Mechanical Technology

Mechanical Technology
Third Edition

D. H. Bacon, MSc, CEng, MIMechE

and

R. C. Stephens, MSc(Eng), CEng, MIMechE

OXFORD AUCKLAND BOSTON JOHANNESBURG MELBOURNE NEW DELHI

Butterworth-Heinemann
Linacre House, Jordan Hill, Oxford OX2 8DP
225 Wildwood Avenue, Woburn, MA 01801-2041
A division of Reed Educational and Professional Publishing Ltd

A member of the Reed Elsevier plc group

First published 1977
Reprinted 1980, 1982, 1983, 1986
Second edition 1990
Reprinted 1991, 1992, 1993, 1994, 1995
Third edition 1998
Reprinted 1999

© D. H. Bacon and R. C. Stephens 1977, 1990, 1998

British Library Cataloguing in Publication Data
Bacon, D. H. (Dennis Henry), 1930 -
Mechanical technology - 3rd ed
1. Mechanics
I. Title II. Stephens, R. C. (Richard Courtney)
620. 1
ISBN 0 7506 3886 9
Library of Congress Cataloguing in Publication Data
Bacon, D. H. (Dennis Henry)
Mechanical technology/D. H. Bacon and R. C. Stephens. – 3rd ed
p. cm
Includes index.

1. Mechanical engineering. I. Stephens, R. C. (Richard Courtney)
II. Title
TII45.B23
CIP 621-dc21 96-10453

Typeset by Laser Words, Madras, India
Printed and bound in Great Britain by Martins the Printers

Contents

Preface

This new edition of *Mechanical Technology* has been substantially revised to reflect the changing policies of the Business and Technician Education Council (BTEC) for courses leading to Higher National qualifications in mechanical engineering. The book is still primarily intended for students in this field, covering the necessary work in stress analysis, dynamics, thermodynamics and fluid mechanics.

Because of the variations in syllabuses, more material has been included than may be required for a particular course and this has the advantage of making the book suitable for students of other disciplines up to degree level who require a fundamental outline of mechanical engineering principles. Such courses embrace management and engineering, systems engineering, computing and engineering, data acquisition and engineering, etc., as well as the traditional non-mechanical courses.

The new material demanded by the BTEC syllabus includes a large extension of the heat transfer work in Chapter 31 and a completely new Chapter 42 on Hydraulic machines. Other small amendments have also been made.

These changes have resulted in a better book and the policy of concise theory presentation with illustrative worked examples and a large number of tutorial examples has been continued.

D. H. Bacon
R. C. Stephens

Acknowledgements

Thermodynamic properties used in this book are taken from:

1. Haywood, R. W., *Thermodynamic Tables in SI(metric) Units*, Cambridge University Press.
2. Rogers, G. F. C. and Mayhew, Y. R., *Thermodynamic and Transport Properties of Fluids; SI Units*, Basil Blackwell, Oxford.

1 Simple stress and strain

1.1 Introduction

When a material is loaded, the force may be resolved into components normal and parallel to any plane within the material. The normal component is termed a *tensile* or *compressive* force and the intensity of loading per unit area is called the *direct stress*, σ. The parallel component is termed a *shear* force and the intensity of loading per unit area is called the *shear stress*, τ.

The distortion of the material due to the direct and shear forces is measured by the *direct strain*, ε, and the *shear strain*, ϕ, respectively.

1.2 Direct stress and strain

Figure 1.1(*a*) shows a piece of material subjected to a tensile force F. If the cross-sectional area of the material is A, then the tensile stress on the cross-section

$$\sigma = \frac{F}{A} \tag{1.1}$$

The basic unit of stress is the N/m^2, termed the *pascal* (Pa) but more practical units are MN/m^2 or N/mm^2.

If the original length of the bar in the direction of F is l, Figure 1.1(*b*), and the extension due to the load is x, then the tensile strain

$$\varepsilon = \frac{x}{l} \tag{1.2}$$

Since strain is merely a ratio of lengths, it has no units.

It will be seen in Figure 1.1(*b*) that the extension in the direction of F is accompanied by a contraction in perpendicular directions and the ratio $\dfrac{transverse\ strain}{longitudinal\ strain}$ is called *Poisson's ratio*, v.

The transverse strain is opposite in direction to the longitudinal strain; thus, if the longitudinal strain is ε, the transverse strain is $-v\varepsilon$.

The value of v for steel is about 0.285 and for brass, about 0.33.

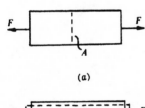

(a)

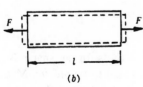

(b)

Figure 1.1

1.3 Shear stress and strain

Figure 1.2(*a*) shows a piece of material subjected to a shear force F. If the cross-sectional area of the material is A, then the shear stress

$$\tau = \frac{F}{A} \tag{1.3}$$

The units of shear stress are the same as for direct stress.

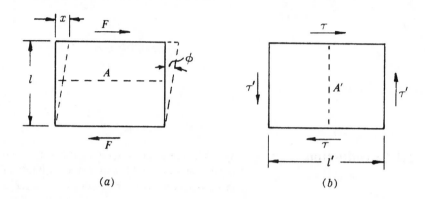

Figure 1.2

If the deformation in the direction of F is x and the distance between the opposite faces is l, then shear strain

$$\phi = \frac{x}{l} \tag{1.4}$$

ϕ is the angular distortion in radians since $\dfrac{x}{l}$ is very small.

1.4 Complementary shear stress

Due to the shear forces, F, a clockwise couple $Fl = \tau Al$ is applied to the material. If the material is to remain in external equilibrium, an equal and opposite couple must be applied by shear stresses τ' induced on perpendicular faces, Figure 1.2(*b*). Thus, for equilibrium,

$$\tau'A'l' = \tau Al$$

But
$$A'l' = Al$$

$$\therefore \tau' = \tau$$

This induced stress is called the *complementary shear stress*.

1.5 Elasticity, plasticity and Hooke's law

If a load is removed from a material and it returns to its former shape, it is said to be *elastic* but if it remains deformed, it is said to be *plastic*. Many engineering materials are elastic up to a certain stress, termed the *elastic limit*, after which they are partly elastic and partly plastic.

Elastic materials obey Hooke's law, which states that the deformation of a material is directly proportional to the load, i.e., the ratio stress/strain is a constant.

For direct stress, the constant of proportionality is called the *modulus of elasticity* (or *Young's modulus*), E.

Thus
$$E = \frac{\sigma}{\varepsilon} = \frac{F/A}{x/l} = \frac{Fl}{Ax} \qquad (1.5)$$

For shear stress, the constant of proportionality is known as the *modulus of rigidity*, G.

Thus
$$G = \frac{\tau}{\phi} = \frac{F/A}{x/l} = \frac{Fl}{Ax} \qquad (1.6)$$

For steel, E is about 205 GN/m^2 (or kN/mm^2) and G is about 82 GN/m^2.

1.6 Factor of safety

The stress at which a material will break is called the ultimate stress but the maximum design stress is considerably less than this to allow for over loading, non-uniformity of stress distribution, faulty workmanship and material, etc.

The ratio $\dfrac{\text{breaking stress}}{\text{design stress}}$ is called the *factor of safety* and may range from about 3 to 10, depending upon the material and nature of the loading.

1.7 Stresses in compound bars

A compound bar is one which is made up of two different materials rigidly fixed together so that there is no relative movement between the ends.

If the bar shown in Figure 1.3 is subjected to a force F, the extension of each part will be the same,

i.e.
$$x_1 = x_2$$

or
$$\frac{\sigma_1 l_1}{E_1} = \frac{\sigma_2 l_2}{E_2} \qquad (1.7)$$

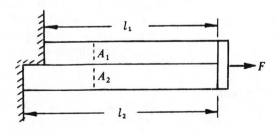

Figure 1.3

Also the load will be carried by the two parts such that

$$F = F_1 + F_2$$
$$= \sigma_1 A_1 + \sigma_2 A_2 \qquad (1.8)$$

σ_1 and σ_2 may be determined from equations (1.7) and (1.8).

If a compound bar undergoes a temperature change, differences in the coefficients of expansion, α, of the two materials coupled with the constraint of no relative movement between the ends will lead to stresses.

Let the bar shown in Figure 1.4(a) be subjected to a temperature rise T. Then material (1) should extend an amount $l_1\alpha_1 T$ and material (2) by an amount $l_2\alpha_2 T$. Since they are constrained to extend the same amount, Figure 1.4(b), material (1) is extended a distance x_1, inducing stress σ_1, and material (2) is compressed a distance x_2, inducing stress σ_2.

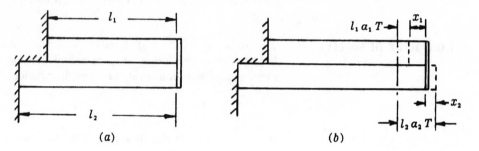

(a)   **(b)**

Figure 1.4

Hence
$$l_1\alpha_1 T + x_1 = l_2\alpha_2 T - x_2$$

or
$$(l_2\alpha_2 - l_1\alpha_1)T = x_1 + x_2 = \frac{\sigma_1 l_1}{E_1} + \frac{\sigma_2 l_2}{E_2} \qquad (1.9)$$

If there is no external force on the bar, the internal forces due to the stresses must balance,

i.e.,
$$\sigma_1 A_1 = \sigma_2 A_2 \qquad (1.10)$$

σ_1 and σ_2 may be determined from equations (1.9) and (1.10).

If there is an external force and a change in temperature, the effects of each can be calculated separately and the results are then combined to obtain the resultant stresses.

Worked examples

    1. *A mild steel rod, 600 mm long, is 25 mm diameter for 150 mm of its length and 50 mm for the rest of its length. It carries an axial tensile pull of 18 kN. With the axial pull applied, the ends of the rod are secured by rigid fixings. Find the temperature through which the rod must be raised to reduce the axial pull by two-thirds.* $\alpha_{\text{steel}} = 11 \times 10^{-6}/°C$; $E_{\text{steel}} = 200 \ GN/m^2$.

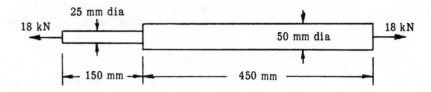

Figure 1.5

$$x = \frac{Fl}{AE} \quad \text{from equation (1.5)}$$

$$\therefore \text{ total extension} = \frac{18 \times 10^3 \times 0.15}{\frac{\pi}{4} \times 0.025^2 \times 200 \times 10^9} + \frac{18 \times 10^3 \times 0.45}{\frac{\pi}{4} \times 0.05^2 \times 200 \times 10^9}$$

$$= 48.1 \times 10^{-6} \text{ m}$$

To reduce the axial pull by two-thirds, the natural extension under a temperature rise T must equal two-thirds of the extension due to the load,

i.e.
$$l\alpha T = \tfrac{2}{3}x$$

i.e.
$$0.6 \times 11 \times 10^{-6}T = \tfrac{2}{3} \times 48.1 \times 10^{-6}$$

$$\therefore T = \underline{4.86°C}$$

2. *A flat steel bar, 10 m long and 10 mm thick tapers from 60 mm at one end to 20 mm at the other. Determine the change in length of the bar when an axial tensile load of 12 kN is applied to it. E = 200 GN/m².*

The arrangement is shown in Figure 1.6 and, from similar triangles, the point of convergence of the tapered sides is at 5 m from the smaller end.

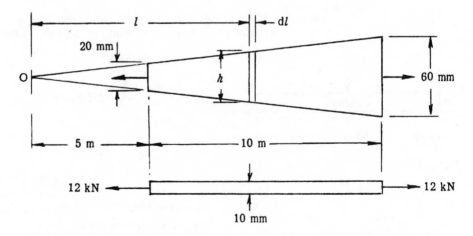

Figure 1.6

In problems where the section of the bar is varying, it is necessary to consider the extension of a small element of the bar and then obtain the overall extension by integration.

From equation (1.5), $x = \dfrac{Fl}{AE}$

Thus the extension of an element of the bar of length dl is given by

$$dx = \frac{F\,dl}{AE}$$

Taking the origin for l at the point O,

$$h = \frac{l}{15} \times 0.06 = 0.004l$$

$$\therefore dx = \frac{12 \times 10^3 dl}{0.01 \times 0.004l \times 200 \times 10^9}$$

$$= 0.001\,5 \frac{dl}{l}$$

$$\therefore x = 0.001\,5 \int_5^{15} \frac{dl}{l}$$

$$= 0.001\,5[\ln l]_5^{15}$$

$$= 0.001\,5 \ln 3$$

$$= 0.001\,648 \text{ m} \quad \text{or} \quad \underline{1.648 \text{ mm}}$$

If, as an approximation, the bar is treated as a uniform bar of width 40 mm,

$$x = \frac{12 \times 10^3 \times 10}{0.01 \times 0.04 \times 200 \times 10^9}$$

$$= 0.001\,5 \text{ m} \quad \text{or} \quad \underline{1.5 \text{ mm}}$$

3. *A steel bar, 20 mm diameter and 50 mm long, is subjected to an axial tensile load of 80 kN. Determine the change in length and diameter. $v = 0.3$.*

Stress on cross-section, Figure 1.7,

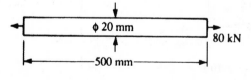

Figure 1.7

$$\sigma = \frac{80 \times 10^3}{\dfrac{\pi}{4} \times 20^2} = 255 \text{ N/mm}^2$$

$$\therefore \text{ axial strain} = \frac{\sigma}{E}$$

$$= \frac{255}{200 \times 10^3} = 1.275 \times 10^{-3}$$

$$\therefore \text{ increase in length} = 1.275 \times 10^{-3} \times 500 = \underline{0.637\,5 \text{ mm}}$$

$$\text{Transverse strain} = 0.3 \times 1.275 \times 10^{-3} = 0.382\,5 \times 10^{-3}$$

$$\therefore \text{ reduction in diameter} = 0.382\,5 \times 10^{-3} \times 20 = \underline{0.007\,65 \text{ mm}}$$

Note: These changes in dimensions will result in an *increase* in volume but it would be difficult to obtain an accurate value for this from the new dimensions. A more satisfactory method is given in Section 8.12.

4. *A wire strand consists of a steel wire 2.7 mm diameter, covered by six bronze wires, each 2.5 mm diameter. If the working stress for the bronze is 60 N/mm², calculate the strength of the strand and the equivalent modulus of elasticity. $E_{steel} = 200$ kN/mm², $E_{bronze} = 85$ kN/mm².*

From equation (1.7), $\dfrac{\sigma_s}{E_s} = \dfrac{\sigma_b}{E_b}$ since the length of each part is the same

$$\therefore \sigma_s = 60 \times \frac{200}{85} = 141.2 \text{ N/mm}^2$$

$$\text{Area of steel} = \frac{\pi}{4} \times 2.7^2 = 5.726 \text{ mm}^2$$

$$\text{Area of bronze} = 6 \times \frac{\pi}{4} \times 2.5^2 = 29.452 \text{ mm}^2$$

$$\therefore \text{ strength of strand} = 141.2 \times 5.726 + 60 \times 29.452 = \underline{2\,575 \text{ N}}$$

From equation (1.5), $$E = \frac{\sigma}{\varepsilon} = \frac{F}{A\varepsilon}$$

For the equivalent modulus of the composite strand, the strain will be that of the steel alone or bronze alone, since both parts stretch by the same amount. Thus, using the given stress in the bronze,

$$\varepsilon = \frac{60}{85 \times 10^3} = 0.705\,9 \times 10^{-3}$$

$$\therefore E = \frac{2\,575}{(5.726 + 29.452) \times 0.705\,9 \times 10^{-3}}$$

$$= \underline{103.7 \times 10^3 \text{ N/mm}^2}$$

5. *A copper tube of mean diameter 120 mm and 6.5 mm thick has its open ends sealed by two rigid plates connected by two steel bolts of 25 mm diameter, initially tensioned to 20 kN at a temperature of 30°C, thus forming a pressure vessel. Determine the stresses in the copper and steel at 0°C.*

$$E_{steel} = 200 \text{ kN/mm}^2 ; \alpha_{steel} = 11 \times 10^{-6}/°C$$

$$E_{copper} = 100 \text{ kN/mm}^2 ; \alpha_{copper} = 18 \times 10^{-6}/°C$$

The arrangement is shown in Figure 1.8.

$$\text{Area of each bolt} = \frac{\pi}{4} \times 25^2 = 491 \text{ mm}^2$$

$$\text{Area of tube} = \pi \times 120 \times 6.5 = 2\,450 \text{ mm}^2$$

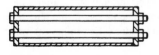

Figure 1.8

$$\text{At 30°C, stress in steel, } \sigma_s = \frac{20 \times 10^3}{491} = 40.73 \text{ N/mm}^2$$

and stress in copper, $\sigma_c = \dfrac{2 \times 20 \times 10^3}{2\,450} = 16.33 \text{ N/mm}^2$

When the temperature is reduced by 30°C,

$$\frac{\sigma_s}{E_s} + \frac{\sigma_c}{E_c} = (\alpha_c - \alpha_s)T \quad \text{from equation (1.9)}$$

i.e. $\dfrac{\sigma_s}{200 \times 10^3} + \dfrac{\sigma_c}{100 \times 10^3} = (18 - 11) \times 10^{-6} \times 30$

or $\sigma_s + 2\sigma_c = 42$ (1)

Also $\sigma_s A_s = \sigma_c A_c \quad \text{from equation (1.10)}$

i.e. $\sigma_s \times 2 \times 491 = \sigma_c \times 2\,450$

or $\sigma_s = 2.5\sigma_c$ (2)

Therefore, from equations (1) and (2),

$$\sigma_s = 23.33 \text{ N/mm}^2$$

and $\sigma_c = 9.33 \text{ N/mm}^2$

Since the coefficient of expansion of the copper is greater than that of the steel, a temperature rise will lead to a compressive stress in the copper and a tensile stress in the steel, and vice versa for a fall in temperature.

Thus resultant stress in steel $= 40.73 - 23.33$

$$= \underline{17.4 \text{ N/mm}^2 \text{(tensile)}}$$

Resultant stress in copper $= 16.33 - 9.33$

$$= \underline{7.0 \text{ N/mm}^2 \text{(compressive)}}$$

Further problems

6. A brass rod 6 mm diameter and 1 m long is joined at one end to a steel rod 6 mm diameter and 1.3 m long. The compound rod is placed in a vertical position with the steel rod at the top and connected at its ends to rigid fixings so that it carries a tensile load of 3.5 kN. A vertically downward load of 1.3 kN is then applied at the junction of the two metals. Calculate the stresses in the steel and brass.

$E_s = 200 \text{ GN/m}^2$; $E_b = 85 \text{ GN/m}^2$.   [153.5 MN/m^2; 107.5 MN/m^2]

7. A steel bar 40 mm diameter and 4 m long is heated through 60°C, after which its ends are firmly fixed. After cooling to normal temperature again, the length of the bar is 1.2 mm less than when at its highest temperature. Determine the force and stress in the cold bar.

$E = 200 \text{ GN/m}^2$ and $\alpha = 11 \times 10^{-6}/°\text{C}$.   [90.5 kN; 72 MN/m^2]

8. The maximum safe compressive stress in a steel punch is limited to 1 GN/m^2 and the punch is used to pierce circular holes in steel plate 20 mm thick.

If the ultimate shearing stress in the plate is 300 MN/m^2, calculate the smallest diameter of hole that can be pierced.   [24 mm]

9. A bar of steel is of rectangular cross-section 30 mm by 20 mm and is subjected to an axial tensile load of 30 kN. If E for steel is 200 GN/m^2 and Poisson's ratio is 0.28, find the change in the cross-sectional dimensions.

[0.021 mm; 0.014 mm]

10. A steel rod ABC, of circular section, transmits an axial pull. AB is 800 mm long and 25 mm diameter, BC is 650 mm long and 20 mm diameter. If the total increase in length is 0.75 mm, determine for the separate parts AB and BC, the changes in (a) length, and (b) diameter. [0.33 mm; 3.1 μm; 0.42 mm; 3.88 μm]

11. A reinforced concrete column is 3 m high and of uniform cross-section 380 mm square. It is reinforced by four 25 mm diameter steel rods symmetrically placed. If the column carries an axial load of 600 kN, determine the stresses in the steel and concrete, the shortening of the column and the energy stored in the column.

$E_s = 200$ GN/m^2; $E_c = 15$ GN/m^2.

[47.5 MN/m^2; 3.56 MN/m^2; 0.713 mm; 214 J]

12. A steel tube, 24 mm external diameter and 18 mm internal diameter, encloses a copper rod 16 mm diameter to which it is rigidly joined at each end. If there is no stress in the compound bar at 15°C, determine the stresses in the rod and tube when the temperature is raised to 200°C.

$E_s = 200$ GN/m^2; $E_c = 90$ GN/m^2; $\alpha_s = 11 \times 10^{-6}$/°C; $\alpha_c = 18 \times 10^{-6}$/°C

[81.3 MN/m^2; 80.0 MN/m^2]

13. A steel bar, 28 mm diameter and 400 mm long, is placed concentrically within a brass tube 400 mm outside diameter and 30 mm inside diameter. The length of the tube exceeds that of the bar by 0.12 mm. Rigid plates are placed on the ends of the tube through which an axial compressive force is applied to the compound bar. Determine the compressive stresses in the bar and tube due to a force of 60 kN.

$E_s = 200$ GN/m^2; $E_b = 100$ GN/m^2. [48.2 MN/m^2; 54.1 MN/m^2]

14. A brass tube 30 mm outside diameter and 5 mm thick has a steel rod 20 mm diameter inside it and joined to it at each end. If at 10°C there is no longitudinal stress, calculate the stress in the rod and the tube when the temperature of the compound bar is raised to 300°C.

For brass: $E = 100$ GN/m^2; $\alpha = 18 \times 10^{-6}$/°C.

For steel: $E = 200$ GN/m^2; $\alpha = 12 \times 10^{-6}$/°C.

[134 N/mm^2, tensile; 107 N/mm^2, compressive]

15. A steel rod of 320 mm^2 cross-sectional area and a coaxial copper tube of 800 mm^2 cross-sectional area are rigidly fixed together at their ends. An axial compressive load of 40 kN is applied to the composite bar and the temperature is then raised by 100°C. Determine the stresses in the steel and copper.

$E_s = 200$ GN/m^2; $E_c = 100$ GN/m^2; $\alpha_s = 12 \times 10^{-6}$/°C; $\alpha_c = 16 \times 10^{-6}$/°C

[11.11 MN/m^2; 45.55 MN/m^2]

2 Shearing force and bending moment

2.1 Introduction

A structural member subject to transverse external forces is called a *beam* and these external forces cause shearing forces and bending moments at each cross-section of the beam. The external forces considered may be due to concentrated or distributed loads, beam mass and supporting reactions.

The *shearing force* at any cross-section of a beam is the algebraic sum of all the external forces to one side of the section. Referring to Figure 2.1(*a*), the shearing force to the left of point C is $W - R_1$ downwards and to the right of point C, it is R_2 upwards. These are of the same magnitude, since $W = R_1 + R_2$.

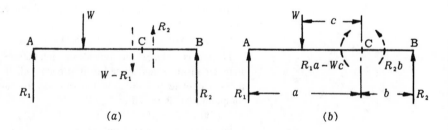

(*a*)   (*b*)

Figure 2.1

The *bending moment* at any cross-section of a beam is the algebraic sum of the moments of all the external forces to one side of the section. Referring to Figure 2.1(*b*), the bending moment to the left of point C is $R_1a - W_c$ clockwise and to the right of point C, it is R_2b anticlockwise. These are the same magnitude since, taking moments about A,

$$R_2(a + b) = W(a - c)$$

or
$$R_2b = Wa - R_2a - Wc = R_1a - Wc$$

2.2 Sign convention

If the Cartesian coordinate system is chosen for beam deflection, Chapter 5, i.e., deflection positive upwards, then the sign convention for shearing force and bending moment is as follows:

(*a*) If the shearing force to the left of a section is upwards (or to the right of a section is downwards), it is regarded as positive.
(*b*) If the bending moment to the left of a section is clockwise (or to the right of a section is anticlockwise), it is regarded as positive.

This convention is illustrated in Figure 2.2.

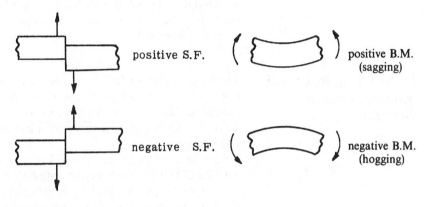

Figure 2.2

2.3 Relation between load intensity, shearing force and bending moment

Consider a small length dx of beam, carrying a uniformly distributed load w per unit length, Figure 2.3. Let the shearing force and bending moment at distance x from the origin be F and M and at distance $x + dx$, $F + dF$ and $M + dM$ respectively.

For equilibrium of forces,

$$F = F + dF + w\,dx$$

$$\therefore\; w = -\frac{dF}{dx} \tag{2.1}$$

For equilibrium of moments about the right-hand end of the element,

$$M + F\,dx - w\,dx \cdot \frac{dx}{2} = M + dM$$

$$\therefore\; F = \frac{dM}{dx}, \text{ neglecting second order terms} \tag{2.2}$$

Equation (2.2) shows that a local maximum or minimum bending moment occurs when the shearing force is zero.

Combining equations (2.1) and (2.2) gives

$$w = -\frac{dF}{dx} = -\frac{d^2M}{dx^2} \tag{2.3}$$

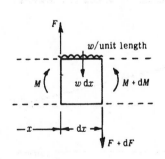

Figure 2.3

Conversely,
$$F = - \int w \, dx$$

and
$$M = \int F \, dx$$

Hence the bending moment at any point is represented by the area of the shearing force diagram up to that point.

Where the load varies in a mathematical manner, the shearing force may be determined by integrating the load and bending moment determined by integrating the shearing force. For irregular loading, these operations may be done graphically.

2.4 Shearing force and bending moment diagrams

These are diagrams which show the variation in shearing force and bending moment along the length of the beam and Figure 2.4 shows four simple examples. Cases (*a*) and (*b*) are cantilevers, in which the support supplies the vertical reaction and the restraining moment; cases (*c*) and (*d*) are simply supported beams.

Concentrated loads are assumed to act at a point but in practice, they must be distributed over a short length of the beam.

In cases where the loading is arbitrary, the loading diagram may be divided into convenient strips (not necessarily of uniform width) and the load represented by each strip is replaced by a concentrated load of the same magnitude placed at the centroid of the strip. Approximate shearing force and bending moment diagrams are then drawn for the concentrated loads, Figure 2.5, and finally, smooth curves are drawn through the points of intersection of the strip boundaries with these approximate diagrams.

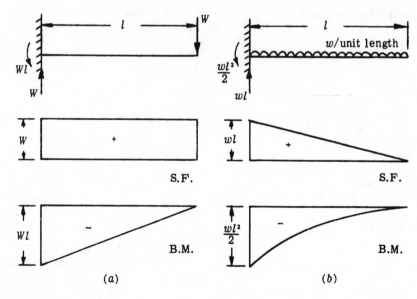

Figure 2.4   (a)–(b)

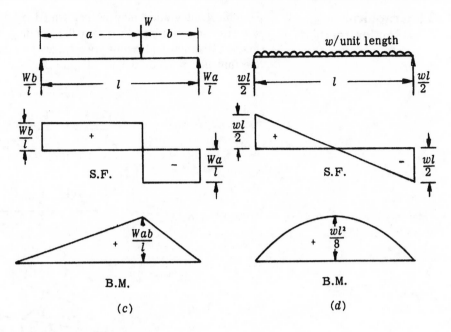

Figure 2.4   (c)–(d)

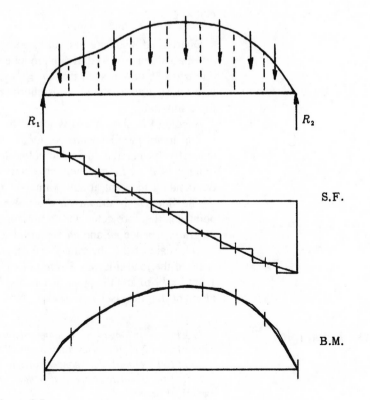

Figure 2.5

2.5 Graphical construction methods

Graphical integration may be performed by area measurement (counting squares) and plotting; to obtain the bending moment diagram from the loading diagram will require two stages and this may become tedious, so the funicular polygon method, Figure 2.6, is often used.

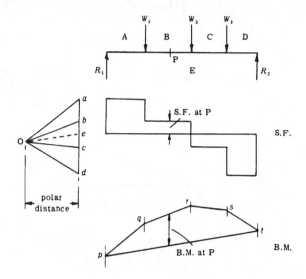

Figure 2.6

Distributed loads are first replaced by equivalent concentrated loads and the spaces on the load diagram are lettered, using Bow's notation. The load line $abcd$ is drawn such that $ab = W_1$, $bc = W_2$, $cd = W_3$ and points a, b, c and d are projected across horizontally to form part of the shearing force diagram.

A pole O is chosen and is joined to a, b, c and d; lines pq, qr, rs and st are drawn parallel with Oa, Ob, Oc and Od in spaces A, B, C and D respectively and the funicular polygon is closed with line tp. This then represents the bending moment diagram and the *vertical* ordinate shows the bending moment at any point.

The line Oe is drawn on the polar diagram parallel with pt and the point e is then projected across horizontally to complete the shearing force diagram, whence ea and ed represent R_1 and R_2 respectively.

The scale of the shearing force diagram is determined by the choice of scale of the load line, say 1 mm $= k_1$ N. If the scale for the beam length is 1 mm $= k_2$ m and the polar distance is l mm, then the scale of the bending moment diagram is 1 mm $= lk_1k_2$ Nm.

Worked examples

1. *A beam 8 m long is freely supported over a span of 6 m and overhangs the right-hand support 2 m. It carries a uniformly distributed load of 20 kN/m together with loads of 160 kN and 130 kN concentrated at 3 m and 8 m from the left-hand support.*

Sketch the shearing force and bending moment diagrams, inserting principal numerical values.

Taking moments about B, Figure 2.7(a):

$$R_a \times 6 = 160 \times 3 - 130 \times 2 + 20 \times 8 \times 2$$

$$= 540$$

$$\therefore R_a = 90 \text{ kN}$$

$$R_b = 160 + 130 + 20 \times 8 - 90$$

$$= 360 \text{ kN}$$

Considering a section at a distance x from A, the shearing forces and bending moments at various points along the beam are as follows:

$\underline{0 < x < 3}$ $F = 90 - 20x$

$$M = 90x - 20x.\frac{x}{2}$$

$$= 90x - 10x^2$$

$\underline{3 < x < 6}$ $F = 90 - 20x - 160$

$$= -70 - 20x$$

$$M = 90x - 20x \cdot \frac{x}{2} - 160(x - 3)$$

$$= -70x - 10x^2 + 480$$

$\underline{6 < x < 8}$ $F = 90 - 20x - 160 + 360$

$$= 290 - 20x$$

$$M = 90x - 20x \cdot \frac{x}{2} - 160(x - 3) + 360(x - 6)$$

$$= 290x - 10x^2 - 1\,680$$

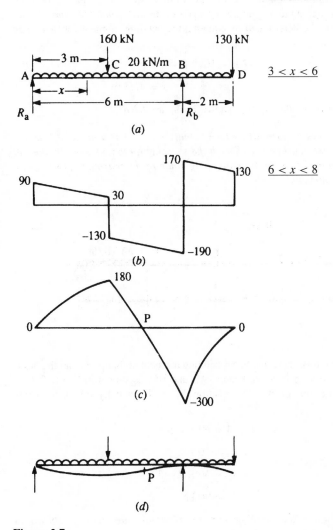

Figure 2.7

The shearing force and bending moment diagrams are shown in Figures 2.7(*b*) and (*c*) respectively.

The maximum positive and negative bending moments are 180 and 300 kNm respectively and it should be noted that these values occur where the shearing force diagram passes through the zero line.

Alternatively, the values of the bending moment may be obtained from the area of the shearing force diagram. Thus, at C,

$$M = \frac{90 + 30}{2} \times 3 \qquad = 180 \text{ kNm}$$

and at B, $$M = 180 - \frac{130 + 190}{2} \times 3 = -300 \text{ kNm}$$

At the point P on the bending moment diagram, Figure 2.7(*c*), the bending moment is zero and this marks the change from positive to negative bending moment, i.e. from sagging to hogging. Such a point is termed a *point of contraflexure* and its position may be determined by equating the bending moment for this range of the beam to zero.

i.e. $$-70x - 10x^2 + 480 = 0$$

from which $$x = 4.33 \text{ m}$$

The deflected form of the beam is shown in Figure 2.7(*d*).

2. *Figure 2.8 shows a beam ABC which is built in at A, has a hinge at B and is supported on a roller bearing at C. Sketch the shearing force and bending moment diagrams and determine the position and magnitude of the maximum positive and negative bending moments.*

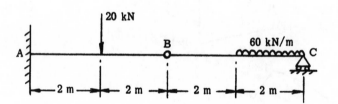

Figure 2.8

Due to the hinge at B, there can be no bending moment at that point and the beam may therefore be split up into two beams, AB and BC, as shown in Figure 2.9(*a*).

From the equilibrium of BC, the reaction R_b is obtained by taking moments about C,

i.e. $$R_b \times 4 = 60 \times 2 \times 1$$

$$\therefore R_b = 30 \text{ kN}$$

and $$R_c = 60 \times 2 - 30$$

$$= 90 \text{ kN}$$

The shearing force and bending moment diagrams are then as shown in Figures 2.8(*b*) and (*c*) respectively.

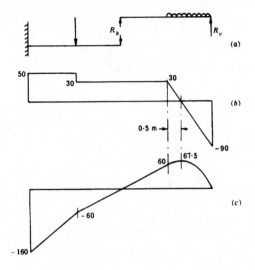

Figure 2.9

From the equilibrium of AB,

$$\text{B.M. at A} = 30 \times 4 + 20 \times 2$$

$$= 160 \text{ kNm}$$

and this is the maximum negative bending moment on the beam.

On span BC, the shearing force will be zero at a point 1.5 m from C. Therefore, maximum bending moment between B and C

$$= 90 \times 1.5 - 60 \times 1.5 \times \frac{1.5}{2} = 67.5 \text{ kNm}$$

and this is the maximum positive bending moment on the beam.

3. *A log of wood, 5 m long and of square cross-section 0.4 m by 0.4 m, floats in a horizontal position in fresh water. It is then loaded at the centre with a weight just sufficient to immerse it completely. Draw the shearing force and bending moment diagrams, stating maximum values. Take the relative density of wood as 0.8.*

$$\text{Volume of log} = 5 \times 0.4 \times 0.4 = 0.8 \text{ m}^3$$

When completely submerged,

$$\text{upthrust of water} = 0.8 \times 10^3 \times 9.81 = 7\,850 \text{ N}$$

since 1 m³ of water has a mass of 1 tonne or 10^3 kg.

$$\text{Weight of log} = 0.8 \times 10^3 \times 9.81 \times 0.8 = 6\,280 \text{ N}$$

Therefore concentrated load required to submerge log

$$= 7\,850 - 6\,280 = 1\,570 \text{ N}$$

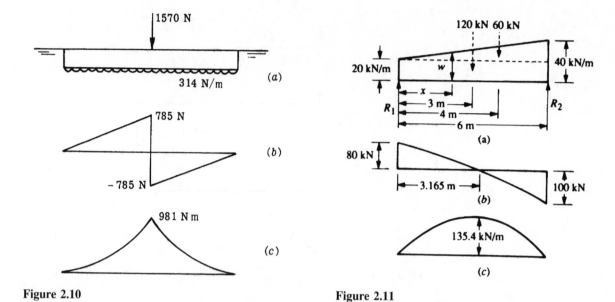

Figure 2.10

Figure 2.11

The log is therefore in equilibrium under a central concentrated load of 1 570 N and an upward uniformly distributed load of 1 570 N, which is equivalent to $\dfrac{1\,570}{5} = 314$ N/m, Figure 2.10(a).

$$\text{S.F. at centre} = 314 \times 2.5 = 785 \text{ N}$$

$$\text{B.M. at centre} = 314 \times 2.5 \times \frac{2.5}{2} = 981 \text{ Nm}.$$

The shearing force and bending moment diagrams are shown in Figures 2.10(b) and (c) respectively.

4. *A simply supported beam of span 6 m carries a distributed load which increases uniformly from 20 kN/m at one end to 40 kN/m at the other end. Sketch the shearing force and bending moment diagrams for the beam and determine the position and magnitude of the maximum bending moment.*

The loading diagram for the beam is shown in Figure 2.11(a). Dividing the load system into a rectangle and a triangle, the loads of 120 kN and 60 kN act at the centroids of these areas, respectively 3 m and 4 m from R_1.

Taking moments about R_1:

$$120 \times 3 + 60 \times 4 = R_2 \times 6$$

$$\therefore R_2 = 100 \text{ kN}$$

$$R_1 = 120 + 60 - 100 = 80 \text{ kN}$$

At a section distance x from R_1 the intensity of loading,

$$w = 20 + \frac{x}{6} \times 20 = 20 \left(1 + \frac{x}{6} \right) \text{ kN/m}$$

$$\therefore F = -\int w\,dx$$

$$= -20\left(x + \frac{x^2}{12}\right) + A$$

When $x = 0$, $\qquad F = R_1 = 80$ kN

$$\therefore A = 80$$

$$\therefore M = \int F\,dx$$

$$= -20\left(\frac{x^2}{2} + \frac{x^3}{36}\right) + 80x + B$$

When $x = 0$, $\qquad M = 0, \therefore B = 0$

The shearing force and bending moment diagrams are obtained by plotting the equations for F and M for various values of x, Figures 2.11(b) and (c).

At the point of maximum bending moment, $F = 0$

i.e. $\qquad -20\left(x + \frac{x^2}{12}\right) + 80 = 0$

from which $\qquad x = 3.165$ m

Hence $\qquad M_{max} = -20\left(\frac{3.165^2}{2} + \frac{3.165^3}{36}\right) + 80 \times 3.165$

$$= \underline{135.4 \text{ kNm}}$$

Further problems

5. A beam 6 m long is freely supported at each end and carries a point load of 70 kN at a distance of 1.5 m from the left-hand end, together with a uniformly distributed load of 15 kN/m from the centre of the span to the right-hand end. Determine the position and magnitude of the maximum B.M.

[95.6 kNm at 1.5 m from L.H. end]

6. A beam 6 m long overhangs its two supports symmetrically and is loaded with three equal loads, one at each end and one at mid-span. Determine the distance between the supports in order that the B.M. at mid-span shall be equal to that at each support.

Draw the S.F. and B.M. diagrams and find the points of contraflexure.

[4.8 m; 1.8 m from each end]

7. A beam 9 m long is freely supported over a span of 7 m and overhangs the right-hand support 2 m. It carries a uniformly distributed load of 25 kN/m together with loads of 150 kN, 70 kN and 120 kN concentrated at 3 m, 5 m and 8 m respectively from the left-hand support.

Draw the S.F. and B.M. diagrams and calculate the value of the maximum bending moment. [395.5 kNm under 150 kN load]

8. A beam ABCD is simply supported at B and C, 6 m apart, and the overhanging parts AB and CD are 2 m and 4 m long respectively. The beam carries a uniformly distributed load of 60 kN/m between A and C and there is a concentrated load of 40 kN at D.

Draw the S.F. and B.M. diagrams. Calculate the position and magnitude of the maximum bending moment between B and C.   [130 kNm at 2.89 m from B]

9. A horizontal beam AD, 10 m long, carries a uniformly distributed load of 200 N/m, together with a concentrated load of 500 N at the left-hand end A. The beam is supported at a point B, 1 m from A, and at C which is in the right-hand half of the beam and x m from the end D.

Determine the value of x if the mid-point of the beam is a point of contraflexure and for this arrangement, draw the B.M. diagram. Locate any other points of contraflexure.

[3 m; B.M. at supports, 600 and 900 Nm; B.M. at
3.75 m from A, 156.25 Nm; 2.5 m from A]

10. Two slings are used to raise a concrete pile of length l and uniform cross-section, the pile remaining horizontal during the lift. Determine the most suitable positions for the slings if the B.M. due to its own weight is to be kept as small as possible. For this arrangement, draw the S.F. and B.M. diagrams.

[0.207 l from ends]

11. A horizontal beam 7 m long rests on two supports A and B, 4 m apart, with a 1 m length overhanging the right-hand support B. The beam carries a uniformly distributed load of 20 kN/m from A to the right-hand end and a uniformly distributed load w kN/m on the part to the left of A. If there is a point of contraflexure 0.5 m to the left of B, find the value of w and the position of the other point of contraflexure. Sketch the bending moment diagram for the beam and find the value of the maximum B.M. between the points of contraflexure.

[35 kN/m; 2 m from A, 5.625 kNm at 2.75 m from A]

12. Determine the position and magnitude of the maximum B.M. for a simply supported beam of length l carrying a distributed load which increases uniformly from zero at one end to w N/m at the other.

[$l/\sqrt{3}$ m from end of zero load; $wl^2/9\sqrt{3}$ Nm]

13. A horizontal beam 4 m long is freely supported at its ends and carries a distributed load which increases uniformly from zero at each end to a maximum of 60 kN/m at the centre. Draw the S.F. and B.M. diagrams.

[Max S.F., 60 kN at each end; max B.M., 80 kNm at centre]

14. A simply supported beam with a span of 4 m carries a distributed load which varies uniformly from 30 kN/m at one end to 90 kN/m at the other. Determine the position and magnitude of the maximum bending moment. Sketch the S.F. and B.M. diagrams.

[121 kNm at 2.16 m from lightly loaded end]

3 Bending and shearing stresses

3.1 Bending

When a bending moment is applied to a beam, Figure 3.1, fibres on one side are extended and those on the other side are compressed. Tensile and compressive stresses are thereby induced in the beam which produce a *moment of resistance*, equal and opposite to the applied bending moment.

Between the areas of tension and compression, there is a layer which is unstressed; this is termed the *neutral layer* and its intersection with the cross-section is called the *neutral axis*.

In the theory of bending, which relates the stresses and curvature to the applied bending moment, it is assumed that:

(*a*) the material is homogeneous, elastic (i.e. obeys Hooke's law) and has the same modulus of elasticity in tension and compression;

(*b*) the beam is initially straight and the resulting radius of curvature is large in comparison with the cross-sectional dimensions;

(*c*) stresses are uniform across the width;

(*d*) the section is symmetrical about the plane of bending;

(*e*) a plane cross-section remains plane after bending.

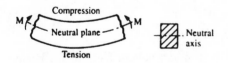

Figure 3.1

3.2 Stresses due to bending

Figure 3.2 shows a beam which is subjected to a bending moment M, acting in a vertical plane through the centroid G. *ab* is a plane cross-section, which moves to *cd* when strained; thus *acdb* represents the strain diagram for the section and since the material is assumed to obey Hooke's law, it is also the stress distribution diagram.

If the stress at a distance y from XX is σ, the stress on an elementary strip of area dA, distance v from XX, is $\dfrac{v}{y}\sigma$.

$$\therefore \text{ force on area } \mathrm{d}A = \frac{v}{y}\sigma \cdot \mathrm{d}A$$

$$\therefore \text{ moment of force about XX} = \frac{v}{y}\sigma \, \mathrm{d}A \cdot v$$

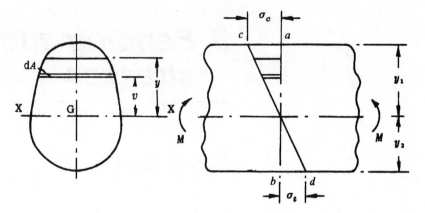

Figure 3.2

$$\therefore \text{ total moment of resistance} = \int \frac{v^2}{y} \sigma \, dA$$

i.e.

$$M = \frac{\sigma}{y} \int v^2 \, dA \quad \text{since} \frac{\sigma}{y} \text{ is a constant}$$

But $\int v^2 \, dA$ is the second moment of area of the section about XX and is denoted by I (see Appendix A).

Thus

$$M = \frac{\sigma}{y} I$$

or

$$\frac{M}{I} = \frac{\sigma}{y} \tag{3.1}$$

The maximum stresses occur at the top and bottom faces of the beam, so that

$$\text{maximum compressive stress, } \sigma_c = \frac{M}{I} y_1$$

and

$$\text{maximum tensile stress, } \sigma_t = \frac{M}{I} y_2$$

3.3 Modulus of section

The maximum stress in a beam is given by

$$\sigma_{max} = \frac{M}{I} y_{max} = \frac{M}{I/y_{max}}$$

The quantity I/y_{max} is called the *modulus of section* or *elastic modulus* and is denoted by Z.

Thus

$$\sigma_{max} = \frac{M}{Z} \tag{3.2}$$

For a rectangular section, breadth b and depth d,

$$Z = \frac{bd^3}{12} \Big/ \frac{d}{2} = \frac{bd^2}{6}$$

For a circular section, diameter d,

$$Z = \frac{\pi d^4}{64} \Big/ \frac{d}{2} = \frac{\pi d^3}{32}$$

For rolled steel structural sections, values of I and Z are tabulated in the structural steel handbooks; an extract from such tables is shown on page 545.

3.4 Position of neutral axis

Since there are no external longitudinal forces acting on the beam, the sum of the internal longitudinal forces must be zero. On the elementary area dA, Figure 3.2, the longitudinal force is $(\sigma/y)v\,dA$ and the total longitudinal force on the section is $\int(\sigma/y)v\,dA$. Thus $\int(\sigma/y)v\,dA = 0$ or $\int v\,dA = 0$ since σ/y is a constant, but $\int v\,dA$ is the first moment of area of the section about the neutral axis and this can only be zero if the neutral axis passes through the centroid of the section.

3.5 Radius of curvature

If the radius of curvature of the neutral axis is R, Figure 3.3, the length of the neutral plane is $R\theta$. The extension of a layer distant y from the neutral plane is $(R + y)\theta - R\theta$ and the strain at this plane,

$$\varepsilon = \frac{(R + y)\theta - R\theta}{R\theta} = \frac{y}{R}$$

Thus the stress

$$\sigma = E\varepsilon = \frac{Ey}{R}$$

so that

$$\frac{\sigma}{y} = \frac{E}{R} \qquad (3.3)$$

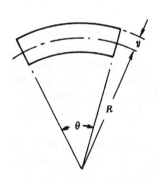

From equations (3.1) and (3.3),

$$\frac{M}{I} = \frac{\sigma}{y} = \frac{E}{R} \qquad (3.4)$$

Figure 3.3

3.6 Combined bending and direct stresses

Materials are frequently subjected to both bending and direct stresses and the resultant stresses are obtained by the algebraic addition of the stresses due to the two causes considered separately.

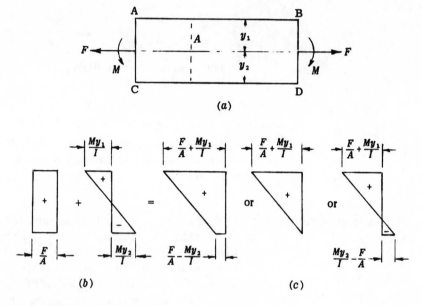

Figure 3.4

If the material shown in Figure 3.4(a) is subjected to a bending moment M and an axial load F, the stress along the edge AB,

$$\sigma_1 = \frac{F}{A} + \frac{M y_1}{I}$$

and the stress along the edge CD,

$$\sigma_2 = \frac{F}{A} - \frac{M y_2}{I}$$

Figure 3.4(b) shows the stress diagrams due to the direct load and the bending moment and Figure 3.4(c) shows the resultant stress diagrams; the stress σ_1 is tensile but the stress σ_2 may be tensile, zero or compressive, depending on whether the bending stress is less than, equal to or greater than the direct stress.

A common example of combined bending and direct stresses occurs when a short member is subjected to an eccentric load. If the load is F and its eccentricity to the longitudinal axis is e, the bar is effectively subjected to an axial load F and a bending moment Fe, as shown in Figure 3.5.

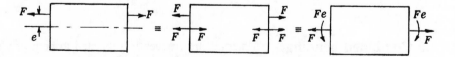

Figure 3.5

The maximum stresses are then given by

$$\sigma = \frac{F}{A} \pm \frac{Fe}{Z}$$

If there is to be no change in the nature of the stress in the section

$$\frac{F}{A} \geq \frac{Fe}{Z}$$

or $$e \leq \frac{Z}{A} \tag{3.5}$$

For a rectangular section, $b \times d$, $Z = \dfrac{bd^2}{6}$, so that $e \leq \dfrac{d}{6}$, i.e., the load must be within the middle third of the bar. For a circular section, diameter d, $Z = \dfrac{\pi}{32}d^3$, so that $e \leq \dfrac{d}{8}$.

Examples arise in the case of structures built of brittle materials such as concrete, masonry, etc., where the compressive load at any section must not be sufficiently eccentric to the axis of that section to give rise to tensile stress.

If the load is eccentric to both axes of symmetry, it may be moved to the axis in two stages, each involving the addition of a moment equal to the force multiplied by the distance moved. Thus the load shown in Figure 3.6 is equivalent to a load on the axis, together with moments Fm in plane XX and Fn in plane YY. The stress at any point, coordinates x and y is then given by

$$\sigma = \frac{F}{A} + \frac{Fm}{I_{YY}} \cdot x + \frac{Fn}{I_{XX}} \cdot y$$

If there is to be no change in the nature of the stress in the section, the load must be within the rhombus $ghjk$ formed by joining the middle third

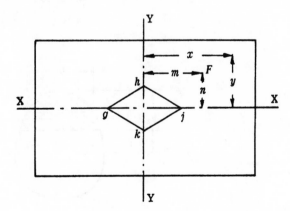

Figure 3.6

points, *gj* being one-third of the horizontal axis and *hk* one-third of the vertical axis.

3.7 Shearing

When a shearing force is applied to a beam, there will be a shear stress on the transverse section and a complementary shear stress on longitudinal sections. The shear stress is not uniform across the transverse section and cannot be obtained by dividing the shearing force by the cross-sectional area. The equilibrium of longitudinal forces is used to determine the complementary shear stress and this is equal to the transverse stress.

It is assumed that the shear stress is uniform across planes parallel to the neutral axis, an assumption which is reasonable if there is no sudden change in width.

3.8 Stresses due to shearing

Figure 3.7 shows an element of a beam which is subjected to a shear force F and a bending moment M, increasing to $M + \mathrm{d}M$ over the length $\mathrm{d}x$.

The force P acting on the end of an element of breadth b and thickness $\mathrm{d}y$

$$= \frac{My}{I} \cdot b\,\mathrm{d}y$$

Similarly,
$$Q = \frac{(M + \mathrm{d}M)y}{I} \cdot b\,\mathrm{d}y$$

Therefore, resultant force on element $= Q - P = \dfrac{\mathrm{d}M}{I} y\,b\,\mathrm{d}y$

∴ total horizontal force on cross-section between AB and CD

$$= \frac{\mathrm{d}M}{I} \int_{y_1}^{y_2} y\,b\,\mathrm{d}y$$

∴ shear stress on CD, $\tau = \dfrac{\dfrac{\mathrm{d}M}{I} \displaystyle\int_{y_1}^{y_2} y\,b\,\mathrm{d}y}{B\,\mathrm{d}x}$

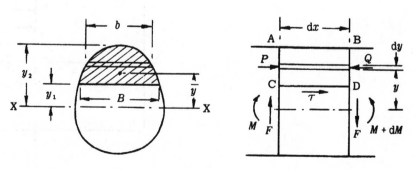

Figure 3.7

But
$$\frac{dM}{dx} = F \quad \text{from equation (2.2)}$$

$$\therefore \ \tau = \frac{F}{IB} \int_{y_1}^{y_2} y\,b\,dy \tag{3.6}$$

If A is the area of the cross-section between the limits y_1 and y_2 and $\overline{y}$ is the distance of the centroid of this area from XX, then

$$\int_{y_1}^{y_2} y\,b\,dy = A\overline{y}$$

so that equation (3.6) can be written in the form

$$\tau = \frac{F}{IB}A\overline{y} \tag{3.7}$$

Equation (3.7) shows that the shear stress is always zero at edges farthest from the neutral axis.

3.9 Application to common cases

(a) Rectangular section of breadth b and depth d, Figure 3.8

At a layer distance h from XX,

$$A = b\left(\frac{d}{2} - h\right), \quad \overline{y} = \frac{1}{2}\left(\frac{d}{2} + h\right), \quad B = b \quad \text{and} \quad I = \frac{bd^3}{12}$$

Hence
$$\tau = \frac{6F}{bd^3}\left(\frac{d^2}{4} - h^2\right) \tag{3.8}$$

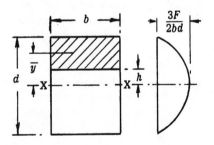

Figure 3.8

This is a parabolic distribution with a maximum at the neutral axis of $\dfrac{3}{2}\dfrac{F}{bd}$, i.e., $\frac{3}{2}$ times the average shear stress.

(b) Circular section, radius r, Figure 3.9

At a layer distance h from XX,

$$\int_h^r y\,b\,dy = \int_h^r y \times 2\sqrt{(r^2 - y^2)}\,dy = \frac{2}{3}(r^2 - h^2)^{3/2}$$

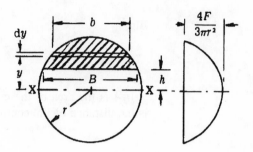

Figure 3.9

$$B = 2\sqrt{(r^2 - h^2)}$$

and

$$I = \frac{\pi r^4}{4}$$

Hence

$$\tau = \frac{4F}{3\pi r^4}(r^2 - h^2) \tag{3.9}$$

This is again a parabolic stress distribution with a maximum at the neutral axis of $\dfrac{4}{3}\dfrac{d}{\pi r^2}$, i.e., $\frac{4}{3}$ times the average shear stress.

(c) I-beam, Figure 3.10

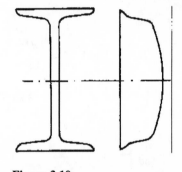

Figure 3.10

When there is an abrupt increase in the width of section, as in an I-beam or T-beam, the shear stress decreases accordingly but the simple theory that this stress is uniform across planes parallel to the neutral axis breaks down. However, the stress in the flange is sufficiently small that this is of little consequence and it is customary to assume that the whole of the shear force is carried by the web (See Example 9).

Worked examples

1. *Figure 3.11 shows a T-section beam which is used as a cantilever, 2 m long, which supports a concentrated load of 100 N at the free end.*

Determine the maximum stress in the beam and calculate also the width of flange required if the maximum stresses in compression and tension are to be in the ratio of 2 to 1.

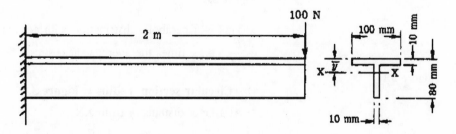

Figure 3.11

Taking moments of area about the top of the flange to find the position of the neutral axis:

$$100 \times 10 \times 5 + 70 \times 10 \times 45 = (100 \times 10 + 70 \times 10)\bar{y}$$

$$\therefore \bar{y} = 21.5 \text{ mm}$$

$$I_{XX} = \frac{100 \times 10^3}{12} + 100 \times 10 \times 16.5^2 + \frac{10 \times 70^3}{12} + 10 \times 70 \times 23.5^2$$

$$= 953\,600 \text{ mm}^4 = 0.953\,6 \times 10^{-6} \text{ m}^4$$

The maximum B.M. occurs at the wall and the maximum stress will be at the bottom of the web, this being the farthest point from the neutral axis.

$$\frac{M}{I} = \frac{\sigma}{y} \quad \text{from equation (3.1)}$$

$$\therefore \sigma = \frac{100 \times 2}{0.953\,6 \times 10^{-6}} \times 58.5 \times 10^{-3}$$

$$= 12.26 \times 10^6 \text{ N/m}^2$$

For the maximum compressive stress to be twice the maximum tensile stress, the neutral axis must be one-third of 80 mm from the upper surface, as shown in Figure 3.12.

Taking moments about the top of the flange:

$$w \times 10 \times 5 + 70 \times 10 \times 45 = (w \times 10 + 70 \times 10) \times \frac{80}{3}$$

$$\therefore w = \underline{59.2 \text{ mm}}$$

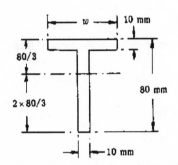

Figure 3.12

2. *Figure 3.13 shows the section of a beam. Find the maximum stress due to bending if a moment of 8 kN m is applied (a) in the plane YY and (b) in the plane XX.*

(*a*) The second moment of area about XX is that of a rectangle, 70 mm wide by 100 mm deep, less that of a circle of diameter 60 mm,

i.e.
$$I_{XX} = \frac{70 \times 100^3}{12} - \frac{\pi}{64} \times 60^4$$

$$= 5.197 \times 10^6 \text{ mm}^4$$

$$= 5.197 \times 10^{-6} \text{ m}^4$$

$$\frac{M}{I} = \frac{\sigma}{y}$$

$$\therefore \sigma = \frac{8 \times 10^3}{5.197 \times 10^{-6}} \times 50 \times 10^{-3}$$

$$= \underline{77 \times 10^6 \text{ N/m}^2}$$

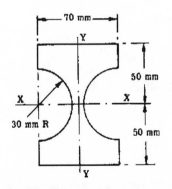

Figure 3.13

(*b*) The second moment of area about YY is that of a rectangle 100 mm wide by 70 mm deep, less that of the two semicircles. The second moment of area of a semicircle about YY is that about the diameter (half the value for a full circle), *less* area × (distance from diameter to centroid)², *plus* area × (distance from centroid to YY)².

Distance of centroid of semicircle from diameter $= \dfrac{4r}{3\pi} = \dfrac{4 \times 30}{3\pi} = 12.74$ mm

$$\therefore I_{YY} = \frac{100 \times 70^3}{12} - 2\left\{\frac{\pi}{128} \times 60^4 - \frac{\pi}{8} \times 60^2 \times 12.74^2 + \frac{\pi}{8} \times 60^2 \times [35 - 12.74]^2\right\}$$

$$= 1.28 \times 10^6 \text{ mm}^4 = 1.28 \times 10^{-6} \text{ m}^4$$

$$\frac{M}{I} = \frac{\sigma}{y}$$

$$\therefore \sigma = \frac{8 \times 10^3}{1.28 \times 10^{-6}} \times 35 \times 10^{-3}$$

$$= \underline{218.5 \times 10^6 \text{ N/m}^2}$$

3. *Select a suitable R.S.J. to support the load system shown in Figure 3.14 if the maximum stress due to bending is not to exceed 120 M N/m².*

Taking moments about R_2,

$$40 \times 1 + 20 \times 4 \times 2 = 4R_1$$

$$\therefore R_1 = 50 \text{ kN}$$

The shearing force diagram for the left-hand part of the beam shows that the shearing force is zero at 2.5 m from R_1 $\left(\text{i.e.} \dfrac{50}{20}\text{m}\right)$

Hence the maximum bending moment, represented by the area of the shearing force diagram up to this section, is given by

$$M = \tfrac{1}{2} \times 50 \times 2.5 = 62.5 \text{ kN m}$$

$$\therefore Z = \frac{M}{\sigma} = \frac{62.5 \times 10^3}{120 \times 10^6} = 0.520\,8 \times 10^{-3} \text{ m}^3$$

$$\text{or} \quad 520.8 \text{ cm}^3$$

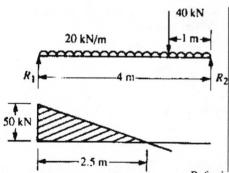

Figure 3.14

Referring to the table for properties of Universal Beams, page 545, the column headed 'elastic modulus' (Z) for bending about axis XX shows that a suitable section would be a 305 mm × 127 mm × 42 kg/m, having a Z value of 520 cm³.

Note: The weight of the beam is 42 × 9.81 = 412 N/m and it may be advisable to check that the given beam section is still suitable when the uniformly distributed load is increased to 20.41 kN/m.

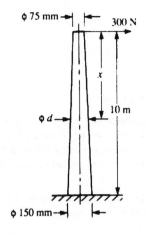

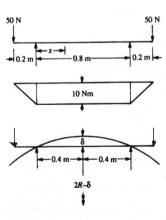

Figure 3.15

4. *A vertical flagstaff 10 m high is of circular section, tapering from 150 mm diameter at the base to 75 mm diameter at the top. A horizontal pull of 300 N is applied at the top. Determine the maximum stress due to bending.*

At a section distance x metres from the top, Figure 3.15.

$$d = 75 + \frac{x}{10} \times 75$$

$$= 7.5(10 + x) \text{ mm} = 0.007\,5(10 + x) \text{ m}$$

B. M. at this section $= 300x$ N m

$$\therefore \sigma = \frac{M}{Z} = \frac{300x}{\dfrac{\pi}{32}\{0.007\,5(10 + x)\}^3}$$

$$= \frac{7.24 \times 10^9 x}{(10 + x)^3}$$

For maximum stress, $\dfrac{d\sigma}{dx} = 0$

from which $x = 5$ m

$$\therefore \sigma_{max} = \frac{7.24 \times 10^9 \times 5}{15^3} = \underline{10.73 \times 10^6 \text{ N/m}^2}$$

5. *A steel rod, 10 mm diameter, is supported symmetrically on two knife-edges at the same level 0.8 m apart and overhangs 0.2 m at each end. Equal loads of 50 N are applied at the two ends.*
Show that the rod bends in the arc of a circle between the supports and find the central deflection. $E = 200$ GN/m².

The arrangement of the beam is shown in Figure 3.16.
Due to symmetry, each reaction is 50 N.
At any section distance x from one support,

$$M = -50(x + 0.2) + 50x = -10 \text{ N m}$$

This is independent of x and so is constant between the supports.

$$\frac{M}{I} = \frac{E}{R}$$

Hence, if M, E and I are constants, R is constant, i.e., the beam bends in the arc of a circle between the supports.

Let the central deflection be δ. Then, from the theorem of intersecting chords of a circle,

$$(2R - \delta) \times \delta = 0.4 \times 0.4$$

δ will be very small in relation to $2R$ so that this equation may be simplified to

$$2R\delta = 0.16 \quad \text{or} \quad \delta = \frac{0.08}{R}$$

Figure 3.16

But

$$R = \frac{EI}{M} = \frac{200 \times 10^9 \times \frac{\pi}{64} \times 0.01^4}{10} = 9.817 \text{ m}$$

$$\therefore \delta = \frac{0.08}{9.817}$$

$$= 8.15 \times 10^{-3} \text{ m} \quad \text{or} \quad \underline{8.15 \text{ mm}}$$

6. *A brick chimney, 30 m high is 2.3 m outside diameter at the top and tapers uniformly to 3.3 m outside diameter at the bottom. The chimney has a mass of 224 tonnes and is 0.675 m thick at the base. If a horizontal wind pressure of 1 kN/m² acts on the projected area of chimney, determine the maximum and minimum stresses on the base.*

The projected area may be divided into a rectangle and a triangle, Figure 3.17, the forces on the two parts, F_1 and F_2, acting at the respective centroids.

$$F_1 = 1 \times 10^3 \times 2.3 \times 30 = 69 \times 10^3 \text{ N}$$

$$F_2 = 1 \times 10^3 \times \tfrac{1}{2} \times 30 = 15 \times 10^3 \text{ N}$$

Therefore moment about base,

$$M = 69 \times 10^3 \times 15 + 15 \times 10^3 \times 10$$

$$= 1\,185 \times 10^3 \text{ Nm}$$

Second moment of area of base section,

$$I = \frac{\pi}{64}(3.3^4 - 1.95^4) = 5.1 \text{ m}^4$$

Therefore bending stress at base,

$$\sigma_b = \frac{M}{I} \times y = \frac{1\,185 \times 10^3}{5.1} \times 1.65$$

$$= 383.5 \times 10^3 \text{ N/m}^2$$

$$\text{Direct stress, } \sigma_d = \frac{W}{A} = \frac{224 \times 10^3 \times 9.81}{\frac{\pi}{4}(3.3^2 - 1.95^2)}$$

$$= 395 \times 10^3 \text{ N/m}^2$$

$\therefore$ maximum compressive stress $= (395 + 383.5) \times 10^3 = \underline{778.5 \times 10^3 \text{ N/m}^2}$

and minimum compressive stress $= (395 - 383.5) \times 10^3 = \underline{11.5 \times 10^3 \text{ N/m}^2}$

7. *A tie-bar 75 mm wide and 25 mm thick sustains an axial load of 100 kN. What depth of metal may safely be removed from one of the narrow sides if the maximum stress over the reduced width may not exceed 100 N/mm² ?*

The bar is shown in Figure 3.18. If a depth x of metal is removed from a narrow side, the load of 100 kN now has an eccentricity $x/2$ from the axis of the remaining metal.

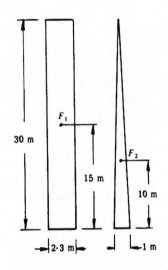

30 m

F_1

15 m

F_2

10 m

2·3 m

1 m

Figure 3.17

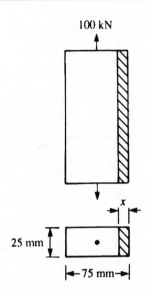

Figure 3.18

Direct stress, $\sigma_d = \dfrac{F}{A}$

$$= \frac{100}{25(75 - x)} = \frac{4}{75 - x} \text{ kN/mm}^2$$

Bending stress, $\sigma_b = \dfrac{M}{Z}$

$$= \frac{100 \times \dfrac{x}{2}}{\dfrac{25(75 - x)^2}{6}}$$

$$= \frac{12x}{(75 - x)^2} \text{ kN/mm}^2$$

The maximum stress in the cross-section is therefore given by

$$\sigma = \frac{4}{75 - x} + \frac{12x}{(75 - x)^2}$$

$$= \frac{300 + 8x}{(75 - x)^2}$$

The maximum depth of metal to be removed leads to a stress of 100 N/mm² or 0.1 kN/mm², so that

$$\frac{300 + 8x}{(75 - x)^2} = 0.1$$

from which $x^2 - 230x + 2\,625 = 0$

Hence $x = \underline{12.05 \text{ mm}}$

8. *A rectangular bar, 80 mm × 40 mm cross-section, is subjected to a pull of W N applied parallel to the axis of the bar but offset 10 mm from each of the axes of the cross-section. If the tensile stress in the material is limited to 150 MN/m², determine the value of W and the stresses at the four corners of the section.*

Obtain the equation of the neutral axis and show its position on a diagram of the cross-section.

The given load is equivalent to a load W on the axis together with moments $10W$ N mm about each of the axes XX and YY, Figure 3.19.

Then direct stress $= \dfrac{W}{A} = \dfrac{W}{80 \times 40 \times 10^{-6}} = 312.5W$ N/m²

Stress at long edge due to bending about XX

$$= \frac{M_{\text{XX}}}{Z_{\text{XX}}} = \frac{10W \times 10^{-3}}{\dfrac{80 \times 40^2}{6} \times 10^{-9}} = 469W \text{ N/m}^2$$

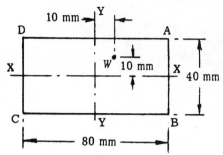

Figure 3.19

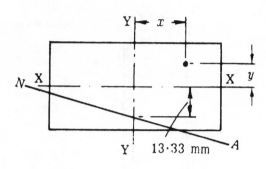

Figure 3.20

Stress at short edge due to bending about YY

$$= \frac{M_{YY}}{Z_{YY}} = \frac{10W \times 10^{-3}}{\dfrac{40 \times 80^2}{6} \times 10^{-9}} = 234.5W \text{ N/m}^2$$

The maximum tensile stress occurs at corner A,

$$\therefore 312.5W + 469W + 234.5W = 150 \times 10^6$$

$$\therefore W = \underline{147.5 \times 10^3 \text{N}}$$

Stress at corner B $= 312.5W - 469W + 234.5W = 78W \qquad = 11.5 \text{ MN/m}^2$

Stress at corner C $= 312.5W - 469W - 234.5W = -391W \qquad = -57.5 \text{ MN/m}^2$

Stress at corner D $= 312.5W + 469W - 234.5W = 547W \qquad = 80.7 \text{ MN/m}^2$

At any point whose coordinates are x and y relative to the axes of the section, Figure 3.20,

$$\sigma = \frac{W}{A} + \frac{M_{XX}}{I_{XX}} \cdot y + \frac{M_{YY}}{I_{YY}} \cdot x$$

$$= \frac{W}{80 \times 40 \times 10^{-6}} + \frac{10W \times 10^{-3}}{\dfrac{80 \times 40^3}{12} \times 10^{-12}} \cdot y + \frac{10W \times 10^{-3}}{\dfrac{40 \times 80^3}{12} \times 10^{-12}} \cdot x$$

$$= W\{312.5 + 23\,450y + 5\,860x\}$$

Along the neutral axis, $\sigma = 0$

i.e. $\qquad\qquad\qquad\qquad 312.5 + 23\,450y + 5\,860x = 0$

from which $\qquad\qquad\qquad\qquad\qquad y = -0.25x - 0.013\,33$

This is the equation of the neutral axis, which is shown in Figure 3.20.

9. *A 120 mm × 300 mm I-beam has flanges 15 mm thick and a web 10 mm thick and at a certain section it is subjected to a shearing force of 200 kN. Draw a diagram showing the distribution of shearing stress across the section and determine the percentage of the shearing force which is carried by the web.*

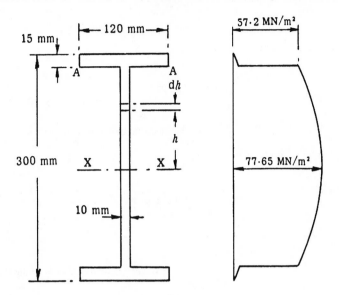

Figure 3.21

Referring to Figure 3.21,

$$I = \left(\frac{120 \times 300^3}{12} - \frac{110 \times 270^3}{12}\right) \times 10^{-12} = 89.5 \times 10^{-6} \text{ m}^4$$

Immediately above section AA,

$$B = 0.12 \text{ m}$$

and $\qquad A\overline{y} = 0.12 \times 0.015 \times 0.142\,5 = 256.5 \times 10^{-6} \text{ m}^3$

$$\therefore \tau = \frac{F}{IB}A\overline{y} \quad \text{from equation (3.7)}$$

$$= \frac{200 \times 10^3 \times 256.5 \times 10^{-6}}{89.5 \times 10^{-6} \times 0.12} = 4.77 \times 10^6 \text{ N/m}^2$$

Immediately below section AA, $B = 0.01$ m and $A\overline{y}$ remains as before.

$$\therefore \tau = 4.77 \times 10^6 \times \frac{0.12}{0.01} = 57.2 \times 10^6 \text{ N/m}^2$$

At a section distance h above XX,

$$A\overline{y} = 0.12 \times 0.015 \times 0.142\,5 + (0.135 - h) \times 0.01 \times \left(\frac{0.135 + h}{2}\right)$$

$$= (0.348 - 5h^2) \times 10^{-3} \text{ m}^3$$

$$\therefore \tau = \frac{200 \times 10^3 \times (0.348 - 5h^2) \times 10^{-3}}{89.5 \times 10^{-6} \times 0.01} = 223.5(0.348 - 5h^2) \times 10^6 \text{ N/m}^2$$

The graph of this equation is shown in Figure 3.21. The maximum value is given by

$$\tau_{max} = 223.5 \times 0.348 \times 10^6 = 77.65 \times 10^6 \text{ N/m}^2 \quad \text{when} \quad h = 0$$

$$\text{S.F. carried by web} = 2 \int_0^{0.135} \tau B \, \mathrm{d}h$$

$$= 2 \int_0^{0.135} 223.5(0.348 - 5h^2) \times 10^6 \times 0.01 \, \mathrm{d}h$$

$$= 4.47 \times 10^6 \left[0.348h - \tfrac{5}{3}h^3 \right]_0^{0.135} = 192 \times 10^3 \text{ N}$$

Therefore percentage of S.F. carried by web

$$= \frac{192}{200} \times 100 = \underline{96\%}$$

If the applied S.F. is assumed to be uniformly distributed over the web area alone,

$$\tau = \frac{200 \times 10^3}{0.27 \times 0.01} = 74 \times 10^6 \text{ N/m}^2$$

which is very near to the actual maximum stress.

Further problems

10. A warehouse floor is to be carried on wooden joists, 100 mm wide and 300 mm deep, simply supported on a span of 3.5 m. If the floor is to carry a load of 20 kN/m^2 and the maximum permissible stress in wood is 8 MN/m^2, calculate the maximum spacing of the joists.   [0.392 m]

11. A cast iron pipe 8 m long, 300 mm outside diameter and 250 mm inside diameter, is simply supported at its ends and is full of water at atmospheric pressure. If the density of water is 1 Mg/m^3 and that of cast iron is 7.8 Mg/m^3, what is the maximum tensile stress in the pipe?   [12.4 MN/m^2]

12. A cantilever 4 m long is of T-section with the web vertical. The section is 150 mm wide by 150 mm deep and the web and flange are 12 mm thick. The cantilever carries a concentrated load of 1.25 kN at a distance of 3 m from the fixed end.

Determine the maximum stress due to bending, neglecting the effect of the weight of the cantilever.   [54 MN/m^2]

13. A horizontal girder of I-section 350 mm by 150 mm is simply supported at the ends of a span of 6 m. The flange thickness is 18 mm whilst that of the web is 10 mm. Calculate the uniformly distributed load which the beam can support including its own weight, if the maximum allowable tensile stress is 120 MN/m^2.   [26.6 kN/m]

14. The section of steel beam is an inverted channel, outside dimensions 220 mm wide by 80 mm deep, thickness of web 10 mm, thickness of vertical flanges 12 mm. The beam is simply supported over a span of 3 m and carries two equal concentrated loads at points distant 0.5 m from each support. Find the value of these loads if the maximum tensile stress is not to exceed 100 MN/m^2.   [7.725 kN]

15. A wooden beam is 80 mm wide and 120 mm deep with a semi-circular groove of 25 mm radius planed out in the centre of each side. Calculate the maximum stress in the section when simply supported on a span of 2 m and loaded with a concentrated load of 400 N at a distance of 0.7 m from one end and a uniformly distributed load of 750 N/m over the whole span.   [2.825 MN/m^2]

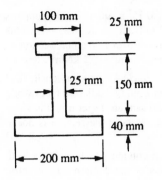

Figure 3.22

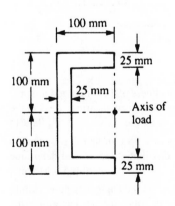

Figure 3.23

16. Figure 3.22 shows the cross-section of a cast-iron beam. Find the position of its neutral axis and the second moment of area about that axis.

What central concentrated load could the beam carry over a simply supported span of 6 m if the stress in the iron is not to exceed 15 MN/m²?

[77 mm from lower edge; 79×10^{-6} m⁴; 10.26 kN]

17. A steel bar, 100 mm diameter and 3 m long, is to be lifted by two vertical slings so that the stress in the bar may be a minimum; calculate that stress. The density of steel is 7.8 Mg/m³.   [1.758 m apart; 1.18 MN/m²]

18. A steel bar, 32 mm wide by 5 mm deep and 2.5 m long, is laid symmetrically on two knife edges at the same level 1.5 m apart.

Show that, when a load W is hung at each end, the part of the bar between the supports bends in a circular arc.

Determine the value of W and the radius of the circular arc if the stress due to bending is not to exceed 55 MN/m². $E = 200$ GN/m².   [14.66 N; 9.09 m]

19. Determine the cross-sectional dimensions of the strongest rectangular section beam which can be cut out of a cylindrical log of wood 300 mm diameter.

[173 mm × 245 mm]

20. A horizontal cantilever 3 m long is of rectangular cross-section 60 mm wide throughout, the depth varying uniformly from 60 mm at the free end to 180 mm at the fixed end. A load of 4 kN acts at the free end. Find the position of the most highly stressed section and the magnitude of the stress at that section.

[Centre section; 41.6 MN/m²]

21. The line of pull in a tension specimen 14 mm diameter is parallel to the axis but displaced from it. Calculate the eccentricity of the pull when the maximum stress is 15% greater than the mean stress.   [0.263 mm]

22. A 250 mm by 150 mm I-section column, thickness of web and flanges 12 mm, is subjected to a thrust of 400 kN at a point on its longer axis 85 mm from its centroid. Determine the maximum tensile and compressive stresses in the column.   [4.4 and 131.3 MN/m²]

23. Figure 3.23 shows the plan view of a short vertical channel. Determine the maximum value of the load if the stress in the steel is not to exceed 80 MN/m².

[136 kN]

24. A chimney, outside diameter 3 m, inside diameter 2.5 m, is to be subjected to a horizontal wind force equal to a uniform load of 2 kN/m of its height. If the density of the masonry is 2.2 Mg/m³, determine the maximum permissible height of the chimney if there is to be no tensile stress at the base.   [29.6 m]

25. A steel chimney is 30 m high, 1 m external diameter and 10 mm thick. It is acted upon by a horizontal wind pressure which is of a uniform intensity of 1 kN/m² of projected area for the lower 15 m and then varies uniformly from 1 kN/m² to 2 kN/m² over the upper 15 m. Calculate the maximum stress in the plates at the base. Steel has a density of 7.8 Mg/m³.   [85.9 MN/m²]

26. A short cast iron column is of hollow section of uniform thickness, 200 mm outside diameter and 125 mm inside diameter. A vertical compressive load acts at an eccentricity of 50 mm from the axis of the column. If the maximum permitted stresses are 80 MN/m² in compression and 16 MN/m² in tension, calculate the greatest allowable load.   [627.5 kN]

27. A short hollow pier, 1.2 m square outside and 0.75 m square inside, supports a vertical point load of 120 kN on a diagonal, 0.69 m from the vertical axis of the pier. Calculate the stresses at the four outside corners of a horizontal section of the pier.   [616.8 kN/m², comp; 136.8 kN/m², comp; 343.2 kN/m², tensile]

28. A tie-bar of rectangular section 75 mm × 25 mm carries an axial tensile load of 100 kN. Metal of thickness 5 mm is now removed from two adjacent faces while the load remains acting along the axis of the original section. Determine the value to which the load must be reduced if the maximum stress in the section is not to exceed the original stress.   [38 kN]

29. Figure 3.24 shows the cross-section of a short column which carries a load, the line of action of which is parallel with the axis of the column but eccentric to both XX and YY. The stresses at A, B and C in MN/m², are respectively 46 compressive, 75 compressive and 12 tensile.

Find the magnitude of the load, the position of its line of action with respect to XX and YY and the stress at point D.

$$I_{XX} = 0.2 \times 10^{-6} \text{ m}^4; I_{YY} = 5 \times 10^{-6} \text{ m}^4; A = 1\,300 \times 10^{-6} \text{ m}^2.$$

[40.95 kN; 1.7 mm from XX; 40.7 mm from YY; 17 MN/m² comp]

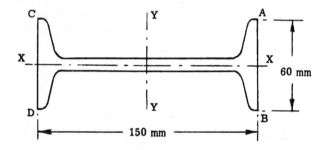

Figure 3.24

30. An I-beam 350 mm × 200 mm has a web thickness of 12.5 mm and a flange thickness of 25 mm. Calculate the ratio of maximum to mean shearing stress in the section and the percentage of the total shear carried by the web.   [3.57; 92.9%]

31. A T-beam has a flange and web each 150 mm × 25 mm and is subjected to a shear force of 200 kN. Sketch the shear stress distribution curve and determine the maximum shear stress.   [65.3 MN/m²]

32. A cantilever of I-section 200 mm × 100 mm has rectangular flanges 10 mm thick and web 75 mm thick. It carries a uniformly distributed load. Determine the length of the cantilever if the maximum bending stress is three times the maximum shearing stress. What is the ratio of the stresses half-way along the cantilever?

[1 m; 1.5]

33. A beam of circular section, diameter D, is subjected to a shearing force F. Find the maximum shearing stress in terms of F and D.

If the beam is simply supported on a span L and carries a central concentrated load, find the ratio L/D if the maximum shearing stress is half the maximum bending stress.   [$16F/3\pi D^2$; $\frac{2}{3}$]

4 Torsion

4.1 Stress due to twisting

When a shaft is subjected to a torque, every section is in a state of shear. The shaft will twist and the shear stress resulting from this strain will produce a moment of resistance, equal and opposite to the applied torque.

In the simple theory applicable to straight, circular, homogeneous, elastic shafts, it is assumed that radial lines remain radial after twisting. The shear strain is therefore proportional to the radius and it therefore follows from Hooke's Law that the stress is also proportional to the radius.

Thus, if the stress at radius r is τ, Figure 4.1,

the stress on an element dA at radius $x = \dfrac{x}{r}\tau$

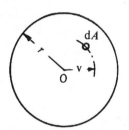

Figure 4.1

$$\therefore \text{ shear force on element} = \frac{x}{r}\tau\,dA$$

$$\therefore \text{ moment of force about O} = \frac{x}{r}\tau\,dA \cdot x$$

$$\therefore \text{ total moment of resistance} = \int \frac{x^2}{r}\tau\,dA$$

$$= \frac{\tau}{r}\int x^2\,dA$$

$\int x^2 dA$ is the polar second moment of area about the axis of the shaft and is denoted by J (see Appendix A).

Thus
$$T = \frac{\tau}{r}J$$

or
$$\frac{T}{J} = \frac{\tau}{r} \tag{4.1}$$

Equation (4.1) gives the stress at the surface of the shaft, where it is a maximum, but the stress at any other radius is proportional to that radius.

For a solid shaft of diameter d, $J = \dfrac{\pi}{32}d^4$; for a hollow shaft of external diameter D and internal diameter d,

$$J = \frac{\pi}{32}(D^4 - d^4).$$

Equation (4.1) can be written as $\tau = \dfrac{T}{J/r}$; the quantity J/r is termed the *modulus of section* and is denoted by Z.

Thus
$$\tau = \frac{T}{Z} \tag{4.2}$$

For a solid shaft,
$$Z = \frac{\pi}{32}d^4 \Big/ \frac{d}{2} \qquad = \frac{\pi}{16}d^3$$

and for a hollow shaft,
$$Z = \frac{\pi}{32}(D^4 - d^4) \Big/ \frac{D}{2} \qquad = \frac{\pi}{16}\frac{D^4 - d^4}{D}$$

It should be noted that values of the section modulus for twisting are not the same as those for bending.

4.2 Angle of twist

The shear strain in the shaft is the angle ϕ, Figure 4.2.

$$\phi = \frac{AB}{l}$$

$$= \frac{r\theta}{l}$$

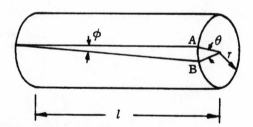

Figure 4.2

Also, from equation (1.6),

$$\phi = \frac{\tau}{G}$$

$$\therefore \frac{\tau}{G} = \frac{r\theta}{l}$$

or
$$\frac{\tau}{r} = \frac{G\theta}{l} \tag{4.3}$$

Combining equations (4.1) and (4.3),

$$\frac{T}{J} = \frac{\tau}{r} = \frac{G\theta}{l} \tag{4.4}$$

Worked examples

1. *A hollow shaft of diameter ratio 3:5 is required to transmit 600 kW at 110 rev/min, the maximum torque being 12 per cent greater than the mean. The shearing stress is not to exceed 60 MN/m² and the twist in a length of 3 m is not to exceed 1°. Calculate the minimum external diameter to satisfy these conditions. G = 80 GN/m².*

$$P = T\omega$$

i.e.

$$600 \times 10^3 = T \times \frac{2\pi}{60} \times 110$$

$$\therefore T = 52.1 \times 10^3 \text{ Nm}$$

$$\therefore \text{maximum torque} = 52.1 \times 10^3 \times 1.12 = 58.3 \times 10^3 \text{ Nm}$$

For the stress condition, $\qquad T = \tau Z$ from equation (4.2)

i.e.

$$58.3 \times 10^3 = 60 \times 10^6 \times \frac{\pi}{16} \frac{D^4 - \left(\frac{3}{5}D\right)^4}{D}$$

from which $\qquad D = 0.178\,3 \text{ m}$

For the twist condition, $\qquad \dfrac{T}{J} = \dfrac{G\theta}{l}$ from equation (4.4)

i.e.

$$\frac{58.3 \times 10^3}{\frac{\pi}{32}\left\{D^4 - \left(\frac{3}{5}D\right)^4\right\}} = \frac{80 \times 10^9 \times 1 \times \frac{\pi}{180}}{3}$$

from which $\qquad D = 0.195\,5 \text{ m}$

The minimum external diameter to satisfy both conditions is therefore <u>0.195 5 m</u>

2. *A shaft runs at 300 rev/min and transmits power from a pulley A at one end to two pulleys B and C which each drive a machine in a workshop. The distance between pulleys A and B is 3 m and that between B and C is 2.4 m. The shaft between A and B is 50 mm diameter and between B and C it is 40 mm diameter.*

If the maximum permissible shear stress in the shaft is 80 MN/m², calculate the maximum power which may be supplied from each of the pulleys B and C, assuming that both machines are in operation at the same time. Also calculate the total angle of twist of one end of the shaft relative to the other when running on full load. G = 80 GN/m².

The arrangement is shown in Figure 4.3.

$$T = \tau Z \quad \text{from equation (4.2)}$$

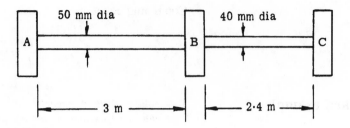

Figure 4.3

Therefore maximum torque transmissible by AB $= 80 \times 10^6 \times \dfrac{\pi}{16} \times 0.05^3$

$$= 1\,964 \text{ Nm}$$

and maximum torque transmissible by BC $= 80 \times 10^6 \times \dfrac{\pi}{16} \times 0.04^3$

$$= 1\,006 \text{ Nm}$$

Therefore maximum power transmissible by AB $= 1\,964 \times \dfrac{2\pi}{60} \times 300$

$$= 61\,700 \text{ W}$$

and maximum power transmissible by BC $= 1\,006 \times \dfrac{2\pi}{60} \times 300$

$$= 31\,600 \text{ W}$$

Therefore maximum power which can be taken from pulley C = <u>31.6 kW</u>

and maximum power which can be taken from pulley B $= 61.7 - 31.6$

$$= \underline{30.7 \text{ kW}}$$

$$\theta = \sum \frac{\tau l}{Gr} \quad \text{from equation (4.3)}$$

$$= \frac{80 \times 10^6}{80 \times 10^9} \left(\frac{3}{0.025} + \frac{2.4}{0.020} \right)$$

$$= 0.24 \quad \text{rad or } \underline{13.75°}$$

3. *Part of a steel tube 24 mm external diameter and 6 mm thick is enlarged to an external diameter of 36 mm. Find the diameter of the bore of the enlarged section so that when the tube is twisted, the maximum shear stresses in both sections are equal.*

If the total length of tube is 1 m, find the length of each part when the total angle of twist is 4° and the maximum shear stress is 75 MN/m².

G = 80 GN/m².

The tube is shown in Figure 4.4.

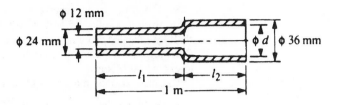

Figure 4.4

For equal stresses under the same torque, the modulus of section of each part must be the same,

i.e.
$$\frac{24^4 - 12^4}{24} = \frac{36^4 - d^4}{36}$$

from which
$$d = \underline{33.2 \text{ mm}}$$

From equation (4.3)
$$\theta = \frac{\tau l}{Gr}$$

Therefore, if the lengths of the 24 mm and 36 mm diameter parts are l_1 and l_2 respectively,

$$\frac{75 \times 10^6}{80 \times 10^9} \left(\frac{l_1}{0.012} + \frac{l_2}{0.018} \right) = 4 \times \frac{\pi}{180}$$

from which
$$1.5l_1 + l_2 = 1.34$$

Also
$$l_1 + l_2 = 1$$

$$\therefore l_1 = \underline{0.68 \text{ m}} \text{ and } l_2 = \underline{0.32 \text{ m}}$$

Further problems

4. What power can be transmitted by a steel shaft 100 mm diameter at a speed of 300 rev/min if the shear stress is limited to 90 MN/m²?

Under these conditions, what would be the angle of twist over a 3 m length of shaft? $G = 80$ GN/m². [555 kW; 3.86⁰]

5. Determine the diameter of a solid steel shaft to transmit 30 MW at a speed of 1 500 rev/min if the angle of twist is limited to 1° for every twenty diameters of length. $G = 80$ GN/m². [303 mm]

6. A torque of 50 kN m is to be transmitted by a hollow shaft of internal diameter half the external diameter. If the maximum shear stress is not to exceed 80 MN/m², calculate the outside diameter of the shaft. What would be the angle of twist over a length of 3 m of the shaft under the above torque? $G = 80$ GN/m².

[150 mm; 2°20′]

7. One quarter of the mass of a solid round shaft, 25 mm diameter, is removed by axial boring. Determine (a) by what percentage the strength of the shaft in torsion is reduced by this boring, (b) the maximum power that could be transmitted by the shaft before and after boring if the maximum allowable shear stress is 80 MN/m² and the speed is 150 rev/min. [6.25%; 3.86 kW; 3.62 kW]

8. A hollow shaft having external and internal diameters of 88 and 50 mm respectively transmits 1.5 MW at 2 000 rev/min. The shaft is connected to another shaft by a flanged coupling having eight bolts on a pitch circle of 180 mm diameter. Determine the minimum permissible diameter of the bolts if the shear stress is not to exceed 50 MN/m^2.

Calculate also the maximum shear stress in the shaft.   [15.9 mm; 59.8 MN/m^2]

9. A hollow shaft is 50 mm outside diameter and 30 mm inside diameter. An applied torque of 1.6 kN m is found to produce an angular twist of 0.4°, measured on a length of 0.2 m of the shaft. Calculate the value of the modulus of rigidity. Calculate also the maximum power which can be transmitted by the shaft at 2 000 rev/min if the maximum allowable shearing stress is 65 MN/m^2.   [86 GN/m^2; 292 kW]

10. A shaft 50 mm in diameter and 0.75 m long has a concentric hole drilled for a portion of its length. Find the maximum length and diameter of the hole so that when the shaft is subjected to a torque of 1.67 kN m, the maximum shearing stress will not exceed 75 MN/m^2 and the total angle of twist will not exceed $1\frac{1}{2}^\circ$.
$G = 80$ GN/m^2.   [27.7 mm; 0.19 m]

11. A solid shaft 50 mm diameter and 1.5 m long is passed through the centre of a hollow shaft of the same material and length, 55 mm and 75 mm inner and outer diameters. The ends of the two shafts are rigidly joined, with the shafts concentric. The composite shaft so formed is used to transmit 375 kW at a speed of 600 rev/min. Find the maximum stresses in the two shafts.

[79.3 MN/m^2; 52.9 MN/m^2]

12. A solid alloy shaft of 50 mm diameter is to be coupled in series with a hollow steel shaft of the same external diameter. Find the internal diameter of the steel shaft if the angle of twist per unit length is to be 75% of that of the alloy shaft.

Determine the speed at which the shafts are to be driven to transmit 200 kW if the limits of shearing stress are to be 55 and 80 MN/m^2 in the alloy and steel respectively. $G_{steel} = 2.2 \times G_{alloy}$.   [39.6 mm; 1 605 rev/min]

5 Deflection of beams

5.1 Introduction

The deflection of a beam depends upon the loading, the dimensions and material of the beam and the method of support. The principal types of beams to be considered are cantilevers, simply supported beams, in which rotation of the ends is unrestrained, and built-in (or *encastré*) beams, where the ends are rigidly built in to walls which restrain the ends to be horizontal but free from longitudinal tension.

5.2 Basic relations

From equation (3.4)

$$\frac{M}{I} = \frac{E}{R} \quad \text{which may be re-arranged to give}$$

$$\therefore \frac{1}{R} = \frac{M}{EI} \tag{5.1}$$

The radius of curvature of any function $y = f(x)$ is expressed by

$$\frac{1}{R} = \frac{\dfrac{d^2 y}{dx^2}}{\left[1 + \left(\dfrac{dy}{dx} \right)^2 \right]^{3/2}}$$

When applied to a deflected beam, $\dfrac{dy}{dx}$ is very small, so that this equation reduces to

$$\frac{1}{R} = \frac{d^2 y}{dx^2} \tag{5.2}$$

Combining equations (5.1) and (5.2) yields

$$\frac{d^2 y}{dx^2} = \frac{M}{EI} \tag{5.3}$$

If the beam is of uniform cross-section and M may be expressed mathematically as a function of x, then the slope of the beam

$$\frac{dy}{dx} = \frac{1}{EI} \int M \, dx + A \qquad (5.4)$$

and the deflection,

$$y = \frac{1}{EI} \int \int M \, dx \, dx + Ax + B \qquad (5.5)$$

Further, from equations (2.1) and (2.2),

$$F = \frac{dM}{dx} \quad \text{and} \quad w = -\frac{dF}{dx}$$

so that

$$\frac{d^3 y}{dx^3} = \frac{F}{EI} \qquad (5.6)$$

and

$$\frac{d^4 y}{dx^4} = -\frac{w}{EI} \qquad (5.7)$$

Equation (5.7) is useful when the load is a function of x, as it may then be difficult to write down an expression for M in terms of x. In such cases, the deflection is obtained by integrating equation (5.7) four times.

Whichever equation is integrated, the resulting deflection will be in accordance with the Cartesian convention used, i.e., y positive upwards. This means that gravitational loading on a horizontal beam will usually produce negative deflections.

Slopes and deflections are inversely proportional to EI, which is termed the *flexural rigidity* of the beam.

5.3 Application of integration method to standard cases

In the following examples, the sign of the bending moment is determined by the convention of Section 2.2.

(a) Cantilever with concentrated end load, Figure 5.1

Taking the origin at the fixed end,

$$\text{B.M. at P} = -W(l - x)$$

$$\therefore \frac{d^2 y}{dx^2} = -\frac{W}{EI}(l - x)$$

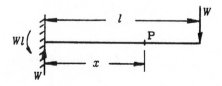

Figure 5.1

$$\therefore \frac{dy}{dx} = -\frac{W}{EI}\left(lx - \frac{x^2}{2}\right) + A$$

When $x = 0$, $\frac{dy}{dx} = 0$, so that $A = 0$

$$\therefore y = -\frac{W}{EI}\left(\frac{lx^2}{2} - \frac{x^3}{6}\right) + B$$

When $x = 0$, $y = 0$, so that $B = 0$.

The maximum slope and deflection occur at the free end, where $x = l$,

i.e.
$$\frac{dy}{dx} = -\frac{Wl^2}{2EI} \tag{5.8}$$

and
$$y = -\frac{Wl^3}{3EI} \tag{5.9}$$

(b) Cantilever with uniformly distributed load, Figure 5.2

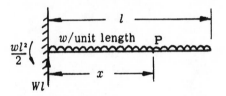

Figure 5.2

Taking the origin at the fixed end,

$$\text{B.M. at P} = -w(l - x)\frac{(l - x)}{2}$$

$$\therefore \frac{d^2y}{dx^2} = -\frac{w}{2EI}(l^2 - 2lx + x^2)$$

$$\therefore \frac{dy}{dx} = -\frac{w}{2EI}\left(l^2x - lx^2 + \frac{x^3}{3}\right) + A$$

When $x = 0$, $\frac{dy}{dx} = 0$, so that $A = 0$

$$\therefore y = -\frac{w}{2EI}\left(\frac{l^2x^2}{2} - \frac{lx^3}{3} + \frac{x^4}{12}\right) + B$$

When $x = 0$, $y = 0$, so that $B = 0$.

The maximum slope and deflection occur at the free end, where $x = l$,

i.e.
$$\frac{\mathrm{d}y}{\mathrm{d}x} = -\frac{wl^3}{6EI}$$
(5.10)

and
$$y = -\frac{wl^4}{8EI}$$
(5.11)

(c) Cantilever with end couple, Figure 5.3

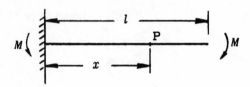

Figure 5.3

Taking the origin at the fixed end,

$$\text{B.M. at P} = -M$$

$$\therefore \frac{\mathrm{d}^2 y}{\mathrm{d}x^2} = -\frac{M}{EI}$$

$$\therefore \frac{\mathrm{d}y}{\mathrm{d}x} = -\frac{Mx}{EI} + A$$

When $x = 0$, $\dfrac{\mathrm{d}y}{\mathrm{d}x} = 0$, so that $A = 0$

$$\therefore y = -\frac{Mx^2}{2EI} + B$$

When $x = 0$, $y = 0$, so that $B = 0$.

The maximum slope and deflection occur at the free end, where $x = l$,

i.e.
$$\frac{\mathrm{d}y}{\mathrm{d}x} = -\frac{Ml}{EI}$$
(5.12)

and
$$y = -\frac{Ml^2}{2EI}$$
(5.13)

(d) Simply supported beam with central concentrated load, Figure 5.4

Taking the origin at the centre,

$$\text{B.M. at P} = \frac{W}{2}\left(\frac{l}{2} - x\right)$$

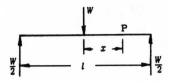

Figure 5.4

$$\therefore \frac{d^2y}{dx^2} = \frac{W}{2EI}\left(\frac{l}{2} - x\right)$$

$$\therefore \frac{dy}{dx} = \frac{W}{2EI}\left(\frac{lx}{2} - \frac{x^2}{2}\right) + A$$

When $x = 0$, $\frac{dy}{dx} = 0$, so that $A = 0$

$$\therefore y = \frac{W}{2EI}\left(\frac{lx^2}{4} - \frac{x^3}{6}\right) + B$$

When $x = \frac{l}{2}$, $y = 0$, so that $B = -\frac{W}{2EI} \cdot \frac{l^3}{24}$

$$\therefore y = \frac{W}{2EI}\left(\frac{lx^2}{4} - \frac{x^3}{6} - \frac{l^3}{24}\right)$$

The maximum slope occurs at the ends, where $x = \frac{l}{2}$,

i.e.
$$\frac{dy}{dx} = \frac{Wl^2}{16EI} \tag{5.14}$$

The maximum deflection occurs at the centre, where $x = 0$

i.e.
$$y = -\frac{Wl^3}{48EI} \tag{5.15}$$

(e) Simply supported beam with uniformly distributed load, Figure 5.5

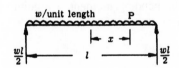

Figure 5.5

Taking the origin at the centre,

$$\text{B.M. at P} = \frac{wl}{2}\left(\frac{l}{2} - x\right) - w\left(\frac{l}{2} - x\right)\frac{\left(\frac{l}{2} - x\right)}{2}$$

$$\therefore \frac{d^2y}{dx^2} = \frac{w}{2EI}\left(\frac{l^2}{4} - x^2\right)$$

$$\therefore \frac{dy}{dx} = \frac{w}{2EI}\left(\frac{l^2x}{4} - \frac{x^3}{3}\right) + A$$

When $x = 0$, $\frac{dy}{dx} = 0$, so that $A = 0$

$$\therefore y = \frac{w}{2EI}\left(\frac{l^2x^2}{8} - \frac{x^4}{12}\right) + B$$

When $x = \frac{l}{2}$, $y = 0$, so that $B = -\frac{w}{2EI} \cdot \frac{5l^4}{192}$

$$\therefore y = \frac{w}{2EI}\left(\frac{l^2x^2}{8} - \frac{x^4}{12} - \frac{5l^4}{192}\right)$$

The maximum slope occurs at the ends, where $x = \dfrac{l}{2}$,

i.e.
$$\frac{dy}{dx} = \frac{wl^3}{24EI} \tag{5.16}$$

The maximum deflection occurs at the centre, where $x = 0$,

i.e.
$$y = -\frac{5wl^4}{384EI} \tag{5.17}$$

5.4 Principle of superposition

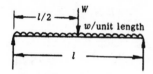

Figure 5.6

The principle of superposition states that if the bending moment, slope, deflection, etc., is directly proportional to the loading, the effect of a combination of loading may be obtained by the algebraic addition of the effects of each load considered separately. This may lead to the rapid solution of problems using the standard cases derived in Section 5.3.

Thus for the loading shown in Figure 5.6, the total deflection at the centre is the sum of the deflections due to the concentrated and distributed loads considered separately,

i.e.
$$y = -\left[\frac{Wl^3}{48EI} + \frac{5}{384}\frac{wl^4}{EI}\right]$$

5.5 Unsymmetrical loading–Macaulay's method

For the loaded beam shown in Figure 5.7, the bending moment at a point between A and C, distance x from A, is $R_1 x$ so that

$$\frac{d^2y}{dx^2} = \frac{R_1}{EI} \cdot x$$

$$\therefore \frac{dy}{dx} = \frac{R_1}{EI} \cdot \frac{x^2}{2} + A_1 \tag{5.18}$$

and
$$y = \frac{R_1}{EI} \cdot \frac{x^3}{6} + A_1 x + B_1 \tag{5.19}$$

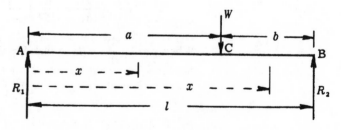

Figure 5.7

When $x = 0$, $y = 0$, so that $B_1 = 0$. No other conditions are available within the range AC so that A_1 must remain unknown at present.

For a point between C and B, keeping the origin at A,

$$\text{bending moment} = R_1 x - W[x - a]$$

$$\therefore \quad \frac{d^2 y}{dx^2} = \frac{1}{EI}\{R_1 x - W[x - a]\} \tag{5.20}$$

$$\therefore \quad \frac{dy}{dx} = \frac{1}{EI}\left\{R_1 \frac{x^2}{2} - W\left[\frac{x^2}{2} - ax\right]\right\} + A_2 \tag{5.21}$$

and $$y = \frac{1}{EI}\left\{R_1 \frac{x^3}{6} - W\left[\frac{x^3}{6} - \frac{ax^2}{2}\right]\right\} + A_2 x + B_2 \tag{5.22}$$

When $x = l$, $y = 0$, so that B_2 can be determined. No other conditions are available within the range CB so that A_2 must also remain unknown at this stage.

However, the slope at C must be the same from equations (5.18) and (5.21) by substituting $x = a$ and the deflection at C must be the same from equations (5.19) and (5.22) by substituting $x = a$. Hence two equations are obtained from which A_1 and A_2 may be calculated. The slopes and deflections in the two ranges are then obtained by using the appropriate equations.

A similar process may be used for several loads but the constants of integration are different for each range and much laborious work is entailed.

If, however, equation (5.20) is integrated to give

$$\frac{dy}{dx} = \frac{1}{EI}\left\{R_1 \frac{x^2}{2} - \frac{W}{2}[x - a]^2\right\} + A_2 \tag{5.23}$$

and $$y = \frac{1}{EI}\left\{R_1 \frac{x^3}{6} - \frac{W}{6}[x - a]^3\right\} + A_2 x + B_2 \tag{5.24}$$

then equating (5.18) and (5.23) at $x = a$ gives $A_1 = A_2$, and equating (5.19) and (5.24) at $x = a$ gives $B_1 = B_2$.

Thus, by this method of integration, the constants are the same for each range of the beam. Further, equations (5.23) and (5.24) may be regarded as applying to the whole beam if the terms involving $[x - a]$ are ignored when $[x - a]$ is negative, i.e., when $x < a$, since these then give the corresponding equations for the range AC. This method, known as *Macaulay's method*, may be applied to any number of loads and it is conventional to use square brackets for terms such as $[x - a]$ which have to be treated in this special manner. The bending moment equation must always be written down at a point within the range remote from the origin so as to embrace all the loads.

For the simple case considered above, $R_1 = \dfrac{Wb}{l}$, so that the general deflection equation for the beam becomes

$$y = \frac{1}{EI}\left\{\frac{Wb}{l}\cdot\frac{x^3}{6} - \frac{W}{6}[x-a]^3\right\} + Ax + B$$

the constants A and B applying to the whole beam and the term $\dfrac{W}{6}[x-a]^3$ being ignored for values of x which make $[x-a]$ negative.

When $x = 0$, $y = 0$, so that $B = 0$, $[x-a]$ being negative for this value of x.

When $x = l$, $y = 0$, so that $A = -\dfrac{Wab}{6EIl}(l+b)$.

The deflection under the load is obtained by putting $x = a$, when

$$y = -\frac{Wa^2b^2}{3EIl} \tag{5.25}$$

At the point of maximum deflection, $\dfrac{dy}{dx} = 0$ but it is first necessary to estimate which range of the beam will contain this point so that terms involving square brackets may be included or ignored as appropriate (see Example 6).

5.6 Distributed loads

A distributed load covering the entire span can be included in the bending moment equation for concentrated loads and is valid for all values of x. If, however, the load is discontinuous, it must extend to the end of the beam remote from the origin, adding, if necessary, a negative load to compensate for an additional load on top of the beam. Thus the load system shown in Figure 5.8(a) must be converted into that in Figure 5.8(b) in order that a continuous equation can be written down for the bending moment at any point, in accordance with the requirements of Macaulay's method.

Considering a section within the range of the beam remote from the origin and taking moments of forces to the left of that section,

$$M = R_1x - \frac{w}{2}[x-a]^2 + \frac{w}{2}[x-b]^2$$

from which

$$y = \frac{1}{EI}\left\{R_1\frac{x^3}{6} - \frac{w}{24}[x-a]^4 + \frac{w}{24}[x-b]^4\right\} + Ax + B$$

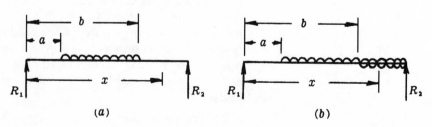

(a)   (b)

Figure 5.8

The terms in square brackets are treated in exactly the same way as for concentrated loads; they must be integrated with respect to the quantities within the brackets and must be ignored when the quantities within the brackets become negative.

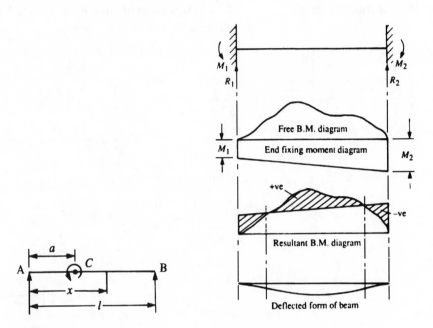

Figure 5.9   Figure 5.10

5.7 Beam with applied couple

Figure 5.9 shows a beam with a couple C applied at a point distance a from A.

As with concentrated loads, the couple only comes in for $x > a$ and so, using Macaulay's method, C must be integrated with respect to $[x - a]$. In order to proceed correctly, it is usual to write the couple term in the form $C[x - a]^0$ (see Example 7).

5.8 Built-in beams

If the ends of a beam are rigidly built in to walls or similar fixings at each end, these fixings exert moments on the beam to maintain the ends horizontal. The bending moment diagram for the beam then consists of the 'free' bending moment diagram (i.e. the bending moment diagram due to the transverse loads only, as if on a simply supported beam) and the bending moment diagram due to the end fixing moments. These are of opposite sign and the resultant bending moment diagram is then given by the difference of the two diagrams, as shown in Figure 5.10.

The bending moment at any point is the *vertical* ordinate on the resultant bending moment diagram, which may be replotted on a horizontal base, if required.

The form of the deflected beam follows from the resultant bending moment diagram; the points at which the bending changes from positive to negative are known as the points of *contraflexure*.

5.9 Standard cases

(a) Central concentrated load, Figure 5.11

Taking the origin at the centre,

$$\text{B.M. at P} = \frac{W}{2}\left(\frac{l}{2} - x\right) - M$$

$$\therefore \frac{d^2y}{dx^2} = \frac{1}{EI}\left\{\frac{W}{2}\left(\frac{l}{2} - x\right) - M\right\}$$

$$\therefore \frac{dy}{dx} = \frac{1}{EI}\left\{\frac{W}{2}\left(\frac{lx}{2} - \frac{x^2}{2}\right) - Mx\right\} + A$$

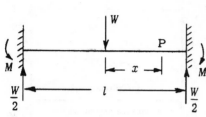

Figure 5.11

When $x = 0$, $\dfrac{dy}{dx} = 0$, so that $A = 0$.

When $x = \dfrac{l}{2}$, $\dfrac{dy}{dx} = 0$, so that $M = \dfrac{Wl}{8}$ (5.26)

$$\therefore \frac{dy}{dx} = \frac{W}{2EI}\left(\frac{lx}{4} - \frac{x^2}{2}\right)$$

$$\therefore y = \frac{W}{2EI}\left(\frac{lx^2}{8} - \frac{x^3}{6}\right) + B$$

When $x = \dfrac{l}{2}$, $y = 0$, so that $B = -\dfrac{Wl^3}{192EI}$

$$\therefore y = \frac{W}{2EI}\left(\frac{lx^2}{8} - \frac{x^3}{6} - \frac{l^3}{96}\right)$$

The maximum deflection occurs at the centre, where $x = 0$,

i.e. $y = -\dfrac{Wl^3}{192EI}$ (5.27)

(b) Uniformly distributed load, Figure 5.12

Taking the origin at the centre,

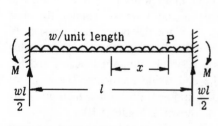

Figure 5.12

$$\text{B.M. at P} = \frac{wl}{2}\left(\frac{l}{2} - x\right) - w\left(\frac{l}{2} - x\right)\frac{\left(\frac{l}{2} - x\right)}{2} - M$$

$$\therefore \frac{d^2y}{dx^2} = \frac{1}{EI}\left\{\frac{w}{2}\left(\frac{l^2}{4} - x^2\right) - M\right\}$$

$$\therefore \frac{dy}{dx} = \frac{1}{EI}\left\{\frac{w}{2}\left(\frac{l^2x}{4} - \frac{x^3}{3}\right) - Mx\right\} + A$$

When $x = 0$, $\dfrac{dy}{dx} = 0$, so that $A = 0$

When $x = \dfrac{l}{2}$, $\dfrac{dy}{dx} = 0$, so that $M = \dfrac{wl^2}{12}$ $\qquad$ (5.28)

$$\therefore \frac{dy}{dx} = \frac{w}{2EI}\left(\frac{l^2 x}{12} - \frac{x^3}{3}\right)$$

$$\therefore y = \frac{w}{2EI}\left(\frac{l^2 x^2}{24} - \frac{x^4}{12}\right) + B$$

When $x = \dfrac{l}{2}$, $y = 0$, so that $B = -\dfrac{wl^4}{384EI}$

$$\therefore y = \frac{w}{2EI}\left(\frac{l^2 x^2}{24} - \frac{x^4}{12} - \frac{l^4}{192}\right)$$

The maximum deflection occurs at the centre, where $x = 0$,

i.e. $\qquad\qquad\qquad\qquad y = -\dfrac{wl^4}{384EI}$ $\qquad$ (5.29)

(c) Single concentrated load not at centre, Figure 5.13

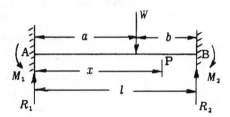

Figure 5.13

Taking the origin at A and using Macaulay's method,

$$\text{B.M. at P} = R_1 x - W[x - a] - M_1$$

$$\therefore \frac{d^2 y}{dx^2} = \frac{1}{EI}\{R_1 x - W[x - a] - M_1\}$$

$$\therefore \frac{dy}{dx} = \frac{1}{EI}\left\{R_1\frac{x^2}{2} - \frac{W}{2}[x - a]^2 - M_1 x\right\} + A \qquad (1)$$

When $x = 0$, $\dfrac{dy}{dx} = 0$, so that $A = 0$, the term $[x - a]$ being negative and hence ignored.

$$\therefore y = \frac{1}{EI}\left\{R_1\frac{x^3}{6} - \frac{W}{6}[x - a]^3 - M_1\frac{x^2}{2}\right\} + B \qquad (2)$$

When $x = 0$, $y = 0$, so that $B = 0$, the term $[x - a]$ again being negative.
When $x = l$, $\dfrac{dy}{dx} = 0$ and $y = 0$.

Inserting these conditions in equations (1) and (2) gives

$$R_1 = \frac{Wb^2}{l^3}(l + 2a) \quad \text{and} \quad M_1 = \frac{Wab^2}{l^2}$$

and, by symmetry,

$$R_2 = \frac{Wa^2}{l^3}(l + 2b) \quad \text{and} \quad M_2 = \frac{Wa^2b}{l^2}$$

The deflection under the load is obtained by putting $x = a$, when

$$y = -\frac{Wa^3b^3}{3EIl^3} \tag{5.30}$$

It will be noted that the reactions R_1 and R_2 are not in the ratio b/a as for a simply supported beam; this is due to the unequal fixing moments M_1 and M_2.

5.10 Area-moment method

Figure 5.14 shows part of a beam AB which, when loaded, deflects to A'B'. The points A and B are at distances x_1 and x_2 from an origin O, the slopes at these points are θ_1 and θ_2 and the deflections are y_1 and y_2 respectively.

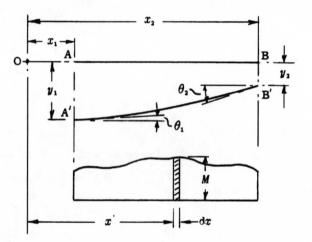

Figure 5.14

The sketch above shows the bending moment diagram for AB, the bending moment at a point distant x from O being M.

From equation (5.3),

$$\frac{\mathrm{d}^2 y}{\mathrm{d}x^2} = \frac{M}{EI} \tag{5.31}$$

$$\therefore \frac{\mathrm{d}y}{\mathrm{d}x} = \frac{1}{EI}\int M\,\mathrm{d}x$$

Thus, the change of slope between A and B,

$$\theta_2 - \theta_1 = \frac{1}{EI} \int_{x_1}^{x_2} M \, dx \qquad (5.32)$$

$$= \frac{1}{EI} \times \text{area of B.M. diagram between A and B}$$

Multiplying equation (5.31) by x gives

$$x\frac{d^2 y}{dx^2} = \frac{Mx}{EI}$$

Integrating both sides between the limits x_1 and x_2 gives

$$\left[x\frac{dy}{dx} - y \right]_{x_1}^{x_2} = \frac{1}{EI} \int_{x_1}^{x_2} Mx \, dx \qquad (5.33)$$

$$= \frac{1}{EI} \times \text{moment of area of B.M. diagram about O}$$

The origin O is chosen so as to make $x\dfrac{dy}{dx}$ zero at both limits, so that the left-hand side then gives the difference of deflection between A and B.

Note: Beams with distributed loads require the parabola data given on page 61.

5.11 Application of area-moment method to standard cases

(a) Cantilever with concentrated end load, Figure 5.15

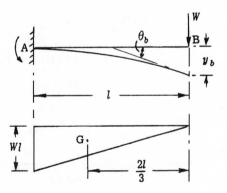

Figure 5.15

$$\theta_b - \theta_a = \frac{1}{EI} \left(-\tfrac{1}{2}Wl \times l \right)$$

i.e.

$$\theta_b = -\frac{Wl^2}{2EI}$$

since $\theta_a = 0$

Taking the origin at B,

$$\left[x\frac{dy}{dx} - y\right]_0^l = \frac{1}{EI}\left(-\tfrac{1}{2}Wl \times l \times \tfrac{2}{3}l\right)$$

i.e.
$$(0 - 0) - (0 - y_b) = y_b = -\frac{Wl^3}{3EI}$$

(b) Cantilever with uniformly distributed load, Figure 5.16

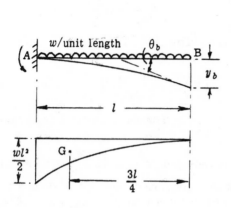

Figure 5.16   **Figure 5.17**

$$\theta_b - \theta_a = \frac{1}{EI}\left(-\tfrac{1}{3}\frac{wl^2}{2} \times l\right)$$

i.e.
$$\theta_b = -\frac{wl^3}{6EI} \text{ since } \theta_a = 0$$

Taking the origin at B,

$$\left[x\frac{dy}{dx} - y\right]_0^l = \frac{1}{EI}\left(-\tfrac{1}{3}\frac{wl^2}{2} \times l \times \tfrac{3}{4}l\right)$$

i.e.
$$(0 - 0) - (0 - y_b) = y_b = -\frac{wl^4}{8EI}$$

(c) Simply supported beam with central concentrated load, Figure 5.17

$$\theta_b - \theta_a = \frac{1}{EI} \times \tfrac{1}{2}\frac{Wl}{4} \times \frac{l}{2}$$

i.e.
$$\theta_b = \frac{Wl^2}{16EI} \text{ since } \theta_a = 0$$

Taking the origin at B,

$$\left[x\frac{dy}{dx} - y\right]_0^{l/2} = \frac{1}{EI} \times \tfrac{1}{2}\frac{Wl}{4} \times \frac{l}{2} \times \frac{2}{3}\frac{l}{2}$$

i.e.
$$(0 - y_a) - (0 - 0) = \frac{Wl^3}{48EI}$$

or
$$y_a = -\frac{Wl^3}{48EI}$$

(d) Simply supported beam with uniformly distributed load, Figure 5.18

$$\theta_b - \theta_a = \frac{1}{EI} \times \frac{2}{3}\frac{wl^2}{8} \times \frac{l}{2}$$

i.e.
$$\theta_b = \frac{wl^3}{24EI} \text{ since } \theta_a = 0$$

Taking the origin at B,

$$\left[x\frac{dy}{dx} - y\right]_0^{l/2} = \frac{1}{EI} \times \frac{2}{3}\frac{wl^2}{8} \times \frac{l}{2} \times \frac{5}{8}\frac{l}{2}$$

i.e.
$$(0 - y_a) - (0 - 0) = \frac{5}{384}\frac{wl^4}{EI}$$

or
$$y_a = -\frac{5}{384}\frac{wl^4}{EI}$$

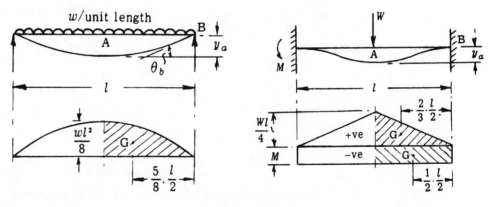

Figure 5.18   Figure 5.19

(e) Built-in beam with central concentrated load, Figure 5.19

$$\theta_b - \theta_a = 0$$

$$\therefore \frac{1}{2}\frac{Wl}{4} \times \frac{l}{2} - M\frac{l}{2} = 0$$

$$\therefore M = \frac{Wl}{8}$$

Taking the origin at B,

$$\left[x\frac{dy}{dx} - y\right]_0^{l/2} = \frac{1}{EI}\left(\frac{1}{2}\frac{Wl}{4} \times \frac{l}{2} \times \frac{2}{3}\frac{l}{2} - M\frac{l}{2} \times \frac{1}{2}\frac{l}{2}\right)$$

i.e. $$(0 - y_a) - (0 - 0) = \frac{1}{EI}\left(\frac{Wl^3}{48} - \frac{Wl^3}{64}\right)$$

$$\therefore y_a = -\frac{Wl^3}{192EI}$$

(f) Built-in beam with uniformly distributed load, Figure 5.20

$$\theta_b - \theta_a = 0$$

$$\therefore \frac{2}{3}\frac{wl^2}{8} \times \frac{l}{2} - M\frac{l}{2} = 0$$

$$\therefore M = \frac{wl^2}{12}$$

Taking the origin at B,

$$\left[x\frac{dy}{dx} - y\right]_0^{l/2} = \frac{1}{EI}\left(\frac{2}{3}\frac{wl^2}{8} \times \frac{l}{2} \times \frac{5}{8}\frac{l}{2} - M\frac{l}{2} \times \frac{1}{2}\frac{l}{2}\right)$$

i.e. $$(0 - y_a) - (0 - 0) = \frac{1}{EI}\left(\frac{5wl^4}{384} - \frac{wl^4}{96}\right)$$

$$\therefore y_a = -\frac{wl^4}{384EI}$$

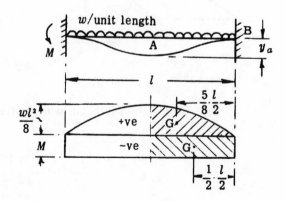

Figure 5.20

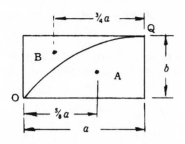

Figure 5.21

Properties of the parabola

The bending moment diagrams for distributed loads are parabolic and the essential properties of the parabola (Figure 5.21) are as follows:

Area of A $= \frac{2}{3}ab$

Area of B $= \frac{1}{3}ab$

Distance of centroid of A from O $= \frac{5}{8}a$

Distance of centroid of B from Q $= \frac{3}{4}a$

Worked examples

1. *A cantilever of length l carries a uniformly distributed load w over the entire length and is propped at a point l/4 from the free end, the level of the prop being adjusted so that there is no deflection at the free end.*

Derive a formula for the reaction at the prop and for the deflection of the beam at the prop.

The downward deflection at the free end due to w alone, shown in Figure 5.22(b), is given by

$$y = -\frac{wl^4}{8EI} \quad \text{from equation (5.11)}$$

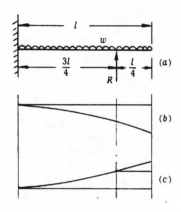

Figure 5.22

The upward deflection at the free end due to the prop force R alone, shown in Figure 5.22(c), is made up of the deflection at the prop, together with the slope of the prop multiplied by the overhanging length $l/4$, this part of the beam being straight.

Thus

$$y = \frac{R\left(\frac{3l}{4}\right)^3}{3EI} + \frac{R\left(\frac{3l}{4}\right)^2}{2EI} \times \frac{l}{4} \quad \text{from equations (5.9) and (5.8)}$$

$$= \frac{27}{128}\frac{Rl^3}{EI}$$

If there is to be resultant deflection at the free end, then

$$-\frac{wl^4}{8EI} + \frac{27}{128}\frac{Rl^3}{EI} = 0$$

from which

$$R = \frac{16}{27}wl$$

The deflection at the prop is the upward deflection at the prop due to R, less the downward deflection at that point due to w, derived in Section 5.3(b),

i.e.

$$y = \frac{R\left(\frac{3l}{4}\right)^3}{3EI} - \frac{w}{2EI}\left[\frac{l^2\left(\frac{3l}{4}\right)^2}{2} - \frac{l\left(\frac{3l}{4}\right)^3}{3} + \frac{\left(\frac{3l}{4}\right)^4}{12}\right]$$

$$= \frac{wl^4}{12EI} - \frac{513}{6\,144}\frac{wl^4}{EI} = -\frac{wl^4}{6\,144EI}$$

2. *Calculate the deflection at the free end of the cantilever shown in Figure 5.23(a).*

The total deflection is made up of (i) the deflection at B (y_1), (ii) the slope at B multiplied by the distance BC (y_2), (iii) the further deflection due to bending of BC (y_3).

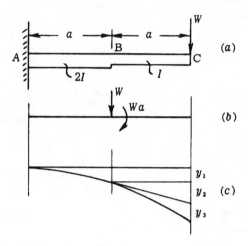

Figure 5.23

In calculating y_1 and y_2 (but not y_3), it is convenient to move the load to B, adding a moment Wa at this point to compensate for this movement.

The equivalent system is shown in Fig. 5.23(b) and the slope and deflection at B can then be written down, using the formulae derived in Sections 5.3(a) and (c). y_3 represents the further deflection due to bending of BC due to the load at C relative to the slope and deflection at B; this is given by equation (5.9).

$$\therefore y_1 = -\frac{Wa^3}{3E(2I)} - \frac{Wa \cdot a^2}{2E(2I)}$$

$$= -\frac{5}{12}\frac{Wa^3}{EI}$$

$$y_2 = \left(-\frac{Wa^2}{2E(2I)} - \frac{Wa \cdot a}{E(2I)}\right) \times a$$

$$= -\frac{3}{4}\frac{Wa^3}{EI}$$

and

$$y_3 = -\frac{Wa^3}{3EI}$$

$$\therefore y = -\frac{Wa^3}{EI}\left(\frac{5}{12} + \frac{3}{4} + \frac{1}{3}\right)$$

$$= -\frac{3}{2}\frac{Wa^3}{EI}$$

This problem may also be solved by the area-moment method. Since I is not constant, it is necessary to express equation (5.33) as

$$\left[x\frac{dy}{dx} - y\right]_{x_1}^{x_2} = \frac{1}{E} \times \text{moment of area of } \frac{M}{I} \text{ diagram about O}$$

The $\frac{M}{I}$ diagram for the cantilever is shown in Figure 5.24(b), which may conveniently be broken down into the two parts shown in Figure 5.24(c).

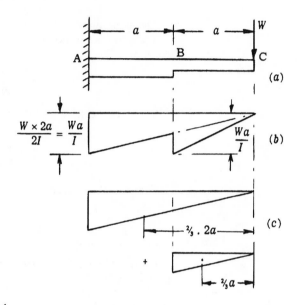

Figure 5.24

Taking moments about C,

$$\left[x\frac{dy}{dx} - y\right]_{0}^{2a} = \frac{1}{E}\left(-\frac{1}{2}\frac{Wa}{I} \times 2a \times \tfrac{2}{3} \cdot 2a - \frac{1}{2}\frac{Wa}{2I} \times a \times \tfrac{2}{3}a\right)$$

i.e. $(0 - 0) - (0 - y_c) = -\frac{1}{E}\left(\frac{4Wa^3}{3I} + \frac{Wa^3}{6I}\right)$

$$\therefore y_c = -\underline{\frac{3Wa^3}{2EI}}$$

3. *A uniform horizontal beam 6 m long, with a rectangular cross-section 50 mm wide by 250 mm deep, is simply supported at the ends and also at the middle by a vertical wire which stretches 0.5 μm/N. When the beam is unloaded, there is no tension in the wire.*

Calculate the deflection at the centre of the beam and the tension in the wire when the beam carries a uniformly distributed load of 20 kN/m over the entire length. $E = 200\ GN/m^2$.

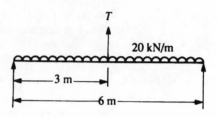

Figure 5.25

The arrangement of the beam is shown in Figure 5.25.

The downward deflection at the centre due to the uniformly distributed load is given by

$$y = -\frac{5wl^4}{384EI} \quad \text{from equation (5.17)}$$

The upward deflection at the centre due to the tension T in the wire is given by

$$y = \frac{Tl^3}{48EI} \quad \text{from equation (5.15)}$$

Therefore the residual deflection at the centre

$$= -\frac{5wl^4}{384EI} + \frac{Tl^3}{48EI}$$

Under a tension T, the centre of the beam moves down a distance $0.5 \times 10^{-6}T$, i.e. the stretch of the wire.

Hence $\quad -\dfrac{5wl^4}{384EI} + \dfrac{Tl^3}{48EI} = -0.5 \times 10^{-6}\ T$

i.e. $\quad -\dfrac{5 \times 20 \times 10^3 \times 6^4}{384} + \dfrac{T \times 6^3}{48} = -0.5 \times 10^{-6} \times 200 \times 10^9 \times \dfrac{50 \times 250^3}{12 \times 10^{12}} T$

from which $\qquad\qquad\qquad T = \underline{30.65 \times 10^3 N}$

$$\text{Central deflection} = -0.5 \times 10^{-6} \times 30.65 \times 10^3$$

$$= \underline{-0.015\,33\ \text{m}}$$

4. *A horizontal cantilever, 1.5 m long, tapers in section from 200 mm deep by 75 mm wide at the fixed end to 75 mm square at the extreme end and carries an end load of 3 kN. Calculate the deflection at the loaded end. E = 14 GN/m².*

The simplest solution will be obtained by taking the origin at the point of convergence of the top and bottom faces of the beam which, by similar triangles, will be found to be 0.9 m beyond the free end of the cantilever, Figure 5.26.

Then at a section distance x from O,

$$M = -3 \times 10^3(x - 0.9)$$

The depth of the section at this point,

$$d = \frac{x}{2.4} \times 0.2 = \frac{x}{12}\text{m}$$

$$\therefore I = \frac{0.075}{12}\left(\frac{x}{12}\right)^3 = 3.62 \times 10^{-6}x^3 \text{ m}^4$$

$$\therefore \frac{d^2y}{dx^2} = \frac{M}{EI} = \frac{-3 \times 10^3(x - 0.9)}{14 \times 10^9 \times 3.62 \times 10^{-6}x^3}$$

$$= -0.059\,2(x^{-2} - 0.9x^{-3})$$

$$\therefore \frac{dy}{dx} = -0.059\,2(-x^{-1} + 0.45x^{-2} + A)$$

When $x = 2.4$ m, $\frac{dy}{dx} = 0$, so that $A = 0.338\,6$

$$\therefore y = -0.059\,2(-\ln x - 0.45x^{-1} + 0.338\,6x + B)$$

When $x = 2.4$ m, $y = 0$, so that $B = 0.250\,4$

When $x = 0.9$ m,

$$y = -0.059\,2\left(-\ln 0.9 - \frac{0.45}{0.9} + 0.338\,6 \times 0.9 + 0.250\,4\right)$$

$$= -0.009\,53 \text{ m} \quad \text{or} \quad \underline{-9.53 \text{ mm}}$$

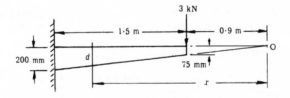

Figure 5.26   **Figure 5.27**

5. *A simply supported beam of span 2l carries a distributed load which varies uniformly from zero at each end to a maximum of w_0 per unit length at the centre. Obtain an expression for the central deflection.*

Taking the origin at the centre, Figure 5.27, the intensity of loading at a distance x from the centre,

$$w = w_0\left(1 - \frac{x}{l}\right)$$

$$\therefore \frac{d^4y}{dx^4} = -\frac{w}{EI}$$

$$= -\frac{w_0}{EI}\left(1 - \frac{x}{l}\right) \quad \text{from equation (5.7)}$$

$$\therefore \quad \frac{d^3 y}{dx^3} = -\frac{w_0}{EI}\left(x - \frac{x^2}{2l} + A\right)$$

When $x = 0$, the S.F. $= 0$, so that $A = 0$

$$\therefore \quad \frac{d^2 y}{dx^2} = -\frac{w_0}{EI}\left(\frac{x^2}{2} - \frac{x^3}{6l} + B\right)$$

When $x = l$, the B.M. $= 0$, so that $B = -\frac{l^2}{3}$

$$\therefore \quad \frac{dy}{dx} = -\frac{w_0}{EI}\left(\frac{x^3}{6} - \frac{x^4}{24l} - \frac{l^2 x}{3} + C\right)$$

When $x = 0$, $\dfrac{dy}{dx} = 0$, so that $C = 0$

$$\therefore \quad y = -\frac{w_0}{EI}\left(\frac{x^4}{24} - \frac{x^5}{120l} - \frac{l^2 x^2}{6} + D\right)$$

When $x = l$, $y = 0$, so that $D = \dfrac{2l^4}{15}$

When $x = 0$, $y = -\dfrac{2}{15}\dfrac{w_0 l^4}{EI}$

6. *A beam of length 8 m is simply supported at its ends and carries two concentrated loads of 20 kN and 40 kN respectively 2 m and 6 m from the left-hand end together with a distributed load of 15 kN/m on the 4 m between the concentrated loads.*

Calculate the deflection at the centre and also the position and magnitude of the maximum deflection, $E = 200$ GN/m² and $I = 0.18 \times 10^{-3}$ m⁴.

Taking moments about R_2, Figure 5.28,

$$8R_1 = 20 \times 6 + 40 \times 2 + 15 \times 4 \times 4$$

$$\therefore R_1 = 55 \text{ kN}$$

Taking the origin at R_1, the distributed load must be continued to the far end and a negative load added to compensate, as shown dotted.

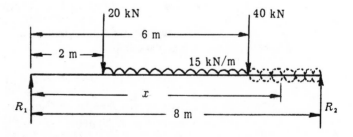

Figure 5.28

Considering a section in the extreme range of the beam from the origin and taking moments of all forces to the left of the section,

$$\frac{d^2y}{dx^2} = \frac{10^3}{EI}\left\{55x - 20[x-2] - 40[x-6] - \frac{15}{2}[x-2]^2 + \frac{15}{2}[x-6]^2\right\}$$

$$\therefore \frac{dy}{dx} = \frac{10^3}{EI}\left\{\frac{55}{2}x^2 - 10[x-2]^2 - 20[x-6]^2 - \frac{5}{2}[x-2]^3 + \frac{5}{2}[x-6]^3 + A\right\}$$

and

$$y = \frac{10^3}{EI}\left\{\frac{55}{6}x^3 - \frac{10}{3}[x-2]^3 - \frac{20}{3}[x-6]^3 - \frac{5}{8}[x-2]^4 \right.$$

$$\left. + \frac{5}{8}[x-6]^4 + Ax + B\right\}$$

When $x = 0$, $y = 0$, so that $B = 0$ since all terms in square brackets are negative and hence ignored.

When $x = 8$, $y = 0$, so that $A = -390$, all terms in square brackets being positive and hence included.

When $x = 4$, $y = \dfrac{10^3}{EI}\left\{\dfrac{55}{6} \times 4^3 - \dfrac{10}{3} \times 2^3 - \dfrac{5}{8} \times 2^4 - 390 \times 4\right\}$,

other terms being negative

$$= -\frac{10^3 \times 1\,010}{200 \times 10^9 \times 0.18 \times 10^{-3}}$$

$$= -0.028\,05 \text{ m or } \underline{-28.05 \text{ mm}}$$

At the point of maximum deflection $\dfrac{dy}{dx} = 0$ and it will be assumed that this point lies within the range $2 < x < 6$, so that terms involving $[x-6]$ must be ignored.

Hence

$$\frac{55}{2}x^2 - 10[x-2]^2 - \frac{5}{2}[x-2]^3 - 390 = 0$$

which reduces to

$$x^3 - 13x^2 - 4x + 164 = 0$$

By Newton's method or plotting, $\qquad x = 4.06$ m

$$\therefore y_{max} = \frac{10^3}{EI}\left\{\frac{55}{6} \times 4.06^3 - \frac{10}{3} \times 2.06^3 - \frac{5}{8} \times 2.06^4 - 390 \times 4.06\right\}$$

$$= -\frac{10^3 \times 1\,011}{200 \times 10^9 \times 0.18 \times 10^{-3}}$$

$$= \underline{-0.028\,08 \text{ m}}$$

If it is assumed that the maximum deflection occurs either to the left of the 20 kN load or to the right of the 40 kN load and the terms in square brackets are ignored or included to correspond, the solution for x will not be within the assumed range, thus indicating that a false assumption has been made.

In all cases of simply supported beams, the point of maximum deflection will be very close to the centre of the beam.

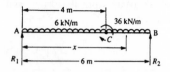

Figure 5.29

7. *Determine the slope and deflection at the point C for the beam shown in Figure 5.29 if EI = 20 MN m².*

Draw the bending moment diagram for the beam.

Taking moments about R_2:

$$6 \times 6 \times 3 = 36 + 6R_1$$

$$\therefore R_1 = 12 \text{ kN}$$

Considering a section in the extreme range of the beam from the origin and taking moments to the left of that section,

$$\frac{d^2y}{dx^2} = \frac{10^3}{EI}\{12x + 36[x-4]^0 - 3x^2\}$$

$$\therefore \frac{dy}{dx} = \frac{10^3}{EI}\{6x^2 + 36[x-4] - x^3 + A\}$$

and

$$y = \frac{10^3}{EI}\left\{2x^3 + 18[x-4]^2 - \frac{x^4}{4} + Ax + B\right\}$$

When $x = 0$, $y = 0$ so that $B = 0$ since the term in square brackets is negative and hence ignored.

When $x = 6$, $y = 0$ so that $A = -30$, the term in square brackets being positive and hence included.

When $x = 4$,

$$\frac{dy}{dx} = \frac{10^3}{20 \times 10^6}\{6 \times 4^2 - 4^3 - 30\}$$

$$= \underline{10^{-4} \text{ rad}}$$

This is positive and shows that the beam slopes upwards at this point, looking from the origin.

When $x = 4$, $y = \dfrac{10^3}{20 \times 10^6}\{2 \times 4^3 - \dfrac{4^4}{4} - 30 \times 4\}$

$$= -2.8 \times 10^{-3} \text{ m} \quad \text{or} \quad \underline{-2.8 \text{ mm}}$$

The bending moment

$$EI\frac{d^2y}{dx^2} = 12x + 36[x-4]^0 - 3x^2$$

and the bending moment diagram may be plotted by substituting values for x in this equation, again ignoring the term in square brackets when $x < 4$.

At $x = 2$, $\qquad\qquad M = 12 \times 2 - 3 \times 2^2 = 12$ kNm

Immediately at the left of C,

$$M = 12 \times 4 - 3 \times 4^2 = 0$$

Immediately to the right of C,

$$M = 12 \times 4 + 36 - 3 \times 4^2 = 36 \text{ kNm}$$

The bending moment diagram is shown in Figure 5.30.

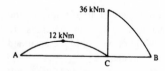

Figure 5.30

8. *A horizontal I-beam, built-in at both ends and 8 m long, carries a uniformly distributed load of 12 kN/m and a central concentrated load of 40 kN. If the bending stress is limited to 75 MN/m² and the deflection is not to exceed 2.5 mm, find the depth of section required. E = 200 GN/m² .*

Due to the concentrated load alone,

$$M = \frac{Wl}{8} \quad \text{from equation (5.26)}$$

$$= \frac{40 \times 8}{8} = 40 \text{ kN m}$$

Due to the distributed load alone,

$$M = \frac{wl^2}{12} \quad \text{from equation (5.28)}$$

$$= \frac{12 \times 8^2}{12} = 64 \text{ kN m}$$

Hence total end fixing moment $= 104$ kN m

Free bending moment at centre due to concentrated load

$$= \frac{Wl}{4} = \frac{40 \times 8}{4} = 80 \text{ kN m}$$

Free bending moment at centre due to distributed load

$$= \frac{wl^2}{8} = \frac{12 \times 8^2}{8} = 96 \text{ kN m}$$

Hence total free bending moment at centre $= 176$ kN m.

The combined bending moment diagram for the two loads is as shown in Figure 5.31 and the maximum bending moment to which the beam is subjected is 104 kN m.

Hence maximum stress, $\qquad\qquad\qquad \sigma = \frac{M}{I} \cdot y$

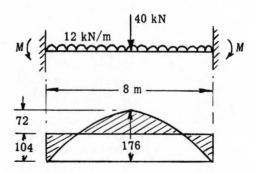

Figure 5.31

i.e.
$$75 \times 10^6 = \frac{104 \times 10^3}{I} \cdot y \qquad (1)$$

Central deflection due to concentrated load

$$= -\frac{Wl^3}{192EI} \quad \text{from equation (5.27)}$$

$$= -\frac{40 \times 10^3 \times 8^3}{192 \times 200 \times 10^9 I} = -\frac{0.533\,3}{10^6 I}\,\text{m}$$

Central deflection due to distributed load

$$= -\frac{wl^4}{384EI} \quad \text{from equation (5.29)}$$

$$= -\frac{12 \times 10^3 \times 8^4}{384 \times 200 \times 10^9 I} = -\frac{0.64}{10^6 I}\,\text{m}$$

$$\therefore \frac{0.533\,3}{10^6 I} + \frac{0.64}{10^6 I} = 0.002\,5$$

from which
$$I = 469.2 \times 10^{-6}\ \text{m}^4$$

Substituting in equation (1),
$$y = 721 \times 469.2 \times 10^{-6} = 0.338\ \text{m}$$

$$\therefore \quad \text{depth of section} = 0.676\ \text{m or } \underline{676\ \text{mm}}$$

9. *Figure 5.32 shows a built-in beam, for which the flexural rigidity EI = 200 MN m². Determine the deflection at each concentrated load and draw the bending moment diagram for the beam.*

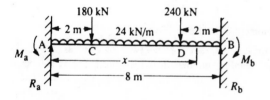

Figure 5.32

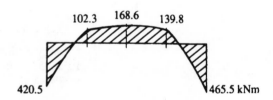

Figure 5.33

Taking the origin at A and considering a section in the extreme range of the beam,

$$\frac{d^2 y}{dx^2} = \frac{10^3}{EI}\left\{R_a x - 180[x-2] - 240[x-6] - 24\frac{x^2}{2} - M_a\right\}$$

$$\therefore \frac{dy}{dx} = \frac{10^3}{EI}\left\{R_a\frac{x^2}{2} - 90[x-2]^2 - 120[x-6]^2 - 4x^3 - M_a x + A\right\} \qquad (1)$$

When $x = 0$, $\dfrac{dy}{dx} = 0$ so that $A = 0$, all terms in square brackets being negative and hence ignored.

$$\therefore y = \frac{10^3}{EI}\left\{R_a\frac{x^3}{6} - 30[x-2]^3 - 40[x-6]^3 - x^4 - M_a\frac{x^2}{2} + B\right\} \qquad (2)$$

When $x = 0$, $y = 0$ so that $B = 0$, all terms in square brackets again being negative.

$\dfrac{dy}{dx}$ and y are also zero when $x = 8$ m, so that, substituting in equations (1) and (2),

$$4R_a - M_a = 721$$

and $$8R_a - 3M_a = 1\,021.5$$

Hence $$M_a = 420.5 \text{ kN m} \quad \text{and} \quad R_a = 285.4 \text{ kN}$$

EI is given as 200 MN m^2 so that equations (1) and (2) become

$$\frac{dy}{dx} = 5 \times 10^{-6}\{142.7x^2 - 90[x - 2]^2 - 120[x - 6]^2 - 4x^3 - 420.5x\}$$

and $$y = 5 \times 10^{-6}\{47.57x^3 - 30[x - 2]^3 - 40[x - 6]^3 - x^4 - 210.3x^2\}$$

When $x = 2$ m, $y = 2.39 \times 10^{-3}$ m = <u>2.39 mm</u>

When $x = 6$ m, $y = 2.55 \times 10^{-3}$ m = <u>2.55 mm</u>

The bending moment,

$$M = EI\frac{d^2 y}{dx^2} = 285.4x - 180[x - 2] - 240[x - 6] - 12x^2 - 420.5 \text{ kN m}$$

The bending moment diagram may therefore be plotted by substituting values for x in this equation, again ignoring terms in which the quantities in brackets are negative.

The diagram is shown in Figure 5.33, values of x at 2 m intervals having been used.

Note that for problems such as this, the values of the reactions and fixing moments may be obtained by summing the values for each of the concentrated loads and the distributed load considered separately (see Sections 5.9(b) and (c)).

Further problems

10. A cantilever consists of a steel tube 3 m long, 120 mm outside diameter and 6 mm thick. Calculate the load which, acting 1.8 m from the fixed end, will give a deflection of 2.5 mm at the free end. $E = 200$ GN/m^2.   [450 N]

11. A cantilever of length l with a concentrated load W at the free end is propped at a distance a from the fixed end to the same level as the fixed end. Find (a) the load on the prop, (b) the distance of the point of inflexion from the fixed end.

$$\left[\frac{W}{2a}(3l - a); \ \frac{a}{3}\right]$$

12. A cantilever of length l supports a load W uniformly distributed along its length. The cantilever is propped to the level of the fixed end at a distance $\frac{3}{4}l$ from that end. Determine the load on the prop.   [0.593 W]

13. A cantilever of uniform section and 6 m long is maintained horizontal at one end and supported by a rigid column at a distance of 4 m from the fixed end. The beam carries a load of 80 kN midway between the fixed end and the column and a load of 15 kN at the free end. Determine the force on the column.   [51.25 kN]

14. A propped cantilever of length l is securely fixed at one end and freely supported at the other. It is subjected to a bending couple M in the vertical plane containing the axis of the beam, applied about an axis 0.75 l from the fixed end. Determine the end fixing moment and the reaction at the support.

$$[0.406\ M;\ 1.406\ M/l]$$

15. A wooden post 6 m high is 50 mm square for the upper 3 m and 100 mm square for the lower 3 m. Find the deflection at the top due to a horizontal pull of 40 N at that point, applied in a direction parallel to one edge of the section. $E = 10$ GN/m^2.   [99.36 mm]

16. A simply supported beam of T-section has a horizontal flange at the top 100 mm wide and 10 mm thick and a vertical web 10 mm thick and 50 mm deep. The span is 1 m and the central load causes a maximum stress of 120 MN/m^2. Calculate the central deflection if $E = 200$ GN/m^2.   [1.11 mm]

17. A wooden beam is 240 mm wide and 80 mm deep. It is supported at each end of a span of 4 m and carries concentrated loads of 1 kN each at distances of 1.2 m from each end. Calculate the deflection under the loads and at the centre. $E = 14$ GN/m^2.   [12.05 mm; 14.75 mm]

18. A beam AB, 8 m long, rests symmetrically on supports C and D, 4 m apart. A load of 40 kN is applied at each of the ends A and B. Calculate the deflection relative to the supports (a) at the ends A and B, (b) at the centre of CD. $EI = 10$ MN m^2.   [42.67 mm; 16 mm]

19. A uniform beam of length l is simply supported at its ends and carries a concentrated load W at a distance $l/3$ from one end. Calculate the deflection (a) under the load, (b) at the centre, (c) at the point of maximum deflection.

$$\left[0.016\,46\frac{Wl^3}{EI};\ 0.017\,75\frac{Wl^3}{EI};\ 0.017\,94\frac{Wl^3}{EI}\right]$$

20. Calculate the deflection at the centre of a simply supported beam of span l carrying a uniformly distributed load w per unit length over the central portion equal to one half of the span.

$$\left[\frac{19wl^4}{2\,048EI}\right]$$

21. A simply supported beam of span l carries a uniformly distributed load w per unit length extending for a length $l/3$ from one end. Determine the deflection at mid-span and at the point of maximum deflection.

$$\left[\frac{wl^4}{311EI};\ \frac{wl^4}{305EI}\right]$$

22. A simply supported beam 6 m long carries concentrated loads of 48 kN and 40 kN at points 1 m and 3 m respectively from one end.

Determine the position and magnitude of the maximum deflection. $E = 200$ GN/m^2 and $I = 85 \times 10^{-6}$ m^4.   [2.87 m from end; 16.75 mm]

23. A simply supported beam, 8 m long, carries a concentrated load of 20 kN at 2 m from the left-hand end and a uniformly distributed load of 5 kN/m over the right-hand half of the beam.

Calculate the deflection (a) under the concentrated load, (b) at the centre, (c) at the point of maximum deflection. $EI = 12 \times 10^6$ Nm^2.

[17.2 mm; 23.3 mm; 23.35 mm at 3.917 m from the left-hand end]

24. A 300 mm $\times$ 125 mm I-beam is built-in at the ends of a span of 6 m and carries a uniformly distributed load of 24 kN/m throughout its length. What is the greatest bending stress in the beam?

By how much per cent is the maximum bending stress increased if the right-hand end becomes free in direction but remains supported at the same level?

I for the beam section $= 86.5 \times 10^{-6}$ m^4 and $E = 200$ GM/m^2.

[124.8 MN/m^2; 50%]

25. Determine the central deflection for the beam shown in Figure 5.34. $EI = 20$ MN m^2.

Draw the bending moment diagram for the beam.   [0.862 5 mm]

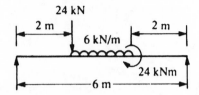

Figure 5.34

26. A simply supported beam of span 5 m carries a load which varies uniformly from 20 kN/m at one end to 50 kN/m at the other end. Find the magnitude of the maximum bending moment.

If the depth of the beam is 0.4 m and the maximum bending stress is 100 MN/m^2, find the central deflection.   [109.6 kN m; 6.48 mm]

27. A horizontal beam, built-in each end, has a clear span of 4.5 m, and carries loads of 50 kN at 1.5 m and 70 kN at 2.5 m from its left-hand end. Calculate the fixing moments and the value and position of the maximum bending moment.

[67.5 kN m; 60.5 kN m; 67.5 kN m at L.H. end]

28. A beam of uniform section, $I = 185 \times 10^{-6}$ m^4, span 6 m, is fixed horizontally at each end. It carries a point load of 120 kN at 3.6 m from one end. Neglecting the weight of the beam itself, find (a) the fixing moments; (b) the reactions; (c) the position and magnitude of the maximum deflection. $E = 200$ GN/m^2.

[69.12 kN m; 103.68 kN m; 42.25 kN; 77.8 kN;
3.35 mm at 0.275 m from mid-span]

29. A fixed-ended beam of span 9 m carries a uniformly distributed load of 15 kN/m (including its own weight) and two equal point loads of 200 kN at the third points of the span. Assuming rigid end-fixing, find the fixing moments and the deflection at the centre. $EI = 210$ MN m^2.   [518 kN m; 6.84 mm]

30. A beam, 4 m long, is rigidly built-in at each end and carries a uniformly distributed load of 6 kN/m from the centre to one end. Find the reactions, end-fixing moments, central deflection and maximum deflection. $EI = 10^6$ N m^2.

[2.25 kN; 9.75 kN; 2.5 kN m; 5.5 kN m; 2 mm;
2.06 mm at 2.23 m from unloaded end]

31. A 250 mm $\times$ 112.5 mm steel beam, $I = 47.6 \times 10^{-6}$ m^4, is used as a horizontal beam with fixed ends and a clear span of 3 m. Calculate the load which can be applied at one-third span if the bending stress is limited to 120 MN/m^2.

[103 kN]

32. A steel joist, 400 mm $\times$ 150 mm, $I = 283 \times 10^{-6}$ m^4, 6 m long, is fixed horizontally at each end and carries loads W and $2W$ at 2 m and 4 m respectively from one end. Draw the bending moment diagram for the beam, stating the values at the principal sections. Find the maximum value of W if the bending stress must not exceed 120 MN/m^2.   [End fixing moments 1.778 W and 2.222 W; 76.5 kN]

6 Struts

6.1 Introduction

A structural member which is subject to a compressive force is called a *strut*. Members which have large cross-sectional area compared with their length will fail by direct compression but those which are slender will fail by buckling before the yield point is reached.

The treatment here is confined to initially straight, homogeneous struts subjected to axial loading applied through the centroids of the ends. Such struts will remain straight until the critical load is reached, when buckling will occur. Any increase in load will cause the strut to collapse and any decrease will cause it to straighten. The value of the critical load depends upon the end-fixing conditions and upon the *slenderness ratio* which is defined as the effective length divided by the least radius of gyration of the section, i.e., slenderness ratio $= l/k$.

Sections other than circles and squares will have two different axes about which buckling may occur and buckling will obviously occur first about the weaker axis.

The principal end-fixing conditions are:

(*a*) pinned or hinged at each end,
(*b*) rigidly built in at each end,
(*c*) built in at one end and free at the other,
(*d*) built in at one end and pinned at the other.

These are shown in Figure 6.1.

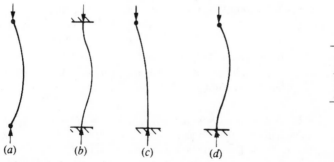

(*a*)   (*b*)   (*c*)   (*d*)   (*a*)   (*b*)

Figure 6.1   **Figure 6.2**

For the purpose of analysis, the pin-ended strut, Figure 6.1(*a*), is regarded as the basic case and other forms of buckling are derived from this. Thus, the middle half of the fixed-ended strut, Figure 6.2(*a*), is of the same form as the pin-ended strut shown in Figure 6.1(*a*); hence, if its actual length is L, it will buckle at the same load that would buckle a pin-ended strut of length $L/2$. The effective length is therefore regarded as $L/2$ when compared with the basic case.

Similarly, the strut shown in Figure 6.2(*b*) is equivalent to half a pin-ended strut, so that, if the actual length is L, the effective length is $2L$ when compared with the pin-ended strut.

The various types are tabulated below; in each case L represents the actual length of the strut.

Case	Both ends pinned	Both ends fixed	One end fixed, other end free	One end fixed, other end pinned
Effective length, l	L	$\dfrac{L}{2}$	$2L$	$\sim 0.7L$

6.2 Euler's Theory

The fundamental mathematical theory, upon which more practical formulae are based, is known as Euler's theory. This ignores the effect of direct compression and assumes that the crippling load depends only on failure due to buckling.

The basic case is shown in Figure 6.3.

If the deflection at distance x from one end is y, then bending moment,

$$EI\frac{d^2 y}{dx^2} = -Wy$$

Note: If the strut is turned into the horizontal position, as for a beam, M is positive when y is negative and vice versa. Hence the bending moment is always of opposite sign to the deflection.

Thus $$\frac{d^2 y}{dx^2} + \frac{W}{EI}y = 0$$

or $$\frac{d^2 y}{dx^2} + \mu^2 y = 0 \quad \text{where } \mu = \sqrt{\left(\frac{W}{EI}\right)}$$

The solution (see Appendix B) is

$$y = A\cos\mu x + B\sin\mu x$$

When $x = 0$, $y = 0$, so that $A = 0$

When $x = l$, $y = 0$, so that $B\sin\mu l = 0$

B cannot be zero, otherwise y is zero for all values of x, i.e. the strut has not buckled.

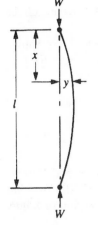

Figure 6.3

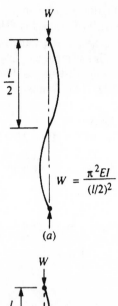

(a)

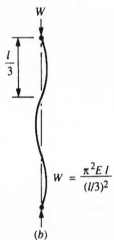

(b)

Figure 6.4

6.3 Validity of Euler's Theory

Hence $\qquad \sin \mu l = 0$

$$\therefore \mu l = 0, \pi, 2\pi, 3\pi, \text{etc.}$$

μl cannot be zero, so that real values of W are given by

$$\frac{W}{EI}l^2 = \pi^2, 4\pi^2, 9\pi^2, \text{etc.}$$

or $\qquad W = \dfrac{\pi^2 EI}{l^2}, \dfrac{4\pi^2 EI}{l^2}, \dfrac{9\pi^2 EI}{l^2}, \text{etc.} \qquad (6.1)$

The lowest value corresponds with the mode of buckling indicated by Figure 6.3. Higher values correspond with the modes of buckling shown in Figure 6.4(a) and (b), these cases arising when the strut is prevented from buckling in the fundamental mode.

Since $W = \sigma A$ and $I = Ak^2$, the formula $W = \dfrac{\pi^2 EI}{l^2}$ may be written

$$\sigma A = \frac{\pi^2 EAk^2}{l^2}$$

or $\qquad \sigma = \dfrac{\pi^2 E}{\left(\dfrac{l}{k}\right)^2} \qquad (6.2)$

Thus the stress at failure is inversely proportional to the square of the slenderness ratio, l/k.

The buckling loads or stresses for the other forms of end fixing shown in Figure 6.1 may be obtained by substituting the effective length for l.

As the length of the strut decreases, the critical load increases and, if short enough, the strut will fail by direct compression before the Euler load has been reached. If the compressive stress at yield or other criterion of failure is denoted by σ_c then, from equation (6.2),

$$\sigma_c = \frac{\pi^2 E}{\left(\dfrac{l}{k}\right)^2} \quad \text{or} \quad \frac{l}{k} = \sqrt{\left(\frac{\pi^2 E}{\sigma_c}\right)}$$

For mild steel, $E \approx 200$ GN/m^2 and $\sigma_c \approx 300$ MN/m^2, so that l/k is approximately 80. Euler's theory is therefore invalid for mild steel struts where $(l/k) < 80$ but in practice, it should not be used for struts of slenderness ratio less than about 120.

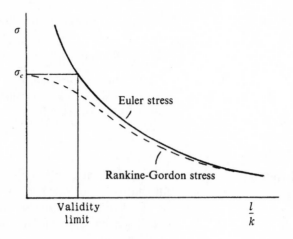

Figure 6.5

Figure 6.5 shows that relation between σ and l/k and the dotted line shows the Rankine–Gordon relation (Section 6.4), which makes allowance for the effect of direct compression. These curves converge at high slenderness ratios.

6.4 The Rankine–Gordon relation

This relation, based on experimental work, allows for the effect of direct compression and suggests that the strut will fail at a load given by

$$\frac{1}{W} = \frac{1}{W_c} + \frac{1}{W_e}$$

where W_c is the load which would cause failure in direct compression and W_e is the Euler load.

Thus
$$W = \frac{W_c W_e}{W_e + W_c} = \frac{W_c}{1 + \dfrac{W_c}{W_e}}$$

$$= \frac{\sigma_c A}{1 + \dfrac{\sigma_c A}{\pi^2 E I / l^2}} = \frac{\sigma_c A}{1 + \dfrac{\sigma_c}{\pi^2 E}\left(\dfrac{l}{k}\right)^2} = \frac{\sigma_c A}{1 + c\left(\dfrac{l}{k}\right)^2} \quad (6.3)$$

or
$$\sigma = \frac{\sigma_c}{1 + c\left(\dfrac{l}{k}\right)^2} \quad (6.4)$$

where c is a constant for any given material.

Values of c could be determined from separate measurements of σ_c and E but experiment shows discrepancies between predicted and measured loads and c is therefore determined directly by experiment. For mild steel, for which $\sigma_c \approx 300$ MN/m^2, $c = \dfrac{1}{7\,500}$ and for cast iron, for which $\sigma_c \approx$ 540 MN/m^2, $c = \dfrac{1}{1\,600}$.

Other empirical relations are available for failure loads on struts but for practical design work, reference should be made to BS449, which tabulates safe working stresses for all slenderness ratios and for different grades of steel.

6.5 Load factor

The load which a strut or column can carry is not proportional to the ultimate stress or yield stress for the material since it also depends on the slenderness ratio and so, instead of dividing the stress by a factor of safety, the buckling load is divided by a *load factor* to obtain the safe working load.

Worked examples

1. *A straight bar is 1.2 m long and is simply supported at its ends in a horizontal position. When loaded at the centre with a concentrated load of 90 N, the central deflection is found to be 5 mm.*

If placed vertically and loaded along its axis, what load would cause it to buckle, according to Euler's theory?

From equation (5.15),
$$y = -\frac{Wl^3}{48EI}$$

i.e.
$$0.005 = \frac{90 \times 1.2^3}{48EI}$$

from which
$$EI = 648 \text{ Nm}^2$$

When used as a strut,
$$W = \frac{\pi^2 EI}{l^2} \quad \text{from equation (6.1)}$$

$$= \frac{\pi^2 \times 648}{1.2^2} = \underline{4\,440 \text{ N}}$$

2. *A cast iron column 200 mm external diameter is 20 mm thick and 4.5 m long. Assuming that it is rigidly fixed at each end, calculate the safe load by Rankine's formula, using a load factor of 4.*

For cast iron, $\sigma_c = 550$ MN/m^2 and $c = \dfrac{1}{1\,600}$.

$$A = \frac{\pi}{4}(D^2 - d^2) = \frac{\pi}{4}\left(\frac{200^2 - 160^2}{10^6}\right) = 0.011\,33 \text{ m}^2$$

$$I = \frac{\pi}{64}(D^4 - d^4) = \frac{\pi}{64}\left(\frac{200^4 - 160^4}{10^{12}}\right) = 46.4 \times 10^{-6} \text{ m}^4$$

$$k = \sqrt{\left(\frac{I}{A}\right)} = \sqrt{\left(\frac{46.4 \times 10^{-6}}{0.011\,33}\right)} = 0.064 \text{ m}$$

Effective length, $l = \dfrac{L}{2} = \dfrac{4.5}{2} = 2.25$ m

Slenderness ratio, $\dfrac{l}{k} = \dfrac{2.25}{0.064} = 35.16$

Therefore, from equation (6.3),

$$W = \frac{1}{4} \times \frac{\sigma_c A}{1 + c \left(\dfrac{l}{k}\right)^2}$$

$$= \frac{1}{4} \times \frac{550 \times 10^6 \times 0.011\,33}{1 + \dfrac{1}{1\,600} \times 35.16^2}$$

$$= \underline{879 \times 10^3 \text{ N}}$$

For cast iron, the validity limit for Euler's theory would be about 43 and so that theory would be quite inappropriate.

3. *A mild steel strut, 8 m long, is fabricated from two flat plates 300 mm wide by 10 mm thick and two channel sections 100 mm wide by 200 mm deep, as shown in Figure 6.6. The second moments of area of a single channel about its centroid G are: $I_{XX} = 26.20 \times 10^6$ mm⁴ and $I_{YY} = 1.57 \times 10^6$ mm⁴. The cross-sectional area is 4 400 mm².*

Assuming pinned ends, determine the critical load (a) by Euler's theory, and (b) by the Rankine–Gordon relation.

$E = 200$ kN/mm² , $c = 1/7\,500$ and $\sigma_c = 320$ N/mm² .

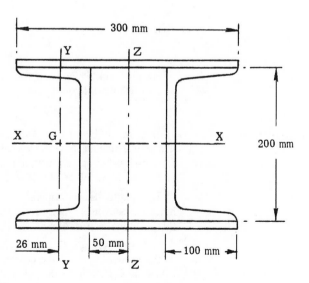

Figure 6.6

For the complete section

$$A = 2(4\,400 + 300 \times 10)$$

$$= 14\,800 \text{ mm}^2$$

$$I_{XX} = 2 \times 26.20 \times 10^6 + 2 \left(\frac{300 \times 10^3}{12} + 300 \times 10 \times 105^2 \right)$$

$$= 118.6 \times 10^6 \text{ mm}^4$$

$$I_{ZZ} = 2(1.57 \times 10^6 + 4\,400 \times 76^2) + 2 \times \frac{10 \times 300^3}{12}$$

$$= 99.0 \times 10^6 \text{ mm}^4$$

Least radius of gyration of section,

$$k = \sqrt{\left(\frac{I_{ZZ}}{A} \right)} = \sqrt{\left(\frac{99.0 \times 10^6}{14\,800} \right)} = 81.7 \text{ mm}$$

Slenderness ratio, $\dfrac{l}{k} = \dfrac{8}{0.081\,7} = 97.9$

(a) $$W = \frac{\pi^2 EA}{\left(\dfrac{l}{k} \right)^2} \quad \text{from equation (6.2)}$$

$$= \frac{\pi^2 \times 200 \times 10^3 \times 14\,800}{97.9^2} = \underline{3.05 \times 10^6 \text{ N}}$$

(b) $$W = \frac{\sigma_c A}{1 + c \left(\dfrac{l}{k} \right)^2} \quad \text{from equation (6.3)}$$

$$= \frac{320 \times 14\,800}{1 + \dfrac{1}{7\,500} \times 97.9^2} = \underline{2.08 \times 10^6 \text{ N}}$$

For this strut, which is approaching the validity limit for Euler's theory, there is an appreciable difference between the results given by the two theories.

Further problems

4. A straight bar of alloy 1 m long and 12.5 mm by 5 mm in section is mounted in a strut testing machine and loaded axially until it buckles. Using the Euler formula for pin-ended struts, calculate the maximum central deflection before the material yields at a stress of 280 MN/m$^2 \cdot E = 75$ GN/m^2.   [0.15 m]

5. Using the Euler formula, calculate the buckling load for a strut 3 m long and hinged at both ends. The strut is of T-section; the flange is 100 mm wide by 10 mm thick and the web is 70 mm deep by 10 mm thick. $E = 200$ GN/m^2.   [209 kN]

6. A steel strip 0.64 m long is arranged as a simply supported beam on a span of 0.5 m and loaded at the centre, when it is found that the central deflection is 0.11 mm/N. Determine the value of EI for the strip and hence calculate the Euler critical load when the strip is tested as a strut with direction-free ends.   [570 N]

7. Calculate the Euler critical load for a flat steel strip, 75 mm long, 12 mm wide and 0.25 mm thick, used as a strut with built-in ends. $E = 200$ GN/m^2.   [21.9 N]

8. A hollow cast-iron column with fixed ends carries an axial load of 1 MN. The column is 4.5 m long and has an external diameter of 250 mm. Find the thickness of metal required, using the Rankine formula. The constant c for cast iron is $1/1\,600$ for pinned ends and the working stress is 80 MN/m^2.   [28 mm]

9. A pin-ended tubular steel strut 2.25 m long has outer and inner diameters of 38 mm and 33 mm respectively. Determine the crippling loads given by the Euler and Rankine formulae. The yield stress is 325 MN/m^2, the Rankine constant is $1/7\,500$ and $E = 200$ GN/m^2.

For what length of strut does the Euler formula cease to apply?

[17.24 kN; 17.20 kN; 0.98 m]

10. In the experimental determination of the buckling loads for 12.5 mm diameter steel struts of various lengths, two of the values obtained were: (i) length 0.5 m, load 9.25 kN, (ii) length 0.2 m, load 25 kN.

(*a*) Determine whether either of these results conform with the Euler critical load.

(*b*) Assuming that both values conform with the Rankine formula, find the two constants. $E = 200$ GN/m^2.   [301.5 MN/m^2; 1/8 540]

7 Strain energy

7.1 Introduction

The work done in distorting an elastic material is retained in the material as strain energy and this work can be recovered if the material is subsequently unloaded. It is assumed that straining forces, moments and torques are gradually applied; if suddenly applied, the material will overshoot its equilibrium position and then oscillate about that position. Internal molecular friction, or *hysteresis*, will cause the oscillation to die away, leaving the material, and energy stored, in the same state as if the material had been gradually strained.

7.2 Strain energy due to direct stress

Let a tensile force F be gradually applied to an elastic material of cross-sectional area A and length l; the graph of force against extension will then be a straight line, as shown in Figure 7.1.

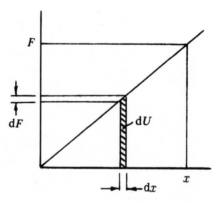

Figure 7.1

An increment of force dF causes an increment of extension dx and the work done is represented by the area dU. Thus the total work done by the force F is represented by the area under the graph.

i.e.   strain energy, $U = \frac{1}{2}Fx$

But $x = \dfrac{Fl}{AE}$, so that $U = \dfrac{1}{2}F\dfrac{FL}{AE} = \dfrac{F^2l}{2AE}$ (7.1)

Alternatively, since $\dfrac{F}{A}$ is the stress, σ, $\qquad U = \dfrac{\sigma^2 Al}{2E}$

But Al is the volume of the material, so that

$$U = \dfrac{\sigma^2}{2E} \times \text{ volume} \qquad (7.2)$$

7.3 Impact loading

If the load is not gradually applied but is suddenly applied or dropped from a height, the stress, strain and strain energy are all increased. These high values are transient and internal damping of the subsequent oscillation will reduce the values to those obtained under static loading.

Let a mass m be dropped from a height h on to a collar at the lower end of a rod of length l and cross-sectional area A, Figure 7.2, producing an instantaneous extension x and instantaneous stress σ.

Then loss of potential energy of mass = gain in strain energy of rod

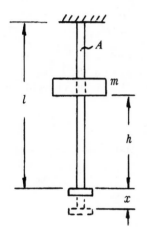

Figure 7.2

i.e. $\qquad mg(h + x) = \dfrac{\sigma^2}{2E} \times \text{volume}$

But $\qquad x = \dfrac{\sigma l}{E}$

so that $\qquad mg\left(h + \dfrac{\sigma l}{E}\right) = \dfrac{\sigma^2}{2E} \times Al$

from which $\qquad \sigma = \dfrac{mg}{A} \pm \sqrt{\left(\left(\dfrac{mg}{A}\right)^2 - \dfrac{2mgEh}{Al}\right)}$ (7.3)

The positive sign relates to the stress at the point of maximum extension and the negative sign relates to the stress at the end of the rebound.

When $h = 0$ and the load is suddenly applied,

$$\sigma = \dfrac{2mg}{A}$$

i.e., the maximum stress is twice that due to a static load.

When h is large compared with the extension x,

$$\sigma = \sqrt{\left(\dfrac{2mgEh}{Al}\right)}$$

7.4 Strain energy due to a uniform moment or torque

If a uniform moment M is gradually applied to a uniform beam of length l, causing an angle of bending ϕ, Figure 7.3,

$$\text{strain energy} = \text{work done} = \tfrac{1}{2}M\phi$$

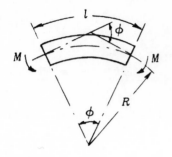

Figure 7.3

But, from Figure 7.3,

$$\phi = \frac{l}{R}$$

and

$$\frac{1}{R} = \frac{M}{EI}$$

Hence

$$U = \tfrac{1}{2}M \cdot \frac{Ml}{EI} = \frac{M^2l}{2EI} \tag{7.4}$$

Similarly, if a uniform torque T is gradually applied to a uniform shaft of length l, causing an angle of twist θ,

$$\text{strain energy} = \tfrac{1}{2}T\theta$$

But

$$\theta = \frac{Tl}{GJ}$$

$$\therefore U = \tfrac{1}{2}T \cdot \frac{Tl}{GJ} = \frac{T^2l}{2GJ} \tag{7.5}$$

The torque applied to a shaft is usually constant along its length but it is rare for the bending moment to remain constant along the length of a beam. It is then necessary to sum the energy stored in small elements of the beam, as shown in the following section.

7.5 Strain energy due to a variable bending moment

From equation (7.4), the strain energy stored in a short length of beam dx is given by

$$dU = \frac{M^2 dx}{2EI}$$

and hence, for the whole beam,

$$U = \int_0^l \frac{M^2 dx}{2EI} \tag{7.6}$$

In most cases, the cross-section of the beam is uniform, so that this may be written

$$U = \frac{1}{2EI} \int_0^l M^2 dx$$

The appropriate expression for M in terms of x must be substituted before the strain energy can be calculated.

7.6 Castigliano's theorem

Figure 7.4 shows an elastic body subjected to a number of forces $F_1, F_2, \ldots, F_n$. Let the resultant deflections at the load points *in the directions of the loads* be $y_1, y_2, \ldots, y_n$. If the forces are applied gradually (static loading), the total strain energy,

$$U = \tfrac{1}{2}F_1 y_1 + \tfrac{1}{2}F_2 y_2 + \cdots + \tfrac{1}{2}F_n y_n \tag{7.7}$$

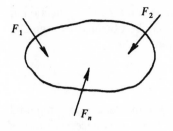

Figure 7.4

Let F_1 increase to $F_1 + \delta F_1$ and let y_1 increase to $y_1 + \delta y_1$, y_2 to $y_2 + \delta y_2, \ldots, y_n$ to $y_n + \delta y_n$ due to this increase in F_1. Then the increase in strain energy,

$$\delta U = \tfrac{1}{2}\delta F_1 \delta y_1 + F_1 \delta y_1 + F_2 \delta y_2 + \cdots + F_n \delta y_n$$

Neglecting the first term, which is of the second order of small quantities, this becomes

$$\delta U = F_1 \delta y_1 + F_2 \delta y_2 + \cdots + F_n \delta y_n \tag{7.8}$$

The total strain energy, $U + \delta U$ for the gradually applied forces $F_1 + \delta F_1, F_2, \ldots, F_n$, causing deflections $y_1 + \delta y_1, y_2 + \delta y_2, \ldots, y_n + \delta y_n$ is given by

$$U + \delta U = \tfrac{1}{2}(F_1 + \delta F_1)(y_1 + \delta y_1) + \tfrac{1}{2}F_2(y_2 + \delta y_2) + \cdots + \tfrac{1}{2}F_n(y_n + \delta y_n)$$

and neglecting again the term $\tfrac{1}{2}\delta F_1 \delta y_1$, this reduces to

$$U + \delta U = \tfrac{1}{2}F_1 y_1 + \tfrac{1}{2}\delta F_1 y_1 + \tfrac{1}{2}F_1 \delta y_1 + \tfrac{1}{2}F_2 y_2 + \tfrac{1}{2}F_2 \delta y_2$$
$$+ \cdots + \tfrac{1}{2}F_n y_n + \tfrac{1}{2}F_n \delta y_n \tag{7.9}$$

Subtracting equation (7.7),

$$\delta U = \tfrac{1}{2}\delta F_1 y_1 + \tfrac{1}{2}F_1 \delta y_1 + \tfrac{1}{2}F_2 \delta y_2 + \cdots + \tfrac{1}{2}F_n \delta y_n \tag{7.10}$$

Multiplying equation (7.10) by 2 and substracting equation (7.8),

$$\delta U = \delta F_1 y_1$$

or
$$y_1 = \frac{\delta U}{\delta F_1}$$

In the limit as $\delta F_1 \to 0$, this becomes

$$y_1 = \frac{\partial U}{\partial F_1} \tag{7.11}$$

Similarly,
$$y_2 = \frac{\partial U}{\partial F_2}, \cdots, y_n = \frac{\partial U}{\partial F_n}$$

The partial derivatives arise because the deflection at each point is a function of all the forces.

Equation (7.11) shows that the deflection in the direction of any force due to all the applied forces is the partial derivative of the *total* strain energy of the system with respect to the force considered.

It may also be shown that the rotation θ at any point at which a couple is applied is the partial derivative of the total strain energy with respect to the couple considered.

7.7 Application to deflection of beams and curved bars

For simple cases of a body subjected to a single force F, the deflection at the load point, in the direction of the force, may be obtained by equating the strain energy to the work done by the force,

i.e.
$$U = \frac{1}{2EI} \int_0^l M^2 \mathrm{d}x = \tfrac{1}{2} F y \qquad (7.12)$$

In cases where there are several forces or where the deflection is required in a direction other than in the direction of a force, it is necessary to use Castigliano's theorem.

$$y = \frac{\partial U}{\partial F} = \frac{\partial}{\partial F} \left(\frac{1}{2EI} \int_0^l M^2 \mathrm{d}x \right)$$

$$= \frac{1}{2EI} \int_0^l \frac{\partial}{\partial F} (M^2) \, \mathrm{d}x$$

$$= \frac{1}{EI} \int_0^l M \frac{\partial M}{\partial F} \, \mathrm{d}x \qquad (7.13)$$

It is legitimate to differentiate M^2 before integrating as the differentiation and integration are with respect to different variables.

If the deflection is required in a direction where there is no applied force, an imaginary force must be applied in that direction. The deflection is then obtained in terms of this force, which is subsequently made zero (see Example 4).

7.8 Variable torque

If the torque varies along the length of a shaft, the energy stored in a short length $\mathrm{d}x$ is given by

$$\mathrm{d}U = \frac{T^2 \mathrm{d}x}{2GJ}$$

and hence, for the whole shaft,

$$U = \int_0^l \frac{T^2 \mathrm{d}x}{2GJ} \qquad (7.14)$$

$$= \frac{1}{2GJ} \int_0^l T^2 \mathrm{d}x \quad \text{for a uniform shaft}$$

Castigliano's theorem is equally applicable to torques or couples. Thus, if U is the total strain energy in a body due to various causes, the rotation θ at any point at which a torque T is applied is given by

$$\theta = \frac{\partial U}{\partial T}$$

Worked examples

1. *A metal bar, 40 mm diameter and 1.2 m long, has a collar fitted at the lower end. It is suspended from the upper end and a mass of 2 000 kg is gradually lowered on to the collar, producing an extension in the bar of 0.25 mm. Find the height from which this load could be dropped on to the collar if the maximum tensile stress in the bar is not to exceed 100 MN/m^2.*

$$E = \frac{Wl}{Ax}$$

$$= \frac{2\,000 \times 9.81 \times 1.2}{\frac{\pi}{4} \times 0.04^2 \times 0.000\,25}$$

$$= 74.9 \times 10^9 \text{ N/m}^2$$

When the mass is dropped on to the collar from a height h,

$$mg\left(h + \frac{\sigma l}{E}\right) = \frac{\sigma^2}{2E} \times \text{volume} \quad \text{from Section 7.3}$$

i.e. $2\,000 \times 9.81 \left(h + \dfrac{100 \times 10^6 \times 1.2}{74.9 \times 10^9}\right) = \dfrac{(100 \times 10^6)^2}{2 \times 74.9 \times 10^9} \times 1.2 \times \dfrac{\pi}{4} \times 0.04^2$

i.e. $\qquad\qquad\qquad h + 0.001\,6 = 0.005\,13$

$$\therefore h = 0.003\,53 \text{ m or } \underline{3.53 \text{ mm}}$$

2. *A simply supported beam of length 1.5 m carries a vertical load of 10 kN at the centre. One half of the beam has a diameter of 50 mm and the other half 75 mm. Calculate the central deflection. $E = 200 \ GN/m^2$.*

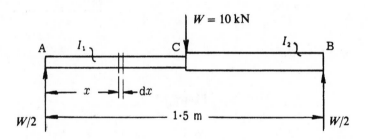

Figure 7.5

The bending moment at a point distance x from the left-hand support, Figure 7.5,

$$= \frac{W}{2}x$$

$$\therefore U_{AC} = \frac{1}{2EI_1} \int_0^{l/2} \left(\frac{W}{2}x\right)^2 dx \quad \text{from equation (7.6)}$$

$$= \frac{W^2}{8EI_1}\left[\frac{x^3}{3}\right]_0^{l/2}$$

$$= \frac{W^2 l^3}{192EI_1}$$

Similarly $\qquad U_{BC} = \dfrac{W^2 l^3}{192EI_2}$

∴ total strain energy,

$$U = \frac{W^2 l^3}{192E} \left(\frac{1}{I_1} + \frac{1}{I_2} \right) = \tfrac{1}{2}\, Wy \quad \text{from equation (7.12)}$$

$$\therefore y = \frac{W l^3}{96E} \left(\frac{1}{I_1} + \frac{1}{I_2} \right)$$

$$= \frac{10 \times 10^3 \times 1.5^3}{96 \times 200 \times 10^9} \left(\frac{64}{\pi \times 0.050^4} + \frac{64}{\pi \times 0.075^4} \right)$$

$$= 0.006\,86 \text{ m or } \underline{6.86 \text{ mm}}$$

Note that the sign of M is of no consequence since this is subsequently squared.

3. *A spring is made of steel rod of diameter d, bent to the form shown in Figure 7.6. Determine the stiffness of the spring if d = 6 mm, r = 40 mm and l = 100 mm. E = 200 kN/mm².*

The spring may be imagined to be cut in half and held rigidly at C, Figure 7.7. The total strain energy is then twice that in the half spring considered.

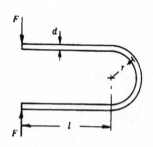

Figure 7.6

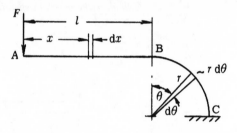

Figure 7.7

For AB, $\qquad M = Fx$

$$\therefore U_{AB} = \frac{1}{2EI} \int_0^l (Fx)^2 \mathrm{d}x$$

$$= \frac{F^2}{2EI} \left[\frac{x^3}{3} \right]_0^l = \frac{F^2 l^3}{6EI}$$

For BC, $\qquad M = F(l + r \sin \theta)$

$$\therefore U_{BC} = \frac{1}{2EI} \int_0^{\pi/2} [F(l + r \sin \theta)]^2 r \, \mathrm{d}\theta$$

$$= \frac{F^2 r}{2EI} \int_0^{\pi/2} (l^2 + 2lr \sin \theta + r^2 \sin^2 \theta) \, \mathrm{d}\theta$$

$$= \frac{F^2 r}{2EI} \left[l^2 \theta - 2lr \cos \theta + \frac{r^2}{2} \left(\theta - \frac{\sin 2\theta}{2} \right) \right]_0^{\pi/2}$$

$$= \frac{F^2 r}{2EI} \left(\frac{\pi l^2}{2} + 2lr + \frac{\pi r^2}{4} \right)$$

$$\therefore \text{ total strain energy, } U = \frac{F^2}{24EI}[4l^3 + 6\pi l^2 r + 24lr^2 + 3\pi r^3] \times 2$$

$$= \tfrac{1}{2}Fy$$

$$\therefore \text{ stiffness, } \quad S = \frac{F}{y} = \frac{6EI}{4l^3 + 6\pi l^2 r + 24lr^2 + 3\pi r^3}$$

$$= \frac{6 \times 200 \times 10^3 \times \dfrac{\pi}{64} \times 6^4}{4 \times 100^3 + 6\pi \times 100^2 \times 40 + 24 \times 100 \times 40^2 + 3\pi \times 40^3}$$

$$= \underline{4.47 \text{ N/mm}}$$

Figure 7.8

4. *A bar of constant section, second moment of area I, is bent as shown in Figure 7.8 and fixed at one end. Find the horizontal and vertical deflections at the free end.*

Since there is no force in the vertical direction at the free end, a force Q, of zero magnitude, must be applied there, Figure 7.9.

Then, for AB, $\qquad\qquad M = Px, \qquad \dfrac{\partial M}{\partial P} = x \quad$ and $\quad \dfrac{\partial M}{\partial Q} = 0$

For BC, $\qquad\qquad M = Pa + Qx, \quad \dfrac{\partial M}{\partial P} = a \quad$ and $\quad \dfrac{\partial M}{\partial Q} = x$

For the horizontal deflection,

$$y_h = \frac{\partial U}{\partial P} = \frac{1}{EI}\int M \frac{\partial M}{\partial P}\,dx$$

$$= \frac{1}{EI}\left[\int_0^a (Px)x\,dx + \int_0^b (Pa + Qx)a\,dx\right]$$

$$= \frac{Pa^2}{EI}\left(\frac{a}{3} + b\right) \quad \text{since } Q = 0$$

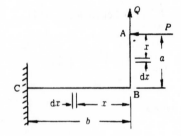

Figure 7.9

For the vertical deflection,

$$y_v = \frac{\partial U}{\partial Q} = \frac{1}{EI}\int M \frac{\partial M}{\partial Q}\,dx$$

$$= \frac{1}{EI}\left[\int_0^a (Px)0\,dx + \int_0^b (Pa + Qx)x\,dx\right]$$

$$= \frac{Pab^2}{2EI} \quad \text{since } Q = 0$$

Note that although Q can be made zero at the integration stage, it cannot be made zero earlier, otherwise no expression for $\dfrac{\partial M}{\partial Q}$ will be obtained.

5. *Figure 7.10 shows a steel rod, 10 mm diameter, with one end firmly fixed. The rod is bent into the form of three-quarters of a circle and the free end is constrained by guides to move in a vertical direction. If the mean radius to which the rod is bent*

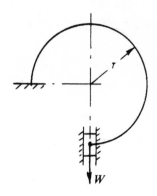

Figure 7.10

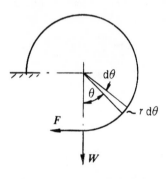

Figure 7.11

is 150 mm, determine the vertical deflection of the free end when a load of 100 N is gradually applied there. $E = 200 \ kN/mm^2$.

In order to constrain the slider to move vertically, a horizontal force must be applied to the slider by the guide.

Let this force be F, Figure 7.11. Then, since there is no horizontal movement in the direction of F, $\dfrac{\partial U}{\partial F} = 0$, an equation which will determine F in terms of W.

At a point subtending an angle θ with the vertical,

$$M = Wr \sin \theta - Fr(1 - \cos \theta)$$

$$\frac{\partial M}{\partial W} = r \sin \theta$$

$$\frac{\partial M}{\partial F} = -r(1 - \cos \theta)$$

$$\frac{\partial U}{\partial F} = 0, \qquad \therefore \ \int M \frac{\partial M}{\partial F} dx = 0$$

i.e.
$$\int_0^{\frac{3\pi}{2}} \{Wr \sin \theta - Fr(1 - \cos \theta)\} \times -r(1 - \cos \theta) \times r \, d\theta = 0$$

i.e.
$$W \int_0^{\frac{3\pi}{2}} (\sin \theta - \sin \theta \cos \theta) \, d\theta = F \int_0^{\frac{3\pi}{2}} (1 - 2 \cos \theta + \cos^2 \theta) \, d\theta$$

i.e.
$$W \left[-\cos \theta + \tfrac{1}{4} \cos 2\theta \right]_0^{\frac{3\pi}{2}} = F \left[\theta - 2 \sin \theta + \frac{\theta}{2} + \tfrac{1}{4} \sin 2\theta \right]_0^{\frac{3\pi}{2}}$$

from which
$$F = \frac{2W}{9\pi + 8}$$

The vertical deflection is given by

$$y = \frac{\partial U}{\partial W} = \frac{1}{EI} \int M \frac{\partial M}{\partial W} dx$$

$$= \frac{1}{EI} \int_0^{\frac{3\pi}{2}} \{Wr \sin \theta - Fr(1 - \cos \theta)\} \times r \sin \theta \times r \, d\theta$$

$$= \frac{r^3}{EI} \int_0^{\frac{3\pi}{2}} \{W \sin^2 \theta - F(\sin \theta - \sin \theta \cos \theta)\} d\theta$$

$$= \frac{r^3}{EI} \left[\frac{W}{2} (\theta - \tfrac{1}{2} \sin 2\theta) - F(-\cos \theta + \tfrac{1}{4} \cos 2\theta) \right]_0^{\frac{3\pi}{2}}$$

$$= \frac{r^3}{EI} \left(\frac{3\pi}{4} W - \frac{F}{2} \right)$$

$$= \frac{Wr^3}{EI} \left(\frac{3\pi}{4} - \frac{1}{9\pi + 8} \right)$$

$$= \frac{100 \times 150^3}{200 \times 10^3 \times \dfrac{\pi}{64} \times 10^4} \times 2.333$$

$$= \underline{8.0 \text{ mm}}$$

6. *A cantilever, lying in a horizontal plane, is in the shape of a quadrant of a circle, of radius 150 mm. One end is firmly fixed and a vertical force of 200 N is applied at the other end. Find the vertical deflection of the free end.*

The cantilever is made from circular rod, 10 mm diameter, for which $E = 200$ kN/mm^2 and $G = 80$ kN/mm^2.

Referring to Figure 7.12, the force F may first be moved to the point A by adding a moment $F \times a$ to compensate and then to B by adding a moment $F \times b$ to compensate.

The moment $F \times a$ causes bending of the bar at B and the moment $F \times b$ causes twisting of the bar at B.

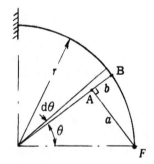

Figure 7.12

Hence

$$M = Fa = Fr \sin \theta$$

and

$$T = Fb = Fr(1 - \cos \theta)$$

$$\therefore U = \frac{1}{2EI} \int_0^l M^2 \, dx + \frac{1}{2GJ} \int_0^l T^2 \, dx$$

$$= \frac{1}{2EI} \int_0^{\frac{\pi}{2}} (Fr \sin \theta)^2 \cdot r \, d\theta + \frac{1}{2GJ} \int_0^{\frac{\pi}{2}} (Fr[1 - \cos \theta])^2 \cdot r \, d\theta$$

i.e. $\dfrac{F\delta}{2} = \dfrac{F^2 r^3}{2} \left\{ \dfrac{1}{EI} \int_0^{\frac{\pi}{2}} \sin^2 \theta \, d\theta + \dfrac{1}{GJ} \int_0^{\frac{\pi}{2}} (1 - 2 \cos \theta + \cos^2 \theta) d\theta \right\}$

$$\therefore \delta = Fr^3 \left\{ \frac{1}{EI} \cdot \frac{\pi}{4} + \frac{1}{GJ} \left(\frac{\pi}{2} - 2 + \frac{\pi}{4} \right) \right\}$$

$$= 200 \times 150^3 \left[\frac{1}{200 \times 10^3 \times \dfrac{\pi}{64} \times 10^4} \frac{\pi}{4} + \frac{1}{80 \times 10^3 \times \dfrac{\pi}{32} \times 10^4} \left(\frac{3\pi}{4} - 2 \right) \right]$$

$$= 0.625(0.008 + 0.004\,54)$$

$$= \underline{8.46 \text{ mm}}$$

Further problems

7. A ring mass of 25 kg encircles a bar and falls through a distance h before striking a stop fixed to the bottom of the bar which hangs from a rigid support. The bar is of steel 25 mm diameter and 3 m long. What must be the value of h if the maximum extension given to the bar is 0.63 mm? $E = 200$ GN/m^2.

[25.9 mm]

8. A mass of 10 kg falls through a height of 150 mm and then commences to stretch a steel bar 18 mm diameter and 0.9 m long. Determine (*a*) the maximum stress induced in the bar, (*b*) the maximum elongation of the bar, (*c*), the energy stored in the bar at the point of maximum extension. $E = 200$ GN/m^2.

[160.3 MN/m^2; 0.721 mm; 14.78 J]

9. A simply supported beam length *l* carries a central concentrated load *W*. Determine the total strain energy in the beam due to bending and hence obtain an expression for the deflection under the load.

$$\left[\frac{W^2 l^3}{96EI}; \frac{Wl^3}{48EI}\right]$$

10. A simply supported beam of span *l* carries a uniformly distributed load *w* per unit length. Determine the total strain energy in the beam due to bending.

$$\left[\frac{w^2 l^5}{240EI}\right]$$

11. A steel tube 56 mm outside diameter and 50 mm internal diameter is fixed vertically on a rigid base. At a distance of 0.9 m from the base the tube is bent into a quadrant of a circle of radius 0.6 m and a vertical load of 2 kN is applied at the free end. Calculate the vertical deflection at the load. $E = 200$ GN/m^2.

[28.1 mm]

12. A steel ring of rectangular cross-section 7.5 mm wide by 5 mm thick has a mean diameter of 300 mm. A narrow radial saw cut is made and tangential separating forces of 5 N each are applied at the cut in the plane of the ring. Determine the additional separation due to these forces. $E = 200$ GN/m^2.   [3.83 mm]

13. A steel spring of uniform section is shown in Figure 7.13. Derive an expression for the vertical movement of the free end due to the vertical force *F*.

[23.92 Fr^3/EI]

14. The stiff frame shown in Figure 7.14 is supported on a smooth surface and loaded at the centre of the span. Derive an expression for the deflection at the load point.   [0.317 5 Wa^3/EI]

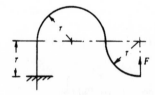

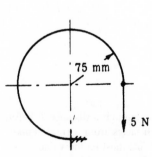

Figure 7.13

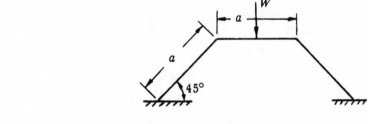

Figure 7.14

Figure 7.15

15. The steel spring shown in Figure 7.15 is rigidly fixed at the lower end and a vertical force of 5 N is applied at the free end. If the section of the spring is 12 mm by 3 mm thick, calculate the vertical and horizontal displacements of the free end. $E = 200$ GN/m^2.   [3.65 mm; 0.195 mm]

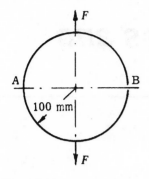

Figure 7.16

16. The ring shown in Figure 7.16 is made of flat steel strip 20 mm by 3 mm. It is cut at the point B and a pull F is applied to the ring along a diameter perpendicular to AB. If the maximum tensile stress due to F is 125 MN/m^2, find the increase in the opening at B. $E = 200$ GN/m^2.   [14.9 mm]

17. A steel tube, having outside and inside diameters of 60 mm and 45 mm respectively, is bent in the form of a quadrant 2 m radius. One end is rigidly attached to a horizontal base-plate to which a tangent to that end is perpendicular and the free end supports a vertical load of 500 N. Determine the vertical and horizontal deflection of the free end.   [36.1 mm; 23.0 mm]

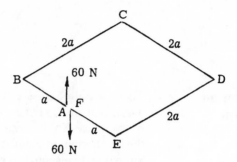

Figure 7.17

18. A 10 mm diameter steel rod is bent to form a square with sides $2a = 40$ mm long. The ends meet at the mid-point of one side and are separated by equal and opposite forces of 60 N applied in a direction perpendicular to the plane of the square as shown in perspective in Figure 7.17. Calculate the amount by which they will be out of alignment. $E = 200$ GN/m^2; $G = 80$ GN/m^2.   [9.78 mm]

19. A cantilever forming a circular arc in plan, and subtending $\pi/3$ rad at the centre, has a circular cross-section of 50 mm diameter and the radius of the centre line is 0.75 m. Find the maximum deflection when a load of 400 N is acting at the free end. $E = 200$ GN/m^2; $G = 80$ GN/m^2.   [2.75 mm]

8 Complex stress and strain

8.1 Stresses on an oblique section

The material of cross-sectional area A shown in Figure 8.1 is subjected to a tensile load F, giving a direct stress $\sigma = F/A$. To determine the stresses on the plane XY inclined at an angle θ to the transverse section, consider the equilibrium of the part ABYX. The area of XY is $A \sec \theta$ so that, if the normal and tangential forces on XY are N and T respectively,

$$N = F \cos \theta \quad \text{and} \quad T = F \sin \theta$$

Therefore the direct stress on XY,

$$\sigma_\theta = \frac{N}{A \sec \theta} = \frac{F}{A} \cos^2 \theta = \sigma \cos^2 \theta \tag{8.1}$$

and the shear stress on XY,

$$\tau_\theta = \frac{T}{A \sec \theta} = \frac{F}{A} \sin \theta \cos \theta = \frac{\sigma}{2} \sin 2\theta \tag{8.2}$$

The maximum value of σ_θ occurs when $\theta = 0$ and is equal to the applied stress σ. The maximum value of τ_θ occurs when $\theta = 45°$ and is equal to $\sigma/2$. Thus if a material has an ultimate shear stress of less than half the ultimate direct stress (tensile or compressive), then failure under *direct* load will occur due to *shear* stress on the oblique plane.

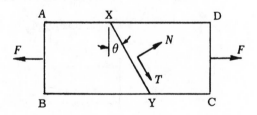

Figure 8.1

8.2 Material subjected to two perpendicular direct stresses

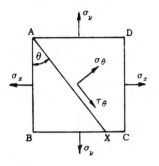

Figure 8.2

Figure 8.2 shows an element of material which is subjected to two perpendicular tensile stresses σ_x and σ_y. To obtain the direct and shear stresses, σ_θ and τ_θ respectively, on plane AX inclined at angle θ to AB, consider the equilibrium of the wedge ABX and let the material be of unit thickness.

Equating *forces* normal to AX.

$$\sigma_\theta \times \mathrm{AX} = \sigma_x \times \mathrm{AB} \cos\theta + \sigma_y \times \mathrm{BX} \sin\theta$$

$$\therefore \ \sigma_\theta = \sigma_x \cos^2\theta + \sigma_y \sin^2\theta$$

$$= \frac{\sigma_x + \sigma_y}{2} + \frac{\sigma_x - \sigma_y}{2}\cos 2\theta \qquad (8.3)$$

The maximum and minimum values of σ_θ are σ_x and σ_y (whichever is greater and lesser respectively) when $\theta = 0$ and $90°$.

Equating *forces* parallel to AX,

$$\tau_\theta \times \mathrm{AX} = \sigma_x \times \mathrm{AB} \sin\theta - \sigma_y \times \mathrm{BX} \cos\theta$$

$$\therefore \ \tau_\theta = (\sigma_x - \sigma_y)\sin\theta\cos\theta$$

$$= \frac{\sigma_x - \sigma_y}{2}\sin 2\theta \qquad (8.4)$$

The maximum value of τ_θ is $\dfrac{\sigma_x - \sigma_y}{2}$ when $\theta = 45°$. It is also apparent that if $\sigma_x = \sigma_y$, τ_θ is zero for all values of θ.

Equations (8.3) and (8.4) may be represented graphically as shown in Figure 8.3. From the pole P, set off $\mathrm{PA} = \sigma_x$ and $\mathrm{PB} = \sigma_y$.

Then $\qquad \mathrm{PO} = \dfrac{\sigma_x + \sigma_y}{2} \quad$ and $\quad \mathrm{OA} = \dfrac{\sigma_x - \sigma_y}{2}$

If σ_x or σ_y is negative, PA or PB must be set off in the opposite direction.

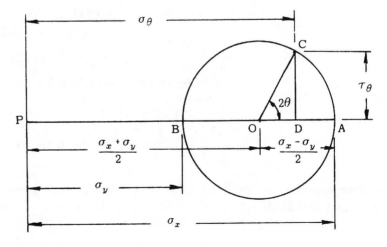

Figure 8.3

Draw a circle, centre O with AB as diameter and draw OC at angle 2θ to OA.

Then
$$PD = \frac{\sigma_x + \sigma_y}{2} + \frac{\sigma_x - \sigma_y}{2} \cos 2\theta = \sigma_\theta$$

and
$$CD = \frac{\sigma_x - \sigma_y}{2} \sin 2\theta \qquad = \tau_\theta$$

This is a simplified form of *Mohr's stress circle* described in Section 8.5.

8.3 Material subjected to shear stress

Figure 8.4 shows an element of material of unit thickness which is subjected to applied and complementary shear stresses τ.

Equating *forces* normal to AX for the equilibrium of ABX,

$$\sigma_\theta \times AX = \tau \times AB \sin \theta + \tau \times BX \cos \theta$$

$$\therefore \sigma_\theta = \tau \sin 2\theta \qquad\qquad (8.5)$$

The maximum value of σ_θ is τ when $\theta = 45°$.

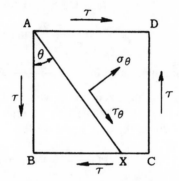

Figure 8.4   **Figure 8.5**

Equating *forces* parallel to AX,

$$\tau_\theta \times AX = -\tau \times AB \cos \theta + \tau \times BX \sin \theta$$

$$\therefore \tau_\theta = -\tau \cos 2\theta \qquad\qquad (8.6)$$

The maximum value of τ_θ is τ when $\theta = 0$ and $90°$ and when $\theta = 45°$, τ_θ is zero.

Equations (8.5) and (8.6) show that the application of shear stress produces tensile and compressive stresses on planes at 45° to the shear planes equal in magnitude to the applied stress, as shown in Figure 8.5.

8.4 The general case of two-dimensional stress

Figure 8.6 shows an element of material which is subjected to perpendicular tensile stresses and applied and complementary shear stresses.

Combining the stresses obtained in Sections (8.2) and (8.3),

$$\sigma_\theta = \frac{\sigma_x + \sigma_y}{2} + \frac{\sigma_x - \sigma_y}{2} \cos 2\theta + \tau \sin 2\theta \qquad\qquad (8.7)$$

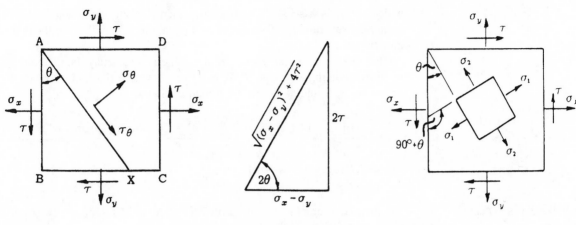

Figure 8.6   **Figure 8.7**   **Figure 8.8**

and
$$\tau_\theta = \frac{\sigma_x - \sigma_y}{2} \sin 2\theta - \tau \cos 2\theta \qquad (8.8)$$

σ_θ is a maximum or minimum when $\dfrac{d\sigma_\theta}{d\theta} = 0$

i.e., when $-(\sigma_x - \sigma_y) \sin 2\theta + 2\tau \cos 2\theta = 0$

or
$$\tan 2\theta = \frac{2\tau}{\sigma_x - \sigma_y} \qquad (8.9)$$

From Figure 8.7, it will be seen that

$$\cos 2\theta = \frac{\sigma_x - \sigma_y}{\sqrt{((\sigma_x - \sigma_y)^2 + 4\tau^2)}} \quad \text{and} \quad \sin 2\theta = \frac{2\tau}{\sqrt{((\sigma_x - \sigma_y)^2 + 4\tau^2)}}$$

so that the maximum and minimum values of σ_θ are given by

$$\sigma_\theta = \frac{\sigma_x + \sigma_y}{2} + \frac{\sigma_x - \sigma_y}{2} \cdot \frac{\sigma_x - \sigma_y}{\sqrt{((\sigma_x - \sigma_y)^2 + 4\tau^2)}} + \tau \cdot \frac{2\tau}{\sqrt{((\sigma_x - \sigma_y)^2 + 4\tau^2)}}$$

$$= \tfrac{1}{2}[(\sigma_x + \sigma_y) \pm \sqrt{((\sigma_x - \sigma_y)^2 + 4\tau^2)}] \qquad (8.10)$$

The planes on which these stresses act are given by equation (8.9),

i.e. $\theta = \tfrac{1}{2} \tan^{-1} \dfrac{2\tau}{\sigma_x - \sigma_y}$ and $\tfrac{1}{2} \tan^{-1} \dfrac{2\tau}{\sigma_x - \sigma_y} + 90°$

These planes of maximum and minimum direct stress are mutually perpendicular and are known as the *principal planes*; the direct stresses which act on them are the *principal stresses*.

Substituting for $\cos 2\theta$ and $\sin 2\theta$ in equation (8.8) shows that *the shear stress on principal planes is zero*.

Figure 8.8 shows the principal planes and stresses in the general stress system, the principal stresses being denoted by σ_1 and σ_2. It is often obvious which stress is associated with each principal plane but when in doubt, numerical substitution of values of θ in equation (8.7) will make the determination.

The maximum shear stress in the general case may be determined by differentiating equation (8.8) but from equation (8.4), it is given by $(\sigma_1 - \sigma_2)/2$, acting at 45° to the principal planes, i.e.

$$\tau_{\max} = \frac{\frac{1}{2}[(\sigma_x+\sigma_y)+\sqrt{((\sigma_x-\sigma_y)^2+4\tau^2)}]-\frac{1}{2}[(\sigma_x+\sigma_y)-\sqrt{((\sigma_x-\sigma_y)^2+4\tau^2)}]}{2}$$

$$= \frac{1}{2}\sqrt{((\sigma_x - \sigma_y)^2 + 4\tau^2)} \tag{8.11}$$

8.5 Mohr's stress circle

An alternative graphical method of determination of principal stresses and planes is given by Mohr's stress circle.

For the stress system shown in Figure 8.9, choose a pole P, Figure 8.10, and set off $PA = \sigma_x$ and $PB = \sigma_y$. The shear stress on the plane of σ_x results in an anticlockwise couple, which is considered negative while the shear on the plane of σ_y results in a clockwise (positive) couple. Set off $AM = -\tau$ and $BN = +\tau$ and join MN to cut PBA at O. Draw a circle of centre O and radius OM and draw OC such that angle $MOC = 2\theta$. Let angle MOD be β.

It will be seen that $PO = \dfrac{\sigma_x + \sigma_y}{2}$ and $OA = \dfrac{\sigma_x - \sigma_y}{2}$.

$$\text{Radius } OM = \sqrt{(OA^2 + AM^2)} = \sqrt{\left(\left(\frac{\sigma_x - \sigma_y}{2}\right)^2 + \tau^2\right)} = OC$$

$$PD = PO + OC\cos(2\theta - \beta)$$

$$= PO + (OM\cos\beta)\cos 2\theta + (OM\sin\beta)\sin 2\theta$$

$$= \frac{\sigma_x + \sigma_y}{2} + \frac{\sigma_x - \sigma_y}{2}\cos 2\theta + \tau\sin 2\theta = \sigma_\theta$$

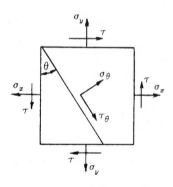

Figure 8.9

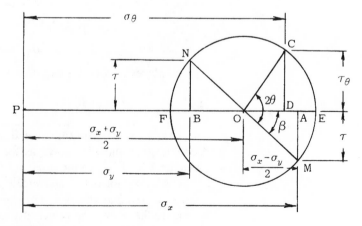

Figure 8.10

$$CD = OC \sin(2\theta - \beta)$$

$$= (OM \cos \beta) \sin 2\theta - (OM \sin \beta) \cos 2\theta$$

$$= \frac{\sigma_x + \sigma_y}{2} \sin 2\theta - \tau \sin 2\theta = \tau_\theta$$

When τ_θ is zero, PC lies on PE and $2\theta = \beta$ or $\beta + 180°$.

Then $\qquad$ PE = PO + OE

$$= \frac{\sigma_x + \sigma_y}{2} + \sqrt{\left(\left(\frac{\sigma_x - \sigma_y}{2}\right)^2 + \tau^2\right)}$$

$$= \tfrac{1}{2}\left[(\sigma_x + \sigma_y) + \sqrt{((\sigma_x - \sigma_y)^2 + 4\tau^2)}\right]$$

and $\qquad$ PF = PO − OF

$$= \frac{\sigma_x + \sigma_y}{2} - \sqrt{\left(\left(\frac{\sigma_x - \sigma_y}{2}\right)^2 + \tau^2\right)}$$

$$= \tfrac{1}{2}\left[(\sigma_x + \sigma_y) - \sqrt{((\sigma_x - \sigma_y)^2 + 4\tau^2)}\right]$$

By comparison with equation (8.10), PE and PF represent the principal stresses, σ_1 and σ_2 and the principal planes are inclined at $\beta/2$ and $\beta/2 + 90°$ to the plane of σ_x.

If the applied shear stress τ is zero, then AM = BN = 0 and the diagram simplifies to that shown in Figure 8.3.

8.6 Simple applications of principal stresses

(a) Combined thrust and twisting

Figure 8.11 shows a shaft of radius r which is subjected to a torque T and an axial thrust F. The direct stress due to F is $\dfrac{F}{A}$ and the shear stress at the surface of the shaft due to T is Tr/J, so that the maximum direct stress,

$$\sigma = \tfrac{1}{2}\left[\frac{F}{A} + \sqrt{\left(\left(\frac{F}{A}\right)^2 + 4\left(\frac{Tr}{J}\right)^2\right)}\right] \qquad \text{from equation (8.10)}$$

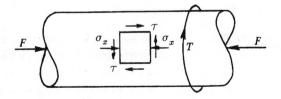

Figure 8.11

and the maximum shear stress,

$$\tau_{\max} = \tfrac{1}{2} \sqrt{\left(\left(\frac{F}{A}\right)^2 + 4\left(\frac{Tr}{J}\right)^2\right)} \qquad \text{from equation (8.11)}$$

(b) Combined bending and twisting

Figure 8.12 shows a shaft of radius r which is subjected to a torque T and a bending moment M. The direct stress at the surface of the shaft due to M is $\dfrac{Mr}{I}$ and the shear stress at the surface due to T is $\dfrac{Tr}{J}$, so that the maximum direct stress,

$$\sigma = \tfrac{1}{2} \left[\frac{Mr}{I} + \sqrt{\left(\left(\frac{Mr}{I}\right)^2 + 4\left(\frac{Tr}{J}\right)^2\right)} \right]$$

element here

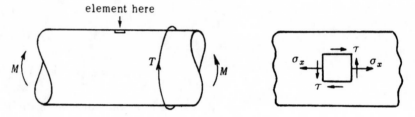

Figure 8.12

Substituting $r = \dfrac{d}{2}$, $I = \dfrac{\pi}{64}d^4$ and $J = \dfrac{\pi}{32}d^4$, this can be rearranged to give

$$\sigma = \frac{16}{\pi d^3}\{M + \sqrt{(M^2 + T^2)}\} \tag{8.12}$$

The maximum shear stress,

$$\tau_{\max} = \tfrac{1}{2} \sqrt{\left(\left(\frac{Mr}{I}\right)^2 + 4\left(\frac{Tr}{J}\right)^2\right)}$$

which can be rearranged to give

$$\tau_{\max} = \frac{16}{\pi d^3} \sqrt{(M^2 + T^2)} \tag{8.13}$$

8.7 Principal strains

If an element is subjected to perpendicular stresses σ_x and σ_y and there is no shear stress on these planes, σ_x and σ_y are the principal stresses, from Section 8.4, and the strains in these directions are the principal strains.

If these strains are ε_x and ε_y, Figure 8.13,

then

$$\varepsilon_x = \frac{\sigma_x}{E} - \nu \frac{\sigma_y}{E} \tag{8.14}$$

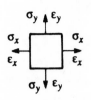

Figure 8.13

and
$$\varepsilon_y = \frac{\sigma_y}{E} - v\frac{\sigma_x}{E} \tag{8.15}$$

where v is Poisson's ratio (Section 1.2).

If the principal strains are measured, then, by rearrangement of equations (8.14) and (8.15), the principal stresses are given by

$$\sigma_x = \frac{E}{1 - v^2}(\varepsilon_x + v\varepsilon_y) \tag{8.16}$$

and
$$\sigma_y = \frac{E}{1 - v^2}(\varepsilon_y + v\varepsilon_x) \tag{8.17}$$

8.8 Strain on an inclined plane

Figure 8.14 shows an element of material ABCD which distorts to A'BC'D' under the action of the principal stresses σ_x and σ_y. The principal strains are ε_x and ε_y and the strain on BD, inclined at angle θ to BC is ε_θ.

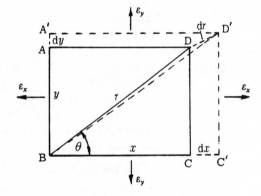

Figure 8.14

Since these strains are very small, the change in length of BD may be taken as DD', i.e. $\varepsilon_\theta = dr/r$,

$$r^2 = x^2 + y^2$$

$$\therefore 2r\,dr = 2x\,dx + 2y\,dy$$

$$\therefore \frac{dr}{r} = \frac{dx}{x}\left(\frac{x}{r}\right)^2 + \frac{dy}{y}\left(\frac{y}{r}\right)^2$$

i.e.
$$\varepsilon_\theta = \varepsilon_x \cos^2\theta + \varepsilon_y \sin^2\theta$$

$$= \frac{\varepsilon_x + \varepsilon_y}{2} + \frac{\varepsilon_x - \varepsilon_y}{2}\cos 2\theta \tag{8.18}$$

This result may be obtained by Mohr's strain circle, Figure 8.15, constructed in a similar manner to the stress circle.

If the direction of the principal planes is unknown, the magnitude and direction of the principal strains and planes may be determined by measuring the strain in three different directions by means of strain gauge rosettes, Section 8.9, and then solving for the three unknown quantities ε_x, ε_y and θ.

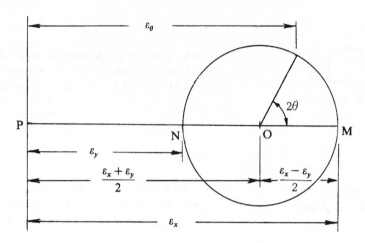

Figure 8.15

8.9 Electric resistance strain gauges

An electric resistance strain gauge consists of a fine wire arranged as shown in Figure 8.16(a) and fixed to a paper backing. This is then cemented to the surface to be investigated and distortion in the direction of the gauge axis changes the electrical resistance of the wire, which is measured by a Wheatstone bridge circuit. Strain in a lateral direction does not affect the resistance of the wire.

The *gauge factor*, defined as the ratio $\dfrac{\text{fractional change in resistance}}{\text{fractional change in length}}$, is approximately 2, this being calibrated by the manufacturer, and strain bridges are available calibrated directly in strain.

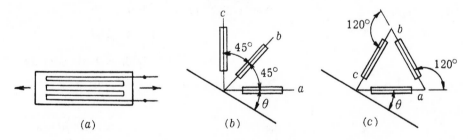

Figure 8.16

To determine the principal strains and planes, strain rosettes are used. These consist of three strain gauges which are glued to the surface with their axes inclined either at 45° intervals, as in Figure 8.16(b), or at 120° intervals, as in Figure 8.16(c).

From equation (8.18), $\varepsilon_\theta = \dfrac{\varepsilon_x + \varepsilon_y}{2} + \dfrac{\varepsilon_x - \varepsilon_y}{2} \cos 2\theta$

which may be written $\quad \varepsilon_\theta = m + n\cos 2\theta \quad$ where $m = \dfrac{\varepsilon_x + \varepsilon_y}{2}$

$$\text{and} \quad n = \frac{\varepsilon_x - \varepsilon_y}{2}$$

For the 45° rosette, $\quad \varepsilon_a = m + n\cos 2\theta$

$$\varepsilon_b = m + n\cos 2(\theta + 45°) = m - n\sin 2\theta$$

and $\qquad\qquad \varepsilon_c = m + n\cos 2(\theta + 90°) = m - n\cos 2\theta$

For the 120° rosette, $\varepsilon_a = m + n\cos 2\theta$

$$\varepsilon_b = m + n\cos 2(\theta + 120°) = m - \frac{n}{2}\cos 2\theta + \frac{\sqrt{3}n}{2}\sin 2\theta$$

and $\qquad \varepsilon_c = m + n\cos 2(\theta + 240°) = m - \frac{n}{2}\cos 2\theta - \frac{\sqrt{3}n}{2}\sin 2\theta$

For either arrangement, the equations are sufficient to solve for m, n and θ and the principal strains are then given by

$$\varepsilon_x = m + n \quad \text{and} \quad \varepsilon_y = m - n$$

The principal stresses are obtained by substitution in equations (8.16) and (8.17).

8.10 Graphical determination of principal strains

The analytical determination of principal strains from strain rosette readings is tedious and these may be determined more readily, and with sufficient accuracy, from a graphical construction, as follows.

For the strain rosette shown in Figure 8.17, draw a vertical line QQ, Figure 8.18, to represent zero strain and then draw three further vertical

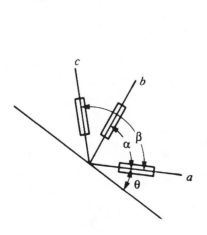

Figure 8.17

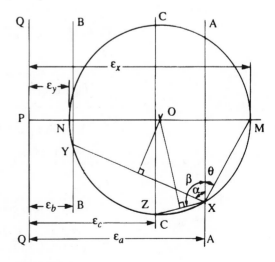

Figure 8.18

lines AA, BB and CC at horizontal distances representing ε_a, ε_b and ε_c respectively from QQ.

Take any point X on AA and through X, draw a line at angle α, anti-clockwise, to intersect BB at Y. Similarly, draw another line through X at angle β to intersect CC at Z. Bisect XY and XZ and draw perpendiculars to intersect at O, which is then the centre of the strain circle passing through X, Y and Z.

The principal strains, ε_x and ε_y, are measured from the point P to the circumference of the strain circle at M and N, as in Figure 8.15, and the angle θ of the principal axis to gauge a is the angle between XM and AA.

8.11 Relation between E, G and v

Figure 8.19(a) shows an element of material subjected to shear stress τ. From Section 8.3, it was seen that this results in direct tensile and compressive stresses τ on an element inclined at 45° to the planes of shear and hence, for the material with modulus of elasticity E, modulus of rigidity G and Poisson's ratio v,

$$\text{strain on the diagonal BD} = \tau E + v\frac{\tau}{E} \tag{8.19}$$

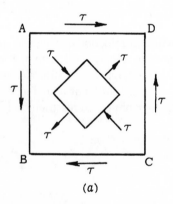
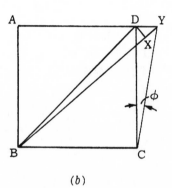

(a) (b)

Figure 8.19

Also, in Figure 8.19(b),

$$\text{strain on BD} = \frac{XY}{BD} = \frac{DY/\sqrt{2}}{\sqrt{2}CD} = \frac{\phi}{2} = \frac{\tau}{2G} \tag{8.20}$$

Hence, from equations (8.19) and (8.20),

$$\frac{\tau}{E}(1+v) = \frac{\tau}{2G}$$

from which $$E = 2G(1+v) \tag{8.21}$$

8.12 Volumetric strain and bulk modulus

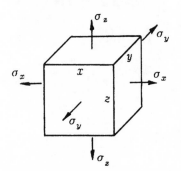

Figure 8.20

Figure 8.20 shows an element of material subjected to principal stresses σ_x, σ_y and σ_z

Strain on side x,

$$\varepsilon_x = \frac{\sigma_x - v(\sigma_y + \sigma_z)}{E} \tag{8.22}$$

Similarly,

$$\varepsilon_y = \frac{\sigma_y - v(\sigma_x + \sigma_z)}{E} \tag{8.23}$$

and

$$\varepsilon_z = \frac{\sigma_z - v(\sigma_x + \sigma_y)}{E} \tag{8.24}$$

New length of side $x = x(1 + \varepsilon_x)$

New length of side $y = y(1 + \varepsilon_y)$

New length of side $z = z(1 + \varepsilon_z)$

$$\therefore \text{ new volume} = xyz(1 + \varepsilon_x)(1 + \varepsilon_y)(1 + \varepsilon_z)$$

$$= xyz(1 + \varepsilon_x + \varepsilon_y + \varepsilon_z)$$

if products of strains are neglected

$$\text{Volumetric strain} = \frac{\text{change in volume}}{\text{original volume}}$$

i.e.

$$e_v = \frac{xyz(1 + \varepsilon_x + \varepsilon_y + \varepsilon_z) - xyz}{xyz}$$

$$= \varepsilon_x + \varepsilon_y + \varepsilon_z \tag{8.25}$$

In the special case of equal stresses, such as with fluid pressure,

$$\sigma_x = \sigma_y = \sigma_z = \sigma$$

and

$$\varepsilon_z = \varepsilon_y = \varepsilon_z = \frac{\sigma}{E}(1 - 2v)$$

so that

$$\varepsilon_v = \frac{3\sigma}{E}(1 - 2v) \tag{8.26}$$

For such a case, the *bulk modulus (K)* is defined as the ratio

$$\frac{\text{stress}}{\text{volumetric strain}}$$

i.e.

$$K = \frac{\sigma}{\varepsilon_v}$$

$$\therefore \varepsilon_v = \frac{\sigma}{K} \tag{8.27}$$

Hence, from equations (8.26) and (8.27),

$$\frac{3\sigma}{E}(1 - 2v) = \frac{\sigma}{K}$$

from which
$$E = 3K(1 - 2v) \qquad (8.28)$$

In the case of unequal stresses,

$$\varepsilon_v = \varepsilon_x + \varepsilon_y + \varepsilon_z$$

$$= \frac{\sigma_x - v(\sigma_y + \sigma_z)}{E} + \frac{\sigma_y - v(\sigma_x + \sigma_z)}{E} + \frac{\sigma_z - v(\sigma_x + \sigma_y)}{E}$$

$$= \frac{1 - 2v}{E}(\sigma_x + \sigma_y + \sigma_z)$$

$$= \frac{\sigma_x + \sigma_y + \sigma_z}{3K} \qquad (8.29)$$

8.13 Strain energy due to two principal stresses

Consider an element of material, of unit thickness, subjected to principal stresses σ_x and σ_y, Figure 8.21.

$$\text{Force in } x \text{ direction} = \sigma_x \times y \times 1 = \sigma_x y$$

$$\text{Strain in } x \text{ direction} = \frac{\sigma_x}{E} - v\frac{\sigma_y}{E} = \frac{1}{E}(\sigma_x - v\sigma_y)$$

$$\therefore \text{ extension in } x \text{ direction} = \frac{x}{E}(\sigma_x - v\sigma_y)$$

$$\therefore \text{ work done in } x \text{ direction} = \tfrac{1}{2} \times \sigma_x y \times \frac{x}{E}(\sigma_x - v\sigma_y)$$

$$= \frac{xy}{2E}(\sigma_x^2 - v\sigma_x\sigma_y)$$

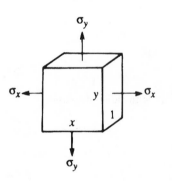

Figure 8.21

Similarly,

$$\text{work done in } y \text{ direction} = \frac{xy}{2E}(\sigma_y^2 - v\sigma_x\sigma_y)$$

Hence,
$$\text{total work done} = \frac{xy}{2E}(\sigma_x^2 + \sigma_y^2 - 2v\sigma_x\sigma_y)$$

This is equal to the energy stored, or strain energy, U. Since xy represents the volume of the element,

$$\text{strain energy per unit volume} = \frac{1}{2E}(\sigma_x^2 + \sigma_y^2 - 2v\sigma_x\sigma_y) \qquad (8.30)$$

8.14 Theories of elastic failure

In simple cases of direct stress, elastic failure is assumed to have occurred when the stress reaches the elastic limit stress for the material. In cases of complex stress, however, stresses on planes perpendicular to that of the maximum principal stress may affect the point at which failure occurs. This depends on whether the smaller stresses are of the same or opposite sign to that of the greater stress and whether the material is brittle or ductile.

In the case of brittle materials, failure is found to occur when the maximum principal stress reaches the elastic limit, regardless of the value or nature of the stresses on perpendicular planes. For ductile materials, however, the stresses on planes perpendicular to that of the maximum principal stress affect the value of the maximum principal stress at which failure occurs, and one theory suggests that failure occurs when the maximum *shear* stress reaches that at the elastic limit in the case of simple direct stress.

Another theory suggests that failure occurs when the energy stored in the material reaches that at the elastic limit in simple direct stress, irrespective of the method of stressing.

Let the principal stresses be σ_x and σ_y, where $\sigma_x > \sigma_y$, and let the stress at the elastic limit, or other criterion of failure, in simple tension be σ_0.

Maximum principal stress theory

Failure is assumed to occur when the greater principal stress, σ_x, reaches σ_0, the lesser stress, σ_y, having no effect,

i.e.
$$\sigma_x = \sigma_0 \tag{8.31}$$

This theory is only found to be applicable to brittle materials.

Maximum shear stress theory

Failure is assumed to occur when the maximum shear stress reaches that at failure in simple tension. The maximum shear stress is half the difference of the greatest and least principal stresses, from equation (8.4) but in the case of two dimensional stress, the third principal stress $\sigma_z = 0$. Thus the maximum shear stress is either $\dfrac{\sigma_x - \sigma_y}{2}$ or $\dfrac{\sigma_x - 0}{2}$, whichever is greater.

In the case of like stresses, $\sigma_x - 0 > \sigma_x - \sigma_y$ but for unlike stresses, where σ_y is negative, $\sigma_x - \sigma_y > \sigma_x - 0$.

Hence, for like stresses, failure occurs when

$$\sigma_x = \sigma_0 \tag{8.32}$$

and for unlike stresses, failure occurs when

$$\sigma_x - \sigma_y = \sigma_0 \tag{8.33}$$

Maximum strain energy theory

Failure is assumed to occur when the strain energy due to σ_x and σ_y reaches that in simple tension at a stress σ_0, i.e. when $\quad \dfrac{1}{2E}(\sigma_x^2 + \sigma_y^2 - 2v\sigma_x\sigma_y) = \dfrac{\sigma_0^2}{2E}\quad$ from equation (8.30)

or $\qquad\qquad \sigma_x^2 + \sigma_y^2 - 2v\sigma_x\sigma_y = \sigma_0^2 \qquad\qquad (8.34)$

The shear stress theory and strain energy theory both find experimental support for ductile materials and will be found to give similar results.

Worked examples

1. *The principal stresses at a point under two-dimensional stress are 80 MN/m² tension and 50 MN/m² compression. Calculate, from first principles, the resultant stress on a plane inclined at 30° to the line of action of the tensile stress.*

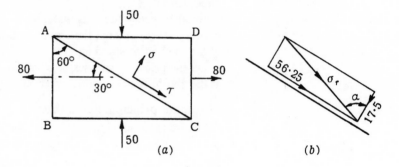

Figure 8.22

Equating forces normal to AC, Figure 8.22(a),

$$\sigma \times \text{AC} = 80\ \text{AB} \cos 60° - 50\ \text{BC} \sin 60°$$

$$\therefore \sigma = 80 \cos^2 60° - 50 \sin^2 60°$$

$$= 80 \times \tfrac{1}{4} - 50 \times \tfrac{3}{4} = -17.5\ \text{MN/m}^2$$

Equating forces parallel to AC,

$$\tau \times \text{AC} = 80\ \text{AB} \sin 60° + 50\ \text{BC} \cos 60°$$

$$\therefore \tau = 80 \cos 60° \sin 60° + 50 \sin 60° \cos 60°$$

$$= \frac{80 + 50}{2} \sin 120°$$

$$= 65 \times \frac{\sqrt{3}}{2} = 56.25\ \text{MN/m}^2$$

From Figure 8.22(b), the resultant stress,

$$\sigma_r = \sqrt{(17.5^2 + 56.25^2)} = \underline{58.9\ \text{MN/m}^2}$$

and $\qquad\qquad \alpha = \tan^{-1}\dfrac{56.25}{17.5} \qquad = \underline{72°48'}$

2. *A thin cylinder with closed ends has an internal diameter of 50 mm and a wall thickness of 2.5 mm. It is subjected to an axial pull of 10 kN and an axial torque of 500 N m while under an internal pressure of 6 MN/m². Determine the principal stresses in the tube and the maximum shear stress.*

The arrangement is shown in Figure 8.23(a).

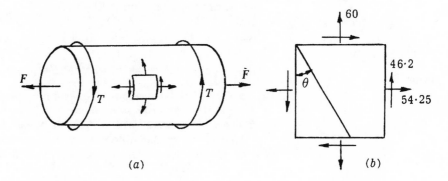

(a)   (b)

Figure 8.23

Due to internal pressure,

$$\sigma_c = \frac{pd}{2t} \quad \text{from equation (9.1)}$$

$$= \frac{6 \times 10^6 \times 50 \times 10^{-3}}{2 \times 2.5 \times 10^{-3}} = 60 \times 10^6 \text{ N/m}^2$$

$$\sigma_l = \frac{\sigma_c}{2} = 30 \times 10^6 \text{ N/m}^2 \quad \text{from equation (9.2)}$$

Due to the axial load,

$$\sigma_l = \frac{F}{A} = \frac{10 \times 10^3}{\pi \times 52.5 \times 2.5 \times 10^{-6}} = 24.25 \times 10^6 \text{ N/m}^2$$

∴ total axial stress $= (30 + 24.25) \times 10^6 \qquad = 54.25 \times 10^6 \text{ N/m}^2$

Due to the torque, $\tau = \dfrac{\text{torque}}{\text{mean radius} \times \text{area}}$ since the tube is thin

$$= \frac{500}{26.25 \times \pi \times 52.5 \times 2.5 \times 10^{-9}} = 46.2 \times 10^6 \text{ N/m}^2$$

The stresses on an element of the tube are shown in Figure 8.23(b). From equation (8.10), the principal stresses are given by

$$\sigma = \tfrac{1}{2}\{(54.25 + 60) \pm \sqrt{((54.25 - 60)^2 + 4 \times 46.2^2)}\}$$

$$= \underline{103.4 \quad \text{and} \quad 10.8 \text{ MN/m}^2}$$

From equation (8.4),

$$\tau_{max} = \frac{103.4 - 10.8}{2} = \underline{46.3 \text{ MN/m}^2}$$

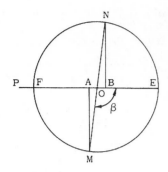

Figure 8.24

From equation (8.9),

$$\theta = \tfrac{1}{2}\tan^{-1}\frac{2 \times 46.2}{54.25 - 60} = \tfrac{1}{2}(180° - 86°26') = \underline{46°47'}$$

Alternatively, using Mohr's stress circle, Figure 8.24, set off PA = 54.25, PB = 60 and AM = BN = 46.2 MN/m². Draw a circle of centre O to pass through M and N, cutting the base line at E and F.

Then $\sigma_1 = PE = 103.4$ MN/m² $\theta = \dfrac{\beta}{2} = 46°47'$

$$\sigma_2 = PF = 10.8 \text{ MN/m}^2$$

3. *On a plane passing through a point in a stressed material, there is a tensile stress of 60 N/mm² and a shearing stress of 40 N/mm². On another plane at 45° to the first plane, there is a tensile stress of 120 N/mm² and an unknown shearing stress.*
Find (a) the principal stresses, (b) the maximum shearing stress, (c) the positions of the principal planes, (d) the principal strains. E = 200 kN/mm² and v = 0.3.

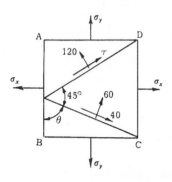

Figure 8.25

In Figure 8.25, let σ_x and σ_y be the principal stresses. Then, from equations (8.3) and (8.4),

$$\sigma_\theta = \frac{\sigma_x + \sigma_y}{2} + \frac{\sigma_x - \sigma_y}{2}\cos 2\theta$$

$$= m + n\cos 2\theta$$

where $m = \dfrac{\sigma_x + \sigma_y}{2}$ and $n = \dfrac{\sigma_x - \sigma_y}{2}$

and $\tau_\theta = n\sin 2\theta$

$$\therefore\ 60 = m + n\cos 2\theta \tag{1}$$

$$40 = n\sin 2\theta \tag{2}$$

and $120 = m + n\cos 2(\theta + 45°) = m - n\sin 2\theta$ (3)

From equations (2) and (3), $m = 160$

so that from equation (1), $n\cos 2\theta = -100$ (4)

Therefore, from equations (2) and (4),

$$n = \sqrt{((-100)^2 + 40^2)} = 107.6$$

and $\tan 2\theta = \dfrac{40}{-100}$

from which $\theta = \tfrac{1}{2}(180° - 21°48') = \underline{79°6'}$

$$\sigma_x = m + n = 160 + 107.6 \qquad = \underline{267.6 \text{ N/mm}^2}$$

and $\sigma_y = m - n = 160 - 107.6$ $= \underline{52.4 \text{ N/mm}^2}$

$$\tau_{max} = \frac{\sigma_x - \sigma_y}{2} = \frac{267.6 - 52.4}{2} \qquad = \underline{107.6 \text{ N/mm}^2}$$

$$\varepsilon_x = \frac{\sigma_x}{E} - \nu\frac{\sigma_y}{E} = \frac{267.6 - 0.3 \times 52.4}{200 \times 10^3} = \underline{1.26 \times 10^{-3}}$$

and

$$\varepsilon_y = \frac{\sigma_y}{E} - \nu\frac{\sigma_x}{E} = \frac{52.4 - 0.3 \times 267.6}{200 \times 10^3} = \underline{-0.139 \times 10^{-3}}$$

4. *In a two-dimensional strain system, the following readings were taken with a 45° strain rosette:*

$$\varepsilon_{0°} = 0.4 \times 10^{-3}, \varepsilon_{45°} = 0.4 \times 10^{-3}, \varepsilon_{90°} = 0.1 \times 10^{-3}$$

Determine the magnitude and directions of the principal strains. If E = 200 kN/mm² and ν = 0.3, find the principal stresses.

From Section 8.9, $\varepsilon_a = m + n \cos 2\theta = 0.4 \times 10^{-3}$ (1)

$\varepsilon_b = m - n \sin 2\theta = 0.4 \times 10^{-3}$ (2)

and $\varepsilon_c = m - n \cos 2\theta = 0.1 \times 10^{-3}$ (3)

Hence, from equations (1) and (3), $m = 0.25 \times 10^{-3}$

Equations (1) and (2) then become $n \sin 2\theta = -0.15 \times 10^{-3}$

and $n \cos 2\theta = 0.15 \times 10^{-3}$

Therefore $n = \sqrt{((-0.15 \times 10^{-3})^2 + (0.15 \times 10^{-3})^2)} = 0.212 \times 10^{-3}$

$$\varepsilon_x = m + n = \underline{0.462 \times 10^{-3}}$$

and $\varepsilon_y = m - n = \underline{0.038 \times 10^{-3}}$

$$\tan 2\theta = \frac{n \sin 2\theta}{n \cos 2\theta} = \frac{-0.15 \times 10^{-3}}{0.15 \times 10^{-3}} = -1$$

Since the sine term is negative and the cosine term is positive, the angle 2θ lies in the fourth quadrant, i.e., $2\theta = 315°$ or $\theta = 157.5°$.

The directions of the principal planes in relation to the strain rosette are shown in Figure 8.26.

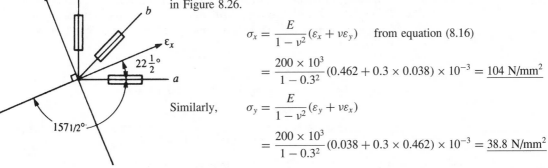

$$\sigma_x = \frac{E}{1 - \nu^2}(\varepsilon_x + \nu\varepsilon_y) \quad \text{from equation (8.16)}$$

$$= \frac{200 \times 10^3}{1 - 0.3^2}(0.462 + 0.3 \times 0.038) \times 10^{-3} = \underline{104 \text{ N/mm}^2}$$

Similarly, $\sigma_y = \frac{E}{1 - \nu^2}(\varepsilon_y + \nu\varepsilon_x)$

$$= \frac{200 \times 10^3}{1 - 0.3^2}(0.038 + 0.3 \times 0.462) \times 10^{-3} = \underline{38.8 \text{ N/mm}^2}$$

Figure 8.26

Alternatively, the principal strains may be determined graphically by the method shown in Section 8.10.

From the starting line QQ, Figure 8.27, draw parallel lines AA, BB and CC at distances representing 0.4, 0.4 and 0.1 millistrain respectively. Take any point X on AA and draw through this point lines at 45° and 90° anticlockwise to intersect BB and CC respectively at points Y and Z (in this example, AA and BB coincide and so Y coincides with X).

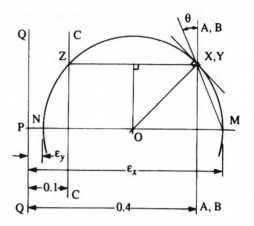

Figure 8.27

The perpendicular bisectors of these lines intersect at O, which is the centre of the Mohr's strain circle, passing through X, Y and Z. The principal strains are then given by PM and PN.

The angle θ of ε_x relative to gauge a is the angle between XM and AA and is positive, relative to AA, i.e. the principal axis lies between gauges a and b at an angle of $22\frac{1}{2}^\circ$.

5. *A bar of metal 20 mm diameter is tested in tension and extends 3.1×10^{-9} m/N on a gauge length of 200 mm. Another bar of the same metal and same diameter is tested in torsion and twists $0.009\,12°/N$ m on a gauge length of 200 mm. Calculate the values of E, G and v.*

A bar of this material and 20 mm diameter is subjected to an axial compressive load of 60 kN together with a lateral pressure of 80 MN/m² applied to the circumference. Determine the change in length on a 200 mm gauge length and the change in diameter.

$$E = \frac{Fl}{Ax} = \frac{1 \times 0.2}{\frac{\pi}{4} \times 0.02^2 \times 3.1 \times 10^{-9}} = \underline{205 \times 10^9 \text{ N/m}^2}$$

$$G = \frac{Tl}{J\theta} = \frac{1 \times 0.2}{\frac{\pi}{32} \times 0.02^4 \times \left(0.009\,12 \times \frac{\pi}{180}\right)} = \underline{80 \times 10^9 \text{ N/m}^2}$$

$$E = 2G(1 + v) \quad \text{from equation (8.21)}$$

$$\therefore v = \frac{E}{2G} - 1 = \frac{205}{2 \times 80} - 1 = \underline{0.28}$$

$$\text{Axial stress} = \frac{60 \times 10^3}{\frac{\pi}{4} \times 0.02^2} = 191 \times 10^6 \text{ N/m}^2$$

$$\therefore \text{ axial strain} = \frac{191 \times 10^6 - 2 \times 0.28 \times 80 \times 10^6}{205 \times 10^9} \quad \text{from equation (8.22)}$$

$$= 0.712 \times 10^{-3}$$

$$\therefore \text{ decrease in length} = 200 \times 0.712 \times 10^{-3} = \underline{0.142\,4 \text{ mm}}$$

$$\text{Lateral strain} = \frac{80 \times 10^6 - 0.28 \times 80 \times 10^6 - 0.28 \times 191 \times 10^6}{205 \times 10^9}$$

$$= 0.02 \times 10^{-3}$$

$$\therefore \text{ decrease in diameter} = 20 \times 0.02 \times 10^{-3} = \underline{0.000\,4 \text{ mm}}$$

6. *Determine the change in volume of a steel bar 80 mm square in section and 1.2 m long when subjected to an axial compressive load of 20 kN. E = 200 kN/mm^2 and G = 80 kN/mm^2.*

$$\sigma_x = \frac{20 \times 10^3}{80^2} = 3.125 \text{ N/mm}^2$$

$$\sigma_y = \sigma_z = 0$$

As in Example 5,
$$\nu = \frac{E}{2G} - 1 = \frac{200}{2 \times 80} - 1 = 0.25$$

$$\therefore \varepsilon_x = \frac{3.125}{200 \times 10^3} = \frac{3.125}{200} \times 10^{-3}$$

$$\varepsilon_y = \varepsilon_z = -0.25 \times \frac{3.125}{200} \times 10^{-3}$$

$$\therefore \varepsilon_v = \varepsilon_x + \varepsilon_y + \varepsilon_z \quad \text{from equation (8.25)}$$

$$= \frac{3.125}{200} \times 10^{-3}(1 - 2 \times 0.25) = \frac{3.125}{400} \times 10^{-3}$$

Alternatively,
$$E = 2G(1 + \nu) \quad \text{from equation (8.21)}$$

and
$$E = 3K(1 - 2\nu) \quad \text{from equation (8.28)}$$

Eliminating ν between these equations gives

$$K = \frac{EG}{3(3G - E)}$$

$$= \frac{200 \times 80 \times 10^6}{3(3 \times 80 - 200) \times 10^3} = \frac{400}{3} \times 10^3 \text{ N/mm}^2$$

$$\therefore \varepsilon_v = \frac{\sigma_x + \sigma_y + \sigma_z}{3K} \quad \text{from equation (8.29)}$$

$$= \frac{3.125}{400 \times 10^3}$$

$$\therefore \text{ change in volume} = \frac{3.125}{400 \times 10^3} \times 80^2 \times 1\,200$$

$$= \underline{60.3 \text{ mm}^3}$$

7. *A compressed air cylinder, 1.8 m long and 0.6 m internal diameter has a wall thickness of 10 mm. Find the increase in volume when it is subjected to an internal pressure of 4 MN/m². E = 200 GN/m² and ν = 0.3.*

Referring to Figure 8.28,

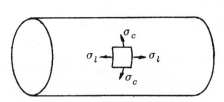

Figure 8.28

$$\sigma_c = \frac{pd}{2t} \quad \text{from equation (9.1)}$$

$$= \frac{4 \times 10^6 \times 0.6}{2 \times 0.01}$$

$$= 120 \times 10^6 \text{ N/m}^2$$

$$\sigma_l = \frac{\sigma_c}{2} \quad \text{from equation (9.2)}$$

$$= 60 \times 10^6 \text{ N/m}^2$$

The volumetric strain is the sum of the strain in the axial direction and the strains on two perpendicular diameters. The diametral strain is the same as the circumferential strain, since the diameter and circumference each expand in the same proportion.

Neglecting the effect of radial stress, which is relatively insignificant,

$$\varepsilon_d = \varepsilon_c = \frac{\sigma_c}{E} - \nu\frac{\sigma_l}{E}$$

$$= \frac{(120 - 0.3 \times 60) \times 10^6}{200 \times 10^9} \qquad = 0.51 \times 10^{-3}$$

$$\varepsilon_l = \frac{\sigma_l}{E} - \nu\frac{\sigma_c}{E}$$

$$= \frac{(60 - 0.3 \times 120) \times 10^6}{200 \times 10^9} \qquad = 0.12 \times 10^{-3}$$

$$\therefore \varepsilon_v = \varepsilon_l + 2\varepsilon_d$$

$$= (0.12 + 2 \times 0.51) \times 10^{-3} \qquad = 1.14 \times 10^{-3}$$

$$\therefore \text{ increase in volume} = \frac{\pi}{4} \times 0.6^2 \times 1.8 \times 1.14 \times 10^{-3} = \underline{0.000\,58 \text{ m}^3}$$

8. *A shaft having a diameter of 100 mm is subjected to a bending moment of 6.5 kN m in addition to the torque which it transmits. Find the maximum allowable torque if (a) the material is brittle and failure is assumed to occur when the maximum direct stress reaches 120 MN/m² ; (b) the material is ductile and failure is assumed to occur when the maximum shear stress reaches that at the elastic limit stress of 300 MN/m² in pure tension.*

From Section 8.6(b), the maximum and minimum principal stresses in a shaft subjected to combined bending and twisting are given by

$$\sigma = \frac{16}{\pi d^3}[M \pm \sqrt{(M^2 + T^2)}] = \frac{16}{\pi \times 0.1^3}[6.5 \times 10^3 \pm \sqrt{((6.5 \times 10^3)^2 + T^2)}]$$

$$= 5\,093[6.5 \times 10^3 \pm \sqrt{(42.25 \times 10^6 + T^2)}]$$

(a) Maximum principal stress, $\sigma_x = 5\,093[6.5 \times 10^3 + \sqrt{(42.25 \times 10^6 + T^2)}]$

which is to be equal to 120 MN/m^2.

Hence $\qquad\qquad\qquad\qquad$ $T = \underline{15.77 \text{ kN m}}$

(b) Maximum principal stress, $\sigma_x = 5\,093[6.5 \times 10^3 + \sqrt{(42.25 \times 10^6 + T^2)}]$

and minimum principal stress, $\sigma_y = 5\,093[6.5 \times 10^3 - \sqrt{(42.25 \times 10^6 + T^2)}]$

Since $\sqrt{(42.25 \times 10^6 + T^2)} > 6.5 \times 10^3$, σ_y will be negative and hence

$$\text{maximum shear stress} = \frac{\sigma_x - \sigma_y}{2}$$

From the maximum shear stress theory of failure,

$$\sigma_x - \sigma_y = \sigma_0 \quad \text{from equation (8.33)}$$

$$\therefore 5\,093[6.5 \times 10^3 + \sqrt{(42.25 \times 10^6 + T^2)}]$$

$$-5\,093[6.5 \times 10^3 - \sqrt{(42.25 \times 10^6 + T^2)}]$$

$$= 300 \times 10^6$$

from which $\qquad\qquad\qquad$ $T = \underline{28.7 \text{ kN m}}$

9. *At a point in a stressed material, the direct stresses on two perpendicular planes are 100 N/mm^2 and 40 N/mm^2, both tensile, and a shear stress on these planes of 20 N/mm^2. The material is ductile and the yield stress in simple tension is 320 N/mm^2. Determine the factor of safety on this yield stress according to*

(a) the maximum shear stress theory,
(b) the maximum strain energy theory. $v = 0.3$.

The principal stresses are given by

$$\sigma = \tfrac{1}{2}\{(100 + 40) \pm \sqrt{((100 - 40)^2 + 4 \times 20^2)}\} \quad \text{from equation (8.10)}$$

$$= 106 \quad \text{and} \quad 43 \text{ N/mm}^2$$

If F is the factor of safety, the working stress is $\dfrac{320}{F}$.

(a) The principal stresses are of the same type, i.e., both tensile, and so, from the maximum shear stress theory,

$$\sigma_x = \sigma_0 \quad \text{from equation (8.32)}$$

i.e. $\qquad\qquad\qquad\qquad$ $106 = \dfrac{320}{F}$

from which $\qquad\qquad\qquad$ $F = \underline{3.02}$

(b) From the maximum strain energy theory,

$$\sigma_x^2 + \sigma_y^2 - 2v\sigma_x\sigma_y = \sigma_0^2 \quad \text{from equation (8.34)}$$

i.e. $$106^2 + 43^2 - 2 \times 0.3 \times 106 \times 43 = \left(\frac{320}{F}\right)^2$$

from which $$F = \underline{2.96}$$

Further problems

10. A thin cylinder, 300 mm diameter and 3 mm thick is subjected to an internal pressure of 3.5 MN/m^2 and an axial tensile force of 100 kN. Determine the normal and shear stresses on a plane inclined at 30° to the axis of the cylinder.

[161 MN/m^2; 22.6 MN/m^2]

11. At a point in a stressed material, the major principal stress is 140 MN/m^2 tension and the maximum shear stress is 80 MN/m^2. Determine (a) the minor principal stress, (b) the direct stress on the plane of maximum shear stress, (c) the stresses on a plane inclined at 30° to the plane on which the major principal stress acts.   [20 MN/m^2; 60 MN/m^2; 100 MN/m^2; 69.3 MN/m^2]

12. The principal stresses at a point in a material are 50 MN/m^2 tension and 30 MN/m^2 tension. Determine, for a plane inclined at 40° to the plane on which the former stress acts, (a) the normal and shear stresses, (b) the magnitude and inclination of the resultant stress.

[41.73 MN/m^2; 9.85 MN/m^2; 42.8 MN/m^2 at 13°18′ to the normal stress]

13. The principal stresses at a point in a material are 60 MN/m^2, both tensile. Determine the normal and shear stresses on a plane inclined at an angle of $\tan^{-1} 0.25$ to the plane on which the maximum principal stress acts.

[57.7 MN/m^2; 9.4 MN/m^2]

14. Direct stresses of 80 MN/m^2 tension and 60 MN/m^2 compression are applied at a point in a material, on planes at right angles to one another. If the maximum direct stress in the material is limited to 100 MN/m^2 tension, what shear stress may be applied to the given planes and what will then be the maximum shearing stress at the point?   [56.56 MN/m^2; 90 MN/m^2]

15. On a plane in a stressed material, there is a tensile stress of 100 MN/m^2 and a shearing stress of 55 MN/m^2. On a plane making an angle of 30° anti-clockwise to this plane, there is a tensile stress of 20 MN/m^2 and an unknown shearing stress. Find the position of the principal planes and the magnitude of the principal stresses.

[69°54′ and 159°54′; +120.3 and −49.7 MN/m^2]

16. A thin cylinder 75 mm diameter and 5 mm thick is subjected to an internal pressure of 5.5 MN/m^2 and also to a torque of 1.6 kN m. Determine the maximum and minimum principal stresses and the maximum shearing stress in the tube.

[64.3 and −2.5 MN/m^2; 33.4 MN/m^2]

17. A hollow shaft 200 mm outside diameter and 125 mm inside diameter is subjected to a bending moment of 43 kN m and a torque of 65 kN m. Calculate the maximum shearing stress in the shaft.   [58.6 MN/m^2]

18. A shaft 80 mm diameter is simply supported in bearings 0.6 m apart. A flywheel of mass 500 kg is mounted midway between the bearings and the shaft transmits 30 kW at 360 rev/min. Calculate the principal stresses and maximum shear stress on the shaft at the ends of a vertical and horizontal diameter in the plane of the flywheel.   [18.1, 10.78, 7.9, 7.9 MN/m^2]

19. A solid circular shaft is subjected to a bending moment M and a torque T. At a point on the circumference, the maximum principal stress is numerically four times the minimum principal stress. Determine the ratio M/T and the angle between the plane of the maximum principal stress and the plane of the bending stress.   [3/4; 26°30′]

20. A hollow shaft 100 mm external diameter and 50 mm internal diameter is subjected to an axial thrust of 50 kN while transmitting 600 kW at 500 rev/min. What bending moment may be applied to the shaft if the greater principal stress is not to exceed 100 MN/m^2? What will then be the value of the smaller principal stress?   [4.85 kN m; 38.85 MN/m^2]

21. A solid shaft 200 mm diameter transmits 2 MW at 250 rev/min and is subjected to a bending moment of 50 kN m. Calculate the maximum permissible end thrust on the shaft if the maximum shearing stress is limited to 80 MN/m^2]
                    [2 MN]

22. Find the dimensions of a hollow steel shaft, internal diameter = 0.6 × diameter, to transmit 150 kW at a speed of 250 rev/min if the shearing stress is not to exceed 70 MN/m^2.

If a bending moment of 2.7 kN m is now applied to the shaft, find the speed at which it must be driven to transmit the same power for the same value of the maximum shearing stress.   [78.2 mm; 46.9 mm; 283 rev/min]

23. The principal stresses at a point in a material are 160 and 40 MN/m^2, both tensile. If $E = 200$ GN/m^2 and $\nu = 0.28$, find the strain in a direction inclined at 30° to that of the greater principal stress. In what direction is the strain zero?
                    [0.552 × 10^{-3}; 79°50′ to direction of 160 MN/m^2 stress]

24. A strain rosette fixed to a point in a stressed material gave the following readings:

Gauge No.	Direction relative to Gauge No. 1	Strain
1	0°	+423 × 10^{-6}
2	45°	+542 × 10^{-6}
3	90°	+ 82 × 10^{-6}

If $E = 200$ GN/m^2 and $\nu = 0.3$, find the magnitude of the principal stresses and their directions relative to Gauge No. 1.
                    [124 MN/m^2; 20.8 MN/m^2; 150°23′; 60°23′]

25. The following strains were recorded with a 120° strain rosette:

$$e_0 = +716 \times 10^{-6} \quad e_{120°} = +539 \times 10^{-6} \quad e_{240°} = +155 \times 10^{-6}$$

Find (a) the principal strains, (b) the principal stresses.
    $E = 200$ GN/m^2 and $\nu = 0.3$.
                    [+801 × 10^{-6}; +139 × 10^{-6}; +185.2 MN/m^2; +83.4 MN/m^2]

26. In a strain rosette experiment, the three strain gauge measurements taken over a small area were: $\varepsilon_{0°} = 400 \times 10^{-6}$, $\varepsilon_{30°} = 150 \times 10^{-6}$, $\varepsilon_{75°} = -40 \times 10^{-6}$.

What reading would have been recorded had a strain gauge been placed at 45° to the 0° line?   [32 × 10^{-6}]

27. A uniform bar 9 m long carries an axial tensile load of 200 kN. Determine the increase in volume of the bar if $G = 80$ GN/m^2 and $\nu = 0.25$.   [4 500 mm^3]

28. An axial compressive load of 500 kN is applied to a metal bar 50 mm square in section. The contraction on a 200 mm gauge length is 0.55 mm and the increase in thickness is 0.045 mm. Find the values of E and ν.

If a uniform lateral pressure of 80 MN/m^2 is applied to the four sides of the bar in addition to the axial load, find the contraction on the 200 mm gauge length and the change in thickness.   [72.7 GN/m^2; 0.327; 0.406 mm; 0.007 9 mm]

29. A solid steel sphere 400 mm diameter is subjected to a uniform hydraulic pressure of 3.5 MN/m^2. Determine the decrease in volume if $E = 200$ GN/m^2 and $\nu = 0.3$.   [703.5 mm^3]

30. In tests on a steel bar 25 mm diameter, a tensile load of 50 kN produced an extension of 0.099 4 mm on a gauge length of 200 mm and a torque of 200 N m produced an angle of twist of 0.925° on a gauge length of 250 mm. Calculate the value of Poisson's ratio for the steel.   [0.27]

31. A circular shaft 0.1 m diameter is subjected to combined bending and twisting moments, the bending moment being three times the twisting moment. If the direct tension yield-point of the material is 350 MN/m^2 and the factor of safety on yield is to be 4, calculate the allowable twisting moment by the following theories of elastic failure: (*a*) maximum principal stress theory; (*b*) maximum shearing stress theory.   [2 790 Nm; 2 715 N m]

32. Three exactly similar specimens of mild steel tube are 40 mm external diameter and 32 mm internal diameter. One of these is tested in tension and reaches the limit of proportionality at an axial tensile load of 90 kN. The second is tested in simple torsion. The third is also tested in torsion, but with a uniform bending moment of 350 N m applied throughout the test. Assuming maximum shear stress to be the criterion of elastic failure, estimate the torque at which the two torsion specimens should fail.   [737 N m; 648.6 N m]

33. A thin cylinder, 0.6 m diameter, is to withstand an internal pressure of 1.7 N/mm^2. Calculate the necessary thickness for a factor of safety of 3 if the criterion of failure is the maximum strain energy and the yield point in pure tension is 240 N/mm^2. $\nu = 0.3$.   [6.21 mm]

9 Cylinders

9.1 Stresses in thin cylindrical shells

A thin cylinder is one in which the thickness is sufficiently small in relation to the diameter for the stress across the thickness to be reasonably uniform. When such a cylinder is subjected to internal pressure, tensile stresses σ_c and σ_l are set up in the circumferential and longitudinal directions respectively. If the internal diameter is d, the length is l, the thickness of the plate is t and the internal pressure is p, then the force tending to separate the top and bottom halves, Figure 9.1(a), is $p \times d \times l$. This is resisted by the stress σ_c acting on an area $2 \times t \times l$ (neglecting the strength of the ends) so that, for, equilibrium,

$$pdl = 2\sigma_c tl$$

$$\therefore \sigma_c = \frac{pd}{2t} \tag{9.1}$$

The force tending to separate the right and left-hand ends, Figure 9.1(b), is $p \times \dfrac{\pi}{4}d^2$. This is resisted by the stress σ_l acting on an area $\pi \times d \times t$, so that, for equilibrium,

$$p\frac{\pi}{4}d^2 = \sigma_l \pi dt$$

$$\therefore \sigma_l = \frac{pd}{4t} \tag{9.2}$$

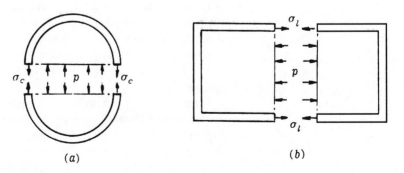

(a)   (b)

Figure 9.1

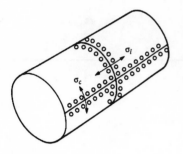

Figure 9.2

The circumferential stress acts on a longitudinal section and the longitudinal stress acts on a circumferential section, Figure 9.2. If the shell is made up of plates riveted together, then if the efficiency of the longitudinal joints, i.e., the ratio strength of joint/strength of undrilled plate, is η_l, the circumferential stress at the joint is given by

$$\sigma_c = \frac{pd}{2t\eta_l} \tag{9.3}$$

Similarly, if the efficiency of the circumferential joints is η_c, the longitudinal stress at the joint is given by

$$\sigma_l = \frac{pd}{4t\eta_c} \tag{9.4}$$

The efficiency of the circumferential joints therefore need only be half that of the longitudinal joints for equal joint stress.

Away from the riveted joints, the stresses are as given by equations (9.1) and (9.2).

9.2 Stress in thin spherical shells

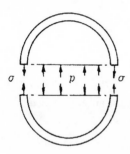

Figure 9.3

If the internal diameter is d, the thickness of the plate is t and the internal pressure is p, then the force tending to separate the two halves, Figure 9.2, is $p \times (\pi/4)d^2$. This is resisted by the stress σ acting on an area πdt so that, for equilibrium,

$$p\frac{\pi}{4}d^2 = \sigma\pi dt$$

$$\therefore \sigma = \frac{pd}{4t} \tag{9.5}$$

If the shell is made up from riveted plates and the efficiency of the joints is η, then

$$\sigma = \frac{pd}{4t\eta} \tag{9.6}$$

9.3 Lamé's Theory

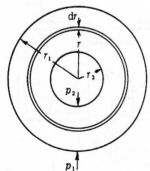

Figure 9.4

When a cylinder is too thick for the simple assumption that the radial stress is negligible and the circumferential stress is therefore uniform across the section, i.e., when t is greater than about $0.1d$, Lamé's theory is used to determine the stress distribution.

If the cylinder is long in comparison with its diameter, the longitudinal stress and strain are assumed to be uniform across the thickness of the cylinder wall.

Figure 9.4 shows a cylinder of external and internal radii r_1 and r_2 subjected to external and internal pressures p_1 and p_2 respectively, and Figure 9.5 shows an element of the cross-section at radius r, subtending an angle $d\theta$ at the centre. The radial and circumferential stresses on the element are σ_r and σ_c respectively and so, equating radial forces on unit axial length,

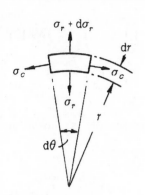

Figure 9.5

$$(\sigma_r + \mathrm{d}\sigma_r)(r + \mathrm{d}r)\,\mathrm{d}\theta = \sigma_r r\,\mathrm{d}\theta + 2\sigma_c\,\mathrm{d}r\frac{\mathrm{d}\theta}{2}$$

which reduces to $\qquad \sigma_r\,\mathrm{d}r + r\,\mathrm{d}\sigma_r = \sigma_c\,\mathrm{d}r$

or $\qquad\qquad\qquad \sigma_r + r\frac{\mathrm{d}\sigma_r}{\mathrm{d}r} = \sigma_c \qquad\qquad (9.7)$

Let the longitudinal stress be σ_l and the longitudinal strain ε_l.

Then $\qquad\qquad\qquad \varepsilon_l = \frac{\sigma_l}{E} - v\left(\frac{\sigma_r + \sigma_c}{E}\right) \qquad\qquad (9.8)$

Since ε_l and σ_l are assumed to be constant across the section,

$$\sigma_r + \sigma_c = \text{constant} = 2a \qquad\qquad (9.9)$$

Substituting $\sigma_c = 2a - \sigma_r$ in equation (9.7) and multiplying throughout by r,

$$2\sigma_r r + r^2\frac{\mathrm{d}\sigma_r}{\mathrm{d}r} - 2ar = 0$$

or $\qquad\qquad\qquad \frac{\mathrm{d}}{\mathrm{d}r}(\sigma_r r^2 - ar^2) = 0$

Integrating, $\qquad\qquad \sigma_r r^2 - ar^2 = \text{constant} = b$

$$\therefore\ \sigma_r = a + \frac{b}{r^2} \qquad\qquad (9.10)$$

and, substituting in equation (9.9),

$$\sigma_c = a - \frac{b}{r^2} \qquad\qquad (9.11)$$

These are Lamé's equations and substitution of the relevant boundary conditions enables the constants a and b to be determined and hence the radial and circumferential stresses at any point. Tensile stresses are regarded as positive and compressive stresses as negative.

9.4 Thick cylinder subject to internal pressure only

Let the gauge pressure be p. Then the boundary conditions are $\sigma_r = -p$ when $r = r_2$ and $\sigma_r = 0$ when $r = r_1$.

Thus, from equation (9.10), $\qquad -p = a + \frac{b}{r_2^2}$

and $\qquad\qquad\qquad\qquad 0 = a + \frac{b}{r_1^2}$

from which $\qquad a = \frac{r_2^2}{r_1^2 - r_2^2}p \quad$ and $\quad b = \frac{-r_1^2 r_2^2}{r_1^2 - r_2^2}p$

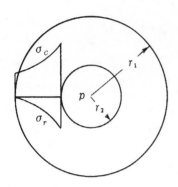

Figure 9.6

Therefore the radial stress
$$\sigma_r = \frac{pr_2^2}{r_1^2 - r_2^2}\left(1 - \frac{r_1^2}{r^2}\right) \qquad (9.12)$$

and the circumferential stress
$$\sigma_c = \frac{pr_2^2}{r_1^2 - r_2^2}\left(1 + \frac{r_1^2}{r^2}\right) \qquad (9.13)$$

The distribution of these stresses across the section is shown in Figure 9.6.

The maximum value of $\qquad \sigma_r = -p$

and the maximum value of $\qquad \sigma_c = p\dfrac{r_1^2 + r_2^2}{r_1^2 - r_2^2} \qquad (19.14)$

9.5 Longitudinal and shear stresses

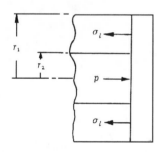

Figure 9.7

The uniform longitudinal stress σ_l is determined by the longitudinal equilibrium of the end of the cylinder, Figure 9.7. Thus, for the simple case of internal pressure only,

$$\sigma_l \times \pi(r_1^2 - r_2^2) = p \times \pi r_2^2$$

$$\therefore \sigma_l = \frac{pr_2^2}{r_1^2 - r_2^2} \qquad (9.15)$$

σ_r, σ_c and σ_l are principal stresses, so that the maximum shear stress τ is equal to half the difference between the maximum and minimum principal stresses (see Section 8.2). Thus for internal pressure only, σ_r is compressive (negative) and σ_c and σ_l are tensile (positive), so that τ is either $\frac{1}{2}(\sigma_c - \sigma_r)$ or $\frac{1}{2}(\sigma_l - \sigma_r)$, whichever is greater. Since, however, $\sigma_c > \sigma_l$,

$$\tau = \frac{\sigma_c - \sigma_r}{2}$$

$$= \frac{\left(a - \dfrac{b}{r^2}\right) - \left(a + \dfrac{b}{r^2}\right)}{2} = -\frac{b}{r^2}$$

$$= \frac{pr_1^2 r_2^2}{(r_1^2 - r_2^2)r^2} \quad \text{from equations (9.12) and (9.13)}$$

Thus the maximum shear stress occurs at the inside surface, where $r = r_2$

i.e.
$$\tau_{\text{max}} = \frac{pr_1^2}{r_1^2 - r_2^2} \qquad (9.16)$$

9.6 The Lamé line

By plotting σ_r against $\dfrac{1}{r^2}$ and σ_c against $-\dfrac{1}{r^2}$, equations (9.10) and (9.11) are represented graphically by a *single* straight line, having the same intercept a and slope b. This is shown in Figure 9.8 for the case of internal pressure only where $\sigma_r = -p$ at $r = r_2$ (point A) and $\sigma_r = 0$ at $r = r_1$ (point B).

Extending the line to negative values of $1/r^2$ gives values of σ_c between points C and D. The maximum value of σ_c occurs at point D and this is given by

$$\frac{\sigma_c}{p} = \frac{\dfrac{1}{r_1^2} + \dfrac{1}{r_2^2}}{\dfrac{1}{r_2^2} - \dfrac{1}{r_1^2}} = \frac{r_1^2 + r_2^2}{r_1^2 - r_2^2} \quad \text{as given by equation (9.14)}$$

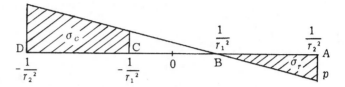

Figure 9.8

$$\text{The ordinate at O} = \frac{p \times \dfrac{1}{r_1^2}}{\dfrac{1}{r_2^2} - \dfrac{1}{r_1^2}} = \frac{p r_2^2}{r_1^2 - r_2^2}$$

From equation (9.15), this is the longitudinal stress, σ_l.

9.7 Compound cylinders

In order to reduce the high circumferential stress at the inside surface of a cylinder under internal pressure, the cylinder can be pre-stressed by shrinking on to it an outer cylinder. The inner diameter of the outer cylinder is made slightly smaller than the outer diameter of the inner cylinder and assembly is achieved by either heating the outer cylinder or cooling the inner one. This puts the inner cylinder in an initial state of compression so that, when the internal pressure is applied, the resulting tensile stress at the inner surface is less than would have been the case in a homogeneous cylinder of the same total thickness.

If the radial pressure between the two cylinders due to shrinking is p_0, the initial stresses in the two parts of the cylinder can be determined and these are then combined with those due to the internal pressure, treating the compound cylinder as homogeneous, to give the resultant stresses.

The combinations of stresses is shown in Figure 9.9.

By this process, the maximum tensile stress in a cylinder of given thickness can be reduced or alternatively, a thinner cylinder can be used for a given maximum stress. The optimum design is that which will give the same maximum stress at the inner surface of each cylinder.

The radial stress distribution in a compound cylinder may be deduced in a similar manner but these stresses are usually relatively unimportant.

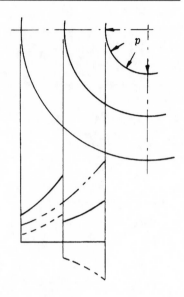

Initial stresses due to
shrinkage pressure p_0 - - - - - - -

Stresses due to internal
pressure p —— - - ——

Resultant stresses   ————————

Figure 9.9

Worked examples

1. *A cylindrical compressed air drum is 2 m in diameter with plates 12.5 mm thick. The efficiencies of the longitudinal and circumferential joints are respectively 85% and 45%. If the tensile stress in the plating is to be limited to 100 MN/m^2, find the maximum safe air pressure.*

For a circumferential stress of 100 MN/m^2,

$$p = \frac{2t\eta_l\sigma_c}{d} \quad \text{from equation (9.3)}$$

$$= \frac{2 \times 12.5 \times \times 10^{-3} \times 0.85 \times 100 \times 10^6}{2} = 106.3 \times 10^3 \text{N/m}^2$$

For a longitudinal stress of 100 MN/m^2,

$$p = \frac{4t\eta_c\sigma_l}{d} \quad \text{from equation (9.4)}$$

$$= \frac{4 \times 12.5 \times 10^{-3} \times 0.45 \times 100 \times 10^6}{2} = 112.5 \times 10^3 \text{ N/m}^2$$

Thus the maximum safe pressure is 106.3 kN/m^2, which will produce a circumferential stress at the joint of 100 MN/m^2 and a longitudinal stress of 94.4 MN/m^2.

2. *A steel pipe 100 mm external diameter and 50 mm internal diameter is subjected to an internal pressure of 14 N/mm^2 and an external pressure of 5.5 N/m^2. Sketch curves showing the distribution of radial and circumferential stress across the section.*

From Lamé's equations, $\sigma_r = a + \dfrac{b}{r^2}$

and $\sigma_c = a - \dfrac{b}{r^2}$

When $r = 25$ mm, $\sigma_r = -14$ N/mm^2

$$\therefore \quad -14 = a + \frac{b}{25^2} \tag{1}$$

When $\qquad r = 50$ mm, $\sigma_r = -5.5$ N/mm^2

$$\therefore \quad -5.5 = a + \frac{b}{50^2} \tag{2}$$

Therefore, from equations (1) and (2),

$$a = -\frac{8}{3} \quad \text{and} \quad b = -\frac{8.5 \times 50^2}{3}$$

$$\therefore \quad \sigma_c = -\frac{8}{3} + \frac{8.5 \times 50^2}{3r^2}$$

When $\quad r = 25$ mm, $\quad \sigma_c = -\frac{8}{3} + \frac{8.5 \times 50^2}{3 \times 25^2} = 8.667$ N/mm^2

When $\quad r = 50$ mm, $\quad \sigma_c = -\frac{8}{3} + \frac{8.5 \times 50^2}{3 \times 50^2} = 0.167$ N/mm^2

The distribution of σ_r and σ_c across the section is shown in Figure 9.10.

Alternatively, using the Lamé line, Figure 9.11, set off $\sigma_r = -14$ N/mm^2 at $1/25^2$ and $\sigma_r = -5.5$ N/mm^2 at 1.50^2. The circumferential stresses of 8.667 and 0.167 N/mm^2 are then given at $-1/25^2$ and $-1/50^2$ respectively.

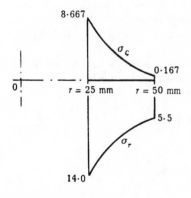

Figure 9.10

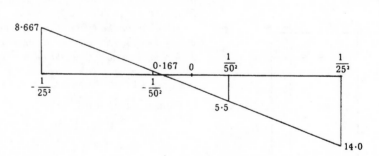

Figure 9.11

3. *A cast iron pipe 150 mm internal diameter and 200 mm external diameter is tested under pressure and breaks at an internal pressure of 48 MN/m^2. Find the safe internal pressure for a pipe of the same material and of the same internal diameter with walls of 40 mm thick, using a factor of safety of 4.*

The maximum stress due to internal pressure only is the circumferential stress at the inside surface and hence the stress at failure is given by

$$\sigma_c = p \frac{r_1^2 + r_2^2}{r_1^2 - r_2^2} \quad \text{From equation (9.14)}$$

$$= 48 \times \frac{100^2 + 75^2}{100^2 - 75^2} = 171.5 \text{ MN/m}^2$$

Therefore, for the second pipe,

$$\frac{171.5}{4} = p \times \frac{115^2 + 75^2}{115^2 - 75^2}$$

from which

$$p = \underline{17.28 \text{ MN/m}^2}$$

Alternatively, these results can be obtained from the Lamé lines for the two cylinders, shown in Figure 9.12(a) and (b) respectively.

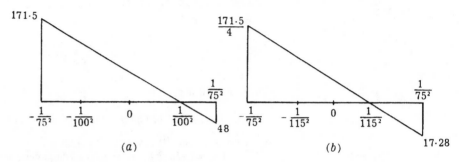

Figure 9.12

4. *The cylinder of a hydraulic ram supported at the open end is 250 mm internal diameter and is required to sustain an internal pressure of 20 MN/m². Calculate the necessary thickness of the wall if the maximum shearing stress is limited to 50 MN/m².*

Allowing for the effect of the longitudinal stress caused by the pressure on the end of the cylinder, calculate the increase in diameter due to the application of the 20 MN/m² pressure. E = 200 GN/m² and ν = 0.28.

Figure 9.13 shows the arrangement of the cylinder.

From equation (9.16),

$$\tau_{\max} = \frac{p r_1^2}{r_1^2 - r^2}$$

i.e.

$$50 = \frac{20 r_1^2}{r_1^2 - 125^2}$$

from which

$$r_1 = \frac{125}{\sqrt{0.6}} = 161.4 \text{ mm}$$

$$\therefore t = 161.4 - 125 = \underline{36.4 \text{ mm}}$$

Figure 9.13

From equation (9.15),

$$\sigma_l = \frac{p r_2^2}{r_1^2 - r_2^2}$$

$$= \frac{20 \times 125^2}{161.4^2 - 125^2} = 30 \text{ MN/m}^2$$

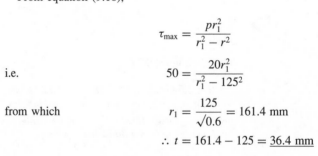

and from equation (9.14),

$$\sigma_c = p\frac{r_1^2 + r_2^2}{r_1^2 - r_2^2}$$

$$= 20 \times \frac{161.4^2 + 125^2}{161.4^2 - 125^2} = 80 \text{ MN/m}^2$$

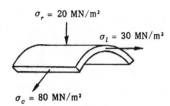

$\sigma_r = 20 \text{ MN/m}^2$

$\sigma_l = 30 \text{ MN/m}^2$

$\sigma_c = 80 \text{ MN/m}^2$

Figure 9.14

The state of stress on an element of the cylinder at the inside surface is shown in Figure 9.14. As in example 7, Chapter 8, the strain on the diameter is the same as the strain on the circumference and this is given by

$$\varepsilon_c = \frac{\sigma_c}{E} + v\frac{\sigma_r}{E} - v\frac{\sigma_l}{E}$$

$$= \frac{(80 + 0.28 \times 20 - 0.28 \times 30) \times 10^6}{200 \times 10^9} = 0.386 \times 10^{-3}$$

Therefore increase in diameter $= 0.386 \times 10^{-3} \times 250 = \underline{0.096\,5 \text{ mm}}$

5. *One steel cylinder is shrunk on to another, the compound cylinder having an inside diameter of 100 mm, an outside diameter of 200 mm and a diameter of 150 mm at the surfaces in contact. If shrinkage produces a radial pressure p_0 at the surfaces in contact, after which the compound cylinder is subjected to an internal pressure of 100 MN/m², find the value of p_0 so that the maximum circumferential stresses in the two cylinders shall be the same and determine the value of this stress.*

(a) Initial stresses due to shrinkage

(i) Inner cylinder:– Boundary conditions are $\sigma_r = -p_0$ when $r = 75$ mm

and $\sigma_r = 0$ when $r = 50$ mm

Hence $\qquad -p_0 = a + \dfrac{b}{75^2}$

and $\qquad 0 = a + \dfrac{b}{50^2}$

from which $\qquad a = -1.8p_0$ and $b = 4\,500p_0$

Therefore, at inner surface, $\quad \sigma_c = -1.8p_0 - \dfrac{4\,500p_0}{50^2} = -3.6p_0 \text{ MN/m}^2$

(ii) Outer cylinder:– Boundary conditions are $\sigma_r = -p_0$ when $r = 75$ mm

and $\sigma_r = 0$ when $r = 100$ mm

Hence $\qquad -p_0 = a + \dfrac{b}{75^2}$

and $\qquad 0 = a + \dfrac{b}{100^2}$

from which $\qquad a = \dfrac{9}{7}p_0$ and $b = -\dfrac{90\,000}{7}p_0$

Therefore, at inner surface, $\quad \sigma_c = \dfrac{9}{7}p_0 + \dfrac{90\,000p_0}{7 \times 75^2} = 3.57p_0 \text{ MN/m}^2$

(b) Stresses due to internal pressure

Boundary conditions are $\sigma_r = -100$ MN/m^2 at $r = 50$ mm

and $\sigma_r = 0$ at $r = 100$ mm

Hence

$$-100 = a + \frac{b}{50^2}$$

and

$$0 = a + \frac{b}{100^2}$$

from which

$$a = \frac{100}{3} \quad \text{and} \quad b = -\frac{10^6}{3}$$

Therefore, at $r = 50$ mm,

$$\sigma_c = \frac{100}{3} + \frac{10^6}{3 \times 50^2} = 166.7 \text{ MN/m}^2$$

and at $r = 75$ mm,

$$\sigma_c = \frac{100}{3} + \frac{10^6}{3 \times 75^2} = 92.6 \text{ MN/m}^2$$

Thus resultant stress at inner surface of inner cylinder

$$= -3.6 p_0 + 166.7 \text{ MN/m}^2$$

and resultant stress at inner surface of outer cylinder

$$= 3.57 p_0 + 92.6 \text{ MN/m}^2$$

Therefore, for equal stresses,

$$-3.6 p_0 + 166.7 = 3.57 p_0 + 92.6$$

from which

$$p_0 = \underline{10.3 \text{ MN/m}^2}$$

Substituting in either of these equations gives the value of the maximum stress in each cylinder as $\underline{129.4 \text{ MN/m}^2}$.

Further problems

6. A cylindrical vessel 0.75 m diameter is to withstand an internal pressure of 2.8 MN/m^2, the maximum allowable stress being 85 MN/m^2. If the efficiency of the riveted joints is 75%, determine (a) the required thickness of the cylinder, (b) the axial stress in the cylinder with this pressure at a point where the full section is undiminished.   [16.46 mm; 31.9 MN/m^2]

7. A cylindrical vessel 1.8 m diameter is subjected to an internal pressure of 1.25 MN/m^2. The plates are 12 mm thick and have an ultimate tensile strength of 450 MN/m^2. If the efficiencies of the longitudinal and circumferential joints are 75% and 50% respectively, calculate the factor of safety.   [3.6]

8. A thin spherical pressure vessel is required to contain 18 000 litres of water at a gauge pressure of 700 kN/m^2. Assuming the efficiency of all riveted joints to be 75 per cent, determine the diameter of the vessel and the thickness of the plate. The stress in the material must not exceed 140 MN/m^2.   [3.248 m, 5.41 mm]

9. A thick cylinder has outer and inner diameters of 400 mm and 250 mm respectively. If the maximum permissible tensile stress is 140 N/mm², calculate the maximum safe internal pressure. What is the circumferential stress at the outer circumference due to this internal pressure?　　　[61.35 N/mm²; 78.65 N/mm²]

10. A thick cylinder having an external diameter of 200 mm and an internal diameter of 100 mm is subjected to an internal pressure of 56 MN/m² and an external pressure of 7 MN/m². Find the maximum direct stress in the cylinder and the change of external diameter. $E = 200$ GN/m² and $\nu = 0.03$.

[74.67 MN/m²; 0.027 8 mm]

11. A steel tube is 18 mm internal diameter, 3 mm thick and 300 mm long. Calculate the safe internal pressure if the maximum stress is not to exceed 150 MN/m² and also the increase in internal volume under this pressure. $E = 200$ GN/m² and $\nu = 1/3.5$　　　[42 MN/m²; 120.6 mm³]

12. A steel cylinder, 1 m inside diameter, is to withstand an internal pressure of 8 MN/m². Calculate the thickness if the maximum shearing stress is not to exceed 35 MN/m².

Calculate the increase in volume due to this internal pressure if the cylinder is 6 m long. $E = 200$ GN/m² and $\nu = \frac{1}{3}$.　　　[69.3 mm; 0.002 835 m³]

13. A thick cylinder, 200 mm internal diameter, is to withstand an internal pressure of 50 MN/m². Find the necessary thickness if the maximum shearing stress is not to exceed 100 MN/m². What will then be the greatest and least hoop stresses in the material?　　　[41.4 mm; 150 and 100 MN/m²]

14. A thick cylinder has a length of 0.25 m and internal and external diameters of 0.1 m and 0.141 4 m respectively. Determine (a) the circumferential and longitudinal stresses at the inner surface when the cylinder is subjected to an internal pressure of 10 MN/m², (b) the increase in internal volume of the cylinder. $E = 200$ GN/m² and $\nu = 0.3$.　　　[30 MN/m²; 10 MN/m²; 628.4×10^{-3} m³]

15. A steel cylinder 200 mm external diameter and 150 mm internal diameter has another cylinder 250 mm external diameter shrunk on to it. If the maximum tensile stress induced in the outer cylinder is 80 N/mm², find the radial compressive stress between the cylinders.

Determine the circumferential stresses at inner and outer diameters of both cylinders and show, by means of a diagram, how these stresses vary with the radius.

[17.6; 80; 62.5; 80.5; 62.8 N/mm²]

16. A steel cylinder of outside diameter 300 mm and inside diameter 250 mm is shrunk on to one having diameters 250 mm and 200 mm, the interference fit being such that under an internal pressure p the inner tensile stress in both cylinders is 84 MN/m². Find the value of p.　　　[36.9 MN/m²]

17. A compound steel cylinder has a bore of 80 mm and an outside diameter of 160 mm, the diameter of the common surface being 120 mm. Find the radial pressure at the common surface which must be provided by shrinkage, if the resultant maximum circumferential tension in the inner cylinder under a superimposed internal pressure of 60 MN/m² is to be half the value of the maximum circumferential tension which would be produced in the inner cylinder if that cylinder alone were subjected to an internal pressure of 60 MN/m².

[6.11 MN/m²]

10 Dynamics

10.1 Introduction

Dynamics involves a study of the motion of systems and the forces associated with the motion, which may be linear, angular or a combination of linear and angular.

10.2 Linear motion

The *velocity*, v, of a body is the rate of change of its displacement s from some reference position with respect to time, t,

i.e.
$$v = \frac{ds}{dt}$$
(10.1)

The *acceleration*, a, of a body is the rate of change of its velocity v with respect to time, t,

i.e.
$$a = \frac{dv}{dt} = \frac{d^2s}{dt^2} = v\frac{dv}{ds}$$
(10.2)

If a body moves with uniform acceleration a such that the velocity changes from u to v in time t while traversing a distance s, then it may be shown that

$$v = u + at$$
(10.3)

$$s = ut + \tfrac{1}{2}at^2$$
(10.4)

and
$$v^2 = u^2 + 2as$$
(10.5)

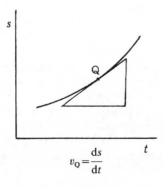

$$v_Q = \frac{ds}{dt}$$

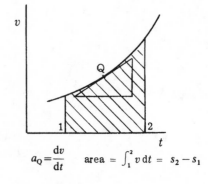

$$a_Q = \frac{dv}{dt} \qquad \text{area} = \int_1^2 v\,dt = s_2 - s_1$$

Figure 10.1

10.3 Angular motion

For a body moving with angular velocity ω and angular acceleration α, turning through an angle θ from some datum position,

$$\omega = \frac{d\theta}{dt} \tag{10.6}$$

and

$$\alpha = \frac{d\omega}{dt} = \frac{d^2\theta}{dt^2} = \omega\frac{d\omega}{d\theta} \tag{10.7}$$

If the angular acceleration is uniform and the angular velocity changes from ω_1 to ω_2 in time t while rotating through an angle θ, then

$$\omega_2 = \omega_1 + \alpha t \tag{10.8}$$

$$\theta = \omega_1 t + \tfrac{1}{2}\alpha t^2 \tag{10.9}$$

and

$$\omega_2^2 = \omega_1^2 + 2\alpha\theta \tag{10.10}$$

10.4 Relations between linear and angular motion

If a wheel of radius r turns through an angle θ rad, the displacement of a point on the circumference,

$$s = \theta r \tag{10.11}$$

If the angular velocity of the wheel is ω and its angular acceleration is α, then by differentiation, the linear velocity and acceleration of points on the circumference are given by

$$v = \omega r \tag{10.12}$$

and

$$a = \alpha r \tag{10.13}$$

10.5 Graphical interpretation of equations of motion

If experimental data relating velocity, distance and time are available, equations (10.1) and (10.2) may be interpreted graphically, as shown in Figure 10.1. Solutions by this method are not restricted to cases of uniform acceleration.

Graphs similar to those in Figure 10.1 may be used for experimental data involving angular motion.

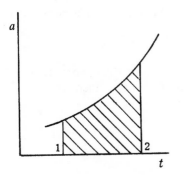

area $= \int_1^2 a\,dt = v_2 - v_1$

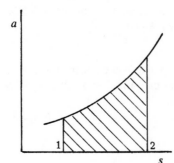

area $= \int_1^2 a\,ds = \dfrac{v_2^2 - v_1^2}{2}$

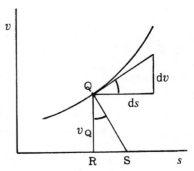

$RS = v_Q\dfrac{dv}{ds} = a_Q$

10.6 Mass, force, weight, momentum

The *mass, m*, of a body is a measure of the amount of matter in the body, this being independent of geographical location.

Force, F, is that necessary to change the state of rest or uniform motion of a body. Unit force is that required to give unit acceleration to unit mass.

The *weight, W*, of a body is the force of attraction between the body and the earth, which varies slightly in different parts of the world due to the variation in the value of *g*.

The *momentum* of a body of mass *m* moving with velocity *v* is the product *mv*. The total momentum of a system in a given direction remains constant unless acted upon by an external force.

10.7 Newton's laws of motion

(i) Every body continues in its state of rest or uniform motion in a straight line unless acted upon by an external force.

(ii) The rate of change of momentum of a body is proportional to the external force and takes place in the direction of the force.

(iii) To every action (force) there is an equal and opposite reaction (force).

From the second law,

$$F \propto \text{ rate of change of momentum}$$

$$\propto \text{ mass } \times \text{ rate of change of velocity}$$

i.e. $F = kma$ where k is a constant.

The units are chosen to make the constant of proportionality unity. In the *Système International d' Unités* (S.I. units), the unit of mass is the kilogramme and the unit of length is the metre. The unit of force is the newton, which is defined as the force necessary to give a mass of 1 kg an acceleration of 1 m/s^2,

i.e. $$F(\text{N}) = m(\text{kg}) \times a(\text{m/s}^2) \tag{10.14}$$

Since a body falling freely under the earth's gravitational pull has an acceleration *g*, the weight

$$W = mg \tag{10.15}$$

g has the standard value of 9.806 65 m/s^2 which, for normal purposes, is taken as 9.81 m/s^2.

10.8 Impulse

The impulse of a constant force F acting for a time t is the product Ft. If this causes a mass m to change its velocity from u to v, then

$$F = ma = m\frac{v - u}{t}$$

or $$Ft = m(v - u) \qquad (10.16)$$

i.e.   impulse = change of momentum.

10.9 Circular motion

Consider a particle of mass m moving in a circular path of radius r with constant speed v, Figure 10.2(a). Let the particle move from P to Q in time dt. Then the magnitude of the velocity is unchanged but the direction has changed and from the relative velocity diagram, Figure 10.2(b), this change is represented by pq

Thus there is an inward or *centripetal* acceleration,

$$a = \frac{pq}{dt} = v\frac{d\theta}{dt} = v\omega$$

But $v = \omega r$, so that

$$a = \omega^2 r \quad \text{or} \quad \frac{v^2}{r} \qquad (10.17)$$

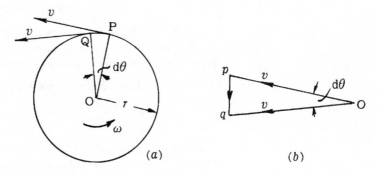

Figure 10.2

The centripetal force necessary to constrain the mass to move in a circular path is given by

$$F = ma = m\omega^2 r \quad \text{or} \quad m\frac{v^2}{r} \qquad (10.18)$$

By Newton's third law, there will be an equal and opposite outward or *centrifugal* reaction to this force, which acts at the centre of rotation.

In problems involving equilibrium of forces, the moving body may be brought to rest by applying a radially outward force, mv^2/r, to the body to convert the dynamic case to the equivalent static case.

10.10 Work, energy and power

If a constant force F moves through a distance s, the work done $= Fs$.

If the force is variable, work done $= \Sigma F \delta s$, which is represented by the area under a force–distance graph. When the force is a function of s, however, this may be expressed in the form

$$\text{work done} = \int_0^s f(s)\,\mathrm{d}s \qquad (10.19)$$

The unit of work is the joule (J), which is the work done by a force of 1 N moving through a distance of 1 m.

Work represents a transfer of energy and thus the energy of a body is a measure of its capacity to do work. Energy may exist in various forms, such as mechanical, thermal and electrical energy but it cannot be created or destroyed; a loss in one form is always accompanied by an equal gain in another form.

Mechanical energy may be of the form of potential, kinetic or strain energy. Potential energy is the energy a body possesses by virtue of its position relative to the earth. If a body of mass m is at a height h above some datum level, the potential energy relative to that datum is given by

$$\text{potential energy} = mgh \qquad (10.20)$$

Kinetic energy is the energy a body possesses due to its velocity. If a force F is applied to a body of mass m such that it attains a velocity v in time t whilst moving a distance s, then

$$\text{kinetic energy} = \text{work done} = F \times s$$

$$= ma \times \frac{v^2}{2a} = \tfrac{1}{2}mv^2 \qquad (10.21)$$

Strain energy is the energy stored in a body when it is deformed. If a force F is applied to a body of stiffness S and causes a deformation x,

$$\text{strain energy} = \text{work done} = \tfrac{1}{2}Fx = \tfrac{1}{2}Sx^2 \qquad (10.22)$$

Power is the work done per unit time or work transfer rate. If a force F moves with velocity v.

$$\text{power } P = Fv. \qquad (10.23)$$

The unit of power is the watt (W), which is 1 Nm/s or 1 J/s.

10.11 Torque and angular acceleration

Figure 10.3 shows a body rotating about O with angular acceleration α due to an applied external torque T.

$$\text{Linear acceleration of particle of mass } \mathrm{d}m = \alpha l$$

$$\therefore \quad \text{force required to accelerate particle} = \mathrm{d}m.\alpha l$$

$$\therefore \quad \text{torque required about O} = \mathrm{d}m.\alpha l^2$$

$$\therefore \quad \text{total torque for whole body} = \int \mathrm{d}m.\alpha l^2$$

$$= \alpha \int \mathrm{d}m l^2$$

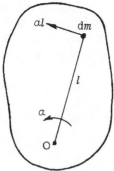

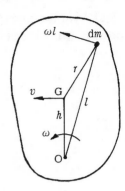

Figure 10.3   Figure 10.4

But $\int \mathrm{d}ml^2$ is the moment of inertia of the body about an axis through O (see Appendix A)

$$\therefore\ T = I_O\alpha \qquad (10.24)$$

10.12 Angular momentum and angular impulse

The angular momentum of a body about an axis is the moment of its momentum about that axis. Figure 10.4 shows a body rotating about O with angular velocity ω.

$$\text{Linear velocity of particle of mass } \mathrm{d}m = \omega l$$

$$\therefore\quad \text{momentum of particle} = \mathrm{d}m.\omega l$$

$$\therefore\quad \text{moment of momentum about O} = \mathrm{d}m.\omega l^2$$

$$\therefore\quad \text{total moment of momentum about O} = \int \mathrm{d}m.\omega l^2 = \omega \int \mathrm{d}ml^2$$

$$= I_O\omega \qquad (10.25)$$

If G is the centre of gravity, then, by the theorem of parallel axes (see Appendix A), $I_O = I_G + mh^2$

$$\therefore\quad \text{moment of momentum about O} = I_G\omega + mh^2\omega$$

$$= I_G\omega + mvh \qquad (10.26)$$

The angular momentum about O is therefore the angular momentum about a parallel axis through G, together with the moment of the linear momentum about O.

The angular momentum of a system about any axis remains constant, unless acted upon by an external torque about that axis.

The angular impulse of a constant torque T acting for a time t is the product Tt. If this causes a body of moment of inertia I to change its angular velocity from ω_1 to ω_2, then $T = I\alpha = I(\omega_2 - \omega_1)/t$, or

$$Tt = I(\omega_2 - \omega_1) \qquad (10.27)$$

i.e.   angular impulse = change of angular momentum

10.13 Angular work, power and kinetic energy

If a constant torque T moves through an angle θ, the work done $= T\theta$.

If the torque is variable, work done $= \Sigma T \delta\theta$, which is represented by the area under a torque–angle graph. When the torque is a function of θ, however, this may be expressed in the form

$$\text{work done} = \int_0^\theta f(\theta)\,d\theta \tag{10.28}$$

If a constant torque T moves with an angular velocity ω, then

$$\text{power } P = T\omega \tag{10.29}$$

Referring to Figure 10.4,

$$\text{kinetic energy of particle of mass } dm = \tfrac{1}{2}dm(\omega l)^2$$

$$\therefore \text{ total kinetic energy of body} = \frac{\omega^2}{2}\int dm.l^2$$

$$= \tfrac{1}{2}I_O\omega^2 \tag{10.30}$$

Alternatively, since $I_O = I_G + mh^2$ (see Appendix A),

$$\text{kinetic energy of body} = \tfrac{1}{2}I_G\omega^2 + \tfrac{1}{2}mh^2\omega^2$$

$$= \tfrac{1}{2}I_G\omega^2 + \tfrac{1}{2}mv^2 \tag{10.31}$$

The total kinetic energy is therefore the kinetic energy due to rotation about G together with the kinetic energy due to the linear velocity of G.

10.14 Equivalent mass of a rotating body

It is sometimes convenient to treat combined linear and angular motion as an equivalent linear problem, the angular effects being allowed for by an equivalent mass added to the actual mass of the body.

Consider the body of mass m shown in Figure 10.5. Let a tangential force F be applied at radius r causing an angular acceleration α.

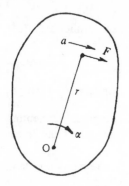

Then
$$Fr = I_O\alpha = mk^2\frac{a}{r}$$

$$\therefore F = m\left(\frac{k^2}{r^2}\right)a \tag{10.32}$$

The quantity $m\left(\dfrac{k^2}{r^2}\right)$ is the equivalent mass of the rotating body referred to the line of action of F.

Figure 10.5

10.15 Acceleration of geared systems

Figure 10.6 shows two gear wheels A and B with moments of inertia I_a and I_b respectively, having a speed ratio $\dfrac{N_b}{N_a} = n$. If a torque T is applied

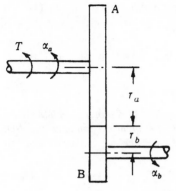

Figure 10.6

to shaft A to give the shafts angular accelerations α_a and α_b, then torque required on B to accelerate $B = I_b\alpha_b = I_b\alpha_a n$ since $\alpha_b/\alpha_a = n$.

$\therefore$ torque required on A to accelerate $B = n \times I_b\alpha_a n = n^2 I_b\alpha_a$

Torque required on A to accelerate $A = I_a\alpha_a$

$\therefore$ torque required on A to accelerate A and B $= I_a\alpha_a + n^2 I_b\alpha_a$

i.e.
$$T = (I_a + n^2 I_b)\alpha_a$$

$$(10.33)$$

The quantity $I_a + n^2 I_b$ is the moment of inertia of the system referred to shaft A and this principle may be extended to any number of gears meshing together.

The tangential force F between the teeth is given by

$$F r_b = I_b \alpha_b$$

or
$$F = \frac{I_b \alpha_b}{r_b}$$

$$(10.34)$$

10.16 Maximum acceleration of vehicles

The maximum acceleration of a vehicle is limited by the friction force between the wheels and the road. This will depend on the normal reactions at the wheels and the coefficient of friction between wheels and road.

Figure 10.7 shows a vehicle of mass m, moving up a gradient inclined at angle θ to the horizontal with acceleration a. The normal reactions at the front and rear wheels are R_f and R_r respectively.

The tractive force $F = \mu R_f$, μR_r or $\mu(R_f + R_r)$, depending on whether the vehicle has front, rear or four-wheel drive. In the latter case, however, the tractive forces at each pair of wheels must be in the ratio of the normal reactions if slipping is to occur simultaneously at all four wheels.

By resolution of forces, $R_f + R_r = mg\cos\theta$ (10.35)

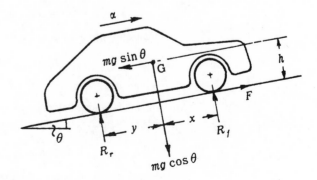

Figure 10.7

and
$$F = mg\sin\theta + ma \qquad (10.36)$$

Taking moments about G, $Fh = R_r y - R_f x \qquad (10.37)$

For maximum retardation of vehicles, the direction of the force F becomes reversed.

Worked examples

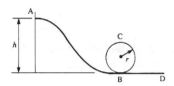

Figure 10.8

1. *A toy car track, shown in Figure 10.8, consists of a sloping section AB, a circular loop-the-loop section BC of radius r and a straight horizontal section BD. Determine the least height of A above B from which the car can start from rest without falling off the track at C.*

In travelling from A to C,

$$\text{loss of PE} = \text{gain of KE}$$

i.e.
$$mg(h - 2r) = \tfrac{1}{2}mv^2$$

$$\therefore v^2 = 2g(h - 2r)$$

For the car to remain in contact with the track at C, the upward centrifugal force must be equal to or greater than the weight

i.e.
$$m\frac{v^2}{r} \geqslant mg$$

i.e.
$$\frac{2g(h - 2r)}{r} \geqslant g$$

from which
$$\underline{h \geqslant 2.5r}$$

2. *A pile of mass $\tfrac{3}{4}t$ can just support a stationary mass of 40 t without subsidence. The mass is removed and the pile is driven to a greater depth by blows of a 2t hammer dropping on to the top of the pile from a height of 1.22m. The hammer does not rebound from the top of the pile.*

Calculate (a) the penetration per blow, assuming that the ground resistance is constant, (b) the energy lost per blow, and (c) the efficiency of the operation.

(*a*) Ground resistance, $R = (40 + \tfrac{3}{4}) \times 10^3 \times 9.81$

$$= 400 \times 10^3 \text{ N}$$

Velocity of impact $= \sqrt{(2gh)}$

$$= \sqrt{(2 \times 9.81 \times 1.22)}$$

$$= 4.89 \text{ m/s}$$

Momentum of hammer before impact $=$ momentum of hammer and pile after impact

i.e. $2 \times 4.89 = \left(2 + \tfrac{3}{4}\right) \times V$ where V is the common velocity

$$\therefore V = 3.56 \text{ m/s}$$

The work done in moving a distance x against a resistance R is equal to the loss of kinetic and potential energy of the system.

i.e. $$400 \times 10^3 x = \tfrac{1}{2} \times 2.75 \times 10^3 \times 3.56^2 + 2.75 \times 10^3 \times 9.81x$$

from which $$x = 0.046\,7 \text{ m or } \underline{46.7 \text{ mm}}$$

(b) Energy lost $=$ loss of potential energy of hammer $-$ work done against ground resistance

$$= 2 \times 10^3 \times 9.81(1.22 + 0.046\,7) - 400 \times 10^3 \times 0.046\,7$$

$$= \underline{6.15 \times 10^6 \text{J}}$$

(c) Efficiency $= \dfrac{\text{work done against ground resistance}}{\text{work done in raising hammer above top of pile}}$

$$= \frac{400 \times 10^3 \times 0.046\,7}{2 \times 10^3 \times 9.81 \times 1.22}$$

$$= 0.78 \quad \text{or} \quad \underline{78\%}$$

3. *A break-down truck of mass 2 300 kg is attached to a car of mass 1 350 kg by a rope 4.5 m long. The initial distance between the fixing points for the rope is 3 m so that the truck can move 1.5 m before the rope tightens and the truck is accelerated from rest under a tractive effort of 2.25 kN.*

Determine the truck speed (a) just before the rope starts to tighten, (b) at the instant when the rope ceases to stretch.

Determine also the impulsive force on the truck during the stretching period if this takes place in 0.1 s.

Ignore frictional losses and the effect of the tractive effort during the tensioning period of 0.1 s.

The arrangement is shown in Figure 10.9.
While taking up the slack in the rope, the acceleration of the truck is given by
$$F = ma$$

i.e. $$2.25 \times 10^3 = 2\,300a$$

$$\therefore a = 0.978 \text{ m/s}^2$$

Therefore, when the truck has moved 1.5 m, its velocity is given by
$$v^2 = u^2 + 2as$$

i.e. $$v = \sqrt{(2 \times 0.978 \times 1.5)} = \underline{1.713 \text{ m/s}}$$

During the tensioning period, the momentum of the system remains constant since equal and opposite forces act on the car and truck.
Hence

total momentum before tensioning $=$ total momentum after tensioning

i.e. $$2\,300 \times 1.715 + 1\,350 \times 0 = (2\,300 + 1\,350) \times V$$

$$\therefore V = \underline{1.08 \text{ m/s}}$$

Impulsive force on truck $=$ rate of change of momentum
$$= \frac{2\,300(1.713 - 1.08)}{0.1}$$

$$= \underline{1\,455 \text{ N}}$$

Figure 10.9

4. *A shaft is being turned in a lathe which is driven by a motor developing 2.25 kW at 1400 rev/min, the speed reduction between the motor and lathe spindle being 10 to 1. The friction torque at the lathe spindle is 17.5 N m. The moment of inertia of the rotating parts of the motor is 0.08 kg m² and that of the lathe spindle and work-piece is 1.2 kg m².*

If the tool is suddenly given an excessively heavy cut which stops the shaft in one revolution, calculate the force on the tool if it is cutting at a radius of 140 mm.

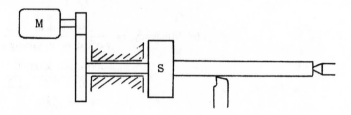

Figure 10.10

Equivalent moment of inertia of spindle S, Figure 10.10,

$$= I_s + I_m \times 10^2 \quad \text{from equation (10.33)}$$

$$= 1.2 + 0.08 \times 100 = 9.2 \text{ kg m}^2$$

The deceleration of the spindle is given by

$$\omega^2 = 2\alpha\theta \quad \text{from equation (10.10)}$$

i.e. $$\alpha = \frac{\left(140 \times \dfrac{2\pi}{60}\right)^2}{2 \times 2\pi} = 17.1 \text{ rad/s}^2$$

$$\text{Torque applied to spindle by motor} = \frac{2.25 \times 10^3}{140 \times \dfrac{2\pi}{60}}$$

$$= 153.5 \text{ N m}$$

The net decelerating torque on the shaft

$$= \text{torque due to tool force, } F + \text{ friction torque } - \text{ driving torque}$$

$$= F \times 0.14 + 17.5 - 153.5$$

$$\therefore 0.14F - 136 = I\alpha = 9.2 \times 17.1$$

from which $$F = \underline{2\,090 \text{ N}}$$

5. *A drum A of mass 200 kg, external diameter 380 mm and radius of gyration 150 mm, rotates on frictionless bearings at 250 rev/min. A stationary drum B of mass 50 kg, external diameter 200 mm and radius of gyration 80 mm, mounted on a frictionless axis parallel to that of A, is pressed into contact with A with a force of 90 N. The coefficient of friction is 0.25.*

Determine (a) the time of slipping and the final speeds of A and B,

(b) the time of slipping if a torque is applied to A to maintain a constant speed of 250 rev/min.

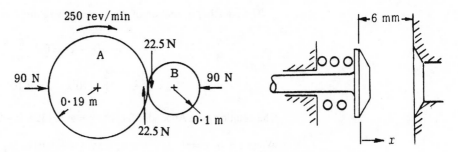

Figure 10.11   **Figure 10.12**

Figure 10.11 shows the drums and the forces acting on them.

If N is the final speed of A, the final speed of B is $N \times \dfrac{0.19}{0.10} = 1.9\,N$

Tangential friction force between cylinders $= 0.25 \times 90 = 22.5\,N$

This force acts so as to decelerate A and accelerate B.

Let the time of slipping be t. Then, applying the equation $T = I\alpha$ to A and B,

$$22.5 \times 0.19 = 200 \times 0.15^2 \left(\frac{250 - N}{t} \right) \times \frac{2\pi}{60}$$

i.e. $250 - N = 9.07t$ (1)

and $22.5 \times 0.10 = 50 \times 0.08^2 \times \dfrac{1.9N}{t} \times \dfrac{2\pi}{60}$

i.e. $N = 35.34t$ (2)

From equations (1) and (2),

$$N = \underline{199 \text{ rev/min}}$$

and $t = \underline{5.62 \text{ s}}$

Speed of B $= 1.9N = \underline{378 \text{ rev/min}}$

When speed of A remains constant at 250 rev/min, equation (2) gives

$$250 = 35.34t$$

from which $t = \underline{7.07 \text{ s}}$

6. *A valve of mass 0.25 kg closes horizontally under the action of a spring. In the closed position, the spring is compressed 12 mm and the maximum opening of the valve is 6 mm. If the spring stiffness is 4 kN/m, find the time required for the valve to close and the velocity with which it strikes the seat.*

In the open position, Figure 10.12, compression of spring $= 12 + 6 = 18$ mm. Therefore, when the valve has moved a distance x m,

$$\text{compression of spring} = 0.018 - x \text{ m}$$

The equation of motion of the valve is therefore $F = ma = m(d^2x/dt^2)$

i.e.
$$4 \times 10^3(0.018 - x) = 0.25\frac{d^2x}{dt^2}$$

$$\therefore \frac{d^2x}{dt^2} + 16 \times 10^3 x = 16 \times 10^3 \times 0.018$$

The solution (see Appendix B) is $x = A\cos 126.4t + B\sin 126.4t + 0.018$

When $t = 0$, $x = 0$, $\therefore A = -0.018$

When $t = 0$, $\dfrac{dx}{dt} = 0$, $\therefore B = 0$

$$\therefore x = 0.018(1 - \cos 126.4t)$$

When $x = 6$ mm $0.006 = 0.018(1 - \cos 126.4t)$

$$\therefore \cos 126.4t = \tfrac{2}{3}$$

$$\therefore t = \underline{0.006\,65 \text{ s}}$$

$$v = \frac{dx}{dt} = 0.018 \times 126.4 \sin 126.4t$$

When $t = 0.066\,5$ s, $v = 0.018 \times 126.4 \sin 126.4 \times 0.006\,65$

$$= 2.275 \sin 48°12'$$

$$= \underline{1.7 \text{ m/s}}$$

7. *A uniform solid cylinder of radius 0.2 m rolls along a horizontal surface with a velocity of 3 m/s when it encounters a step of height 0.1 m, parallel with the axis of the cylinder.*

Determine its velocity after mounting the step.

The moment of inertia of the cylinder about its axis $= \dfrac{mr^2}{2}$ (see Appendix A)

$$= \frac{m \times 0.2^2}{2} = 0.02m \text{ kg m}^2$$

Figure 10.13(a) shows the position immediately before impact at P.

The initial angular velocity, $\omega_1 = \dfrac{v_1}{r} = \dfrac{3}{0.2} = 15$ rad/s

Initial moment of momentum about P

$\qquad$ = angular momentum about O + moment of linear momentum about P

$\qquad = I\omega_1 + mv_1(0.2 - 0.1) = 0.02m \times 15 + m \times 3 \times 0.1 = 0.6m$

Immediately after impact, Figure 10.13(b),

$\qquad$ moment of momentum about P $= I\omega_2 + mv_2 \times 0.2$

$$= 0.02m\left(\frac{v_2}{0.2}\right) + mv_2 \times 0.2 = 0.3mv_2$$

Figure 10.13

During impact, the moment of momentum of the cylinder about P remains constant since the external force through P has no moment about that point.

Hence
$$0.6m = 0.3mv_2$$

$$\therefore v_2 = 2 \text{ m/s}$$

If the velocity on the level, after climbing the step, is v_3, then

$$\text{KE immediately after impact} = \text{KE on level} + \text{gain in PE}$$

i.e.
$$\tfrac{1}{2}mv_2^2 + \tfrac{1}{2}I\omega_2^2 = \tfrac{1}{2}mv_3^2 + \tfrac{1}{2}I\omega_3^2 + mg \times 0.1$$

i.e.
$$\tfrac{1}{2}mv_2^2 + \tfrac{1}{2} \times 0.02m \left(\frac{v_2}{0.2}\right)^2 = \tfrac{1}{2}mv_3^2 + \tfrac{1}{2} \times 0.02m \left(\frac{v_3}{0.2}\right)^2 + mg \times 0.1$$

from which
$$v_2^2 = v_3^2 + \frac{0.2g}{3}$$

Hence
$$v_3 = \sqrt{(4 - 0.654)} = \underline{1.83 \text{ m/s}}$$

8. *One end of a thin uniform rod 0.45 m long is hinged to a rigid support. The rod, which has a mass of 2 kg and initially hangs downwards, is raised through 60° from the vertical and then released. As the rod approaches the end of its travel, it is restrained by a horizontal compression spring situated 0.4 m below the hinge. The stiffness of the spring is 35 kN/m and it is arranged so that the rod just reaches the vertical position at the point of maximum compression.*

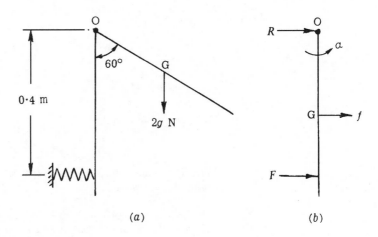

Figure 10.14

Calculate the greatest reaction at the hinge. Where should the spring be placed for this value to be zero?

Let F be the force exerted by the spring at the maximum compression x. When released from the highest position, Figure 10.14(a),

$$\text{loss of potential energy of rod} = \text{gain of strain energy of spring}$$

i.e.
$$mgh = \tfrac{1}{2}Sx^2 = \tfrac{1}{2}\frac{F^2}{S} \text{ since } F = Sx$$

$$\therefore\ 2 \times 9.81 \times \frac{0.45}{2}(1 - \cos 60°) = \tfrac{1}{2} \times \frac{F^2}{35 \times 10^3}$$

from which
$$F = 393 \text{ N}$$

This is the maximum compressive force in the spring and at this instant, Figure 10.14(b), the angular acceleration of the rod is given by taking moments about O,

i.e.
$$F \times 0.4 = I_O\alpha = \frac{ml^2}{3}\alpha$$

$$\therefore\ 393 \times 0.4 = 2 \times \frac{0.45^2}{3}\alpha$$

$$\therefore\ \alpha = 1\,165 \text{ rad/s}^2$$

If the reaction at the hinge is R, the linear motion of the rod is given by

$$F + R = ma = m \times \frac{l}{2}\alpha$$

i.e.
$$393 + R = 2 \times 0.225 \times 1\,165 = 524$$

$$\therefore\ R = \underline{131 \text{ N}}$$

If the spring is at a distance h below O, then

$$Fh = I_O\alpha$$

$$\therefore\ \alpha = \frac{Fh}{I_O} = \frac{3Fh}{ml^2}$$

$$R = ma - F = m\frac{l}{2}\alpha - F$$

$$= \frac{3Fh}{2l} - F$$

Therefore, when $R = 0$,
$$\frac{3h}{2l} = 1$$

$$\therefore\ h = \tfrac{2}{3}l = \tfrac{2}{3} \times 0.45 = \underline{0.3 \text{ m}}$$

Note: The point of impact which produces no reaction at the hinge is called the *centre of percussion*.

9. *The loaded cage of a goods hoist has a mass of 1 200 kg. The rope passes over a drum at the top of the shaft and then to a balance mass of 500 kg. The cage and balance mass move in guides and the friction force at each guide is 500 N. The drum* *has a diameter of 1.5 m, a mass of 600 kg and a radius of gyration of 0.6 m. The maximum acceleration attained is 1.5 m/s², which occurs at a speed of 2.5 m/s. The maximum speed is 4.5 m/s and retardation is at a uniform rate from that speed to zero in the last 4.5 m of travel.*

Determine (a) the power required to drive the drum at the condition of maximum acceleration, (b) the rope tensions during retardation.

Suffices 1 and 2 refer to the cage and balance mass respectively.

Let the rope tensions be F_1 and F_2, Figure 10.15, and let the friction force be f. Then during the acceleration period,

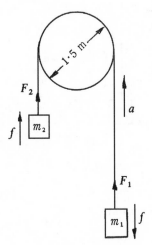

Figure 10.15

$$F_1 = m_1 g + m_1 a + f$$
$$= 1\,200(9.81 + 1.5) + 500$$
$$= 14\,070 \text{ N}$$

and
$$F_2 = m_2 g - m_2 a - f$$
$$= 500(9.81 - 1.5) - 500$$
$$= 3\,660 \text{ N}$$

Therefore, torque on drum,

$$T = I\alpha + (F_1 - F_2)r$$
$$= 600 \times 0.6^2 \times \frac{1.5}{0.75} + (14\,070 - 3\,660) \times 0.75 = 8\,239 \text{ N m}$$

$$P = T\omega = 8\,239 \times \frac{2.5}{0.75} = \underline{27\,460 \text{ W}}$$

During retardation, the deceleration is given by

$$v^2 = 2as$$

i.e.
$$4.5^2 = 2a \times 4.5$$

$$\therefore a = 2.25 \text{ m/s}^2$$

$$\therefore F_1 = m_1 g - m_1 a + f$$
$$= 1\,200(9.81 - 2.25) + 500 = \underline{9\,572 \text{ N}}$$

and
$$F_2 = m_2 g + m_2 a - f$$
$$= 500(9.81 + 2.25) - 500 = \underline{5\,530 \text{ N}}$$

10. *In a double reduction lifting gear, the moments of inertia of the motor armature and pinion, intermediate and drum shafts are 3.5, 45 and 1000 kg m² respectively. The motor runs at 6 times the speed of the intermediate shaft and 6G times the speed of the drum shaft. The drum radius is 0.9 m and the load lifted is 1 t. For a constant motor torque of 550 Nm, find the value of G for maximum acceleration and determine the value of this acceleration*

The arrangement of the drive is shown in Figure 10.16.

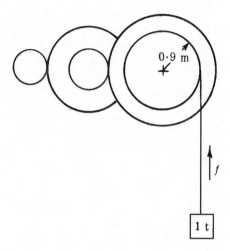

Figure 10.16

Effective moment of inertia of motor

$$= 3.5 + \frac{45}{6^2} + \frac{1\,000}{(6G)^2} \quad \text{from Section 10.15}$$

$$= 4.75 + \frac{27.78}{G^2}$$

If the acceleration of the load is a,

$$\text{rope tension} = 1\,000(9.81 + a)$$

$$\therefore \text{drum torque} = 1\,000(9.81 + a) \times 0.9$$

$$\therefore \text{motor torque} = \frac{900(9.81 + a)}{6G}$$

$$\text{Angular acceleration of drum} = \frac{a}{0.9}$$

$$\therefore \text{angular acceleration of motor} = \frac{a}{0.9} \times 6G$$

Motor torque = torque to accelerate rotating parts + torque to accelerate load

i.e. $\qquad 550 = \left(4.75 + \dfrac{27.78}{G^2}\right) \times \dfrac{6aG}{0.9} + \dfrac{900(9.81 + a)}{6G}$

from which $\quad a = \dfrac{550G - 1\,472}{31.7G^2 + 335}$ $\qquad\qquad\qquad$ (1)

For maximum acceleration, $\dfrac{\mathrm{d}a}{\mathrm{d}G} = 0$

i.e. $\qquad \dfrac{(31.7G^2 + 335) \times 550 - (550G - 1\,472) \times 63.4G}{(31.7G^2 + 335)^2} = 0$

from which $G^2 - 5.35G - 10.56 = 0$

Hence $G = \underline{6.88}$

$$a_{\text{max}} = \frac{550 \times 6.88 - 1472}{31.7 \times 6.88^2 + 335} \quad \text{substituting in equation (1)}$$

$$= \underline{1.26 \text{ m/s}^2}$$

11. *A car of mass 1300 kg, with the engine at full throttle, can travel at 162 km/h on a level road with the engine developing 75 kW. The resistance to motion varies as the square of the speed.*

Determine the time taken for the speed of the car to rise from 72 km/h to 108 km/h at full throttle on an upgrade of 1 in 20, assuming that the engine torque remains constant.

$$1 \text{ km/h} = \frac{1\,000}{3\,600} = \frac{1}{3.6} \text{ m/s}$$

$$\therefore 162 \text{ km/h} = \frac{162}{3.6} = 45 \text{ m/s}$$

$$72 \text{ km/h} = \frac{72}{3.6} = 20 \text{ m/s}$$

and $$108 \text{ km/h} = \frac{108}{3.6} = 30 \text{ m/s}$$

At 162 km/h, the whole of the engine power is absorbed in overcoming the resistance to motion,

i.e. $$\text{tractive force} = \frac{P}{v} = \frac{75 \times 10^3}{45} = k \times 45^2$$

from which $$k = 0.823$$

At a speed v,

$$\text{tractive force} = \text{mass} \times \text{acceleration} + \text{resistance to motion}$$

$$+ \text{component of weight down incline}$$

i.e. $$\frac{75 \times 10^3}{45} = 1\,300\frac{dv}{dt} + 0.823v^2 + \frac{1\,300 \times 9.81}{20}$$

from which $$dt = \frac{1\,300\,dv}{1\,030 - 0.823v^2}$$

$$\therefore t = \int_{20}^{30} \frac{1\,580\,dv}{35.4^2 - v^2}$$

$$= 1\,580 \times \frac{1}{2 \times 35.4} \left[\ln\frac{35.4 + v}{35.4 - v}\right]_{20}^{30}$$

$$= 22.3 \ln 3.37 = \underline{27 \text{ s}}$$

12. *The resistance to motion of a car of mass 800 kg is (54 + 0.8 v²) N, where v is the speed in m/s. The gear ratio between the engine and rear axle is 5.3:1, the moment of inertia of the rotating parts of the engine is 0.3 kg m² and that of the wheels is 1 kg m². The diameter of the wheels is 0.52 m.*

Find the distance travelled by the car up an incline of 1 in 40 whilst accelerating from 25 to 50 km/h assuming that the engine develops a constant torque of 40 N m during this period.

Effective moment of inertia of wheels

$$= 1.0 + 5.3^2 \times 0.3 \quad \text{from Section 10.15}$$

$$= 9.43 \text{ kg m}^2$$

Therefore effective mass of car for acceleration purposes

$$= 800 + \frac{9.43}{0.26^2} \quad \text{from Section 10.14}$$

$$= 939.5 \text{ kg}$$

$$\text{Tractive force} = \frac{40 \times 5.3}{0.26} = 815 \text{ N}$$

Component of weight down incline

$$= \frac{800 \times 9.81}{40} = 196.2 \text{ N}$$

$$\therefore 815 = 939.5v\frac{dv}{ds} + (54 + 0.8v^2) + 196.2$$

from which
$$ds = \frac{939.5v\,dv}{565 - 0.8v^2}$$

$$50 \text{ km/h} = \frac{50}{3.6} = 13.9 \text{ m/s}$$

and
$$25 \text{ km/h} = \frac{25}{3.6} = 6.95 \text{ m/s}$$

$$\therefore s = \int_{6.95}^{13.9} \frac{1\,175\ v\,dv}{706.5 - v^2}$$

$$= -\frac{1\,175}{2}\left[\ln(706.5 - v^2)\right]_{6.95}^{13.9}$$

$$= -587.5\ln 0.779 = \underline{146 \text{ m}}$$

13. *A car is driven by the rear wheels and when it is stationary, 0.55 of its weight is supported on the rear wheels. The height of the centre of gravity above the road is one-fifth of the wheel base. In a test on a level road it was found that the greatest acceleration obtainable without skidding was 3 m/s².*

Calculate the coefficient of friction between the tyres and road. Using the same coefficient of friction, find the steepest gradient which the vehicle could climb.

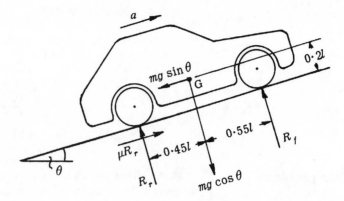

Figure 10.17

Equating forces normal and parallel to the plane when $\theta = 0$, Figure 10.17,

$$R_r + R_f = mg \tag{1}$$

and $$\mu R_r = ma \tag{2}$$

Taking moments about G,

$$\mu R_r \times 0.2l = R_r \times 0.45l - R_f \times 0.55l \tag{3}$$

From equation (3) $$R_f = \frac{0.45 \times 0.2\mu}{0.55} R_r$$

Therefore in equation (1)

$$R_r \times \frac{1 - 0.2\mu}{0.55} = mg$$

Substituting for R_r in equation (2),

$$\mu \times \frac{0.55mg}{1 - 0.2\mu} = ma$$

from which $$\frac{a}{g} = \frac{0.55\mu}{1 - 0.2\mu} = \frac{3}{9.81}$$

$$\therefore \mu = \underline{0.5}$$

At a uniform speed on an inclined road,

$$R_f + R_r = mg \cos \theta \tag{4}$$

$$\mu R_r = mg \sin \theta \tag{5}$$

and $$\mu R_r \times 0.2l = R_r \times 0.45l - R_f \times 0.55l \tag{6}$$

From equation (6), $$R_f = \frac{0.35}{0.55} R_r$$

From equations (4) and (5),

$$\tan\theta = \frac{0.5R_r}{R_f + R_r}$$

$$= \frac{0.5}{\dfrac{0.35}{0.55} + 1} = 0.305\,5$$

$$\therefore \theta = \underline{17°}$$

Further problems

14. A mass of 700 kg falling 0.2 m is used to drive a pile of mass 500 kg into the ground. Assuming there is no rebound, find the common velocity of the driver and pile at the end of the blow and the loss of kinetic energy. If the resistance of the ground is constant, find its value if the pile is driven 75 mm.

[1.155 m/s; 572 J; 22.45 kN]

15. A stationary truck of mass 9 t is set in motion by a shunting locomotive which provides an impulse of 30 kNs. The truck travels freely along a level track for a period of 15 s when it collides with a truck of mass 12 t which is moving at 0.6 m/s in the same direction. After collision, both trucks move on together. The track resistance is 65 N/t. Determine their common speed and the loss of energy at impact.   [1.353 m/s; 8 kJ]

16. A flywheel of mass 50 kg is mounted on a 75 mm diameter shaft in bearings on either side of the wheel. Due to friction, the speed of the flywheel falls from 200 rev/min to 150 rev/min in 14 s with uniform deceleration.

A plain cast iron ring, outside diameter 450 mm, inside diameter 350 mm, thickness 75 mm and density 7.2 Mg/m³ is now bolted concentrically on the side of the wheel and the effect of friction is to reduce the speed uniformly from 200 rev/min to 150 rev/min in 20 s.

Calculate the coefficient of friction and the radius of gyration of the flywheel.

[0.02; 140 mm]

17. A steel bar 75 mm diameter starts from rest and rolls without slipping down a plane inclined at 10° to the horizontal. The mass of the bar is 88 kg. Find the kinetic energy of the bar when it has rolled 6 m down the plane.   [900 J]

18. Two parallel shafts A and B are connected by gear wheels so that A rotates at four times the speed of B. The moments of inertia of the rotating parts on shafts A and B are respectively 160 and 500 kg m². Find the total kinetic energy of the system when B rotates at 200 rev/min and the driving torque required on shaft B to accelerate the system uniformly from rest so that the speed of B is 200 rev/min after 30 s.

If the shafts are 0.75 m apart, what is the tangential force between the teeth of the gears during this acceleration?   [671 kJ; 2.14 kN m; 2.98 kN]

19. A flywheel A of mass 12.5 kg, outside diameter 450 mm and radius of gyration 175 mm, is initially rotating at 300 rev/min and second flywheel B of mass 7.5 kg, outside diameter 380 mm and radius of gyration 150 mm, is mounted on a parallel shaft and is initially stationary.

The two shafts are moved together so that the wheels make circumferential contact. Determine the speeds of the wheels when slipping ceases.

If the coefficient of friction between the wheels is 0.12 and the normal force between them is 60 N, calculate the time taken for slipping to cease.

[185.4 rev/min; 219.5 rev/min; 2.84 s]

20. A rigid beam AB of uniform cross-section and of mass 40 kg is hinged at A to a fixed support and is maintained in a horizontal position by a vertical helical spring attached to B, AB being 1.8 m. A mass of 2.5 kg falls on to the beam with a striking velocity of 3 m/s, the point of impact being 1.2 m from A.

Assuming the mass and beam move together, determine the angular velocity of the beam immediately after impact. [0.192 4 rad/s]

21. Figure 10.18 shows a tilt hammer hinged at O with its head A resting on top of the pile B. The hammer and arm have a mass of 25 kg, the centre of gravity G is 400 mm from O and the radius of gyration about an axis through G parallel to the axis of the pin O is 75 mm. The pile has a mass of 135 kg. The hammer is raised through 45° from the horizontal and released. On striking the pile there is no rebound.

Find the angular velocity of the hammer immediately before impact and the linear velocity of the pile immediately after impact. [5.79 rad/s; 0.343 m/s]

22. A valve is opened by a cam and then released by a trip-gear, when it is closed by a helical spring concentric with the valve stem. The mass of the valve is 3 kg and the maximum opening is 16 mm. The stiffness of the spring is 25 kN/m and the compression when the valve is closed is 20 mm.

Determine (a) the time taken for the valve to close, (b) the velocity at the moment of impact. [0.010 8 s; 2.735 m/s]

23. The table of a machine tool slides on horizontal guides. The table has a mass of 100 kg and the frictional force opposing its motion is 180 N. When the table is moving at 0.9 m/s, the driving mechanism is disconnected and the table is brought to rest by a spring buffer which is initially unstressed.

Calculate the time required for the spring to attain its greatest compression of 40 mm. [0.072 3 s]

24. A uniform disc of radius r, rolling without slipping along a horizontal plane with velocity v, encounters a step of height $3r/4$ perpendicular to its plane of motion. Assuming that no slipping occurs, show that the disc will surmount the step if $v^2 > 4rg$.

25. A uniform rod AB, 0.75 m long and of mass 20 kg, is hinged at A and held in a horizontal position. It is allowed to fall, rotating about A in a vertical plane, until it strikes a horizontal spring C of stiffness 35 kN/m, whose centre line is 0.35 m below the level of A. When the spring force is a maximum, the rod is vertical.

Find (a) the maximum spring force, (b) the maximum force on the hinge A.

[2 270 N; 682 N]

26. The tailboard of a lorry is a uniform rectangle, 1.5 m long by 0.75 m high and has a mass of 27 kg. It is hinged along the bottom edge to the floor of the lorry. Chains are attached to the top corners of the board and to the sides of the lorry so that when the board is in the horizontal position, the chains are inclined at 45° to the horizontal. A tension spring is inserted in each chain to reduce the shock and these are adjusted to prevent the board from dropping below the horizontal. Each spring has a stiffness of 52 kN/m.

Find the greatest force in each spring and the resultant force at the hinges when the board falls freely from the vertical position. [2.27 kN; 3.59 kN]

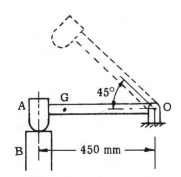

Figure 10.18

27. A shaft A carries rotating parts of moment of inertia 10 kg m^2 and this is geared to another shaft B, whose rotating parts have a moment of inertia of 2 kg m^2. Shaft B rotates at 3 times the speed of shaft A, which is accelerated from rest by a constant torque of 56 N m. Determine the time taken for shaft A to reach a speed of 12 rad/s.   [6 s]

28. A double reduction winding gear hauls a 10 t truck up an incline of 1 in 40 against a frictional resistance of 2 kN. The diameter of the winding drum is 1.6 m. The moments of inertia of the motor shaft, intermediate shaft and drum are 3, 100 and 2 500 kg m^2 respectively and each gear ratio is 5:1.

If the starting torque of the motor is 800 N m, determine the acceleration of the truck.   [0.99 m/s^2]

29. A truck of mass 12 t moves down an incline of 1 in 12. The motion of the truck is controlled by means of a rope attached to the truck and wound round a drum at the top of the incline. The mass of the drum is 2 t and it has a diameter of 2.5 m and a radius of gyration of 0.9 m. A braking torque of 2 700 N m is applied to the drum and the tractive resistance to the motion of the truck is 90 N/t. Calculate the acceleration of the truck down the incline.   [0.504 m/s^2]

30. A winding drum raises a cage of mass 500 kg through a height of 120 m. The winding drum has a mass of 250 kg, a radius of 0.5 m and a radius of gyration of 0.36 m. The mass of the rope is 3 kg/m. The cage has at first an acceleration of 1.5 m/s^2 until a velocity of 9 m/s is reached, after which the velocity is constant until the cage nears the top, when the final retardation is 6 m/s^2.

Find (*a*) the time taken for the cage to reach the top, (*b*) the starting torque on the drum, (*c*) the power at the end of the acceleration period.

[17.08 s; 4 960 N m; 82.14 kW]

31. A lift of mass 900 kg is connected to a rope which passes over a drum 1 m diameter and then to a balance mass of 450 kg. The moment of inertia of the drum is 100 kg m^2 and it is driven through a reduction gear of 25 to 1, of 90 per cent efficiency. Neglecting the inertia of the gears, calculate the motor torque for a lift acceleration of 3 m/s^2. If the maximum output of the motor is 15 kW, what will be the maximum uniform speed of the lift?   [215 N m; 3.06 m/s]

32. A truck of mass 5 t is hauled up an incline of 1 in 15 by a rope parallel to the track. The rope is wound on a drum driven by an electric motor. The drum is 1 m in diameter and has a mass of 1 t and a radius of gyration of 0.4 m. The efficiency of the drive from motor to drum is 88 per cent and the frictional resistance to the motion of the truck is 1.3 kN.

When the truck is moving up the incline with a speed of 3 m/s and an acceleration of 0.3 m/s^2, find the power output from the motor.   [21.4 kW]

33. In a double reduction lifting gear driven by an electric motor, the gear ratio between the motor shaft and the intermediate shaft is 3.5 and between the intermediate shaft and drum shaft, it is 4.5. The moments of inertia of the three shafts are 5, 40 and 500 kg m^2 respectively. On the drum, 1.2 m diameter, is suspended a loaded cage of mass 6 t and a balance mass of 4.5 t which descends as the loaded cage is lifted. Determine the motor torque required to raise the cage with an acceleration of 0.4 m/s^2.   [830 N m]

34. In a double reduction lifting gear, the speed of the motor shaft is G times the speed of the intermediate shaft and G^2 times the speed of the drum shaft. The moments of inertia of the three shafts are 3, 35 and 850 kg m^2 respectively. A load of 800 kg is suspended from the drum which is 1.2 m diameter and the motor exerts a constant torque of 570 N m.

Determine the value of G for maximum acceleration of the load when being raised and calculate this maximum acceleration.   [5.59; 1.535 m/s^2]

35. A car has a mass of 1 t and the moment of inertia of the wheels are together 8.4 kg m^2. The diameter of the wheels is 0.62 m. Find the acceleration on the level when the engine output torque is 100 N m, the overall speed reduction is 14, the resistance to motion is 200 N and the transmission efficiency is 88 per cent.

[3.47 m/s^2]

36. A car of mass 1 t has four wheels, each of radius 0.3 m and moment of inertia 0.8 kg m^2. The rotating parts of the engine have a moment of inertia of 0.5 kg m^2 and the gear ratio, engine to back axle, is 4 to 1.

When the speed of the car is v m/s, the resistance to motion is $(220 + 0.8v^2)$ N. If the tractive effort remains constant at 1 500 N, find the time taken to accelerate from 20 to 30 m/s on a level road.   [15 s]

37. A car has a maximum speed on a level road of 70 m/s, at which speed the engine develops 180 kW. The four road wheels have a rolling radius of 0.3 m, a radius of gyration of 0.2 m and the mass of each wheel is 23 kg. The rotating parts of the engine have a moment of inertia of 1 kg m^2 and the gear ratio, engine to wheels, is 4:1. The total mass of the car is 1 350 kg. The torque output of the engine may be regarded as constant over a wide speed range.

If the resistance to motion varies as the square of the road speed, determine the time taken for the speed to rise from 40 m/s to 60 m/s on a level road.   [27 s]

38. The resistance to motion of a train of mass 550 t is given by $R = 3\,800 + 900v$ where R is in newtons and v is in m/s. If the locomotive exerts a constant tractive force of 50 kN, find the distance travelled and the time taken to accelerate from 32 km/h to 48 km/h on an up gradient of 1 in 200.   [271 s; 3.06 km]

39. A car has a mass of 2 t and the four wheels are each of mass 18 kg, diameter 0.75 m and radius of gyration 0.32 m. The engine develops a torque of 115 N m and the rotating parts have a moment of inertia of 0.47 kg m^2. Calculate the ratio of engine speed to back-axle speed for maximum acceleration up an incline in 1 in 120 against a wind resistance of 180 N.   [25.9]

40. To maintain a uniform speed of 20 m/s on a level road, a car of mass 1.5 t requires 9 kW. The resistance to motion is given by $R = 160 + kv^2$ N where k is a constant and v is the speed.

If, while travelling at 20 m/s, the car starts to climb a uniform slope of 1 in 20, find the time taken for the speed to drop to 12 m/s, assuming that the tractive effort remains constant.   [8.57 s]

41. A flywheel of mass 80 kg and radius of gyration 500 mm is accelerated from rest by a constant torque and reaches a steady speed of 75 rad/s. The friction torque opposing the motion is 10ω where ω is the speed in rad/s at any instant.

Find the value of the applied torque and the time taken to reach a speed of 50 rad/s.   [750 N m; 2.2 s]

42. A vehicle is driven along a horizontal road by the rear wheels. The wheelbase is 3.3 m and the centre of gravity is 0.75 m above the ground and 1.35 m behind the front axle. The coefficient of friction between the tyres and ground is 0.3.

Determine (*a*) the maximum acceleration if the wheels are not to slip; (*b*) the maximum retardation when a braking torque is applied to the rear wheels.

[1.29 m/s^2; 1.126 m/s^2]

43. A car is driven by the rear wheels and when the car is stationary, 0.55 of the mass is supported by the rear wheels. The height of the centre of gravity above the ground is one-fifth of the wheel base. On a level road, the greatest acceleration possible without skidding the wheels is 3 m/s^2.

(*a*) Calculate the coefficient of friction between the tyres and the road.

(*b*) Using this coefficient of friction, find the steepest gradient which the car could climb.   [0.5; $\tan^{-1} 0.305\,5$]

11 Velocity and acceleration diagrams

11.1 Introduction

It is often necessary to determine the velocity and acceleration of points in a mechanism in order to obtain the forces involved. These can be found from velocity and acceleration diagrams, which give the velocity and acceleration of any point relative to any other point for one particular position of the mechanism.

11.2 Velocity of a rigid link

The velocity of one point on a link relative to another must be perpendicular to the axis of the link, otherwise there would be a component along the axis which would involve a change in length. Thus, in the link shown in Figure 11.1, the velocity of B relative to A is given by $v_{ba} = \omega$. AB, perpendicular to AB; this is represented by the vector ab.

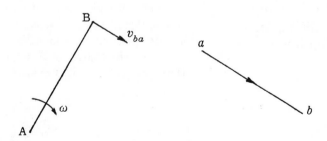

Figure 11.1

If this link is part of a mechanism of several links such as the four-bar mechanism shown in Figure 11.2 and the angular velocity of AB is given, the velocity diagram may be built up as follows:

The velocity of D relative to A is zero since AD is fixed; hence A and D are represented by a single point ad.

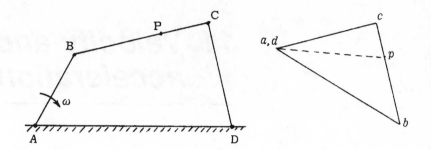

Figure 11.2

The velocity of B relative to A is ω. AB and is perpendicular to AB; this is represented by ab.

The velocity of C relative to B is perpendicular to BC and passes through b while the velocity of C relative to D is perpendicular to CD, passing through d. The intersection of these two vectors gives the point c.

For a point on the mechanism such as P, the corresponding point on the velocity diagram is obtained by proportion,

i.e.
$$\frac{bp}{bc} = \frac{BP}{BC}$$

and the velocity of P relative to the fixed points A and D is given by the vector ap.

The angular velocities of BC and CD are given by

$$\omega_{bc} = \frac{bc}{BC} \quad \text{and} \quad \omega_{cd} = \frac{cd}{CD}$$

11.3 Velocity of a block sliding on a rotating link

If a block B slides with velocity v along a link which rotates about A at rate ω, Figure 11.3, the velocity of the coincident link point B′ is given by ω. AB′, perpendicular to AB′. This is represented by the vector $ab′$. The velocity of B relative to B′ is v, parallel to AB′ and this passes through $b′$. Hence the velocity of B relative to A is represented by ab.

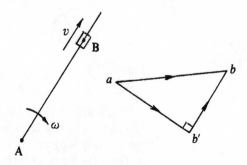

Figure 11.3

11.4 Acceleration of a rigid link

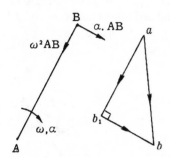

Figure 11.4

The acceleration of one point on a link relative to another has two components, (a) a centripetal component due to the angular velocity of the link, and (b) a tangential component due to the angular acceleration of the link.

If the link AB, Figure 11.4, is rotating with angular velocity ω and angular acceleration α, the centripetal acceleration of B relative to A is ω^2 AB, directed towards A, and the tangential acceleration is αAB, perpendicular to AB. These are represented in the acceleration diagram by vectors ab_1 and b_1b respectively, the resultant acceleration of B relative to A being given by ab.

If this link is part of a mechanism such as the four-bar mechanism shown in Figure 11.5, then A and D are represented by a single point, as in the velocity diagram since there is no relative motion between them.

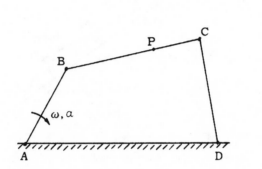

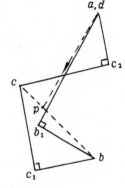

Figure 11.5

The centripetal and tangential accelerations of B relative to A are given by ab_1 and b_1b respectively, as before. The centripetal acceleration of C relative to B is given by v_{cb}^2/BC and is directed towards B. The value of v_{cb} is obtained from the velocity diagram and this acceleration is represented by bc_1. The tangential acceleration of C relative to B is unknown in magnitude but its direction is perpendicular to BC, so that a vector is drawn through c_1 perpendicular to bc_1.

The centripetal acceleration of C relative to D is given by v_{cd}^2/CD and is directed towards D; this is represented by dc_2. The tangential acceleration of C relative to D is again unknown but its direction is perpendicular to CD and is represented by a line through c_2 perpendicular to dc_2. The intersection of the lines through c_1 and c_2 then give the point c.

The acceleration of a point such as P is again obtained by proportion, i.e. $\dfrac{bp}{bc} = \dfrac{\mathrm{BP}}{\mathrm{BC}}$, the absolute acceleration of P being represented by ap.

Also $\alpha_{bc} = \dfrac{cc_1}{\mathrm{BC}}$ and $\alpha_{cd} = \dfrac{cc_2}{\mathrm{CD}}$, only the tangential component of the acceleration of one end of the link relative to the other being relevant.

11.5 Acceleration of a block sliding on a rotating link

Let a block B slide with velocity v and acceleration a along a link which rotates about A with angular velocity ω and angular acceleration α, Figure 11.6, and let the link turn through an angle $d\theta$ in time dt.

The velocity of the coincident link point B′ relative to A changes from ωr, represented by ab_1 to $(\omega + d\omega)(r + dr)$, represented by ab_2. The change in velocity is represented by b_1b_2 which has radial and tangential components $\omega r\, d\theta$ and $\omega\, dr + r\, d\omega$ respectively, neglecting the product $d\omega\, dr$.

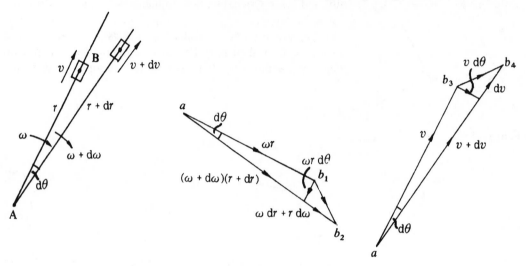

Figure 11.6

The velocity of B relative to the link point B′ changes from v, represented by ab_3, to $v + dv$, represented by ab_4. The change in velocity is represented by b_3b_4 which has radial and tangential components dv and $v\, d\theta$ respectively.

Thus the total change of velocity of B relative to A in the radial direction is $dv - \omega r\, d\theta$ so that

$$\text{radial acceleration} = \frac{dv}{dt} - \omega r \frac{d\theta}{dt}$$

$$= a - \omega^2 r \tag{11.1}$$

This is evidently the outward acceleration, a, of the block, less the centripetal acceleration of the link point B′ relative to A.

The total change of velocity of B relative to A in the tangential direction is $v\, d\theta + \omega\, dr + r\, d\omega$ so that

$$\text{tangential acceleration} = v\frac{d\theta}{dt} + \omega\frac{dr}{dt} + r\frac{d\omega}{dt}$$

$$= v\omega + \omega v + r\alpha$$

$$= \alpha r + 2v\omega \tag{11.2}$$

The term αr represents the tangential acceleration of the link point B′ relative to A and so the term $2v\omega$ represents the tangential acceleration of the block B relative to B′. This is called the *Coriolis component* and arises whenever a block slides along a rotating link or when a link slides through a swivel block. The direction of the Coriolis component is obtained by rotating the sliding velocity vector through 90° in the direction of rotation of the link. Where the acceleration of the link point B′ relative to the block B is required, this direction becomes reversed.

The acceleration diagram for the block is shown in Figure 11.7. The acceleration of B′ relative to A has components $\omega^2 r$ and αr, while the acceleration of B relative to B′ has components a and $2v\omega$. The resultant acceleration of B relative to A is then represented by ab.

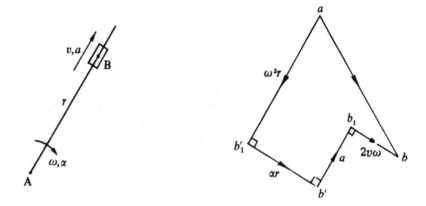

Figure 11.7

11.6 Inertia force on a link

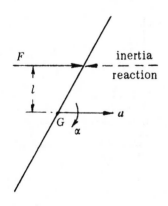

Figure 11.8

Figure 11.8 shows a link of mass m; the linear acceleration of the centre of gravity G is a and the angular acceleration is α. The force necessary to accelerate the link $F = ma$ and the torque necessary, $T = I\alpha = mk^2\alpha$ where k is the radius of gyration of the link about G.

The combined effect of F and T may be obtained by a single force F acting in the direction of the acceleration but offset from it by a perpendicular distance l such that

$$Fl = mk^2\alpha$$

or

$$l = \frac{mk^2\alpha}{ma} = \frac{k^2\alpha}{a} \tag{11.3}$$

The inertia reaction of the link is equal and opposite to the resultant accelerating force F.

Alternatively the link may be replaced by two concentrated masses m_1 and m_2 provided that these masses are equivalent to that of the link, Figure 11.9. The conditions for equivalence are:

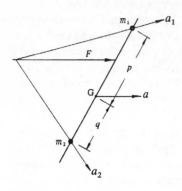

Figure 11.9

(a) the total mass must be the same,

i.e. $$m_1 + m_2 = m \qquad (11.4)$$

(b) the centre of gravity must be at G,

i.e. $$m_1 p = m_2 q \qquad (11.5)$$

(c) the moment of inertia about G of the two-mass system must be the same as that of the link,

i.e. $$m_1 p^2 + m_2 q^2 = mk^2 \qquad (11.6)$$

From equations (11.4), (11.5) and (11.6),

$$m_1 = \frac{q}{p+q}m, \quad m_2 = \frac{p}{p+q}m \quad \text{and} \quad pq = k^2 \qquad (11.7)$$

Either p or q can be chosen arbitrarily and the other distance is then determined from the relation $pq = k^2$.

The directions of the accelerations of m_1 and m_2 are obtained from the acceleration diagram, shown by a_1 and a_2 and the line of action of F must then be parallel to the direction of a and pass through the intersection of the lines of a_1 and a_2, since the accelerating forces on the masses are in the directions of a_1 and a_2. It is then unnecessary to calculate the value of l or to decide on which side of G the force F must act. It is also unnecessary to calculate the magnitudes of m_1 and m_2; it is only their positions which are relevant.

Worked examples

1. *In the mechanism shown in Figure 11.10, the crank OA rotates in a clockwise direction at 120 rev/min, the block B moves along the axis XX and the block D moves*

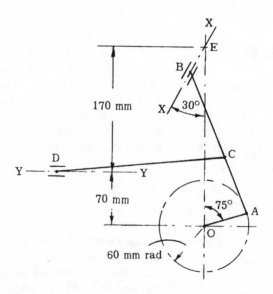

Figure 11.10

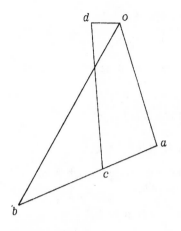

Figure 11.11

along the axis YY. AB and CD are connected by a pin-joint at C. AB = 210 mm, AC = 80 mm and CD = 240 mm. Draw the velocity diagram when the angle EOA is 75°, and find the velocity of D.

If a force of 100 N acting along YY resists the motion of D, determine the forces acting on the pins at the ends of the bar AB and the torque required on the crank OA, neglecting friction.

$$v_a = \omega \cdot \text{OA} = 120 \times \frac{2\pi}{60} \times 0.06 = 0.754 \text{ m/s}$$

In the velocity diagram, Figure 11.11, the absolute velocity of A is represented by oa and the velocity of B relative to A is perpendicular to AB, passing through the point a. The velocity of B relative to the fixed point, O is inclined at 30° to the vertical and passes through the point o. The intersection of these lines through a and o then gives the point b.

The point c on ab is positioned such that ac:ab: :AC:AB. The velocity of D relative to C is perpendicular to DC and the velocity of D relative to the fixed point O is horizontal. Thus the intersection of a line through c perpendicular to DC and a horizontal line through o gives the point d.

From the diagram, $v_d = od = \underline{0.187 \text{ m/s}}$

The reaction at the slider D is perpendicular to YY, Figure 11.12, and so, for a horizontal force of 100 N, the force in CD is 101 N, as determined by a triangle of forces. The forces acting on AB are then the force exerted at C by DC, the reaction at B which is perpendicular to XX and the force through A, which must be concurrent with the other two forces.

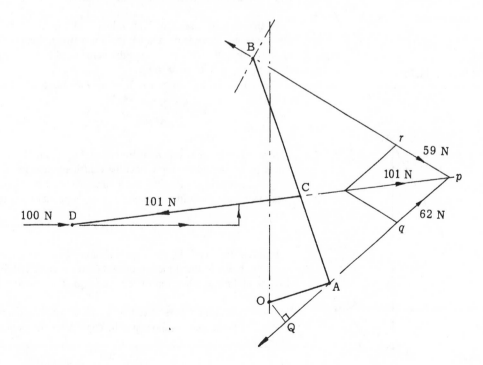

Figure 11.12

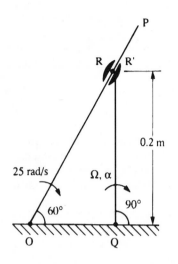

Figure 11.13

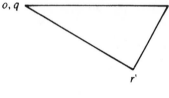

Figure 11.14

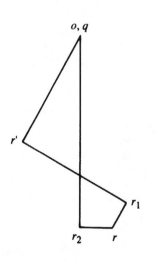

Figure 11.15

From the parallelogram of forces, force at A $= pq = $ <u>62 N</u>

and   force at B $= pr = $ <u>59 N</u>

The torque on OA is determined by equating the work done by this torque to the work done against the resistance at D,

i.e. $$T \times \left(120 \times \frac{2\pi}{60}\right) = 100 \times 0.187$$

$$\therefore \quad T = \underline{1.49 \text{ N m}}$$

Alternatively, the torque on OA is the product of the force through A and the perpendicular distance of its line of action from O, OQ.

Thus $$T = 62 \times 0.024 = \underline{1.49 \text{ N m}}$$

2. *Figure 11.13 shows part of a mechanism in which the link OP, rotating at a constant speed of 25 rad/s, drives the link QR through a sliding block.*

For the position shown, determine the angular velocity and angular acceleration of QR.

Let R$'$ be the point on OP which is coincident with the block R.

Then $v_{r'o} = \omega.\text{OR}' = 25 \times 0.231 = 5.775$ rad/s

and $a_{r'o} = \omega^2.\text{OR}' = 25^2 \times 0.231 = 144.5$ rad/s^2

In the velocity diagram, Figure 11.14, O and Q are represented by the single point oq since there is no relative motion between them. or' represents the velocity of R' relative to O, $r'r$ represents the velocity of R relative to R$'$ and qr represents the velocity of R relative to Q.

From the diagram, velocity of R relative to Q $= 6.67$ m/s. Hence angular velocity of QR,

$$\Omega = \frac{v_{rq}}{\text{QR}} = \frac{6.67}{0.2} = \underline{33.4 \text{ rad/s}}$$

In the acceleration diagram, Figure 11.15, O and Q are again represented by a single point oq and or' represents the centripetal acceleration of R$'$ relative to O.

The acceleration of R relative to R$'$ has two components, a tangential (or Coriolis) component and a sliding component. The tangential component is given by

$$2v_{rr'}\omega = 2 \times 2.89 \times 25 = 144.5 \text{ m/s}^2$$

From Section 11.5, its direction is obtained by rotating the relative velocity vector rr' through 90° in the direction of ω, i.e. to the right of OP. This is shown by $r'r_1$ and the sliding acceleration of R relative to R$'$ is parallel to OP but of unknown magnitude.

The acceleration of R relative to Q also has two components, a centripetal component and a tangential component. The centripetal component is given by

$$\frac{v_{rq}^2}{\text{QR}} = \frac{6.67^2}{0.2} = 222 \text{ m/s}^2$$

This is directed towards Q and is represented by qr_2. The tangential acceleration is perpendicular to QR and passes through r_2. The intersection of this line and the sliding acceleration of R relative to R′ then gives the point r.

Tangential acceleration of R relative to $Q = r_2 r = 35$ m/s²

Therefore angular acceleration of QR,

$$\alpha = \frac{a_{rq\text{(tangential)}}}{\text{QR}} = \frac{35}{0.2} = \underline{175 \text{ rad/s}^2}$$

Since the tangential acceleration of R relative to Q is to the right, the angular acceleration of QR is in the same sense as the angular velocity.

3. *Figure 11.16 shows a quick-return motion in which the driving crank OA rotates at 120 rev/min in a clockwise direction. For the position shown, determine the acceleration of the block D.*

$$v_a = \omega \cdot \text{OA} = 120 \times \frac{2\pi}{60} \times 0.2 \qquad = 2.515 \text{ m/s}$$

$$a_a = \omega^2 \cdot \text{OA} = \left(120 \times \frac{2\pi}{60}\right)^2 \times 0.2 = 31.6 \text{ m/s}^2$$

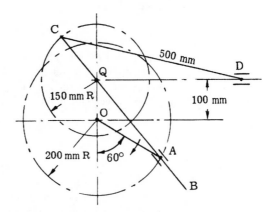

Figure 11.16

Let A′ be the point on QB which is coincident with the block A. Then, in the velocity diagram, Figure 11.17, O and Q are represented by a single point oq since there is no relative motion between them. oa represents the absolute velocity of A, aa' represents the sliding velocity of A′ relative to A and qa' represents the velocity of A′ relative to Q.

$a'q$ is extended to c such that $a'q:qc:: :A'Q:QC$. cd then represents the velocity of C relative to D and the velocity of D relative to the fixed points is horizontal.

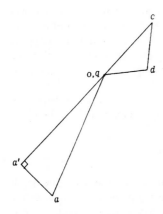

From the diagram, $v_{aa'} = 0.8$ m/s

$v_{a'q} = 2.37$ m/s

and $v_{dc} = 0.9$ m/s

Figure 11.17

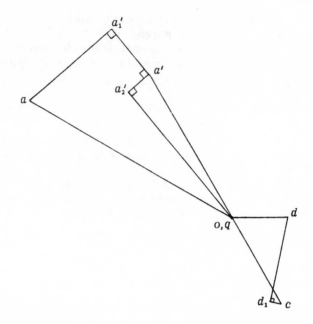

Figure 11.18

In the acceleration diagram, Figure 11.18, the fixed points O and Q are again represented by a single point oq and oa is the absolute acceleration of A. The acceleration of A′ relative to A has two components, a tangential (or Coriolis) component and a sliding component.

The tangential component is given by

$$2v_{a'a}\omega_{qb} = 2 \times v_{a'a} \times \frac{v_{a'q}}{\text{A}'\text{Q}}$$

$$= 2 \times 0.8 \times \frac{2.37}{0.26} = 14.6 \text{ m/s}^2$$

From Section 11.5, the direction of the acceleration of A relative to A′ is obtained by turning the relative velocity vector aa' through 90° in the direction of ω_{qb}, i.e. to the left of QB. The acceleration of A′ relative to A is therefore in the opposite direction, i.e. to the right and this is represented by aa'_1. The sliding acceleration of A′ relative to A is parallel to QB and passes through a'_1 but its magnitude is unknown.

The acceleration of A′ relative to Q also has two components, a centripetal component and a tangential component. The centripetal component is given by

$$\frac{v_{a'q}^2}{\text{A}'\text{Q}} = \frac{2.37^2}{0.26} = 21.6 \text{ m/s}^2$$

Its direction is towards Q and this is represented by qa'_2. The tangential acceleration is perpendicular to QA′ and passes through the point a'_2; the intersection of this line with the sliding acceleration through a'_1 then gives the point a'. $a'q$ is extended to c such that $a'q:qc:$:A′Q:QC. The centripetal acceleration of D relative to C is given by

$$\frac{v_{dc}^2}{DC} = \frac{0.9^2}{0.5} = 1.62 \text{ m/s}^2$$

This is directed towards C and is represented by cd_1. The tangential acceleration of D relative to C is perpendicular to CD and passes through d_1. The acceleration of D relative to the fixed points is horizontal and thus the point d is obtained.

From the diagram, $a_d = od = \underline{7 \text{ m/s}^2}$

4. *Figure 11.19 shows a mechanism in which the crank AB rotates anticlockwise about A at 70 rev/min. The link CD swings about D and is connected to the crankpin B by the link BC. The lengths are: AB, 0.25 m, BC, 1.0 m, CD, 0.75 m. The link BC has a mass of 14 kg, its centre of gravity G is at its mid-point and its radius of gyration about a transverse axis through G is 0.3 m.*

For the position in which AB is at 45° to the horizontal, find the torque required on AB to overcome the inertia of BC.

$$v_b = \omega \cdot AB = 70 \times \frac{2\pi}{60} \times 0.25 = 1.83 \text{ m/s}$$

$$a_b = \omega^2 \cdot AB = \left(70 \times \frac{2\pi}{60}\right)^2 \times 0.25 = 13.4 \text{ m/s}^2$$

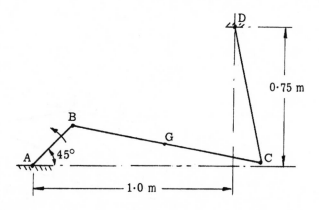

Figure 11.19

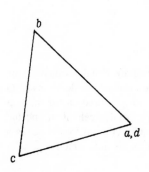

Figure 11.20

In the velocity diagram, Figure 11.20, ab represents the absolute velocity of B and cb and db represent the velocities of B relative to C and D respectively.

In the acceleration diagram, Figure 11.21, ab represents the absolute acceleration of B, bc_1 represents the centripetal acceleration of C relative to B($= v_{cb}^2/CB$) and dc_2 represents the centripetal acceleration of C relative to D($= v_{cd}^2/CD$). The tangential accelerations of C relative to B and D are perpendicular to CB and CD respectively, passing through points c_1 and c_2, and the intersection of these lines gives the point c. The resultant acceleration of C relative to B is then given by bc.

To find the inertia force on BC, its mass is first replaced by an equivalent two-mass system (Section 11.6), with one mass at B and the other at E, where BG × GE = k_G^2, Figure 11.22.

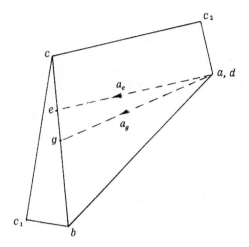

Figure 11.21

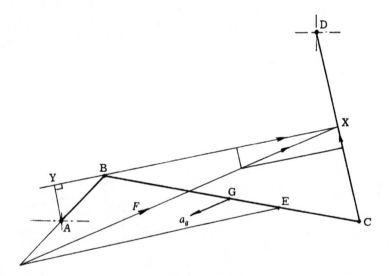

Figure 11.22

Thus
$$GE = \frac{0.3^2}{0.5} = 0.18 \text{ m}$$

The corresponding points in the acceleration diagram are g and e which divide bc in the same proportions as G and E divide BC; the absolute accelerations of G and E are then given by ag and ae respectively. The inertia force on the mass at B is directed along BA since the crank has only centripetal acceleration and the inertia force on the mass at E is parallel to ae. The intersection of these lines is a point on the line of action of the resultant inertia force on the link, which is parallel to ag and opposite in direction.

Inertia force, $\qquad F = ma_g = 14 \times 10.9 = 152.6 \text{ N}$

The forces acting on BC are the inertia force F and the reactions through the joints at B and C. In the absence of friction, the reaction at C is directed along CD and so, from the parallelogram of forces at X,

$$\text{force through B} = 150 \text{ N}$$

$$\text{Therefore torque on A} = 150 \times \text{AY}$$

$$= 150 \times 0.14 = \underline{21 \text{ N m}}$$

Note: It is not necessary to place one of the equivalent masses at B; any positions may be used for m_1 and m_2 provided that the condition $pq = k^2$ is satisfied.

Further problems

5. The end A of a bar AB, Figure 11.23, moves along the vertical path AD and the bar passes through a swivel bearing pivoted at C. When A has a velocity of 1.0 m/s towards D, find the velocity of sliding through the swivel and the angular velocity of the bar. [0.5 m/s; 7.5 rad/s]

6. In the quick return mechanism shown in Figure 11.24, the distance between the fixed centres BC = 90 mm, crank AB = 180 mm, CD = 180 mm and DE = 360 mm with its centre of gravity, G, 90 mm from D. If AB rotates clockwise at 120 rev/min, find the linear velocity of G and the angular velocity of DE. [1.56 m/s; 3.16 rad/s]

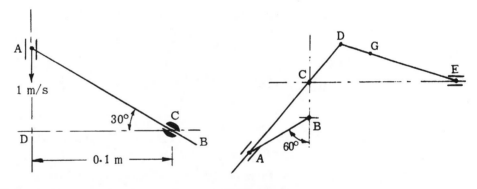

Figure 11.23   **Figure 11.24**

7. Figure 11.25 shows a crank OA, 72 mm long, which rotates anticlockwise about O at 150 rev/min. The bar DBC is pivoted at B which is 150 mm vertically below O. BC = 75 mm. The slider E moves in horizontal guides 30 mm below B and the rod CE is 240 mm long. A horizontal force of 150 N opposes the motion of E.

For the given position, where angle AOB = 120°, find
(*a*) the linear velocity of E and the angular velocity of DBC;
(*b*) the driving torque on the crank OA.   [0.292 m/s; 4.34 rad/s; 2.79 Nm]

8. In the mechanism shown in Figure 11.26, the crank AB is 75 mm long and rotates clockwise at 8 rad/s. BC = 300 mm and BD = DC = DE. Find the velocity and acceleration of pistons C and E. [0.187 5 m/s; 0.6 m/s; 4.16 m/s²; 9.06 m/s²]

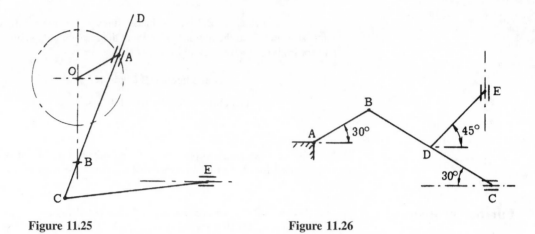

Figure 11.25   Figure 11.26

9. In the mechanism shown in Figure 11.27, the block P reciprocates along the line AB and the crank OC rotates at 240 rev/min. OC = 132 mm, CP = 732 mm, OD = 720 mm and angle COD = 60°. Find the velocity and acceleration of P.

[3.72 m/s; 63.9 m/s²]

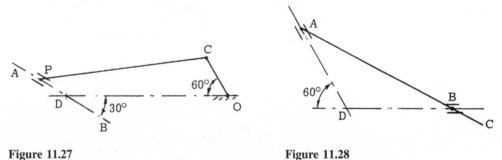

Figure 11.27   Figure 11.28

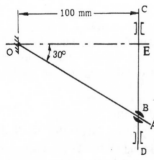

Figure 11.29

10. In the link ABC, Figure 11.28, AB = 600 mm and BC = 225 mm. A and B are attached by pin-joints to the sliding blocks. When BD = 375 mm, A is sliding towards D with a velocity of 6 m/s and a retardation of 150 m/s². Find the acceleration of C and the angular acceleration of AC.   [259 m/s²; 294 rad/s²]

11. The rod OA shown in Figure 11.29 rotates about O and lifts the vertical rod CD by means of the trunnion at B. At the instant when the angle EOA is 30°, OA has an anticlockwise angular velocity of 5 rad/s and zero angular acceleration.

Find the velocity and acceleration of CD at this instant.   [$\frac{2}{3}$ m/s; 2.5 m/s²]

12. In part of a quick-return mechanism shown in Figure 11.30, the crank OA rotates uniformly at 2.5 rad/s. OA = 225 mm and OQ = 300 mm. Determine the angular acceleration of the link QB.   [0.33 rad/s²]

13. In the mechanism shown in Figure 11.31, crank OA rotates at 60 rev/min. The rod EF is pinned to AB at D and slides through a swivel block at Q.

OA = 25 mm, AD = 75 mm, BC = 75 mm, DB = 75 mm and DE = 50 mm.

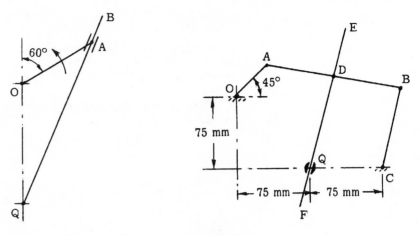

Figure 11.30   **Figure 11.31**

For the position shown, find the velocity and acceleration of E.

[0.208 m/s; 1.06 m/s²]

14. In the mechanism shown in Figure 11.32, the crank AB rotates clockwise at 110 rev./min. AB = 70 mm, CD = 140 mm and BD = 260 mm. BD slides through a swivelling block at E at the lower end of EF and EF slides in vertical guides. For the position shown, find the linear velocity and linear acceleration of F and the angular velocity and angular acceleration of BD.

[0.452 m/s; 2.36 m/s²; 1.395 rad/s; 31.4 rad/s²]

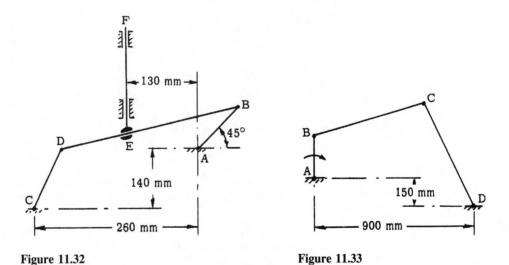

Figure 11.32   **Figure 11.33**

15. In the mechanism shown in Figure 11.33, the crank AB rotates at a uniform speed of 10 rad/s and CD oscillates about the fixed centre D. AB = 225 mm, BC = 600 mm and CD = 600 mm. CD is a uniform thin rod of mass 16 kg. Find the turning moment which must be applied to the crank to accelerate CD for the position where AB is vertical.   [13.8 Nm]

16. Figure 11.34 shows a mechanism in which the crank OP revolves about O at 180 rev/min. The lever AB has a mass of 2.7 kg and its centre of gravity is at the pivot Q. OQ = 125 mm, OP = 50 mm and the radius of gyration of AB about Q is 100 mm.

When the angle POQ is 30°, find the torque on the crankshaft to overcome the inertia of AB.   [3 Nm]

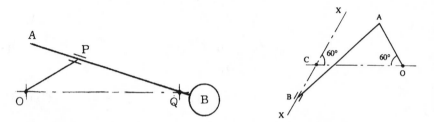

Figure 11.34   **Figure 11.35**

17. In the mechanism shown in Figure 11.35, OA rotates about the fixed centre O and the block B slides along the fixed axis XX. OA = 75 mm, AB = 175 mm and OC = 135 mm.

When the angle AOC is 60°, OA has a clockwise velocity of 7.5 rad/s and an anticlockwise acceleration of 40 rad/s^2. For this position, determine the torque required on OA to overcome the inertia of AB, which is a uniform link of mass 4 kg.   [1.05 Nm]

12 Reciprocating mechanisms

12.1 Introduction

The reciprocating mechanism is of such common application that special methods have been developed to determine velocity, acceleration and inertia forces.

12.2 Piston velocity

The velocity of the crankpin C is perpendicular to OC, Figure 12.1(*a*), and the piston P is constrained to move along OP. Thus, in the velocity diagram, Figure 12.1(*b*), $oc = \omega \cdot OC$, cp is perpendicular to CP and op is parallel to OP.

If PC is produced to intersect the perpendicular to PO at M, then OCM is similar to the velocity diagram ocp.

Therefore
$$\frac{v_p}{v_c} = \frac{op}{oc} = \frac{OM}{OC}$$

$$\therefore v_p = v_c \times \frac{OM}{OC} = \omega \cdot OM \qquad (12.1)$$

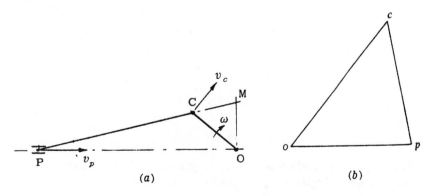

(*a*)   (*b*)

Figure 12.1

12.3 Piston acceleration

Assuming the crank to be rotating at a uniform angular velocity, the acceleration diagram is shown in Figure 12.2. oc is the centripetal acceleration of $C(= \omega^2 \cdot OC)$ and cp_1 is the centripetal acceleration of P relative to $C(= v_{pc}^2/PC)$. $p_1 p$ is the tangential acceleration of P relative to C and op is the acceleration of P relative to O.

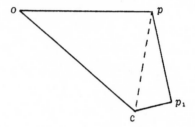

Figure 12.2

12.4 Klein's construction

Draw a circle with diameter PC, Figure 12.3. Extend PC to cut the perpendicular through O at M and draw a circle of centre C and radius CM. Join H to K, H and K being the intersection of the two circles, and let HK intersect PC at L and PO at N.

The quadrilateral OCLN is then similar to the acceleration diagram $ocp_1 p$, Figure 12.2. Since $a_c = \omega^2 OC$,

then
$$a_p = \omega^2 ON \qquad (12.2)$$

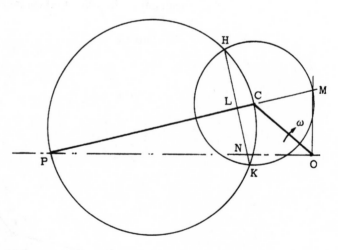

Figure 12.3

12.5 Analytical method for piston velocity and acceleration

The displacement of the piston P, Figure 12.4, from the inner dead centre position is given by

$$x = (l + r) - (r \cos \theta + l \cos \phi)$$

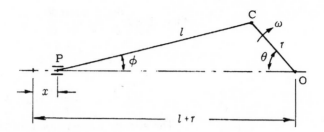

Figure 12.4

But $\quad l\sin\phi = r\sin\theta$

$\therefore\ \sin\phi = \dfrac{r}{l}\sin\theta$

$\qquad\ = \dfrac{\sin\theta}{n}\quad$ where $\quad n = \dfrac{l}{r}$

$\therefore\ \cos\phi = \sqrt{(1-\sin^2\phi)} = \sqrt{\left(1-\left(\dfrac{\sin\theta}{n}\right)^2\right)}$

$\qquad\ \approx 1 - \dfrac{\sin^2\theta}{2n^2}\quad$ since $\dfrac{1}{n}$ is small and higher powers may be neglected

Thus $\qquad x = (l+r) - \left(r\cos\theta + l\left[1 - \dfrac{\sin^2\theta}{2n^2}\right]\right)$

$\qquad\qquad = r(1-\cos\theta) + \dfrac{l\sin^2\theta}{2n^2}$

$\qquad\qquad = r\left(1-\cos\theta + \dfrac{\sin^2\theta}{2n}\right)$

$\therefore\ v_p = \dfrac{dx}{dt} = r\left(\sin\theta + \dfrac{\sin 2\theta}{2n}\right)\dfrac{d\theta}{dt} = \omega r\left(\sin\theta + \dfrac{\sin 2\theta}{2n}\right)\quad(12.3)$

and

$\qquad\qquad a_p = \dfrac{d^2x}{dt^2} = \omega r\left(\cos\theta + \dfrac{\cos 2\theta}{n}\right)\dfrac{d\theta}{dt} = \omega^2 r\left(\cos\theta + \dfrac{\cos 2\theta}{n}\right)$

$$(12.4)$$

The angular velocity Ω of the connecting rod is given by

$$\Omega = \dfrac{d\phi}{dt}$$

Since $\qquad \sin\phi = \dfrac{\sin\theta}{n}$

$$\text{then} \quad \cos\phi \frac{d\phi}{dt} = \frac{\cos\theta}{n}\frac{d\theta}{dt}$$

$$\therefore \Omega = \frac{\cos\theta}{n\cos\phi}\omega$$

$$\approx \frac{\omega\cos\theta}{n} \quad \text{since } \phi \text{ is small for usual values of } n \qquad (12.5)$$

The angular acceleration α of the connecting rod is given by

$$\alpha = \frac{d\Omega}{dt} = -\frac{\omega^2\sin\theta}{n} \qquad (12.6)$$

12.6 Crankshaft torque due to piston force and mass

Let the force acting on the piston in the direction of motion be F_p, Figure 12.5, and the force on the crankpin perpendicular to the crank be F_c. Then, neglecting friction and the change in kinetic energy of the connecting rod,

$$\text{power input} = \text{power output}$$

i.e. $$F_p v_p = F_c v_c$$

$$\therefore F_c = F_p \times \frac{v_p}{v_c} = F_p \times \frac{OM}{OC}$$

$$\therefore \text{crankshaft torque} = F_c \times OC = F_p \times OM \qquad (12.7)$$

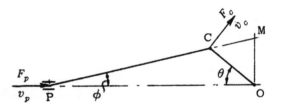

Figure 12.5

If the piston is subjected to a gas pressure p acting on an area A, the force on the piston is pA.

If the mass of the piston is m, the force necessary to accelerate the piston is ma_p, so that the net piston force is given by

$$F_p = pA - ma_p.$$

$$= pA - m\omega^2 r\left(\cos\theta + \frac{\cos 2\theta}{n}\right)$$

Hence crankshaft torque, $$T = \left\{pA - m\omega^2 r\left(\cos\theta + \frac{\cos 2\theta}{n}\right)\right\} \times OM$$

$$(12.8)$$

Equation (12.8) is applicable for all values of θ since $\cos\theta + \dfrac{\cos 2\theta}{n}$ becomes negative when θ exceeds about 77°.

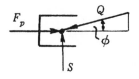

Figure 12.6

The side thrust S on the cylinder and the force in the connecting rod Q may be obtained from the equilibrium of the forces on the small end pin, Figure 12.6.

Thus $\qquad\qquad\qquad\qquad\qquad S = F_p \tan \phi \qquad\qquad$ (12.9)

and $\qquad\qquad\qquad\qquad\qquad Q = F_p \sec \phi \qquad\qquad$ (12.10)

12.7 Crankshaft torque due to connecting rod inertia

The inertia force on a connecting rod may be obtained by use of the two-mass system described in Section 11.6 but a separate acceleration diagram may be avoided by using that provided by Klein's construction, Section 12.4.

This force F, Figure 12.7, has components R and S acting through the crank-pin and piston; the component R, multiplied by the perpendicular distance h from O then gives the inertia torque on the crankshaft.

This process is shown in detail in Example 3.

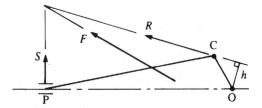

Figure 12.7

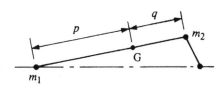

Figure 12.8

However, as an approximation, the connecting rod may be replaced by two masses, one at the crank pin and one at the piston, Figure 12.8, such that $m_1 + m_2 = m$ and $m_1 p = m_2 q$.

The mass m_2, rotating at constant speed, has no inertia effects and the mass m_1 may be added to that of the piston. This will only be an approximation since condition (c) in Section 11.6, i.e. that

$$m_1 p^2 + m_2 q^2 = k_G^2$$

will not be fulfilled.

Worked examples

1. *A horizontal steam engine has a stroke of 600 mm and a cylinder diameter of 225 mm. The piston rod diameter is 50 mm and the length of the connecting rod is 1 200 mm. The reciprocating parts have a mass of 100 kg. When the crank has rotated 60° from the inner dead centre, the steam pressure on the outside of the piston is 560 kN/m² and on the inside is 160 kN/m².*

For this position, calculate the thrust in the connecting road, the thrust on the cross-head guide, the turning moment on the crankshaft and the radial force in the crank when the crankshaft rotates at 240 rev/min.

Crank radius, $r = 0.3$ m $\quad$ and $\quad n = \dfrac{l}{r} = \dfrac{1.2}{0.3} = 4$

Net steam force on piston

$$= 560 \times 10^3 \times \frac{\pi}{4} \times 0.225^2 - 160 \times 10^3 \times \frac{\pi}{4}(0.225^2 - 0.05^2)$$

$$= 16\,200 \text{ N}$$

Inertia force on piston $= m\omega^2 r \left(\cos\theta + \frac{\cos 2\theta}{n} \right)$ from equation (12.4)

$$= 100 \left(240 \times \frac{2\pi}{60} \right)^2 \times 0.3 \left(\cos 60° + \frac{\cos 120°}{4} \right)$$

$$= 7\,100 \text{ N}$$

$\therefore$ net piston force, $F_p = 16\,200 - 7\,100 = 9\,100 \text{ N}$

Referring to Figure 12.5, $1.2 \sin\phi = 0.3 \sin 60°$

$$\therefore \phi = 12°30'$$

$$\frac{\text{OM}}{\sin(\theta + \phi)} = \frac{\text{OC}}{\sin(90° - \phi)}$$

Hence $\qquad \text{OM} = 0.3 \dfrac{\sin 72°30'}{\sin 77°30'} = 0.293 \text{ m}$

Thus thrust in connecting rod, $Q = F_p \sec\phi$ from equation (12.10)

$$= 9\,100 \sec 12°30' = \underline{9\,320 \text{ N}}$$

Side thrust, $S = F_p \tan\phi$ from equation (12.9)

$$= 9\,100 \tan 12°30' = \underline{2\,015 \text{ N}}$$

Turning moment, $T = F_p \times \text{OM}$ from equation (12.7)

$$= 9\,100 \times 0.293 = \underline{2\,670 \text{ Nm}}$$

Referring to Figure 12.9,

$$\alpha = 60° + 12°30'$$

$$= 72°30'$$

$\therefore$ radial force in crank

$$= Q \cos\alpha$$

$$= 9\,320 \cos 72°30'$$

$$= \underline{2\,800 \text{ N}}$$

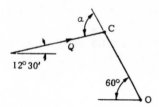

Figure 12.9

2. *A marine engine has a stroke of 200 mm and a connecting rod of length 400 mm, its centre of gravity being at 175 mm from the crank-pin centre and radius of gyration about the centre of gravity being 125 mm. The connecting rod has a mass of 120 kg and the reciprocating mass is 90 kg. The crank rotates at 240 rev/min.*

Determine (a) the crankshaft torque due to the inertia of the reciprocating mass, and (b) the kinetic energy of the connecting rod for a crank angle of 45°.

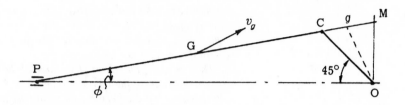

Figure 12.10

By drawing or calculation, as in Example 1,

$$\phi = 10°11' \quad \text{and} \quad OM = 83.3 \text{ mm, Figure 12.10}$$

$$r = 0.1\text{m} \quad \text{and} \quad n = \frac{400}{100} = 4$$

From equation (12.4), $F_p = m\omega^2 r \left(\cos\theta + \frac{\cos 2\theta}{n} \right)$

$$= 90 \times \left(240 \times \frac{2\pi}{60} \right)^2 \times 0.1 \left(\cos 45° + \frac{\cos 90°}{4} \right)$$

$$= 4\,020 \text{ N}$$

$\therefore$ crankshaft torque, $\quad T = F_p \times OM \quad$ from equation (12.7)

$$= 4\,020 \times 0.083\,3$$

$$= \underline{335 \text{ Nm}}$$

Triangle OCM represents the velocity diagram for the mechanism, OC representing the velocity of C relative to O, OM the velocity of P relative to O and CM the velocity of P relative to C.

Thus the velocity of G is represented by Og, where

$$\frac{Cg}{CM} = \frac{CG}{CP}$$

Since $v_c = \omega \cdot OC$, then $\quad v_g = \omega \cdot Og$

$$= 240 \times \frac{2\pi}{60} \times 0.085 = 2.14 \text{ m/s}$$

The direction of v_g is perpendicular to Og.

From equation (12.5), $\quad \Omega = \frac{\omega \cos\theta}{n}$

$$= 240 \times \frac{2\pi}{60} \times \frac{\cos 45°}{4} = 4.44 \text{ rad/s}$$

$\therefore$ total kinetic energy $= \frac{1}{2}mv_g^2 + \frac{1}{2}I\Omega^2$

$$= \frac{1}{2} \times 120 \times 2.14^2 + \frac{1}{2} \times 120 \times 0.125^2 \times 4.44^2$$

$$= \underline{293.5 \text{ J}}$$

3. *An engine of 120 mm stroke has a connecting rod 260 mm long between centres and of mass 1.25 kg. The centre of gravity is 80 mm from the big end centre and when suspended as a pendulum from the gudgeon pin axis, the rod makes 21 complete oscillations in 20 s.*

Determine (a) the radius of gyration of the rod about an axis through the centre of gravity, and (b) the inertia torque exerted on the crankshaft when the crank is 40° from the top dead centre and is rotating at 1 500 rev/min.

The periodic time of a compound pendulum is given by

$$T = 2\pi\sqrt{\left(\frac{k^2 + h^2}{gh}\right)} \quad \text{from equation (18.20)}$$

where k is the radius of gyration about an axis through the c.g. and h is the distance of the c.g. from the point of suspension.

$$h = 260 - 80 = 180 \text{ mm}$$

so that

$$\frac{20}{21} = 2\pi\sqrt{\left(\frac{k^2 + 0.18^2}{9.81 \times 0.18}\right)}$$

from which

$$k^2 = 0.008\,1 \text{ m}^2$$

and

$$k = 0.09 \text{ m} \quad \text{or} \quad \underline{90 \text{ mm}}$$

Figure 12.11 shows the required configuration of the engine. Replacing the connecting rod mass by an equivalent two-mass system with the masses placed at P and X, then

$$PG \times GX = k^2 \quad \text{from equation (11.7)}$$

i.e.

$$GX = \frac{0.008\,1}{0.18} = 0.045 \text{ m} \quad \text{or} \quad 45 \text{ mm}$$

Klein's construction is superimposed on the mechanism, giving the acceleration diagram OCLN. CN represents the acceleration of P relative to C and Og and Ox then give the directions of the accelerations of G and X respectively. The inertia

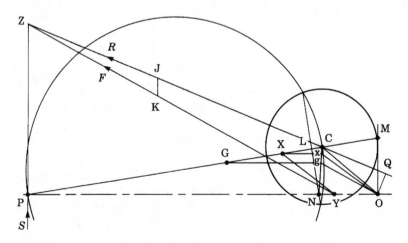

Figure 12.11

force on the mass at X is parallel to Ox and the inertia force on the mass at P acts along PO; these intersect at the point Y.

The resultant inertia force on the connecting rod passes through Y and is parallel to Og; its magnitude is given by

$$F = m\omega^2 \mathrm{O}g$$

$$= 1.25 \times \left(1\,500 \times \frac{2\pi}{60}\right)^2 \times 0.057 = 1\,760 \text{ N}$$

The three forces acting on the rod are the inertia force, F, the side thrust at the piston, S, and the force at the crank pin, R. Forces F and S intersect at Z and so R must also pass through this point. Drawing the triangle of forces at Z, KZ represents F (= 1 760 N) and JZ represents R. This is found to be 1 620 N and the perpendicular distance of the line of action of R from O (OQ) is found to be 15 mm, so that the crankshaft torque,

$$T = R \times \mathrm{OQ}$$

$$= 1\,620 \times 0.015 = \underline{24.3 \text{ Nm}}$$

If, as an approximation, the connecting rod mass is replaced by masses at P and C, that at P is, from Section 12.7, $1.25 \times \dfrac{80}{260} = 0.385$ kg.

The inertia force on this mass is given by

$$F = m\omega^2 r \left(\cos\theta + \frac{\cos 2\theta}{n}\right) \quad \text{from equation (12.4)}$$

$$= 0.385 \times \left(1\,500 \times \frac{2\pi}{60}\right)^2 \times 0.60 \left(\cos 40° + \frac{\cos 80°}{260/60}\right)$$

$$= 459.4 \text{ N}$$

For this configuration,   OM $= 45.6$ mm

Hence inertia torque $= 459.4 \times 0.045\,6$

$$= \underline{20.9 \text{ Nm}}$$

Further problems

4. A vertical single-cylinder engine has a cylinder diameter of 250 mm and a stroke of 450 mm. The reciprocating parts have a mass of 180 kg, the ratio connecting rod/crank radius is 4 and the speed is 360 rev/min. When the crank has turned through an angle of 45° from t.d.c. the net pressure on the piston is 1.05 MN/m². Calculate the crankshaft torque for this position.   [2 390 Nm]

5. A vertical steam engine, 450 mm bore and 750 mm stroke, runs at 240 rev/min. The reciprocating parts of the engine have a mass of 70 kg and the connecting rod is 1.2 m long. When the piston is moving downwards and the crank is 90° beyond t.d.c., the steam pressure above the piston is 800 kN/m² and that below the piston is 120 kN/m².

Determine the instantaneous torque on the crankshaft, neglecting the piston rod area.   [41.5 kNm]

6. An engine mechanism has a 150 mm crank radius and a 375 mm connecting rod with a piston mass of 10 kg. The crank rotates at 300 rev/min. Determine the acceleration of the piston and the crankshaft torque due to piston inertia when the crank is 45° from the o.d.c. position.   [103 m/s^2; 78.5 Nm]

7. A horizontal reciprocating engine has a stroke of 200 mm and a connecting rod length of 400 mm. The effective mass of the reciprocating parts is 40 kg and the crankshaft rotates at 300 rev/min. The cylinder diameter is 120 mm and when the crank makes an angle of 60° with the inner dead centre position, the pressure in the cylinder is 2 bar.

Determine the torque on the crankshaft for this position.   [76.25 Nm]

8. The crankshaft of a vertical single-cylinder engine, stroke 250 mm, rotates at 300 rev/min. The reciprocating parts have a mass of 100 kg. The connecting rod has a mass of 120 kg, it is 450 mm long, the c.g. is 300 mm from the gudgeon-pin axis and the radius of gyration about that axis is 363 mm.

When the crank is 30° from the t.d.c. position and moving downwards, find (a) the reaction at the cylinder walls due to the inertia of the reciprocating parts; (b) the total kinetic energy of the connecting rod.   [1.74 kN; 739 J]

9. A petrol engine of cylinder diameter 100 mm and stroke 120 mm has a piston of mass 1.1 kg and a connecting rod of length 250 mm. When rotating at 2 000 rev/min, the gas pressure is 700 kN/m^2 when the crank is at 20° from the t.d.c. position.

Find (a) the resultant load on the gudgeon pin, (b) the thrust on the cylinder wall.

Determine also the speed at which the gudgeon pin load would be reversed in direction, the other conditions remaining constant.

[2 263 N; 186 N; 2 603 rev/min]

10. A single-cylinder engine has a cylinder diameter of 250 mm, a stroke of 450 mm and runs at 360 rev/min. The reciprocating parts have a mass of 100 kg. The connecting rod is 900 mm long and has a mass of 150 kg. Its centre of gravity is 300 mm from the crank-pin axis and its inertia may be allowed for by apportioning its mass between the two ends.

When the crank has turned through an angle of 40° from the inner dead centre position, the pressure in the cylinder is 1.2 MN/m^2. Calculate the torque on the crankshaft for this position.   [3.653 kNm]

11. The connecting rod of an engine is 0.9 m long between centres, its mass is 25 kg and its centre of gravity is 0.3 m from the crank-pin. The radius of gyration about an axis through the c.g. is 0.375 m and the crank rotates at 300 rev/min. The crank radius is 0.3 m.

Determine the crankshaft torque due to the inertia of the connecting rod when the crank makes an angle of 45° with the t.d.c. position.   [520 Nm]

12. A connecting rod has a mass of 1.125 kg and the distance between the centres of the end bearings is 250 mm. The c.g. is 162.5 mm from the centre of the small end bearing and the moment of inertia about a transverse axis through the c.g. is 0.011 8 kg m^2. The crank radius is 62.5 mm and the speed is 200 rad/s. For a crank angle of 30° past the inner dead centre position, determine (a) the torque required at the crankshaft to accelerate the rod, assuming the rod to be equivalent to two mass particles, one at each end of the rod, (b) the percentage error in this assumption.

[37.1 Nm; 11.8%]

13 Turning moment diagrams

13.1 Introduction

The output torque from a reciprocating engine varies considerably over the working cycle and if the engine is driving a generator or machine which offers a constant resisting torque, the resulting speed will vary because the engine torque is at times greater than or less than the resisting torque. In order to reduce this fluctuation of speed, a flywheel is fitted to the engine to absorb energy at some points in the cycle and release it at others. The inertia of the flywheel required depends on the fluctuation of the energy available from the engine and the fluctuation of speed which is acceptable.

13.2 Crank effort diagrams

If the output torque from the engine is plotted against crank angle, a turning moment or crank effort diagram is obtained. The net area under the graph represents the work done during the cycle and the average height represents the mean torque, which is equal to the resisting torque if the mean speed is to remain constant.

A typical crank effort diagram for a four stroke single cylinder engine is shown in Figure 13.1. At points where the curve cuts the mean torque line,

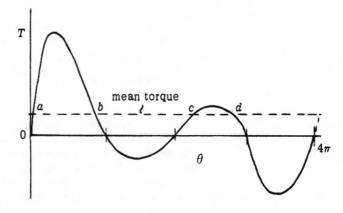

Figure 13.1

the engine speed is constant. Between points a and b, the engine torque is greater than the resisting torque and so the engine speeds up; similarly, between points b and c, the engine torque is less than the resisting torque and so the engine slows down.

A similar situation arises in the case of an electric motor driving a machine which offers a variable resistance. The turning moment then consists of a uniform input torque and a variable resisting torque but the analysis is identical with that for an engine application.

By inspection of the areas of the loops above and below the mean torque line, the points of maximum and minimum speed in the engine cycle may be determined (see Example 1). The difference of energy supplied and energy required between these points is then responsible for the change of speed; this is termed the *fluctuation of energy, U.*

If the moment of inertia of the rotating parts of the system is I and ω_1 and ω_2 are respectively the maximum and minimum speeds during the cycle, the change of kinetic energy of the system is $\frac{1}{2}I\omega_1^2 - \frac{1}{2}I\omega_2^2$, which is brought about by the fluctuation of energy supplied by the engine,

i.e.
$$U = \tfrac{1}{2}I\omega_1^2 - \tfrac{1}{2}I\omega_2^2 \qquad (13.1)$$

The fluctuation of speed, $\omega_1 - \omega_2$, is small in comparison with the mean speed ω and assuming that the variations above and below the mean speed are equal,

$$\omega_1 + \omega_2 = 2\omega$$

Thus equation (13.1) may be written

$$U = \tfrac{1}{2}I(\omega_1 + \omega_2)(\omega_1 - \omega_2)$$
$$= I\omega(\omega_1 - \omega_2)$$
$$\text{or} \quad I\omega^2 \times \frac{\omega_1 - \omega_2}{\omega} \qquad (13.2)$$

The term $\dfrac{\omega_1 - \omega_2}{\omega}$ is called the *coefficient of fluctuation of speed* and the ratio $\dfrac{\text{fluctuation of energy}}{\text{work done per cycle}}$ is called the *coefficient of fluctuation of energy.*

Worked examples

1. *The turning moment diagram for an engine is drawn on a base of crank angle and the mean resisting torque line added. The areas above and below the mean line are $+4400, -1150, +1300, -4550$ mm^2, the scales being 1 mm = 100 Nm torque and 1 mm = 1° of crank angle.*

Find the mass of the flywheel required to keep the speed between 297 and 303 rev/min if its radius of gyration is 0.525 m.

The turning moment diagram is shown in Figure 13.2.

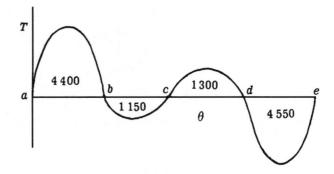

Figure 13.2

1 mm on the torque scale represents 100 Nm

1 mm on the crank angle scale represents $1 \times \dfrac{\pi}{180}$ rad

$\therefore$ 1 mm^2 of area represents $100 \times \dfrac{\pi}{180} = 1.745$ J

Commencing at point a, the engine speeds up to point b, slows down to point c, speeds up to point d and slows down to point e, when the cycle is repeated. Between b and c, less energy is abstracted than is added between a and b and so the engine is running faster at c than at a. It is running faster still at d but then a substantial amount of energy is abstracted between d and e, which restores the speed to that at a. Hence the minimum speed occurs at a and the maximum at d.

Thus the fluctuation of energy between points of minimum and maximum speeds is represented by *either* $4\,400 - 1\,150 + 1\,300$ mm^2 *or* $4\,550$ mm^2, these being equal.

Hence fluctuation of energy $= 4\,550 \times 1.745 = 7\,940$ J

Therefore, from equation (13.1), $7\,940 = \frac{1}{2}I(\omega_1^2 - \omega_2^2)$

$$= \tfrac{1}{2}I(303^2 - 297^2) \times \left(\frac{2\pi}{60}\right)^2$$

$$\therefore \quad I = 402 \text{ kg m}^2$$

$$\therefore \quad 402 = m \times 0.525^2$$

$$\therefore \quad m = \underline{1\,459 \text{ kg}}$$

2. *A motor driving a punching machine exerts a constant torque of 675 Nm on the flywheel, which rotates at an average speed of 120 rev/min. The punch operates 60 times per minute, the duration of the punching operation being $\frac{1}{5}$ s. It may be assumed that during the punching operation, the resisting torque on the flywheel is constant.*

Deduce the value of the resisting torque and find the moment of inertia of the flywheel if the speed variation between maximum and minimum is not to exceed 10 rev/min.

The turning moment diagram is shown in Figure 13.3.

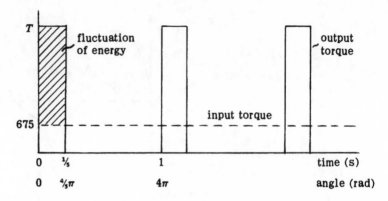

Figure 13.3

Equating the areas under the input and output torques over a period of 1 s,

$$T \times \tfrac{1}{5} = 675 \times 1$$

$$\therefore \quad T = 3\,375 \text{ Nm}$$

Fluctuation of energy $= (3\,375 - 675) \times \dfrac{4\pi}{5}$

$$= 6\,786 \text{ J}$$

$$\therefore \quad 6\,786 = \tfrac{1}{2}I(\omega_1^2 - \omega_2^2) \quad \text{from equation (13.1)}$$

$$= \tfrac{1}{2}I(N_1^2 - N_2^2)\left(\frac{2\pi}{60}\right)^2 \quad \text{where } N_1 \text{ and } N_2 \text{ are the speeds in rev/min}$$

$$= \tfrac{1}{2}I(N_1 + N_2)(N_1 - N_2)\left(\frac{2\pi}{60}\right)^2$$

$$= \tfrac{1}{2}I \times 240 \times 10 \times \frac{\pi^2}{900}$$

$$\therefore \quad I = \underline{516 \text{ kg m}^2}$$

3. *A single cylinder gas engine, working on the four-stroke cycle, develops 11 kW at 300 rev/min. The work done on the gas during the compression stroke is 0.7 times the work done by the gases during the power stroke. The turning moment diagram for the compression stroke may be taken as an isosceles triangle and that for the power stroke as another isosceles triangle. The turning moment during the suction and exhaust strokes is negligible.*

If the mass of the flywheel is 1 500 kg and its radius of gyration is 400 mm, find the coefficient of fluctuation of speed.

The turning moment diagram is shown in Figure 13.4

$$\text{Power} = \text{mean torque} \times \text{speed}$$

$$\therefore \quad \text{mean torque} = \frac{11\,000}{300 \times \dfrac{2\pi}{60}} = 350 \text{ Nm}$$

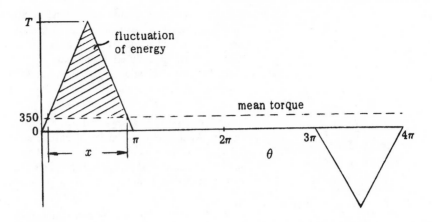

Figure 13.4

Equating areas under actual and mean torque lines,

$$\tfrac{1}{2}T\pi - 0.7 \times \tfrac{1}{2}T\pi = 350 \times 4\pi$$

$$\therefore T = 9\,333 \text{ Nm}$$

From similar triangles, $\dfrac{x}{\pi} = \dfrac{9\,333 - 350}{9\,333} = \dfrac{8\,983}{9\,333}$

$$\therefore x = 3.025 \text{ rad}$$

Therefore, fluctuation of energy $= \tfrac{1}{2} \times 8\,983 \times 3.025$

$$= 13\,582 \text{ J}$$

$$\therefore \quad 13\,582 = I\omega^2 \times \frac{\omega_1 - \omega_2}{\omega} \quad \text{from equation(13.2)}$$

$$= 1\,500 \times 0.4^2 \times \left(300 \times \frac{2\pi}{60}\right)^2 \times \frac{\omega_1 - \omega_2}{\omega}$$

$$\therefore \quad \frac{\omega_1 - \omega_2}{\omega} = 0.0\,573 \quad \text{or} \quad \underline{5.73\%}$$

Further problems

4. The turning-moment diagram for a petrol engine is drawn to the following scales: turning moment 1 mm = 5 Nm; crank angle 1 mm = 1°. The turning-moment diagram repeats itself at every half-revolution of the engine and the areas above and below the mean turning-moment line, taken in order, are 295, 685, 40, 340, 960, 270 mm². The rotating parts are equivalent to a mass of 36 kg at a radius of gyration of 150 mm. Determine the coefficient of fluctuation of speed when the engine runs at 1 800 rev/min.   [0.299%]

5. Distinguish between the functions of the governor and the flywheel of an engine.

A double-acting steam engine runs at 100 rev/min. A curve of the turning moment plotted on a crank angle base showed the following areas alternately above and below the mean turning-moment line: 780, 400, 520, 620, 260, 460, 340, 420 mm². The scales used were 1 mm = 400 Nm and 1 mm = 1° crank angle.

If the total fluctuation in speed is limited to $1\frac{1}{2}$ per cent of the mean speed, determine the mass of the flywheel necessary if the radius of gyration is 1.05 m.

[3 464 kg]

6. A machine press is worked by an electric motor, delivering 2.25 kW continuously. At the commencement of an operation, a flywheel of moment of inertia 50 kg m² on the machine is rotating at 250 rev/min. The pressing operation requires 4.75 kJ of energy and occupies 0.75 s. Find the maximum number of pressings that can be made in 1 h and the reduction in speed of the flywheel after each pressing.

[1 705; 23.5 rev/min]

7. The torque required by a machine rises at a uniform rate from zero to a maximum of 180 Nm and then falls to zero at a uniform rate over one half of a revolution. The remainder of the revolution requires zero torque.

If the torque supplied by the driving motor is constant, find

   (*a*) the power required at a mean speed of 240 rev/min;
   (*b*) the moment of inertia required to keep the machine speed within ±2 rev/min from the mean speed.   [1 130 W; 15.1 kg m²]

8. A single-cylinder four-stroke internal combustion engine develops 30 kW at 300 rev/min. The turning-moment diagram for the expansion and compression strokes may be taken as two isosceles triangles, on bases 0 to π and 3π to 4π radians respectively, and the net work done during the exhaust and inlet strokes is zero. The work done during compression is negative and is one quarter of that during expansion.

Sketch the turning moment diagram for one cycle and find the maximum value of the turning moment during expansion.

If the load remains constant, mark on the diagram the points of maximum and minimum speeds. Also find the moment of inertia, in kg m², of a flywheel to keep the speed fluctuation within ±1.5 per cent of the mean speed.

[10.18 kNm; 444 kg m²]

9. A gas engine develops 22.5 kW at 270 rev/min. It has hit-and-miss governing and there are 125 explosions per minute. The flywheel has a mass of 900 kg and a radius of gyration of 0.675 m. If it is assumed that the work done is identical for each working cycle, that the work done by the gases on the explosion stroke is 2.4 times the work done on the gases during compression stroke, and that the work done on the other two strokes is negligible, find the maximum fluctuation of speed of the flywheel as a percentage of the mean speed.   [4.82 per cent]

10. A shaft fitted with a flywheel rotates at 250 rev/min and drives a machine the resisting torque of which varies in a cyclic manner over a period of three revolutions. The torque rises from 675 Nm to 2 700 Nm in a uniform manner during $\frac{1}{2}$ revolution and remains constant for the following 1 revolution. It then falls uniformly to 675 Nm during the next $\frac{1}{2}$ revolution and remains constant for 1 revolution, the cycle being then repeated.

If the driving torque applied to the shaft is constant and the flywheel has a mass of 450 kg and a radius of gyration of 0.6 m, find the power necessary to drive the machine and the percentage fluctuation of speed.   [44.2 kW; ± 3.58 per cent]

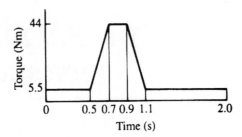

Figure 13.5

11. Figure 13.5 shows the variation with time of the torque required on the driving shaft of a machine during one cycle of operations. The shaft is direct-coupled to an electric motor which exerts a constant torque and runs at a mean speed of 1 500 rev/min. The rotating parts are equivalent to a flywheel of mass 18 kg with a radius of gyration of 250 mm.

Determine: (*a*) the power of the motor, neglecting friction, and (*b*) the percentage fluctuation of speed.   [2.075 kW; 6.275%]

14 Balancing of rotating masses

14.1 Static balance

A shaft is in static balance if the centre of gravity of the masses or rotors which it carries lies on the axis of the shaft; it will then remain in any angular position in which it is placed. If it is not in static balance, it will rotate until the centre of gravity is in its lowest position.

Figure 14.1 shows the end view of a shaft carrying masses m_1, m_2 and m_3 at radii r_1, r_2 and r_3 and angular positions θ_1, θ_2 and θ_3 respectively. For static equilibrium, the moment about O of the gravitational forces on the masses must be zero, i.e., $\Sigma mgr \cos \theta = 0$ and $\Sigma mgr \sin \theta = 0$.

This is satisfied if a vector diagram representing the vectors $m_1 r_1$, $m_2 r_2$ and $m_3 r_3$ is a closed figure, which may be expressed by the vector equation $\Sigma mr = 0$. If the polygon does not close, a mass m is required at a radius r such that the product mr gives the closing vector.

The longitudinal position of the masses is of no consequence.

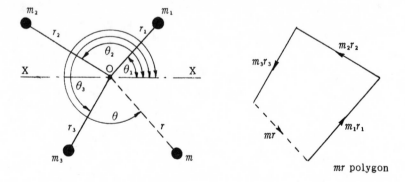

Figure 14.1

14.2 Dynamic balance of masses rotating in the same plane

If all the masses rotate in the same plane, the centrifugal forces on the shaft are balanced if $\Sigma m\omega^2 r = 0$. Since the angular velocity of the shaft is common to all the masses, this reduces to $\Sigma mr = 0$, which is the same condition for static balance.

14.3 Dynamic balance of masses rotating in different planes: Dalby's method

If the masses rotate in different planes, then not only must the centrifugal forces be balanced but the moments of these forces about any plane of revolution must be balanced.

Figure 14.2 shows a shaft carrying masses m_1, m_2 and m_3 at radii r_1, r_2 and r_3, disposed at distances l_1, l_2 and l_3 from a reference plane. The forces $m_1\omega^2 r_1$, $m_2\omega^2 r_2$ and $m_3\omega^2 r_3$ may be transferred to the reference plane by adding couples of magnitude force × distance moved, i.e., $m_1\omega^2 r_1 l_1$, $m_2\omega^2 r_2 l_2$, etc., acting in the planes containing the forces and the shaft axis.

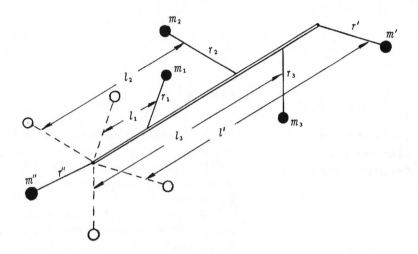

Figure 14.2

These couples are first balanced by drawing a vector polygon representing the couples $m_1 r_1 l_1$, $m_2 r_2 l_2$ and $m_3 r_3 l_3$, drawn in the directions of the respective forces (the term ω^2 is common to all the forces and couples and may therefore be omitted from the calculations). If the polygon does not close, balance is achieved by placing a balance mass m' at a radius r' situated at a distance l' from the reference plane such that $m'r'l'$ is equal to the closing vector, this vector also giving the angular position of the mass.

The mass m' is also transferred to the reference plane; the forces in this plane are then balanced by drawing a vector diagram representing the forces $m_1 r_1$, $m_2 r_2$, $m_3 r_3$ and $m'r'$. If the polygon does not close, balance is achieved by placing a mass m'' at radius r'' such that $m''r''$ is equal to the closing vector, this again giving the angular position of the balance mass.

The couple and force polygons for the shaft are shown in Figure 14.3.

Two balance masses in separate planes of rotation may therefore be required for dynamic balance. The reference plane is normally taken to coincide with the plane of revolution of one of the unknown masses to eliminate the couple due to this mass. If this plane has masses on either side of it, the vectors on one side must be considered positive and those on the other side negative, these being reversed in direction.

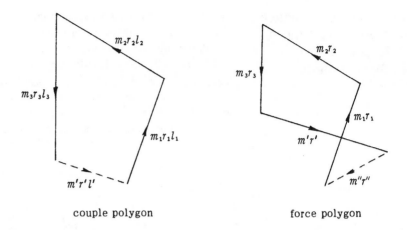

Figure 14.3

14.4 Forces on bearings due to dynamic loading

The forces acting on bearings due to lack of dynamic balance are obtained by finding the *mr* products necessary in the planes of the bearings to balance the shaft. These products, multiplied by ω^2, then give the dynamic loads on the bearings, these being opposite in direction to the forces necessary for balance.

If the shaft is in static balance, it is subjected only to an unbalanced couple. The forces on the bearings are then equal in magnitude and opposite in direction.

Worked examples

1. *A two-cylinder engine has a stroke of 0.5 m. The cranks are set at right-angles and the mass which may be considered to be rotating at each crank pin is 100 kg. The planes of rotation of the cranks are 0.6 m apart and they are symmetrically placed with respect to two flywheels which are 1.5 m apart. Find the magnitude and angular positions of balance masses required in the flywheels at a radius of 0.9 m.*

The arrangement of the planes of rotation of the masses is shown in Figure 14.4. The crank radius is half the stroke, i.e., 0.25 m, and due to the symmetry of the

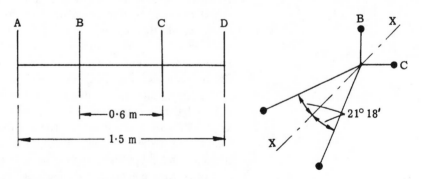

Figure 14.4

system, the balance masses are equal in magnitude and are symmetrically disposed with regard to the axis XX.

Taking the plane A as the reference plane to eliminate the couple due to the force in that plane, the following table is compiled.

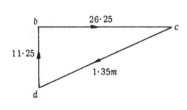

Figure 14.5

Plane	m(kg)	r(m)	mr	l(m)	mrl
A	m	0.9	$0.9m$	0	0
B	100	0.25	25	0.45	11.25
C	100	0.25	25	1.05	26.25
D	m	0.9	$0.9m$	1.5	$1.35m$

Using the data from the mrl column, the couple polygon is drawn, Figure 14.5. From the closing side, $1.35m = 28.52$ kg m^2. Hence $m = 21.1$ kg at an angle of $23°12'$ to the horizontal.

The positions of the balance masses relative to the cranks are as shown in Figure 14.4.

2. A three-throw crankshaft has cranks of 150 mm radius set at 120° to each other and equally spaced with a pitch of 500 mm. The rotating masses at crank radius are: No. 1, 30 kg; No. 2, 40 kg; No. 3, 40 kg. Balance is to be effected by a balance mass in a plane 150 mm outside the plane of crank No. 1 and another in a plane 750 mm outside the plane of crank No. 3; the radii of rotation of the masses are to be 225 mm and 750 mm respectively.

Determine the magnitudes and angular positions of the balance masses.

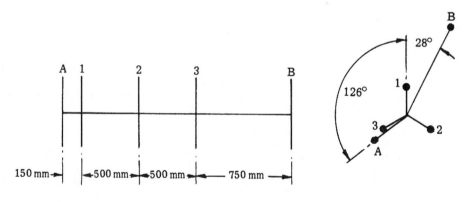

Figure 14.6

The arrangement of the planes of rotation of the masses is shown in Figure 14.6. Taking the plane A as the reference plane to eliminate the couple due to the force in that plane, the following table is compiled.

Plane	m (kg)	r (m)	mr	l (m)	mrl
A	m_a	0.225	$0.225m_a$	0	0
1	30	0.15	4.5	0.15	0.675
2	40	0.15	6.0	0.65	3.9
3	40	0.15	6.0	1.15	6.9
B	m_b	0.75	$0.75m_b$	1.90	$1.425m_b$

Using the data from the mrl column, the couple polygon is drawn, Figure 14.7. From the closing side, $1.425m_b = 5.42$ kg m². Hence $m_b = 3.8$ kg at an angle of 28° to the direction of crank No. 1 and the mr value for plane B is therefore $0.75 \times 3.8 = 2.85$ kg m.

Using the data from the mr column, the force polygon is drawn, Figure 14.8. From the closing side, $0.225m_a = 1.65$ kg m. Hence $m_a = 7.33$ kg at an angle of 126° to the direction of crank No. 1.

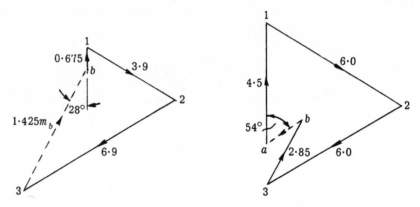

Figure 14.7   Figure 14.8

3. A shaft is supported in two bearings 2.4 m apart and projects 0.6 m beyond the bearings at each end. The shaft carries three pulleys, one at each end and one at the middle of its length. The end pulleys have masses of 90 kg and 50 kg and their

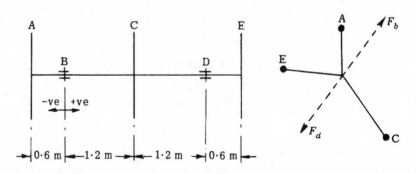

Figure 14.9

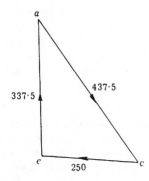

Figure 14.10

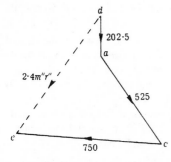

Figure 14.11

centres of gravity are at 3.75 mm and 5 mm respectively from the shaft axis. The centre pulley has a mass of 70 kg and its centre of gravity is 6.25 mm from the shaft axis. If the pulleys are arranged so as to give static balance, determine the dynamic forces on the bearings when the shaft rotates at 300 rev/min.

The arrangement of the planes of rotation is shown in Figure 14.9. Taking the plane B as the reference plane, the following table is compiled, the couple due to the mass in plane A being regarded as negative.

Plane	m (kg)	r (mm)	mr	l (m)	mrl
A	90	3.75	337.5	-0.6	-202.5
B	m'	r'	$m'r'$	0	0
C	70	6.25	437.5	$+1.2$	$+525$
D	m''	r''	$m''r''$	$+2.4$	$+2.4m''r''$
E	50	5	250	$+3.0$	$+750$

Since the shaft is to be in static balance, the mr products for A, C and E must be represented by a closed figure, Figure 14.10. This gives the relative directions of the three masses.

The couple polygon is then drawn, using the values in the mrl column and the directions of A, C and E already found. This is shown in Figure 14.11, the couple due to the mass in plane A being drawn in the opposite direction to the force, since this couple is negative. The out-of-balance couple is represented by de, of magnitude 740 kg mm m.

Thus
$$m''r'' = \frac{740}{2.4} = 308 \text{ kg mm}$$

$$\therefore F = m''r''\omega^2$$

$$= 308 \times 10^{-3} \times \left(300 \times \frac{2\pi}{60}\right)^2 = \underline{304 \text{ N}}$$

Since the forces are balanced, the shaft is subject only to a pure couple and the reaction at B is also 304 N, opposite in direction to that at D.

Further problems

4. A machine has a rotating mass of 80 kg whose centre of gravity is 375 mm from the axis of rotation. What masses must be placed at radii of 450 mm and in planes of revolution 0.5 m and 1.5 m from, and on the same side of, the plane of revolution of the 80 kg mass for complete balance?

If, instead of the balancing masses, there were bearings situated at the same distances from the 80 kg mass carrying the shaft to which it is attached, what would be the dynamic forces on these bearings when the shaft rotates at 120 rev/min?

[100 and 33.3 kg; 7.1 kN and 2.37 kN]

5. The two cranks of a reciprocating engine are at right-angles and the distance between the cylinder centre lines is 0.625 m. The stroke of the piston is 0.6 m and the revolving parts are equivalent to 300 kg at each crank pin. Find the magnitude and angular positions of the balance masses required at a radius of 0.8 m in planes 1.5 m apart, arranged symmetrically with respect to the cylinder centre lines.

[86.2 kg, 22°37′ to axis of symmetry]

6. A rotating shaft carries four masses A, B, C and D attached to it and the mass centres are at 30 mm, 36 mm 39 mm and 33 mm respectively. A, C and D are 7.5 kg, 5 kg and 4 kg, the axial distance between A and B is 0.4 m, that between B and C is 0.5 m and the eccentricities of A and C are at 90° to each other.

Find, for complete balance, (a) the angles between A, B and D; (b) the axial distance between the planes of C and D: (c) the mass B.

[AB $162\frac{1}{2}°$; AD $47\frac{1}{2}°$; 0.505 m; 9.16 kg]

7. A casting has a mass of 225 kg and is mounted on centres 1.15 m apart. It is given static balance by two masses A and B in planes situated 0.5 m and 0.4 m respectively on either side of the plane containing the mass centre. The masses A and B are 10 kg and 12 kg at radii of 0.375 m and 0.45 m respectively and their mass centres are at 90° to each other.

Determine the eccentricity of the mass centre of the casting, its position relative to that of mass A and the forces on the centres when the rotor is run at 50 rev/min.

[29.38 mm; 124°47′; 69.2 N]

8. A shaft supported in bearings 1.2 m apart carries two masses A and B, each of mass 6 kg, placed at distances of 0.3 m and 0.6 m respectively from one of the bearings. Static balance is obtained by the addition of a balance mass of 5 kg at a radius of 200 mm, the position of the balance mass being midway between A and B. If the radii of the mass centres of A and B are 175 mm and 225 mm respectively, find the relative angular positions of the three masses and the magnitude of the unbalanced couple acting on the bearings when the shaft rotates at 100 rev/min.

[132°23′ between A and B; 129°9′ between B and balance mass; 37.1 N m]

9. The out-of-balance of a machine rotor is equivalent to 5 kg at 10 mm radius in one plane A, together with an equal mass at 15 mm radius in a second plane B. AB = 0.375 m and the two radii are at 120°. Find the mass required in a third plane C at a radius of 125 mm and its angular position with respect to the given radii so that there is no resultant out-of-balance force.

Find also the position of C along the axis for the residual couple to be a minimum and the value of this couple when the speed is 500 rev/min.

[0.525 kg; 260° to A, 0.3215 m from A; 51.2 N m]

10. Three rotating masses, A = 14 kg, B = 11 kg and C = 21 kg, are carried by a shaft with centres of gravity 275 mm, 400 mm and 150 mm respectively from the shaft axis. The angular positions of B and C are 60° and 135° respectively from A, measured in the same direction. The distances between the planes of rotation of A and B is 1.35 m and between those of A and C is 3.6 m, B and C being on the same side of A.

Two balance masses are to be fitted, each with its centre of gravity 225 mm from the shaft axis, in planes midway between those of A and B and of B and C. Determine the magnitude and angular position of each balance mass with respect to A.

[36.7 kg between A and B; 187° to A; 29.3 kg between B and C; 310° to A]

15 Friction clutches, bearings and belt drives

15.1 Plate clutches

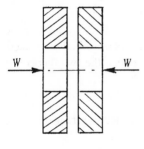

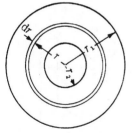

Figure 15.1

A clutch is a device for connecting and disconnecting the drive between two co-axial shafts. In a friction clutch, the two parts are held in contact by springs and the torque is transmitted by friction between the contact surfaces. The drive may then be disconnected by separating the contact surfaces.

In a multi-plate clutch, a number of flat annular plates are held in contact by an axial force normal to the surfaces. Each pair of surfaces carries the full load and each side of each plate is effective, so that the total torque transmitted is then $2n$ times the torque transmitted by one pair of surfaces, where n is the number of plates.

Figure 15.1 shows two surfaces pressed together with an axial force W; the outer and inner radii are r_1 and r_2 respectively and the pressure at a radius r is p.

Normal force on annular ring of radius r and width dr

$$= p \times 2\pi r \, dr$$

$$\therefore \text{ total axial force, } W = 2\pi \int_{r_2}^{r_1} pr \, dr \qquad (15.1)$$

Friction force on ring $= \mu \times 2\pi pr \, dr$, where μ is the coefficient of friction

$\therefore$ moment of friction force about axis $= \mu \times 2\pi pr \, dr \times r$

$$\therefore \text{ total torque transmitted, } T = 2\pi\mu \int_{r_2}^{r_1} pr^2 \, dr \qquad (15.2)$$

Equations (15.1) and (15.2) may be integrated if some assumptions are made regarding the variation of pressure with radius.

For unworn clutches, it is assumed that p is constant, giving

$$W = 2\pi p \int_{r_2}^{r_1} r \, dr \quad = \pi p(r_1^2 - r_2^2) \tag{15.3}$$

and

$$T = 2\pi\mu p \int_{r_2}^{r_1} r^2 \, dr = \tfrac{2}{3}\pi\mu p(r_1^3 - r_2^3)$$

$$= \tfrac{2}{3}\mu W \frac{r_1^3 - r_2^3}{r_1^2 - r_2^2} \tag{15.4}$$

For worn clutches, it is assumed that wear is uniform over the contact area.

Since

$$\text{wear} \propto \text{pressure} \times \text{velocity}$$

$$\propto \text{pressure} \times \text{radius},$$

then

$$pr = c,$$

giving

$$W = 2\pi c \int_{r_2}^{r_1} dr = 2\pi c(r_1 - r_2) \tag{15.5}$$

and

$$T = 2\pi\mu c \int_{r_2}^{r_1} r \, dr = \pi\mu c(r_1^2 - r_2^2) = \mu W \frac{r_1 + r_2}{2} \tag{15.6}$$

The torque transmissible by a worn clutch is less than by an unworn clutch and so this theory should always be used unless otherwise stated. If, however, the ratio $r_2/r_1 > \tfrac{1}{4}$, the difference between the two theories is very small.

15.2 Cone clutches

A cone clutch consists of a single pair of friction faces arranged as the frustum of a cone, as shown in Figure 15.2.

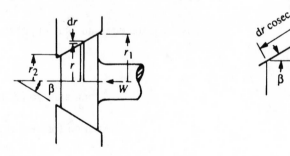

Figure 15.2

Let p be the normal pressure between the surfaces at radius r.

Then normal force on elementary ring $= p \times 2\pi r \, dr \, \text{cosec} \, \beta$

$\therefore$ axial component of this force $= p \times 2\pi r \, dr \, \text{cosec} \, \beta \sin \beta$

$$= p \times 2\pi r \, dr$$

$$\therefore \text{ total axial force, } W = 2\pi \int_{r_2}^{r_1} pr \, dr \qquad (15.7)$$

$$\text{Friction force on ring} = \mu p \times 2\pi r \, dr \operatorname{cosec} \beta$$

$$\therefore \text{ moment of friction force about axis} = \mu p \times 2\pi r \, dr \operatorname{cosec} \beta \times r$$

$$\therefore \text{ total torque transmitted, } T = 2\pi r \operatorname{cosec} \beta \int_{r_2}^{r_1} pr^2 \, dr \quad (15.8)$$

If p is assumed constant,
$$T = \tfrac{2}{3} \mu W \frac{r_1^3 - r_2^3}{r_1^2 - r_2^2} \operatorname{cosec} \beta \qquad (15.9)$$

If pr is assumed constant,
$$T = \mu W \frac{r_1 + r_2}{2} \operatorname{cosec} \beta \qquad (15.10)$$

15.3 Centrifugal clutches

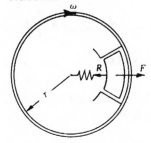

A centrifugal clutch consists of a number of shoes which move outwards in guides due to centrifugal force. The shoes come into contact with a rim and cause the rim to rotate due to friction between shoes and rim. When the speed falls, the shoes are pulled in from the rim by springs.

Figure 15.3 shows a clutch shoe which is subjected to a centrifugal force F and an inward spring force R.

$$\text{Then radial force between shoe and rim} = F - R$$

$$\therefore \text{ friction force between shoe and rim} = \mu(F - R)$$

$$\therefore \text{ friction torque due to each shoe} = \mu(F - R)r$$

If there are n shoes, total clutch torque, $T = n \mu r (F - R)$ (15.11)

Figure 15.3

15.4 Bearings

Figure 15.4(a), (b) and (c) shows three types of plain bearing, a collar bearing, a footstep bearing and a conical pivot respectively. In each case, the friction torque to be overcome is given by an identical equation to that for the corresponding flat plate or conical clutch.

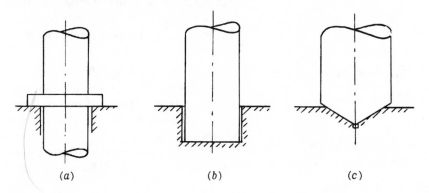

(a)   (b)   (c)

Figure 15.4

In order to distribute the load over a larger area and hence reduce the bearing pressure, a multi-collar bearing, as shown in Figure 15.5, may be used. The thrust pads between each pair of collars are horse-shoe shaped and are adjusted so that each takes an equal share of the axial thrust.

If the axial load is W and there are n collars, each collar carries a load W/n. The friction torque per collar can then be calculated for this load but since the total torque is n times that for a single collar, this is the same as that for one collar carrying the total load.

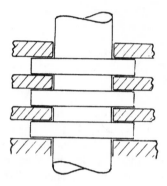

Figure 15.5

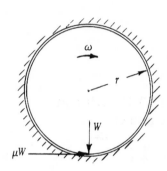

Figure 15.6

15.5 Journal bearings

Figure 15.6 shows a shaft of radius r supported by a concentric bearing. If the radial load between the shaft and the bearing is W,

$$\text{tangential friction force} = \mu W$$

$$\therefore \text{ friction torque, } T = \mu W r$$

and   and power loss $= T\omega = \mu W r\omega$ (15.12)

15.6 Lubricated surfaces

In the foregoing cases of friction torques, it has been assumed that the coefficient of friction, μ, has been constant, which is appropriate to unlubricated surfaces. If the bearing is lubricated, however, the effective coefficient of friction depends on the viscosity of the oil, the pressure between the surfaces, bearing dimensions and the shaft speed.

15.7 Viscosity

When a fluid flows smoothly over a stationary boundary, the layer in contact with the boundary is at rest and subsequent layers move with increasing velocities as the distance from the boundary increases. Thus there is a velocity gradient across the section of flow and a shearing action between adjacent layers.

If the velocity changes by dv in a distance dy perpendicular to the direction of flow, Figure 15.7, the viscous strain rate,

$$\phi = \frac{dv}{dy}$$

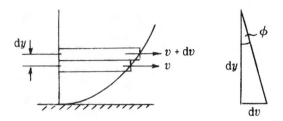

Figure 15.7

The viscous stress τ is the viscous resistance per unit area and the coefficient of viscosity η is defined as the ratio viscous stress/viscous strain rate,

i.e.

$$\eta = \frac{\tau}{\phi} = \frac{\tau}{\mathrm{d}v/\mathrm{d}y}$$

or

$$\tau = \eta \frac{\mathrm{d}v}{\mathrm{d}y} \tag{15.13}$$

In the case of a thin film of lubricant, of thickness y, between two surfaces moving with a velocity v relative to one another, the gradient may be regarded as uniform, so that equation (15.13) becomes

$$\tau = \eta \frac{v}{y} \tag{15.14}$$

The units of η are kg/m s or N s/m^2 (1 poise $= 0.1$ kg/m s).

15.8 Application to bearings

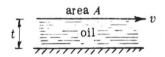

Figure 15.8

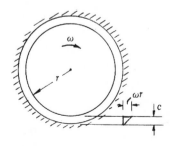

Figure 15.9

(a) Parallel bearings

Figure 15.8 shows two flat surfaces of area A, between which is an oil film of thickness t. Then, if the upper surface is moving at a velocity v relative to the lower surface,

$$\text{viscous force, } F = \text{stress} \times \text{area}$$

$$= \eta \frac{v}{t} \times A$$

$$\text{Power loss} = Fv = \frac{\eta v^2 A}{t} \tag{15.15}$$

(b) Journal bearings

Figure 15.9 shows a journal bearing of radius r and radial clearance c, rotating at ω rad/s.

Then

$$\tau = \eta \frac{v}{c} = \eta \frac{\omega r}{c}$$

∴ viscous force for a length l

$$= \text{stress} \times \text{area}$$

$$= \eta\frac{\omega r}{c} \times 2\pi rl$$

$$= \frac{2\pi\eta\omega r^2 l}{c}$$

∴ viscous torque $= \dfrac{2\pi\eta\omega r^2 l}{c} \times r = \dfrac{2\pi\eta\omega r^3 l}{c}$ $\hspace{2em}$ (15.16)

and $\hspace{3em}$ power loss $= \dfrac{2\pi\eta\omega r^3 l}{c} \times \omega = \dfrac{2\pi\eta\omega^2 r^3 l}{c}$ $\hspace{2em}$ (15.17)

The coefficient of friction is defined by

$$\mu = \frac{\text{viscous friction force}}{\text{bearing load}}$$

$$= \frac{\dfrac{2\pi\eta\omega r^2 l}{c}}{p \times 2rl} = \frac{\pi\eta\omega r}{pc} \hspace{2em} (15.18)$$

where p is the projected bearing pressure given by $\dfrac{\text{bearing load}}{\text{projected bearing area}}$

(c) Collar and footstep bearings

Figure 15.10 shows a collar bearing of external and internal radii r_1 and r_2 respectively. The shaft is supported on an oil film of thickness t and is rotating at ω rad/s.

Consider an annular element of radius x and thickness $\mathrm{d}x$. Then

$$\tau = \eta\frac{v}{t} = \eta\frac{\omega x}{t}$$

Therefore shearing force on element

$$= 2\pi x\,\mathrm{d}x \times \eta\frac{\omega x}{t}$$

Hence torque on element

$$= 2\pi x\,\mathrm{d}x \times \eta\frac{\omega x}{t} \times x$$

$$= \frac{2\pi\eta\omega}{t}x^3\,\mathrm{d}x$$

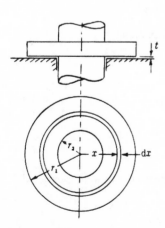

Figure 15.10

$$\text{so that total torque} = \frac{2\pi\eta\omega}{t} \int_{r_2}^{r_1} x^3 \, dx$$

$$= \frac{\pi\eta\omega}{2t}(r_1^4 - r_2^4) \qquad (15.19)$$

$$\text{Power loss} = T\omega = \frac{\pi\eta\omega^2}{2t}(r_1^4 - r_2^4) \qquad (15.20)$$

In the case of a footstep bearing of radius r, Figure 15.4 (b), these equations reduce to

$$T = \frac{\pi\eta\omega r^4}{2t} \qquad (15.21)$$

and

$$P = \frac{\pi\eta\omega^2 r^4}{2t} \qquad (15.22)$$

15.9 Belt drives

When a belt transmits power to a pulley, there is a difference in tension between the tight and slack sides due to the friction force between the belt and pulley.

Consider the section of flat belt shown in Figure 15.11, which has an angle of lap θ on the pulley and is about to slip in the direction of motion, the tight and slack side tensions then being T_1 and T_2 respectively. On a small arc subtending an angle $d\theta$, let T and $T + dT$ be the tensions at the element and R be the radial force between the pulley and belt.

For radial equilibrium,

$$(T + dT)\sin\frac{d\theta}{2} + T\sin\frac{d\theta}{2} = R$$

which reduces to

$$T \, d\theta = R \qquad (15.23)$$

assuming that $\sin\dfrac{d\theta}{2} = \dfrac{d\theta}{2}$ rad and neglecting products of small quantities.

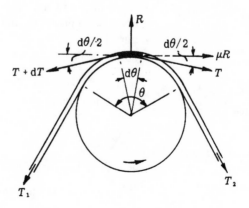

Figure 15.11

For tangential equilibrium,

$$(T + dT)\cos\frac{d\theta}{2} = T\cos\frac{d\theta}{2} + \mu R$$

which reduces to $$dT = \mu R, \tag{15.24}$$

assuming that $\cos\dfrac{d\theta}{2} = 1$.

From equations (15.23) and (15.24),

$$dT = \mu T\, d\theta$$

or $$\frac{dT}{T} = \mu\, d\theta$$

$$\therefore \int_{T_2}^{T_1} \frac{dT}{T} = \int_0^\theta \mu\, d\theta$$

$$\therefore \ln\frac{T_1}{T_2} = \mu\theta$$

or $$\frac{T_1}{T_2} = e^{\mu\theta} \tag{15.25}$$

If the belt is used to transmit power between two pulleys of unequal diameter, the belt will slip first on the pulley with the smaller angle of lap, i.e. on the smaller pulley.

If the belt moves with a velocity v, then

$$\text{power transmitted} = (T_1 - T_2)v = T_1\left(1 - \frac{1}{e^{\mu\theta}}\right)v \tag{15.26}$$

15.10 Centrifugal tension

As the belt passes over the pulley, each element is subjected to centrifugal force, which will increase the tension in the belt. Consider an element of mass m per unit length subtending an angle $d\theta$ and moving with velocity v, Figure 15.12.

$$F = mr\, d\theta \times \frac{v^2}{r} = mv^2\, d\theta \tag{15.27}$$

For radial equilibrium,

$$F = 2T_c\frac{d\theta}{2} = T_c\, d\theta \tag{15.28}$$

where T_c is the centrifugal tension.

Therefore, from equations (15.27) and (15.28),

$$T_c = mv^2 \tag{15.29}$$

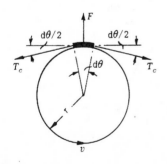

Figure 15.12

This tension is additional to that due to the transmission of power. If allowance is to be made for this, equation (15.23) becomes

$$T \, d\theta = R + F = R + T_c \, d\theta$$

which leads to

$$\frac{dT}{T - T_c} = \mu \, d\theta$$

so that

$$\frac{T_1 - T_c}{T_2 - T_c} = e^{\mu\theta} \tag{15.30}$$

T_1 and T_2 are the total tensions in the belt and $(T_1 - T_c)$ and $(T_2 - T_c)$ are the effective driving tensions.

The power transmitted is still $(T_1 - T_2)v$ but equation (15.26) becomes

$$P = (T_1 - T_c)\left(1 - \frac{1}{e^{\mu\theta}}\right)v \tag{15.31}$$

The maximum power transmissible, *if the belt velocity may be varied*, is given by

$$\frac{d}{dv}[(T_1 - T_c)v] = 0$$

i.e.

$$\frac{d}{dv}[T_1 v - mv^3] = 0$$

from which

$$mv^2 = T_c = \frac{T_1}{3} \tag{15.32}$$

This equation gives the velocity for maximum power, which may then be substituted in equation (15.31) but in most cases, the belt velocity is a constant.

15.11 Initial tension

A belt is assembled on pulleys with an initial tension T_0. To determine the tension required, it is assumed that the belt material is elastic and hence obeys Hooke's law. Since the total length remains constant, the increase in length of the tight side is equal to the decrease in length of the slack side and it therefore follows that the increase in tension on the tight side is equal to the decrease in tension on the slack side.

i.e.

$$T_1 - T_0 = T_0 - T_2$$

or

$$T_0 = \frac{T_1 + T_2}{2} \tag{15.33}$$

In determining the power transmissible by a belt, it is assumed that the tight side tension is at its maximum permissible value and also that the belt is at the point of slipping. These conditions can only occur simultaneously if the initial tension is set correctly from this relationship.

15.12 V-belt drives

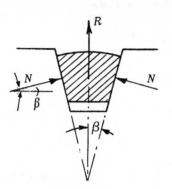

Figure 15.13

For a V-grooved pulley, the normal reaction between the belt and pulley is given by

$$R = 2N \sin \beta, \quad \text{Figure 15.13.}$$

The friction force between the belt and pulley is then

$$2\mu N = \mu R \operatorname{cosec} \beta$$

The coefficient of friction is therefore effectively increased from μ to $\mu \operatorname{cosec} \beta$ and equation (15.25) then becomes

$$\frac{T_1}{T_2} = e^{\mu\theta \operatorname{cosec} \beta} \qquad (15.34)$$

$\mu \operatorname{cosec} \beta$ is the effective, or virtual, coefficient of friction, μ', when compared with an equivalent flat belt.

Worked examples

1. *The rotating parts of a grinding mill have a mass of 2 t and rotate at 120 rev/min. They are supported by a conical bearing of 125 mm outer diameter and 50 mm inner diameter with an included angle of 120°. Assuming $\mu = 0.075$ and that the intensity of pressure varies inversely as the radius, determine the power wasted in friction.*

From equation (15.10), $T = \mu W \dfrac{r_1 + r_2}{2} \operatorname{cosec} \beta$

$$= 0.075 \times (2 \times 10^3 \times 9.81) \times \frac{62.5 + 25}{2 \times 10^3} \operatorname{cosec} 60°$$

$$= 74.3 \text{ N m}$$

$\therefore$ power wasted in fraction $= T\omega$

$$= 74.3 \times 120 \times \frac{2\pi}{60} = \underline{935 \text{ W}}$$

2. *An electric motor drives a co-axial rotor through a single-plate clutch which has two pairs of driving surfaces, each of 275 mm external and 200 mm internal diameter. The total spring load pressing the plates together is 500 N. The motor armature has a mass of 800 kg and a radius of gyration of 260 mm; the rotor has a mass of 1350 kg and a radius of gyration of 225 mm.*

When the motor is running at 1250 rev/min, the current is switched off and the clutch suddenly engaged. Determine the final speed of the motor and rotor, the time taken to reach that speed and the kinetic energy lost during clutch slip. $\mu = 0.35$.

Torque transmitted by clutch, $T = \mu W \dfrac{r_1 + r_2}{2} \times 2$ from equation (15.6)

$$= 0.35 \times 500 \times (0.1375 + 0.100)$$

$$= 41.6 \text{ N m}$$

Let suffices m and r refer to the motor and rotor respectively.

Then $\qquad\qquad\qquad\qquad\quad I_m = 800 \times 0.26^2 = 54.1 \text{ kg m}^2$

and $\qquad\qquad\qquad\qquad\quad I_r = 1\,350 \times 0.225^2 = 68.3 \text{ kg m}^2$

Let N be the final speed in rev/min and t be the time of slipping.

Then, for the motor, $\qquad\qquad T = I_m \alpha_m$

i.e. $\qquad\qquad\qquad\qquad 41.6 = 54.1 \times \dfrac{1\,250 - N}{t} \times \dfrac{2\pi}{60}$

which reduces to $\qquad\qquad 7.34 = \dfrac{1\,250 - N}{t} \qquad\qquad\qquad\qquad (1)$

and for the rotor, $\qquad\qquad T = I_r \alpha_r$

i.e. $\qquad\qquad\qquad\qquad 41.6 = 68.3 \times \dfrac{N}{t} \times \dfrac{2\pi}{60}$

which reduces to $\qquad\qquad 5.82 = \dfrac{N}{t} \qquad\qquad\qquad\qquad\qquad (2)$

Therefore, from equations (1) and (2),

$$N = 552 \text{ rev/min}$$

and $\qquad\qquad\qquad\qquad\qquad t = \underline{95 \text{ s}}$

$$\text{Loss of K.E.} = \tfrac{1}{2} \times 54.1 \times \left(1\,250 \times \frac{2\pi}{60} \right)^2 - \tfrac{1}{2} \times (54.1 + 68.3) \times \left(552 \times \frac{2\pi}{60} \right)^2$$

$$= \underline{259\,000 \text{ J}} \text{ or } \underline{259 \text{ kJ}}$$

3. *A multi-plate friction clutch has to be designed to transmit 75 kW from an engine rotating at 2 000 rev/min. The inner and over diameters are respectively 100 mm and 150 mm, the pressure is to be assumed uniform at 150 kN/m^2 and $\mu = 0.25$. Determine the necessary end thrust and the number of plates required.*

If this clutch is then used to transmit power from a larger engine to a rotor which has a mass of 1 150 kg and a radius of gyration of 200 mm, determine the time required for this rotor to reach 1 500 rev/min from standstill, assuming that the clutch is transmitting the maximum possible torque.

$$P = T\omega$$

i.e. $\qquad\qquad\qquad 75 \times 10^3 = T \times 2\,000 \times \dfrac{2\pi}{60}$

$$\therefore T = 358 \text{ Nm}$$

$$W = \pi p (r_1^2 - r_2^2) \quad \text{from equation (15.3)}$$

$$= \pi \times 150 \times 10^3 \times \dfrac{75^2 - 50^2}{10^6}$$

$$= 1\,474 \text{ N}$$

From equation (15.4), $T = \frac{2}{3}\mu W \frac{r_1^3 - r_2^3}{r_1^2 - r_2^2} \times n$ where n is the number of contact surfaces

i.e $\qquad 358 = \frac{2}{3} \times 0.25 \times 1\,474 \times \dfrac{75^3 - 50^3}{(75^2 - 50^2) \times 10^3} \times n$

from which $\qquad n = 15.35$

Therefore 8 plates are required, since each side of each plate is effective. When used with a larger engine, maximum torque transmissible

$$= \frac{16}{15.35} \times 358 = 373 \text{ Nm}$$

Therefore the equation of motion of the rotor is

$$373 = 1\,150 \times 0.2^2 \alpha$$

$$\therefore \alpha = 8.1 \text{ rad/s}^2$$

$$\therefore t = \frac{1\,500 \times \dfrac{2\pi}{60}}{8.1} = \underline{19.4 \text{ s}}$$

4. *A centrifugal clutch consists of a spider carrying four shoes which are kept from contact with the clutch case until increase in centrifugal force overcomes the spring force and power is transmitted by friction between the shoes and case.*

Determine the mass of each shoe if it is required to transmit 22.5 kW at 750 rev/min with engagement beginning at 75 per cent of the running speed. The inside diameter of the drum is 300 mm and the radial distance of the centre of gravity of each shoe from the shaft axis is 125 mm. $\mu = 0.25$.

$$P = T\omega$$

i.e. $\qquad 22.5 \times 10^3 = T \times 750 \times \dfrac{2\pi}{60}$

$$\therefore T = 286.5 \text{ Nm}$$

From equation (15.11), $\qquad T = n\mu r(F - R)$

At 75% of 750 rev/min, (i.e. 562.5 rev/min)

$$R = F = m\left(562.5 \times \frac{2\pi}{60}\right)^2 \times 0.125$$

Therefore, at 750 rev/min,

$$T = 4 \times 0.25 \times 0.15 \times m\left[\left(750 \times \frac{2\pi}{60}\right)^2 - \left(562.5 \times \frac{2\pi}{60}\right)^2\right] \times 0.125$$

$$= 50.6\,m$$

Hence $\quad m = \dfrac{286.5}{50.6} = \underline{5.66 \text{ kg}}$

5. *A piston, 50 mm diameter and 75 mm long, moves vertically in an open-ended lubricated cylinder with a radial clearance of 0.1 mm. When falling due to its own weight, the piston moves through 30 mm in 4.2 s at a uniform velocity. When a mass of 0.05 kg is added to the piston, it moves with uniform velocity through the same distance in 2.4 s. Calculate the viscosity of the oil.*

If the lubricated area is unwrapped, it becomes equivalent to the parallel bearing considered in Section 15.8(*a*).

Thus viscous force $= \eta \dfrac{v}{t} A$ from equation (15.15)

Let the weight of the piston be W.

Then $\qquad\qquad\qquad W = \eta \times \dfrac{0.03/4.2}{0.000\,1} \times \pi \times 0.05 \times 0.075 = 0.841\eta$

and $\quad W + 0.05 \times 9.81 = \eta \times \dfrac{0.03/2.4}{0.000\,1} \times \pi \times 0.05 \times 0.075 = 1.472\eta$

Hence, by subtraction $\qquad\qquad 0.490\,5 = 0.631\eta$

or $\qquad\qquad\qquad\qquad \eta = \underline{0.777 \text{ Ns/m}^2}$

6. *A journal bearing 50 mm diameter and 90 mm long carries a radial load of 4 500 N and the shaft runs at 1 400 rev/min. Calculate the power absorbed in friction,*
  (a) if the bearing is unlubricated and the coefficient of friction is 0.08;
  (b) if the bearing is lubricated with oil of viscosity 0.065 Ns/m² and there is a radial clearance of 0.025 mm.

(a) $\qquad\qquad P = \mu W r \omega$ from equation (15.12)

$$= 0.08 \times 4\,500 \times 0.025 \times \left(1\,400 \times \frac{2\pi}{60} \right)$$

$$= \underline{1\,317 \text{ W}}$$

(b) $\qquad\qquad P = \dfrac{2\pi\eta\omega^2 r^3 l}{c}$ from equation (15.17)

$$= \frac{2\pi \times 0.065 \times \left(1\,400 \times \dfrac{2\pi}{60} \right)^2 \times 0.025^3 \times 0.09}{0.025 \times 10^{-3}}$$

$$= \underline{494 \text{ W}}$$

Alternatively, $\quad \mu = \dfrac{2\pi\eta\omega r^2 l}{cW}$ from equation (15.18)

$$= \frac{2\pi \times 0.065 \times \left(1\,400 \times \dfrac{2\pi}{60} \right) \times 0.025^2 \times 0.09}{0.025 \times 10^{-3} \times 4\,500} = 0.03$$

$$\therefore P = \frac{0.03}{0.08} \times 1\,317 = \underline{494 \text{ W}}$$

7. *A belt drive transmits power from an electric motor to a machine. The diameter of the pulley on the motor shaft is 150 mm, that on the machine is 200 mm and the centre distance is 600 mm. If the motor speed is 1440 rev/min and the maximum permissible belt tension is 900 N, then the maximum power transmissible is 6 kW.*

It is required to increase the power transmitted to 6.75 kW using the same pulleys, centre distance and motor speed. The belt material is to be treated with a preparation which inceases the coefficient of friction by 10% and in addition, a jockey pulley is to be fitted.

Determine (a) the original coefficient of friction, (b) the new angle of lap.

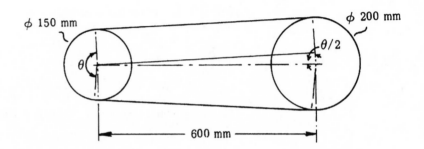

Figure 15.14

The arrangement of the drive is shown in Figure 15.14.

$$\cos \frac{\theta}{2} = \frac{\frac{1}{2}(200 - 150)}{600} = 0.041\,67$$

i.e. angle of lap on smaller pulley, on which the belt would slip first,

$$\theta = 175°14' = 3.055 \text{ rad}$$

$$v = 1\,440 \times \frac{2\pi}{60} \times 0.075 = 11.32 \text{ m/s}$$

$$P = T_1 \left(1 - \frac{1}{e^{\mu\theta}}\right) v \quad \text{from equation (15.26)}$$

i.e.
$$6 \times 10^3 = 900 \left(1 - \frac{1}{e^{3.055\mu}}\right) \times 11.32$$

from which
$$e^{3.055\,\mu} = 2.43$$

$$\therefore \mu = \underline{0.29}$$

New value of $\mu = 1.1 \times 0.29 = 0.319$

$$\therefore 6.75 \times 10^3 = 900 \left(1 - \frac{1}{e^{0.319\,\theta}}\right) \times 11.32$$

from which
$$e^{0.319\theta} = 0.662$$

$$\therefore \theta = 3.39 \text{ rad} = \underline{194°30'}$$

8. *An electric motor running at 1 400 rev/min transmits power by three V-belts, each of 320 mm² cross-sectional area, the total groove angle being 45°. The density of the belt material is 1.6 Mg/m³ and the maximum allowable stress in the belts is 2 MN/m². The angle of lap on the motor pulley is 145° and μ = 0.2. Calculate the maximum power which can be transmitted and the diameter of the motor pulley.*

$$\text{Mass of 1 m of belt} = \frac{320}{10^6} \times 1.6 \times 10^3 = 0.512 \text{ kg}$$

$$T_1 = \frac{320}{10^6} \times 2 \times 10^6 = 640 \text{ N}$$

For maximum power,

$$T_c = mv^2 = \frac{T_1}{3} = 213 \text{ N} \quad \text{from equation (15.32)}$$

$$\therefore \text{ velocity for maximum power} = \sqrt{\frac{T_c}{m}} = \frac{213}{\sqrt{0.512}}$$

$$= 20.4 \text{ m/s}$$

$$\text{Effective coefficient of friction} = \mu \operatorname{cosec} \beta \quad \text{from Section 15.12}$$

$$= 0.2 \operatorname{cosec} 22\tfrac{1}{2}^{\circ}$$

$$= 0.522\,6$$

$$\text{Angle of lap} = 145° \times \frac{\pi}{180} = 2.53 \text{ rad}$$

$$\therefore e^{\mu\theta \operatorname{cosec} \beta} = e^{0.522\,6 \times 2.53}$$

$$= 3.76$$

$$\text{Power transmitted by 3 belts} = (T_1 - T_c)\left(1 - \frac{1}{e^{\mu\theta \operatorname{cosec} \beta}}\right) v \times 3$$

from equation (15.31)

$$= (640 - 213)\left(1 - \frac{1}{3.76}\right) \times 20.4 \times 3$$

$$= 19\,200 \text{ W or } \underline{19.2 \text{ kW}}$$

$$\text{Angular speed of motor pulley, } \omega = 1\,400 \times \frac{2\pi}{60}$$

$$= 146.6 \text{ rad/s}$$

$$\therefore d = \frac{2v}{\omega} = 2 \times \frac{20.4}{146.6}$$

$$= \underline{0.278 \text{ m}}$$

9. *A belt drive connects two pulleys A and B, the centres of which are 4 m apart. The belt has a mass of 1.15 kg/m. Pulley A is 1 m diameter, has a mass of 25 kg and a radius of gyration of 420 mm. Pulley B is 0.5 m diameter, has a mass of*

18 kg and a radius of gyration of 225 mm. When at rest, the tension in the belt is 700 N. Assuming that the belt obeys Hooke's law, determine the tensions in the two portions of the belt between the pulleys when 1.5 kW is being transmitted, the speed of A being 180 rev/min.

Find also the kinetic energy of the belt and pulleys under these conditions.

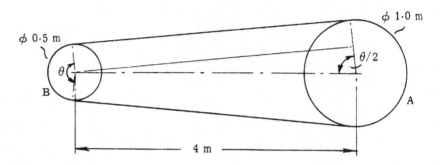

Figure 15.15

In Figure 15.15, $\cos \dfrac{\theta}{2} = \dfrac{\frac{1}{2}(1 - 0.5)}{4} = 0.062\,5$

i.e. angle of lap on smaller pulley,

$$\theta = 172°50' = 3.02 \text{ rad}$$

$$v = \omega r = 180 \times \frac{2\pi}{60} \times 0.5 = 9.425 \text{ m/s}$$

Also $T_1 + T_2 = 2T_0 = 1\,400 \text{ N}$ from equation (15.33)

$$\therefore T_2 = 1\,400 - T_1$$

$$P = (T_1 - T_2)v \quad \text{from equation (15.26)}$$

i.e. $1\,500 = (T_1 - [1\,400 - T_1]) \times 9.425$

from which $T_1 = \underline{779.6 \text{ N}}$

$$T_2 = 1\,400 - 779.6 = \underline{620.4 \text{ N}}$$

Angle of lap on larger pulley $= 2\pi - 3.02 = 3.26$ rad

$$\therefore \text{ length of belt} = 0.25 \times 3.02 + 0.5 \times 3.26 + 2\sqrt{(4^2 - 0.25^2)}$$

$$= 10.365 \text{ m}$$

$$\therefore \text{ mass of belt} = 1.15 \times 10.365 = 11.93 \text{ kg}$$

$$\therefore \text{ K.E. of belt} = \tfrac{1}{2} \times 11.93 \times 9.425^2 = \underline{529 \text{ J}}$$

$$\text{Speed of pulley A} = \frac{9.425}{0.5} = 18.85 \text{ rad/s}$$

$$\text{and speed of pulley B} = \frac{9.425}{0.25} = 37.7 \text{ rad/s}$$

$$\therefore \text{ K.E. of pulleys} = \tfrac{1}{2}I_a\omega_a^2 + \tfrac{1}{2}I_b\omega_b^2$$

$$= \tfrac{1}{2}\{25 \times 0.42^2 \times 18.85^2 + 18 \times 0.255^2 \times 37.7^2\}$$

$$= \underline{1\,434 \text{ J}}$$

Further problems

10. A plate clutch has four pairs of contact surfaces, each of 240 mm external diameter and 120 mm internal diameter. Assuming uniform pressure, find the total spring load pressing the plates together to transmit 25 kW at 1575 rev/min. Take $\mu = 0.3$.

If there are six springs each of stiffness 13 kN/m and each of the contact surfaces has worn away by 1.25 mm, what is the maximum power that can be transmitted at the same speed, assuming uniform wear and the same coefficient of friction?

[1.355 kN; 10.25 kW]

11. Determine the time required to accelerate a wheel of mass 500 kg and radius of gyration 0.2 m from rest to 250 re/min through a single plate clutch of internal and external radii 125 mm and 200 mm, taking μ as 0.3 and the spring load as 600 N. Assume uniform pressure.   [8.81 s]

12. A multi-plate clutch is to transmit 12 kW at 1 500 rev/min. The inner and outer radii of the plates are 50 mm and 100 mm respectively and the maximum spring force is limited to 1 kN. If $\mu = 0.35$, determine the necessary number of pairs of surfaces, assuming uniform wear.

What will be the necessary axial force?   [3; 970 N]

13. A shaft running at a steady speed of 175 rev/min drives a countershaft through a single-plate clutch of external and internal diameters 375 mm and 225 mm respectively. The rotating parts on the countershaft have a mass of 350 kg and a radius of gyration of 250 mm. The axial spring load is 450 N and $\mu = 0.3$. Determine the time required to reach full speed from rest and the energy dissipated due to clutch slip during this time.   [9.9 s; 3.67 kJ]

14. A thrust of 30 kN along the axis of a shaft is taken by a pivot bearing consisting of the frustum of a cone. The outer and inner diameters are 200 mm and 100 mm and the semi-angle of the cone is 60°. The shaft speed is 200 rev/min and $\mu = 0.02$. Assuming uniform pressure, determine (*a*) the magnitude of this pressure, (*b*) the power absorbed in friction.   [1.272 MN/m^2; 1.13 kW]

15. An electric motor drives a machine through a cone clutch of mean diameter 240 mm and included angle 40°. The axial thrust on the clutch is 800 N and $\mu = 0.3$. The moment of inertia of the machine parts is 5 kg m^2.

Determine the time required for the machine shaft to attain its running speed of 300 rev/min from rest, and the energy lost due to clutch slip.   [1.87 s; 2 467 J]

16. A cone clutch is required to transmit 30 kW at 1 200 rev/min. The mean diameter of the bearing surface is 250 mm and the cone angle is 25°. Assuming that $\mu = 0.3$ and the mean pressure is 140 kN/m^2, determine the axial width of the conical bearing surface and the axial load required.   [54.1 mm; 1.32 kN]

17. Two co-axial rotors A and B are connected by a single-plate clutch with two pairs of friction surfaces, each of 300 mm external and 220 mm internal diameter. The total spring load pressing the plates together is 700 N. The masses and radii of gyration of A and B are 1 100 kg, 200 mm and 800 kg, 350 mm respectively. $\mu = 0.3$.

The rotor A is given a speed of 1 200 rev/min while B is stationary and the clutch is then engaged. Determine the time for A and B to reach the same speed, the magnitude of that speed and the kinetic energy lost due to clutch slip. Assume constant wear.   [70 s; 372 rev/min.; 240 kJ]

18. A centrifugal clutch is fitted to a motor shaft to enable it to start without load. As the speed rises, the shoes move radially outward under centrifugal force against the inward pull of springs and press against the inner surface of a pulley. They then take up the drive by friction.

There are four shoes, each of mass 1.35 kg and the centre of gravity of each shoe is at 100 mm radius when it is just touching the pulley, which has a radius of 125 mm. Each spring then exerts a pull of 1.2 kN. $\mu = 0.25$.

Calculate the speed at which the shoes first touch the drum and the power which can be transmitted at 1 200 rev/min.   [900 rev/min; 14.6 kW]

19. A centrifugal clutch has four blocks which slide radially in a spider keyed to the driving shaft and make contact with the internal surface of a drum keyed to the driven shaft. When the clutch is stationary, each block is pulled against a stop by a spring, leaving a radial clearance of 6 mm between the block and drum. The spring pull is then 450 N and the centre of gravity of each block is 200 mm from the axis of the clutch.

If the internal diameter of the drum is 500 mm, the mass of each block is 7 kg, the stiffness of each spring is 35 kN/m and $\mu = 0.3$, find the maximum power the clutch can transmit at 500 rev/min.   [51.7 kW]

20. A bearing consists of a cylinder of outside diameter 150.00 mm which rotates at 100 rev/min in a fixed co-axial hollow cylinder of inside diameter 150.50 mm. The axial length of the surfaces is 0.3 m and the radial clearance is filled with oil of viscosity 0.12 Ns/m^2. Calculate the power dissipated.   [41.5 W]

21. A uniform film of oil 0.1 mm thick separates two 0.1 m diameter discs, one of which has a 10 mm diameter hole at the centre. Calculate the power required to rotate one disc at 5 rev/s relative to the other if the viscosity of the oil is 0.14 kg/ms.
                                                                          [13.57 W]

22. The thrust at the lower end of a vertical shaft is supported by a bearing consisting of a flat disc 100 mm diameter rotating with the shaft and a stationary housing. The disc is separated from the housing by a film of oil 0.25 mm thick and of viscosity 0.13 Ns/m^2. Calculate the power absorbed in friction when the shaft rotates at 800 rev/min, ignoring the side effects.   [35.7 W]

23. Each of the belts detailed below has an angle of lap of 180° and moves with the same velocity:
(a) a flat belt, maximum tension 360 N, $\mu = 0.3$;
(b) a V-belt, included angle 40°, maximum tension 240 N, $\mu = 0.35$.
    Determine which belt can transmit the greater power.
                                                         [(a) 219.7 v; (b) 230.4 v]

24. A ship is pulled through a lock by means of a capstan and rope. The capstan has a diameter of 0.5 m and turns at 30 rev/min. The rope makes three complete turns round the capstan and a pull of 100 N is applied at the free end. $\mu = 0.25$. Find (a) the pull on the ship, (b) the power required to drive the capstan.
                                                                 [11.1 kN; 8.65 kW]

25. A multiple V-belt drive is required to transmit 4 kW. The smaller pulley of the system is 150 mm diameter and rotates at 200 rev/min. The groove angle is

40° and the coefficient of friction between belt and pulley is 0.3. If the maximum tension in any one belt is limited to 800 N and the angle of lap is 135°, determine the number of belts required.   [4]

26. In a belt drive, the angle of lap on the small pulley is 150°. With a belt speed of 20 m/s and a tight side tension of 1 400 N, the greatest power which can be transmitted is 10 kW. What power could be transmitted for the same belt speed and maximum tension if an idler pulley is used to increase the angle of lap to 210°? Allow for centrifugal tension, the mass of the belt being 0.75 kg/m.   [20.21 kW]

27. A pulley is driven by a flat belt, the angle of lap being 120°. The belt is 100 mm wide by 6 mm thick and has a mass of 1 Mg/m^3. If $\mu = 0.3$ and the maximum stress in the belt is limited to 1.5 MN/m^2, find the greatest power the belt can transmit and the corresponding speed of the belt. [6.265 kW; 22.36 m/s]

28. Power is transmitted by a rope drive between two shafts 4.5 m apart. The pulleys are 3 m and 2 m diameter and the groove angle is 40°. If the rope has a mass of 4 kg/m and the maximum tension is 20 kN, determine the maximum power which the rope can transmit and the corresponding speed of the smaller pulley. $\mu = 0.2$.   [446 kW; 390 rev/min]

29. Power is transmitted from a shaft rotating at 250 rev/min by five ropes running in grooves in the periphery of a wheel of effective diameter 1.65 m. The groove angle is 50° and the arc of contact is 180°. The maximum permissible load in each rope is 900 N and its mass is 0.55 kg/m. $\mu = 0.3$. What power can be transmitted under these conditions?   [62 kW]

30. A 4 to 1 speed reduction between two shafts at 2 m centres is provided by four V-belts running on pulleys mounted on the shafts. The effective diameter of the driving pulley is 350 mm and it runs at 740 rev/min. The groove angle is 40° and each belt has a mass of 0.45 kg/m. $\mu = 0.28$. Determine the power that can be transmitted if the tension in each belt is not to exceed 800 N.   [34.3 kW]

31. The drive from an electric motor to a shaft is by three parallel V-belts. The mean diameter of the motor pulley is 150 mm, the motor and shaft speeds are respectively 1 600 and 400 rev/min and the shafts are 1 m apart. Each belt has a mass of 1.4 Mg/m^3 and a cross-sectional area of 800 mm^2; the groove angle is 30° and $\mu = 0.13$.

Find the power that can be transmitted by the drive if the tensile stress in the belt is not to exceed 8 MN/m^2.   [173.6 kW]

32. A rope drive is required to transmit power from a pulley of 1.15 m effective diameter. The ropes each have a mass of 1.2 kg/m, the groove angle is 50° and the angle of lap is 170°. $\mu = 0.3$.

(*a*) If the initial tension in each rope is 900 N, what is the maximum power which can be transmitted per rope? What will then be the load in the rope and its linear speed?

(*b*) If the permissible load is 1.6 kN per rope and this is to be utilized for maximum power, what then should be the initial tension in the rope?

[14 kW; 1.273 kN; 18.8 m/s; 1.13 kN]

16 Gyroscopic motion

16.1 Introduction

Whenever the axis of a rotating body is caused to change direction, a couple is required, called a *gyroscopic couple*. The reaction to this couple is experienced in such cases as aircraft, marine and car engines when changing direction and gyroscopic effects may be usefully employed in navigating systems.

16.2 Gyroscopic couple

Figure 16.1 shows a rotor of polar moment of inertia I rotating at a rate ω about the axis OX, called the *spin axis*. The angular momentum, $I\omega$, is represented by the vector *op*, its direction being given by the forward movement of a corkscrew when turned in the direction of rotation.

If the spin axis is now rotated, *or precessed*, about the perpendicular axis OY at a rate Ω then, after a time dt, the vector *op* has moved to *oq*. The change of momentum is represented by *pq* and if the angle *poq* is dθ, then

$$pq = I\omega\,d\theta$$

Thus, rate of change of momentum $= I\omega\dfrac{d\theta}{dt}$

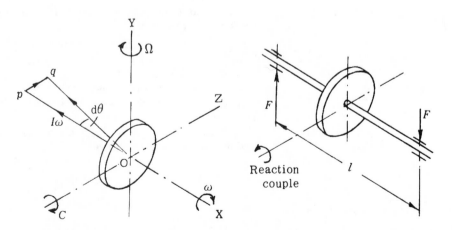

Figure 16.1   Figure 16.2

i.e.   gyroscopic couple, $C = I\omega\Omega$ (16.1)

By applying the corkscrew rule to the vector pq, it will be seen that the couple required to cause precession is clockwise looking along axis OZ, mutually perpendicular to OX and OY.

The reaction couple, which is provided by the shaft bearings, is equal and opposite to the gyroscopic couple. If the force at the bearings is F and the distance between the bearings is l, Figure 16.2, then

$$F = \frac{I\omega\Omega}{l}$$ (16.2)

Worked examples

1. *A pair of locomotive driving wheels and axle have a moment of inertia of 400 kg m^2. The diameter of the wheels is 2 m and the distance between the wheel centres is 1.5 m. When the locomotive is travelling at 100 km/h, defective ballasting causes one wheel to fall 10 mm and rise again in a total time of 0·1 s. If the movement of the wheel takes place with simple harmonic motion, find the gyroscopic reaction on the locomotive.*

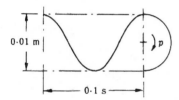

Figure 16.3

Referring to Figure 16.3, the S.H.M. may be considered to be generated by a vector rotating at a uniform angular velocity p (see Section 18.2).

Periodic time of S.H.M. = 0.1 s

$$\therefore\ p = \frac{2\pi}{0.1}\ \text{rad/s}$$

Amplitude of S.H.M., $X = 0.005$ m

Therefore maximum vertical velocity of wheel when descending,

$$v = pX = \frac{2\pi}{0.1} \times 0.005 = \frac{\pi}{10}\ \text{rad/s}$$

Therefore maximum velocity of precession

$$\Omega = \frac{v}{l} = \frac{\pi}{10 \times 1.5} = \frac{\pi}{15}\ \text{rad/s}$$

$$\text{Spin velocity, } \omega = \frac{\text{velocity of loco}}{\text{wheel radius}}$$

$$= \frac{100/3.6}{1} = \frac{100}{3.6}\ \text{rad/s}$$

Therefore gyroscopic couple, $C = I\omega\Omega$ from equation (16.1)

$$= 400 \times \frac{100}{3.6} \times \frac{\pi}{15}$$

$$= \underline{2\,330\ \text{Nm}}$$

Figure 16.4 shows a view of the back of the wheels. If the left-hand wheel falls, the angular momentum vector op moves to oq, and the change of angular

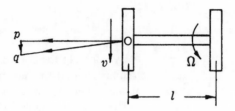

Figure 16.4

momentum, pq, shows that a clockwise couple looking downwards is required to cause precession. The reaction couple is opposite to this, i.e., anticlockwise downwards, and this tends to cause the locomotive to swing to the left.

2. *An electric motor on board ship is arranged with its rotor athwart the ship. Find the maximum load on its bearings due to gyroscopic action if the ship rolls with simple harmonic motion 30° on each side of the vertical and the time for one complete roll is 4 s. The mass of the rotor is 200 kg, its radius of gyration is 240 mm, the bearings are 1.2 m apart and the rotor speed is 3000 rev/min, clockwise viewed from the starboard side.*

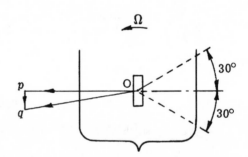

Figure 16.5

Figure 16.5 shows a view from astern of the ship.

Periodic time of S.H.M. $= 4$ s

Therefore angular velocity of vector generating S.H.M.,

$$p = \frac{2\pi}{4} \text{ rad/s}$$

Angular amplitude, $\phi = 30° = \dfrac{\pi}{6}$ rad

Therefore maximum angular velocity during rolling = maximum velocity of precession,

$$\Omega = p\phi = \frac{2\pi}{4} \times \frac{\pi}{6} = \frac{\pi^2}{12} \text{ rad/s}$$

Velocity of spin, $\omega = 3\,000 \times \dfrac{2\pi}{60} = 100\pi$ rad/s

Therefore gyroscopic couple, $C = I\omega\Omega$ from equation (16.1)

$$= 200 \times 0.24^2 \times 100\pi \times \frac{\pi^2}{12}$$

$$= 2\,975 \text{ Nm}$$

Therefore force on bearings, $F = \dfrac{I\omega\Omega}{l}$ from equation (16.2)

$$= \frac{2\,975}{1.2} = \underline{2\,480 \text{ N}}$$

If the ship heels to port, the angular momentum vector moves from *op* to *oq* and hence a clockwise moment looking downwards must be applied by the shaft bearings to cause precession. The reaction couple is therefore anticlockwise looking downwards and tends to turn the ship to port.

3. *A truck with four wheels, each 750 mm diameter, travels on rails round a curve of 75 m radius at a speed of 50 km/h. The rails lie in a horizontal plane and are 1.4 m apart. The total mass of the truck is 5 t and its centre of gravity is midway between the axles, 1.2 m above the rails and midway between them. The moment of inertia of each pair of wheels is 15 kg m^2.*

Determine the load on each rail.

Let $\qquad\qquad\qquad\qquad\qquad\qquad v =$ velocity of truck,

$\qquad\qquad\qquad\qquad\qquad\qquad\qquad r =$ radius of wheels

and $\qquad\qquad\qquad\qquad\qquad\qquad R =$ radius of curve.

Then, referring to Figure 16.6,

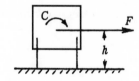

centrifugal force, $F = m\dfrac{v^2}{R}$

$$= \frac{5 \times 10^3}{75} \left(\frac{50}{3.6}\right)^2$$

$$= 12\,850 \text{ N}$$

$\therefore$ overturning couple $= F \times h$

$$= 12\,850 \times 1.2$$

$$= 15\,420 \text{ Nm}$$

Angular velocity of wheels,

$$\omega = \frac{v}{r} = \frac{50}{3.6 \times 0.375} = 37 \text{ rad/s}$$

Velocity of precession,

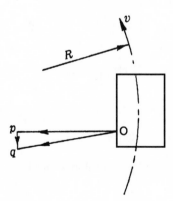

Figure 16.6

$$\Omega = \frac{v}{R} = \frac{50}{3.6 \times 75} = 0.185 \text{ rad/s}$$

Therefore gyroscopic couple due to wheel precession

$$= I\omega\Omega \qquad \text{from equation (16.1)}$$

$$= 2 \times 15 \times 37 \times 0.185 = 205 \text{ Nm}$$

As the truck turns, the angular momentum vector op turns to oq, so that a couple represented by pq is required to cause precession of the wheels. This is clockwise looking backwards, so that the reaction couple is clockwise looking forwards, tending to overturn the truck outwards in the same way as centrifugal force.

Therefore total overturning couple $= 15\,420 + 205 = 15\,625$ Nm

Reaction at rails due to overturning couple $= \dfrac{15\,625}{1.4} = 11\,150$ N

This increases the load on the outer rail and reduces that on the inner rail.

Load on each rail due to dead weight $= \dfrac{mg}{2}$

$$= \frac{5 \times 10^3 \times 9.81}{2} = 24\,550 \text{ N}$$

Therefore resultant load on outer rail $= 24\,550 + 11\,150 = \underline{35\,700 \text{ N}}$

and resultant load on inner rail $= 24\,550 - 11\,150 = \underline{13\,400 \text{ N}}$

4. *A motor cyclist travels at 140 km/h round a curve of 120 m radius. The cycle and rider have a mass of 150 kg and their centre of gravity is 0.7 m above ground level when the machine is vertical. Each wheel is 0.6 m diameter and the moment of inertia about its axis of rotation is 1.5 kg m². The engine has rotating parts whose moment of inertia about their axis of rotation is 0.25 kg m² and it rotates at five times the wheel speed in the same direction.*

Find (a) the angle of banking so that there will be no tendency to side slip, (b) the angle of inclination of the cycle and rider to the vertical.

Let $\qquad\qquad\qquad\qquad\qquad v = $ velocity of cycle,

$\qquad\qquad\qquad\qquad\qquad\qquad r = $ radius of wheels

and $\qquad\qquad\qquad\qquad\qquad R = $ radius of curve

The forces acting on the cycle, Figure 16.7, are

(i) its weight, $\qquad W = mg = 150 \times 9.81 = 1\,472$ N

(ii) the centrifugal force, $\qquad F = m\dfrac{v^2}{R}$

$$= \frac{150}{120}\left(\frac{140}{3.6}\right)^2 = 1\,890 \text{ N}$$

For no tendency to sideslip, the resultant force must be normal to the track,

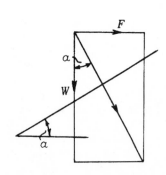

Figure 16.7

i.e. $\qquad\qquad\qquad\qquad \tan\alpha = \dfrac{F}{W} = \dfrac{1\,890}{1\,472} = 1.284$

$$\therefore \alpha = \underline{52°5'}$$

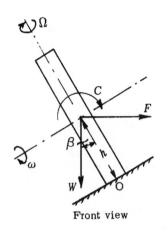

Front view

Figure 16.8

Angular velocity of wheels $= \dfrac{v}{r} = \dfrac{140}{3.6 \times 0.3} = 129.6$ rad/s

Angular velocity of engine $= 5 \times 129.6 = 648$ rad/s

If β is the inclination of the cycle to the vertical, Figure 16.8, the velocity of precession, perpendicular to the axis of spin of the wheels and engine, is given by

$$\Omega = \frac{v}{R}\cos\beta = \frac{140}{3.6 \times 120}\cos\beta = 0.324\cos\beta$$

Therefore gyroscopic couple due to precession of wheels

$$= 2 \times 1.5 \times 129.6 \times 0.324\cos\beta = 125.9\cos\beta$$

and gyroscopic couple due to precession of engine

$$= 0.25 \times 648 \times 0.324\cos\beta = 52.5\cos\beta$$

Since the directions of rotation of the wheels and engine are the same,

total gyroscopic couple, $C = (125.9 + 52.5)\cos\beta = 178.4\cos\beta$ N m

As in Example 3, the gyroscopic reaction tends to overturn the cycle outwards. Therefore, taking moments about the point of contact with the track, O,

$$C + F \times h\cos\beta = W \times h\sin\beta$$

i.e. $178.4\cos\beta + 1\,890 \times 0.7\cos\beta = 1\,472 \times 0.7\sin\beta$

from which $\tan\beta = 1.458$

$$\therefore \beta = \underline{55°33'}$$

Further problems

5. One of the driving axles of a locomotive, with its two wheels, has a moment of inertia of 350 kg m^2. The wheels are 1.85 m diameter and the distance between the planes of the wheels is 1.5 m. When travelling at 100 km/h, the locomotive passes over a defective rail which causes the right-hand wheel to fall 12 mm and rise again in a total time of 0.1 s, the vertical movement of the wheel being with S.H.M.

Find the maximum gyroscopic torque caused and state the effect this has on the locomotive.   [2.64 kN m; loco tends to swing to right]

6. A generator is arranged on board ship with its axis parallel to the longitudinal centre-line of the ship. The revolving parts have a mass of 1 400 kg, a radius of gyration of 400 mm and revolve at 420 rev/min.

Find the magnitude and sense of the gyroscopic couple exerted on the ship when it steams at 36 km/h round a curve of 180 m radius.

[547 Nm; tends to lift bow when turning left if revolving clockwise looking forward]

7. The turbine rotor of a ship has a mass of 30 t, a radius of gyration of 0.6 m and rotates at 2 400 rev/min in a clockwise direction when viewed from aft. The ship pitches through a total angle of 15°, the motion being simple harmonic with a periodic time of 12 s.

Determine the maximum gyroscopic couple on the ship and its effect as the bow rises.   [186.5 kNm; ship turns to starboard]

8. A car travels round a bend of 100 m radius. Each of the four wheels has a moment of inertia of 1.6 kg m^2 and a diameter of 0.6 m. The rotating parts of the engine have a moment of inertia of 0.85 kg m^2, the engine is parallel to the axles and the crankshaft rotates in the same sense as the wheels at three times the speed. The car has a mass of 1 400 kg and its centre of gravity is 0.45 m above road level. The width of the track of the car is 1.5 m.

Determine the limiting speed round the curve if the wheels are not to leave the road surface, which is horizontal.   [142 km/h]

9. A diesel-electric locomotive has two axles 4 m apart. The wheels on each axle are 1.2 m diameter and 1.5 m apart and the moment of inertia of each axle is 70 kg m^2. The rotating parts of the engine and generator have a moment of inertia of 60 kg m^2 and they rotate about the longitudinal axis of the locomotive at 2 200 rev/min in an anticlockwise direction when looking forward. When the speed of the locomotive is 40 km/h, it enters a left-hand curve of 150 m radius. Find the change in the vertical reactions on each wheel due to gyroscopic action.

[front outer, +192 N; front inner, +64 N;
rear outer, −64 N; rear inner, −192 N]

10. A car has a mass of 1.5 t and the centre of gravity is 0.95 m above road level. The moment of inertia of the two front wheels is 10 kg m^2 and that of the two rear wheels is 15 kg m^2. The moment of inertia of the rotating parts of the engine is 2 kg m^2 and the gear ratio from engine to wheels is 10 to 1. The engine rotates in a clockwise direction when viewed from the front of the car. The wheel diameter is 0.64 m and when the car is travelling at 80 km/h, it enters a right-hand curve of 150 m radius.

Determine the magnitude and sense of the couples on the car due to (a) centrifugal effects, (b) gyroscopic effects of wheel rotation, (c) gyroscopic effects of engine rotation.   [4.69 kNm, tending to overturn car outwards;
257 N m, tending to overturn car outwards; 206 N m, tending to lift nose]

11. A motor cycle and rider have a total mass of 320 kg, with the centre of gravity 0.525 m above ground level. The wheels each have a mass of 9 kg, a radius of gyration of 0.225 m and a rolling radius of 0.3 m. The rotating parts of the engine have a mass of 12 kg, a radius of gyration of 0.075 m and rotate in the same sense as the road wheels at 3.5 times the speed.

The machine travels round a banked curve of 60 m radius at 160 km/h. Determine the angle of banking necessary for the cycle to ride normal to the track.

[73°45′ to horizontal]

12. A solo motor cycle, complete with rider, has a mass of 225 kg, the centre of gravity being 0.6 m above ground level. The moment of inertia of each road wheel is 1 kg m^2 and the rolling diameter is 0.6 m. The engine crankshaft rotates, in the same sense as the wheels, at 5 times the speed of the wheels. The rotating parts of the engine are equivalent to a flywheel whose moment of inertia is 0.2 kg m^2.

Determine the heel-over angle required when the unit is travelling at 100 km/h in a curve of radius 60 m.   [54°36′]

13. A gyroscope turn indicator for an aeroplane consists of a uniform disc, 50 mm diameter, which rotates at 3 000 rev/min. Its bearings are carried in a frame which is free to turn in trunnions, the centre-line of the trunnions being at right angles to, and 5 mm above, the axis of rotation of the disc. On a straight course, the plane of the disc is vertical and at right angles to the fore-and-aft centre-line of the plane. Find the angle through which the frame will tilt when the speed is 240 km/h and the course is altered to a circular arc of 300 m radius.   [24°]

17 Gear trains

17.1 Introduction

Gears are used to transmit rotary motion from one shaft to another. The shafts may be parallel or inclined to one another and their speed ratio is determined by the numbers of teeth on the gears. Spur gears have teeth which are parallel with the shaft axis but helical gears have teeth which are cut on a helix; this gives a smoother and quieter drive as the engagement of mating teeth is gradual instead of instantaneous. The force on the helical tooth has a component in the axial direction which must be resisted by a thrust bearing; alternatively, double helical gears may be used in which the axial forces in the two parts balance each other.

17.2 Gear teeth definitions

Figure 17.1 shows two gears in mesh; the smaller one is termed the *pinion* and larger one the *wheel* or *spur*.

The *pitch circle diameters* (p.c.d) are the diameters of discs which would transmit the same velocity ratio by friction as the gear wheels and the pitch point is the point of contact of the two pitch circles.

Let the numbers of teeth on the pinion and wheel be respectively t and T, the pitch circle diameters d and D and the speeds ω and Ω.

Then
$$\frac{\Omega}{\omega} = \frac{d}{D} = \frac{t}{T} \qquad (17.1)$$

The *circular pitch* (p) is the distance between a point on one tooth and the corresponding point on the next tooth, measured along the pitch circle,

i.e.
$$p = \frac{\pi d}{t} = \frac{\pi D}{T} \qquad (17.2)$$

The *diametral pitch* (P) is the number of teeth per mm of p.c.d.,

i.e.
$$P = \frac{t}{d} = \frac{T}{D} = \frac{\pi}{p} \qquad (17.3)$$

The *module* (m) is the number of millimetres of p.c.d. per tooth,

i.e.
$$m = \frac{d}{t} = \frac{D}{T} = \frac{1}{P} \qquad (17.4)$$

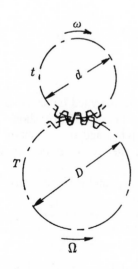

Figure 17.1

The *centre distance* $= \dfrac{d + D}{2}$ (17.5)

Gear teeth are normally of involute profile, the involute curves being generated from a *base circle* and the angle which the generator makes with the common tangent to the pitch circles is termed the *pressure angle* (ψ) since the generator is the line of contact between the mating teeth, Figure 17.2.

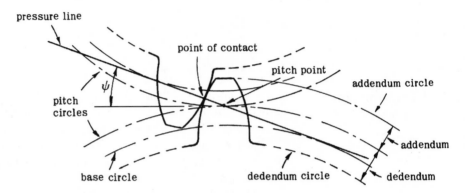

Figure 17.2

The radial depth of the tooth above the pitch circle is termed the *addendum* and that below the pitch circle the *dedendum*. The diameter of the dedendum circle is smaller than that of the base circle to give clearance to the tips of the mating teeth and that part of the tooth profile below the base circle is non-involute.

The *working depth* is the sum of the addenda of two mating teeth.

Standard gear teeth proportions are:

$$\text{addendum} = 1/P = m$$

$$\text{dedendum} = 1.25/P = 1.25\,m$$

$$\text{pressure angle} = 20°$$

17.3 Simple gear trains

A simple gear train is one in which all the gears are mounted on separate shafts. Figure 17.3(*a*) shows a train of two wheels and Figure 17.3(*b*) shows a train of three wheels. In the first case, the driving and driven shafts rotate in opposite directions and in the second case, they rotate in the same direction.

Let N_a and N_b be the speeds of rotation of wheels A and B respectively. Then, allowing for changes in direction of rotation,

$$\frac{N_a}{N_b} = -\frac{T_b}{T_a} = -\frac{D_b}{D_a} \quad \text{in the first case}$$

and

$$\frac{N_a}{N_b} = \frac{T_b}{T_a} = \frac{D_b}{D_a} \quad \text{in the second case}$$

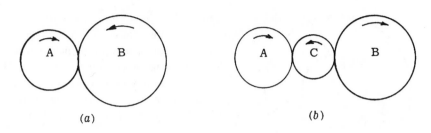

Figure 17.3

The idler C does not affect the velocity ratio between the driving and driven shafts but reverses the direction.

17.4 Compound gear trains

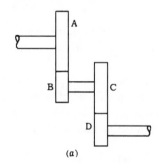

(a)

Figure 17.4

A compound gear train has more than one wheel mounted on a shaft, a simple example being shown in Figure 17.4(a). For this train, allowing for changes of direction,

$$\frac{N_a}{N_b} = -\frac{T_b}{T_a} \quad \text{and} \quad \frac{N_c}{N_d} = -\frac{T_d}{T_c}$$

Hence

$$\frac{N_a N_c}{N_b N_d} = \frac{N_a}{N_d} = \frac{T_b T_d}{T_a T_c}$$

If the input and output shafts are mounted co-axially, as shown in Figure 17.4(b), this is referred to as a co-axial or reverted train.

The gear ratio

$$\frac{N_a}{N_d} = \frac{T_b T_d}{T_a T_c}$$

as before but the equation

$$D_a + D_b = D_c + D_d$$

must also be satisfied to give equal centre distances.

If the pitch of each pair of wheels is the same, then

$$T_a + T_b = T_c + T_d$$

but the pitches of the two pairs may be made different to achieve a required velocity ratio and centre distance. Thus, if the modules of A and B are m_1 and those of C and D are m_2, then

$$(T_a + T_b)m_1 = (T_c + T_d)m_2 \tag{17.6}$$

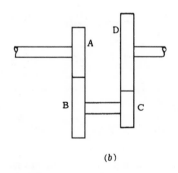

(b)

17.5 Epicyclic gear trains

An alternative method of obtaining concentric motion of input and output shafts is to use an epicyclic gear train. This design can give a high velocity ratio within a small volume, leading to considerable saving in weight and it also has the capacity of giving a change of velocity ratio by locking or freeing one member of the gear train.

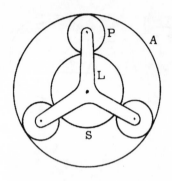

Figure 17.5

Figure 17.5 shows a simple epicyclic train in which A is an annular wheel having internal teeth, P is a planet wheel, L is an arm, star or spider carrying pins on which the planets can rotate freely and S is a sun wheel which rotates about the same axis as A. The input and output shafts may be connected to any two of the sun, annulus or arm.

Solution of an epicyclic gear problem is one of relative motion and a simple tabular method is used which involves three steps.

	Number of revs. of wheels			
	L	A	P	S
(i) Rotate whole gear train $+m$ revs	$+m$	$+m$	$+m$	$+m$
(ii) Fix arm and rotate annulus $+n$ revs	0	$+n$	$+n \cdot \dfrac{T_a}{T_p}$	$-n \cdot \dfrac{T_a}{T_s}$
(iii) Add lines (i) and (ii)	m	$m+n$	$m+n \cdot \dfrac{T_a}{T_p}$	$m-n \cdot \dfrac{T_a}{T_s}$

Line (iii) gives a general relation between the speeds of the various members and the constants m and n can then be solved from the given speeds of those members.

In many applications, one of the wheels is fixed so that the speed of that wheel is zero. This can be simply achieved by giving all members $+1$ rev in line (i) and then giving that fixed wheel -1 rev in line (ii) while the arm is held fixed. The resultant speed of the fixed wheel in line (iii) will then be zero.

If, as an example, the annulus is fixed, then, in tabular form:

	L	A	P	S
(i) Rotate whole gear train $+1$ rev	$+1$	$+1$	$+1$	$+1$
(ii) Fix arm and rotate annulus, -1 rev	0	-1	$-\dfrac{T_a}{T_p}$	$+\dfrac{T_a}{T_s}$
(iii) Add lines (i) and (ii)	1	0	$1-\dfrac{T_a}{T_p}$	$1+\dfrac{T_a}{T_s}$

17.6 Torques in gear trains

Figure 17.6 shows a gear train in which A is the input shaft, B is the output shaft and C is the casing, usually connected to the annulus.

If C_a, C_b and C_c are the external torques applied *to* the train, then C_a is in the *same* direction as the rotation N_a and C_b is in the *opposite* direction to the rotation N_b.

If there is no loss of power in the gears, the net power supplied to the train is zero.

i.e.
$$C_a N_a + C_b N_b = 0$$

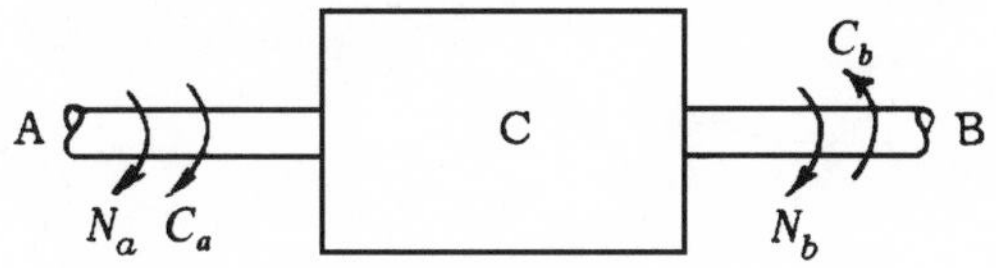

Figure 17.6

If the direction of N_a is taken as positive, then either N_b or C_b will be negative.

The magnitude and direction of the torque on C will be determined from the equilibrium of the whole train, which is represented by the equation

$$C_a + C_b + C_c = 0 \qquad (17.7)$$

Worked examples

1. *Two shafts A and D in the same line are geared together through an intermediate parallel shaft carrying wheels B and C which mesh with wheels on A and D respectively. Wheels A and B have a module of 4 and wheels C and D have a module of 9. The number of teeth on any wheel is to be not less than 15, the speed of D is to be about but not greater than 1/12 the speed of A and the ratio of each reduction is the same. Find suitable wheels, the actual reduction and the distance of the intermediate shaft from shafts A and D.*

The arrangement of the gears is as shown in Figure 17.4(b).

From equation (17.6),

$$(T_a + T_b)m_1 = (T_c + T_d)m_2$$

i.e.
$$T_a + T_b = 2.25(T_c + T_d) \qquad (1)$$

$$\frac{T_a}{T_b} = \frac{T_c}{T_d} \approx \frac{1}{\sqrt{12}}$$

$$\approx 0.288\,5 \qquad (2)$$

From equations (1) and (2) $\qquad T_b = 2.25\,T_d \qquad (3)$

and $\qquad\qquad\qquad\qquad T_a = 2.25\,T_c \qquad (4)$

$T = D/m$ and since m for C and D is 9 compared with 4 for wheels A and B, the smallest number of teeth will be on wheel C.

Thus the lowest tooth numbers to satisfy equation (4) is

$$T_c = \underline{16} \quad \text{and} \quad T_a = \underline{36}$$

From equation (2), $\qquad\qquad T_b \geq \dfrac{T_a}{0.285\,5} \geq 124.8 \qquad (5)$

The lowest tooth numbers to satisfy equations (3) and (5) are

$$T_b = \underline{126} \quad \text{and} \quad T_d = \underline{56}$$

$$\text{Actual gear ratio} = \frac{T_a}{T_b} \times \frac{T_c}{T_d}$$

$$= \frac{36}{126} \times \frac{16}{56} = \frac{1}{\underline{12.25}}$$

$$\text{Centre distance} = \frac{D_a + D_b}{2}$$

$$= (T_a + T_b) \times \frac{m_1}{2}$$

$$= (36 + 126) \times \frac{4}{2} = \underline{324 \text{ mm}}$$

2. *In the epicyclic gear shown in Figure 17.7, the driving wheel A has 14 teeth and the fixed annular wheel C, 100 teeth. The ratio of tooth numbers on wheels E and D is 98:41. If 2 kW at 1 200 rev/min is supplied to wheel A, find the speed and direction of rotation of E and the fixing torque required at C.*

$$T_b = \frac{T_c - T_a}{2}$$

$$= \frac{100 - 14}{2}$$

$$= 43$$

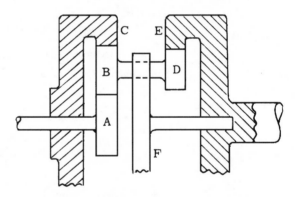

Figure 17.7

	A	B, D	C	E	F
(i) Rotate whole gear train +1 rev	+1	+1	+1	+1	+1
(ii) Fix arm and rotate C −1 rev	$+\dfrac{100}{14}$	$-\dfrac{100}{43}$	−1	$-\dfrac{100}{43} \cdot \dfrac{41}{98}$	0
(iii) Add lines (i) and (ii)	8.143	−1.326	0	0.027 1	1
Multiply by $\dfrac{1\,200}{8.143}$	1 200			3.99	

The speed of E is therefore <u>3.99 rev/min</u>, in the same direction as A.

$$C_a = \frac{2 \times 10^3}{1\,200 \times \dfrac{2\pi}{60}} = 15.92 \text{ Nm}$$

From equation (17.6), $C_a N_a + C_e N_e = 0$

$$\therefore C_e = -\frac{15.92 \times 1\,200}{3.99}$$

$$= -4\,785 \text{ N m, opposite in direction to } C_a.$$

The directions of the torques are similar to those shown in Figure 17.6 and so, from equation (17.7),

$$C_a + C_c + C_e = 0$$

$$15.92 + C_c - 4\,785 = 0$$

i.e.

$$\therefore C_c = \underline{4\,769 \text{ N m}}, \text{ in the same direction as } C_a.$$

3. *In the epicyclic gear train shown in Figure 17.8, A rotates at 1000 rev/min clockwise while E rotates at 500 rev/min anticlockwise. Determine the speed and direction of rotation of the annulus D and the shaft F. All gears are of the same pitch and the numbers of teeth are A, 30, B, 20, and E, 80.*

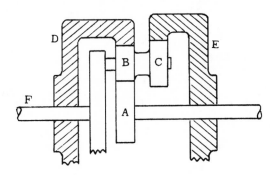

Figure 17.8

Since diameters are proportional to numbers of teeth and all wheels are of the same pitch,

$$T_d = T_a + 2T_b$$

$$= 30 + 2 \times 20 = 70$$

$$T_e = T_a + T_b + T_c$$

i.e.

$$T_c = 80 - 30 - 20 = 30$$

	F	A	D	E	B, C
(i) Rotate whole gear train $+m$ revs	$+m$	$+m$	$+m$	$+m$	$+m$
(ii) Fix arm and rotate wheel A $+n$ revs	0	$+n$	$-n \cdot \dfrac{30}{70}$	$-n \cdot \dfrac{30}{20} \cdot \dfrac{30}{80}$	$-n \cdot \dfrac{30}{20}$
(iii) Add lines (i) and (ii)	m	$m+n$	$m - \frac{3}{7}n$	$m - \frac{9}{16}n$	$m - \frac{3}{2}n$

Taking clockwise rotation as positive,

$$\text{speed of A} = +1\,000 = m + n$$

and
$$\text{speed of E} = -500 \;\; = m - \tfrac{9}{16}n$$

$$\therefore \; m = +40 \text{ and } n = +960$$

$$\therefore \; \text{speed of D} = +40 - \tfrac{3}{7} \times 960$$

$$= -371.4 \text{ rev/min}$$

and
$$\text{speed of F} = +40 \text{ rev/min}$$

4. *In the epicyclic gear shown in Figure 17.9, the gear B has 120 teeth externally and 100 teeth internally. The driver A has 20 teeth and the arm E is connected to the driven shaft. Gear D has 60 teeth. If A revolves at +100 rev/min and D revolves at +27 rev/min, find the speed of the arm, E.*

If D is now fixed and A transmits a torque of 10 N m at +100 rev/min, what will be the available torque on the arm E, assuming 96 per cent efficiency of transmission?

$$T_c = \frac{T_b - T_d}{2} = \frac{100 - 60}{2} = 20$$

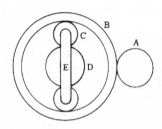

Figure 17.9

Wheel A is not a member of the epicyclic train and must therefore be treated separately.

$$\text{Speed of A} = +100 \text{ rev/min}$$

$$\therefore \; \text{speed of B} = -100 \times \frac{20}{120}$$

$$= -\frac{100}{6} \text{ rev/min}$$

	B	C	D	E
(i) Rotate whole gear $+m$ rev	$+m$	$+m$	$+m$	$+m$
(ii) Fix arm and rotate wheel $B + n$ rev	$+n$	$+n \cdot \dfrac{100}{20}$	$-n \cdot \dfrac{100}{60}$	0
(iii) Add lines (i) and (ii)	$m + n$	$m + 5n$	$m - \tfrac{10}{6}n$	m

$$\text{Speed of B} = -\frac{100}{6} = m + n$$

and
$$\text{speed of D} = +27 = m - \tfrac{10}{6}n$$

$$\therefore \; m = -0.292 \text{ and } n = -16.4$$

$$\therefore \; \text{speed of E} = m = -0.29 \text{ rev/min}$$

When D is fixed,
$$m + n = -\frac{100}{6}$$

and
$$m - \tfrac{10}{6}n = 0$$

$$\therefore \; m = -\frac{1\,000}{96} \quad \text{and} \quad n = -\frac{100}{16}$$

$$\therefore \text{ speed of E } = m = -\frac{1\,000}{96}\text{rev/min}$$

The ideal output torque is given by

$$C_a N_a + C_e N_e = 0$$

$$\therefore C_e = \frac{-10 \times 100}{-\dfrac{1\,000}{96}} = 96 \text{ Nm}$$

Since, however, the gearing efficiency is only 96%,

$$\text{available torque} = 0.96 \times 96 = \underline{92.16 \text{ Nm}}$$

Further problems

5. Two parallel shafts X and Y are to be connected by toothed wheels; wheels A and B form a compound pair which can slide along, but rotate with, shaft X; wheels C and D are rigidly attached to shaft Y and the compound pair may be moved so that A engages with C or B with D.

Shaft X rotates at 640 rev/min and the speeds of shaft Y are to be 340 rev/min exactly, and 240 rev/min as nearly as possible. Using a module of 12 for all wheels, find the minimum distance between the shaft axes, suitable tooth numbers for the wheels and the lower speed of Y.

[294 mm; A, 13; B, 17; C, 36; D, 32; 231.1 rev/min]

6. The first and third shafts of a double reduction gear are in line and a total reduction of approximately 10 to 1 is required. The module of the high speed pair is to be 5, that of the low speed pair is to be 8 and no wheel is to have fewer than 20 teeth. Obtain a suitable value for the centre distance between the first and second shafts and the numbers of teeth on the wheels to satisfy the above conditions. What is the actual gear ratio? [340 mm; A, 32; B, 104; C, 20; D, 65; 1:9.5]

7. Figure 17.10 shows an epicyclic gear train in which the wheel A is held stationary and the arm E is rotated at 300 rev/min. The compound planet B, C rotates freely on a pin carried by the arm. The numbers of teeth are as follows:

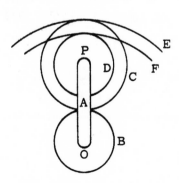

Figure 17.10

A, 50; B, 20; C, 40; D, 30.

Find the speed and direction of wheel D. [−700 rev/min]

8. In an epicyclic gear of the sun and planet type, the p.c.d. of the internally toothed ring is to be as nearly as possible 220 mm and the teeth are to have a module of 4. When the ring is stationary, the spider, which carries three planets, is to make one revolution for every five of the driving spindle carrying the sun wheel. Determine suitable numbers of teeth for all the wheels and the exact p.c.d. of the ring.

If a torque of 12 Nm is applied to the shaft carrying the sun wheel, what torque will be required to keep the ring stationary? [56, 21, 14; 224 mm; 48 Nm]

9. In the epicyclic gear shown in Figure 17.11, the pinion B and the internal wheels E and F are mounted independently on the spindle O while C and D form a compound wheel which rotates on the pin P attached to the arm A. The wheels B, C and D have 15, 30 and 25 teeth respectively, all of the same pitch.

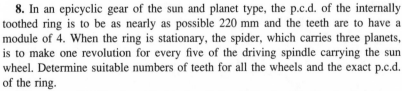

Figure 17.11

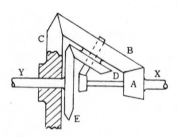

Figure 17.12

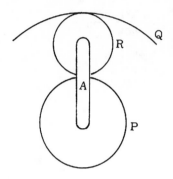

Figure 17.13

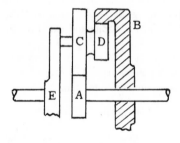

Figure 17.14

(*a*) If wheel E is fixed, what is the ratio of the speed of F to that of B?

(*b*) If wheel B is fixed, what are the ratios of the speeds of E and F to that of A?

[1:56; 6:5; 33:28]

10. Figure 17.12 shows an epicyclic gear in which the wheel P, having 45 teeth of 15 mm pitch, is geared with Q through the intermediate wheel R at the end of the arm A. When P is rotating at 63 rev/min in a clockwise direction and A is rotating at 9 rev/min, also in a clockwise direction, Q is required to rotate at 21 rev/min anticlockwise. Find the necessary numbers of teeth on Q and R and the p.c.d. of Q.

[81; 18; 386.9 mm]

11. An epicyclic gear consists of a sun wheel which has 24 teeth, planet wheels which have 28 teeth and an internally toothed annulus which is held stationary. Neglecting friction, find the torque required to hold the annulus fixed when 9 kW is being transmitted, the sun wheel rotating at 700 rev/min.

If the teeth have a module of 4, what is the diameter of the circle traced out by the centres of the planets? [409 N m; 208 mm]

12. An epicyclic gear consists of two sun wheels S_1 and S_2 with 24 and 28 teeth respectively, engaging with a compound planet with 26 and 22 teeth. S_1 is keyed to the driven shaft and S_2 is a fixed wheel co-axial with the driven shaft. The planet is carried on an arm fixed to the driving shaft. Find the velocity ratio of the gear.

If 750 W is transmitted when the output speed is 100 rev/min, what torque is required to hold S_2? [−2.61:1; 98:7 N m]

13. An epicyclic train has a sun-wheel with 30 teeth and two planet wheels of 50 teeth, the latter meshing with the internal teeth of a fixed annulus. The input shaft, carrying the sun wheel, transmits 4 kW at 300 rev/min. The output shaft is connected to an arm which carries the planet wheels. Find the speed of the output shaft and the torque transmitted if the overall efficiency is 95%.

If the annulus is rotated independently, what should be its speed if the output shaft is to rotate at 10 rev/min? [56.25 rev/min: 645 N m; −56.9 rev/min]

14. In the epicyclic gear shown in Figure 17.13, the input shaft A runs at 12 000 rev/min and the annular wheel B is fixed. Find the speed of the output shaft E and the speed of the planets relative to the spindle on which they are mounted. The tooth numbers are A, 15; B, 81; C, 41; D, 25.

[1 216 rev/min; −3 946 rev/min]

15. In the gear shown in Figure 17.14, the wheel C is fixed and shaft X rotates at 650 rev/min. Determine the speed of shaft Y. The numbers of teeth are A, 18; B, 55; C, 60; D, 24; E, 28. [9.75 rev/min]

18 Free vibrations

18.1 Introduction

When an elastic system is displaced from its equilibrium position, the internal restoring force (and hence the acceleration) is proportional to the displacement and is directed towards the equilibrium position. Thus the body oscillates with simple harmonic motion, the rate being known as the natural frequency. In practice, internal damping will oppose the vibration, which will soon cease unless maintained by external excitement.

18.2 Simple harmonic motion

If a point moves in a circular path with uniform velocity, the component of the motion along a diameter of the circle satisfies the conditions for simple harmonic motion. If the radius of the circle is X, Figure 18.1, and the radius OP rotates at a rate ω rad/s, then measuring time from the point B, the angle BOP $= \omega t$.

The linear velocity and acceleration of P are respectively ωX and $\omega^2 X$ in the directions shown in Figure 18.1, and the components of these quantities along AB give the corresponding velocity and acceleration of the point Q.

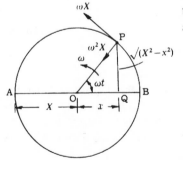

Figure 18.1

Thus
$$x = X \cos \omega t \tag{18.1}$$

$$v = \omega X \sin \omega t$$
$$= \omega \sqrt{(X^2 - x^2)} \tag{18.2}$$

and
$$a = \omega^2 X \cos \omega t$$
$$= \omega^2 x \tag{18.3}$$

Thus the acceleration of Q is proportional to the displacement from the mid-position and is always directed towards that point, so that the motion of Q is simple harmonic.

Equations (18.1), (18.2) and (18.3) give merely numerical relations between displacement, velocity and acceleration without regard to direction.

The maximum displacement, X is termed the *amplitude* of the oscillation.

The maximum velocity is ωX and occurs at the mid-position when $x = 0$.

The maximum acceleration is $\omega^2 X$ and occurs at the end of the stroke.

The periodic time, T, is the time taken for one complete oscillation,

i.e.
$$T = \frac{2\pi}{\omega}$$

But from equation (18.3), $\omega = \sqrt{\left(\frac{a}{x}\right)}$, so that

$$T = 2\pi\sqrt{\left(\frac{x}{a}\right)} \quad \text{or} \quad 2\pi\sqrt{\left(\frac{\text{displacement}}{\text{acceleration}}\right)} \tag{18.4}$$

The frequency, n, is the number of complete cycles per unit time,

i.e.
$$n = \frac{1}{T} = \frac{1}{2\pi}\sqrt{\left(\frac{a}{x}\right)} \tag{18.5}$$

The unit of frequency is the hertz (Hz), which is 1 cycle/s.

Simple harmonic motion also occurs in angular vibrations. If the angular amplitude of the motion is ϕ and the angular displacement, velocity and acceleration are θ, Ω and α respectively,

then
$$\theta = \phi \cos \omega t \tag{18.6}$$

$$\Omega = \omega\phi \sin \omega t = \omega\sqrt{(\phi^2 - \theta^2)} \tag{18.7}$$

$$\alpha = \omega^2\phi \cos \omega t = \omega^2\theta \tag{18.8}$$

$$T = 2\pi\sqrt{\left(\frac{\theta}{\alpha}\right)} \tag{18.9}$$

and
$$n = \frac{1}{2\pi}\sqrt{\left(\frac{\alpha}{\theta}\right)} \tag{18.10}$$

18.3 Linear motion of an elastic system

Consider the elastic system represented by the spring of stiffness S and the body of mass m shown in Figure 18.2. If the mass is given a displacement x from the equilibrium position, the restoring force due to the spring stiffness is Sx. When released, this force gives the mass an acceleration a which is given by

$$Sx = ma$$

Thus the acceleration is proportional to the displacement and is always directed towards the equilibrium position, so that the mass moves with simple harmonic motion.

Thus the periodic time,

$$T = 2\pi\sqrt{\left(\frac{x}{a}\right)} = 2\pi\sqrt{\left(\frac{m}{s}\right)} \tag{18.11}$$

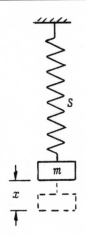

Figure 18.2

If the static deflection of the spring under the action of gravity on the mass is δ, then

$$\delta = \frac{mg}{S}$$

so that

$$\frac{m}{S} = \frac{\delta}{g}$$

$$\therefore T = 2\pi\sqrt{\left(\frac{\delta}{g}\right)} \tag{18.12}$$

The frequency

$$n = \frac{1}{2\pi}\sqrt{\left(\frac{g}{\delta}\right)} \text{ and if } \delta \text{ is measured in metres,}$$

$$g = 9.81$$

so that

$$n = \frac{1}{2.006\,5\sqrt{\delta}} \approx \frac{1}{2\sqrt{\delta}}\text{Hz} \tag{18.13}$$

18.4 Angular motion of an elastic system

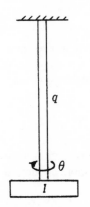

Figure 18.3

Consider the elastic system represented by the rod of torsional stiffness q and the rotor of moment of inertia I shown in Figure 18.3. If the rotor is given an angular displacement θ from the equilibrium position, the restoring torque due to the rod stiffness is $q\theta$. When released, this torque gives the rotor an acceleration α which is given by

$$q\theta = I\alpha$$

Thus the acceleration is proportional to the displacement and is always directed towards the equilibrium position, so that the rotor moves with simple harmonic motion.

Thus the periodic time, $T = 2\pi\sqrt{\left(\frac{\theta}{\alpha}\right)}$

$$= 2\pi\sqrt{\left(\frac{I}{q}\right)} \tag{18.14}$$

If the length of the rod is l, the polar second moment of area of the cross-section is J and the modulus of rigidity of the material is G, then

$$q = \frac{GJ}{l} \quad \text{from equation (4.4)}$$

Thus

$$T = 2\pi\sqrt{\left(\frac{Il}{GJ}\right)} \tag{18.15}$$

and

$$n = \frac{1}{2\pi}\sqrt{\left(\frac{GJ}{Il}\right)} \tag{18.16}$$

18.5 Effect of mass of spring and inertia of shaft

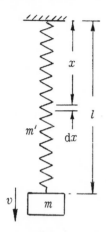

Figure 18.4

Let the mass of the spring be m', Figure 18.4. Then mass of element of length $dx = m'\dfrac{dx}{l}$. If the element is situated at a distance x from the support and the velocity of the free end is v, then

$$\text{velocity of element} = \frac{x}{l}v$$

$$\therefore \text{ K.E. of element} = \tfrac{1}{2}m'\frac{dx}{l}\left(\frac{x}{l}v\right)^2$$

$$= \tfrac{1}{2}m'\frac{v^2}{l^3}x^2 dx$$

$$\therefore \text{ total K.E. of spring} = \tfrac{1}{2}m'\frac{v^2}{l^3}\int_0^l x^2 dx$$

$$= \tfrac{1}{2}\frac{m'}{3}v^2 \qquad (18.17)$$

The mass of the spring is therefore equivalent to a mass $\dfrac{m'}{3}$ added to the concentrated mass m at the free end.

It can similarly be shown that, in the case of angular motion, the moment of inertia of the rod, I', is equivalent to a body of moment of inertia $I'/3$ added to the rotor at the free end.

18.6 The motion of a pendulum

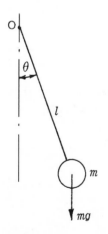

Figure 18.5

Figure 18.5 shows a simple pendulum consisting of a concentrated mass m at the end of a string of length l. If the string is given an angular displacement θ from the vertical, the restoring moment about O is $mg\,l\sin\theta$ and when released, the angular acceleration is given by

$$mg\,l\sin\theta = I_O\alpha = ml^2\alpha$$

Thus α is proportional to $\sin\theta$ and the motion is therefore not simple harmonic. If, however, θ is small, $\sin\theta \approx \theta$ and the equation of motion then becomes

$$mg\,l\theta = ml^2\alpha$$

from which

$$\frac{\theta}{\alpha} = \frac{l}{g}$$

whence

$$T = 2\pi\sqrt{\left(\frac{l}{g}\right)} \qquad (18.18)$$

The pendulum thus gives approximate S.H.M. provided that θ is kept small.

If the mass is not concentrated at a point, as shown in Figure 18.6, the pendulum becomes *compound*. If the distance between the point of

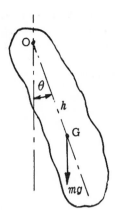

Figure 18.6

suspension O and the centre of gravity G is h, and the pendulum is displaced through a small angle θ, restoring moment about O $\approx mgh\theta$. When released, the angular acceleration is given by

$$mgh\theta = I_O\alpha$$

$$\therefore \frac{\theta}{\alpha} = \frac{I_O}{gh}$$

whence

$$T = 2\pi\sqrt{\left(\frac{I_O}{mgh}\right)} \qquad (18.19)$$

By the theorem of parallel axes (see Appendix A),

$$I_O = I_G + mh^2$$
$$= m(k_G^2 + h^2)$$

so that the periodic time may be expressed in the form

$$T = 2\pi\sqrt{\left(\frac{k_G^2 + h^2}{gh}\right)} \qquad (18.20)$$

18.7 Two-rotor torsional system

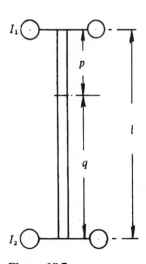

Figure 18.7

Figure 18.7 shows a shaft carrying rotors of moments of inertia I_1 and I_2 at its ends. When set in torsional oscillation, the rotors will move in opposite directions at any instant and there will be a stationary point (called a node) somewhere between the two rotors.

Let the node be at distances p and q from I_1 and I_2 respectively. Then treating each as a rotor on the end of a shaft attached to a fixed point,

$$n_1 = \frac{1}{2\pi}\sqrt{\left(\frac{GJ}{I_1 p}\right)} \qquad (18.21)$$

and

$$n_2 = \frac{1}{2\pi}\sqrt{\left(\frac{GJ}{I_2 q}\right)} \qquad (18.22)$$

The frequency of each rotor must be the same, so that

$$I_1 p = I_2 q$$

or

$$\frac{p}{q} = \frac{I_2}{I_1} \qquad (18.23)$$

Also

$$p + q = l \qquad (18.24)$$

p and q may be determined from equations (18.23) and (18.24) and the frequency of oscillation is then obtained by substitution in either equation (18.21) or equation (18.22).

Worked examples

1. *In a mechanism, a crosshead moves in a straight guide with simple harmonic motion. At distances of 125 mm and 200 mm from its mean position, the crosshead has velocities of 6 and 3 m/s respectively. Determine (a) the amplitude of the motion, (b) the maximum velocity, and (c) the periodic time.*
   If the crosshead has a mass of 0.2 kg, what is the maximum inertia force?

From equation (18.2), $v = \omega\sqrt{(X^2 - x^2)}$

$$\therefore 6 = \omega\sqrt{(X^2 - 0.125^2)}$$

and $\qquad 3 = \omega\sqrt{(X^2 - 0.200^2)}$

$$\therefore \omega = 33.3 \text{ rad/s}$$

and $\qquad\qquad X = \underline{0.219\,5 \text{ m}}$

$\qquad v_{max} = \omega X \qquad = 33.3 \times 0.219\,5 = \underline{7.3 \text{ m/s}}$

$$T = \frac{2\pi}{\omega} \qquad = \frac{2\pi}{33.3} = \underline{0.188 \text{ s}}$$

$\qquad a_{max} = \omega^2 X \qquad = 33.3^2 \times 0.219\,5 = 243 \text{ m/s}$

$$\therefore F_{max} = 0.2 \times 243 = \underline{48.7 \text{ N}}$$

2. *In the system shown in Figure 18.8, the upper spring has a stiffness of 1 000 N/m and the lower spring 500 N/m. The suspended mass is 0.5 kg. Find the natural frequency of vertical vibration.*

If a force F is applied to the compound spring, Figure 18.8,

$$\text{extension of upper spring} = \frac{F}{1\,000}$$

$$\text{and extension of lower spring} = \frac{F}{500}$$

Therefore effective stiffness of compound spring

$$= \frac{F}{\dfrac{F}{1\,000} + \dfrac{F}{500}}$$

$$= \frac{500}{1.5} \text{ N/m}$$

$$n = \frac{1}{T} = \frac{1}{2\pi}\sqrt{\left(\frac{S}{m}\right)} \qquad \text{from equation (18.11)}$$

$$= \frac{1}{2\pi}\sqrt{\left(\frac{500}{1.5 \times 0.5}\right)}$$

$$= \underline{4.11 \text{ Hz}}$$

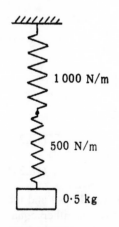

1 000 N/m

500 N/m

0·5 kg

Figure 18.8

3. *A thin uniform bar of mass m is placed on two identical discs which rotate with equal but opposite speeds. Show that if the centre of gravity of the bar is displaced from the mid-position, the bar will oscillate with a periodic time given by*

$$T = 2\pi \sqrt{\left(\frac{l}{\mu g}\right)}$$ *where μ is the coefficient of friction between the discs and the bar and 2 l is the distance between the centres of the two discs.*

Let the centre of gravity of the bar be displaced a distance x from the midposition, Figure 18.9.

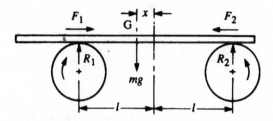

Figure 18.9

Taking moments about R_2,

$$mg(l + x) = R_1 \times 2l$$

$$\therefore R_1 = \frac{mg}{2l}(l + x)$$

Similarly,

$$R_2 = \frac{mg}{2l}(l - x)$$

If F_1 and F_2 are the friction forces exerted on the bar by the discs, then

$$F_1 = \mu R_1 = \frac{\mu mg}{2l}(l + x) \quad \text{and} \quad F_2 = \mu R_2 = \frac{\mu mg}{2l}(l - x)$$

Therefore resultant force $F = \dfrac{\mu mg}{2l}(l + x) - \dfrac{\mu mg}{2l}(l - x) = \dfrac{\mu mgx}{l}$

The acceleration of the bar is then given by

$$a = \frac{F}{m} = \frac{\mu gx}{l}$$

Hence

$$T = 2\pi \sqrt{\left(\frac{\text{displacement}}{\text{acceleration}}\right)}$$

$$= 2\pi \sqrt{\left(\frac{x}{\frac{\mu gx}{l}}\right)} = 2\pi \sqrt{\left(\frac{l}{\mu g}\right)}$$

4. *A large gear wheel has a mass of 2 t and is suspended from a knife-edge so that it is free to swing in a vertical plane at right-angles to the gear axis. If the point of suspension is 0.7 m from the gear axis and the periodic time is 2.25 s, determine (a) the moment of inertia of the gear about its axis, and (b) the minimum possible periodic time if the point of suspension can be moved.*

From equation (18.20),
$$T = 2\pi\sqrt{\frac{k_G^2 + h^2}{gh}}$$

i.e.
$$2.25 = 2\pi\sqrt{\frac{k_G^2 + 0.7^2}{9.81 \times 0.7}}$$

from which
$$k_G^2 = 0.392$$

$$\therefore I_G = mk_G^2$$

$$= 2 \times 10^3 \times 0.392 = \underline{784 \text{ kg m}^2}$$

For minimum periodic time, $\dfrac{\mathrm{d}T}{\mathrm{d}h} = 0$

i.e.
$$\frac{\mathrm{d}}{\mathrm{d}h}\frac{k_G^2 + h^2}{h} = 0$$

from which
$$h = k_G = 0.626 \text{ m}$$

$$\therefore T_{\min} = 2\pi\sqrt{\frac{2k_G}{g}}$$

$$= 2\pi\sqrt{\frac{2 \times 0.626}{9.81}} = \underline{2.245 \text{ s}}$$

5. *A uniform bar AB, 2.5 m long and mass 100 kg, is supported on a hinge at one end A and on a spring support at the other end B so that it can vibrate in a vertical plane. The stiffness of the spring is 20 kN/m and when in static equilibrium, the bar is horizontal.*

The end B of the bar is depressed 10 mm and then released. Calculate (a) the frequency of the vibrations and (b) the maximum angular acceleration of the bar.

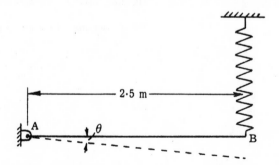

Figure 18.10

If the beam is depressed through a small angle θ, Figure 18.10,

$$\text{extension of spring} = 2.5\theta \text{ m}$$

$$\therefore \text{ restoring force of spring} = 2.5\theta \times 20 = 50\theta \text{ kN}$$

$$\therefore \text{ restoring moment about A} = 50\theta \times 2.5 = 125\theta \text{ kN m}$$

$$\therefore 125 \times 10^3 \theta = I_A \alpha = \frac{ml^2}{3}\alpha$$

$$= \frac{100 \times 2.5^2}{3}\alpha$$

$$\therefore \frac{\alpha}{\theta} = 600 \tag{1}$$

$$\therefore n = \frac{1}{2\pi}\sqrt{(600)} = \underline{3.9 \text{ Hz}}$$

When extension of spring is 10 mm,

$$\theta = \frac{0.010}{2.5} = \frac{1}{250} \text{ rad}$$

$$\therefore \alpha = \frac{600}{250} \quad \text{from equation (1)}$$

$$= \underline{2.4 \text{ rad/s}^2}$$

6. *A vertical steel wire, 2 mm diameter and 2 m long, is fixed at its upper end and at the lower end, a solid steel cylinder is attached centrally so that its axis is horizontal. The cylinder is 75 mm diameter and its density is 7.8 Mg/m³. Find the length of the cylinder to give 0.6 torsional oscillation per second.*

Calculate the amplitude of the vibrations when the maximum shearing stress is 120 MN/m². G = 80 GN/m².

For a solid cylinder of length L and radius R, the moment of inertia about the axis of rotation AB, Figure 18.11, is given by

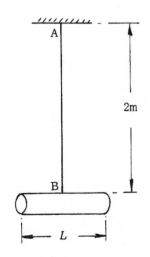

Figure 18.11

$$I = m\left(\frac{L^2}{12} + \frac{R^2}{4}\right)$$

$$= \frac{\rho\pi R^2 L}{12}(L^2 + 3R^2)$$

$$= \frac{7.8 \times 10^3 \pi \times 0.037\,5^2 L(L^2 + 3 \times 0.037\,5^2)}{12}$$

$$= 2.87L(L^2 + 0.004\,22)$$

For the wire $J = \frac{\pi d^4}{32}$

$$= \frac{\pi \times 2^4 \times 10^{-12}}{32} = 1.571 \times 10^{-12} \text{ m}^4$$

$$n = \frac{1}{2\pi} \sqrt{\left(\frac{GJ}{Il}\right)} \quad \text{from equation (18.16)}$$

i.e.

$$0.6 = \frac{1}{2\pi} \sqrt{\left(\frac{80 \times 10^9 \times 1.571 \times 10^{-12}}{2.87L(L^2 + 0.004\,22) \times 2}\right)}$$

i.e. $L(L^2 + 0.004\,22) = 0.001\,54$

By trial or plotting, $L = \underline{0.103\,5\ \text{m}}$

From equation (4.4), $\dfrac{\tau}{r} = \dfrac{G\theta}{l}$

$$\therefore \theta = \frac{120 \times 10^6 \times 2}{80 \times 10^9 \times 0.001}$$

$$= \underline{3\ \text{rad}} \quad \text{or} \quad \underline{172°}$$

7. *A rotor has a mass of 225 kg and a radius of gyration of 0.4 m. It is bolted between the ends of two shafts, one of which is 75 mm diameter, 0.9 m long and the other is 65 mm diameter, 0.45 m long. The other ends of the shafts are rigidly fixed in position. Find the frequency of natural torsional vibration of the rotor. $G = 80\ GN/m^2$.*

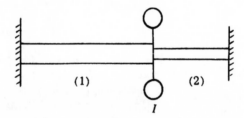

Figure 18.12

The arrangement is shown in Figure 18.12.

From equation (4.5), $\dfrac{T}{J} = \dfrac{G\theta}{l}$

Therefore when the rotor is twisted through an angle θ,

$$\text{restoring torque, } T = \frac{GJ_1\theta}{l_1} + \frac{GJ_2\theta}{l_2}$$

Hence $G\theta \left(\dfrac{J_1}{l_1} + \dfrac{J_2}{l_2}\right) = I\alpha$

$$\therefore n = \frac{1}{2\pi} \sqrt{\left(\frac{\alpha}{\theta}\right)} = \frac{1}{2\pi} \sqrt{\left(\frac{G}{I}\left(\frac{J_1}{l_1} + \frac{J_2}{l_2}\right)\right)}$$

$$= \frac{1}{2\pi} \sqrt{\left(\frac{80 \times 10^9}{225 \times 0.4^2} \times \frac{\pi}{32}\left(\frac{75^4}{0.9} + \frac{65^4}{0.45}\right) \times 10^{-12}\right)}$$

$$= \underline{20.3\ \text{Hz}}$$

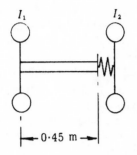

Figure 18.13

8. *The flywheel of an engine driving a dynamo has a mass of 135 kg and a radius of gyration of 250 mm. The armature has a mass of 100 kg and a radius of gyration of 200 mm. The driving shaft is 450 mm long and 50 mm diameter and a spring coupling is incorporated at one end, having a stiffness of 28 kN m/rad. Determine the natural frequency of torsional vibration of the system. G = 80 GN/m².*

The arrangement is shown in Figure 18.13.

The torsional stiffness of a shaft is given by

$$q = \frac{T}{\theta} = \frac{GJ}{l}$$

$$\therefore l = \frac{GJ}{q}$$

Thus the spring coupling is equivalent to a length of shaft of diameter 50 mm given by

$$l = \frac{80 \times 10^9 \times \dfrac{\pi}{32} \times 0.05^4}{28 \times 10^3} = 1.753 \text{ m}$$

Therefore the actual shaft and coupling is equivalent to a uniform shaft of length $0.45 + 1.753 = 2.203$ m, as shown in Figure 18.14.

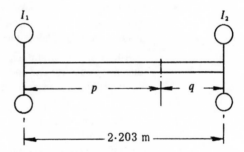

Figure 18.14

Hence, from equations (18.23) and (18.24),

$$\frac{p}{q} = \frac{I_2}{I_1} = \frac{100 \times 0.20^2}{135 \times 0.25^2} = 0.474$$

and

$$p + q = 2.203$$

$$\therefore p = 0.708 \text{ m} \quad \text{and} \quad q = 1.495 \text{ m}$$

Therefore, from equation (18.21),

$$n = \frac{1}{2\pi} \sqrt{\left(\frac{80 \times 10^9 \times \dfrac{\pi}{32} \times 0.05^4}{135 \times 0.25^2 \times 0.708} \right)}$$

$$= \underline{14.45 \text{ Hz}}$$

Further problems

9. A mass of 25 kg is suspended from a spring of stiffness 14 kN/m and vibrates with an amplitude of 12 mm. Find the periodic time, the velocity and acceleration when displaced 8 mm from the equilibrium position and the time taken to move from this position to the position of maximum displacement.

[0.266 s; 0.212 m/s; 4.48 m/s^2; 0.035 6 s]

10. A body of mass 5 kg oscillates in a horizontal straight line with S.H.M. The amplitude of the oscillation is 0.9 m and the maximum horizontal force acting on the body is 300 N. Determine (a) the frequency of the oscillations, (b) the velocity of the body at a point 0.45 m from the mid-position, (c) the time taken for the body to travel 0.45 m from one extreme end of the oscillation.

[1.3 Hz; 6.363 m/s; 0.128 s]

11. A particle moving with S.H.M. performs 10 complete oscillations per minute and its speed, when at a distance of 200 mm from the centre of oscillation, is $\frac{3}{5}$ of the maximum speed. Find the amplitude, the maximum acceleration and the speed of the particle when it is 150 mm from the centre of oscillation.

[0.25 m; 0.274 2 m/s^2; 0.209 4 m/s]

12. A rough horizontal table moves horizontally with S.H.M., the period being 3 s and the maximum speed 1.2 m/s. A small heavy mass is placed on the table. Find the least coefficient of friction if the mass does not slide on the table throughout the motion.   [0.256]

13. A 2 kg mass is hung from the end of a helical spring and is set vibrating vertically. The mass makes 100 complete oscillations in 55 s. Determine the stiffness of the spring. Also calculate the maximum amplitude of vibration if the mass is not to leave the hook during its motion.   [261 N/m; 75.2 mm]

14. A spring of stiffness 200 N/m has a mass of 0.75 kg. A mass of 5 kg is attached to the free end and set in vertical motion. Find the frequency of oscillation (a) neglecting the mass of the spring; (b) allowing for the mass of the spring.

[1.005 Hz; 0.982 Hz]

15. A horizontal shaft, supported in bearings at the ends, deflects at the centre by 0.005 mm per 100 N of load applied there. When a wheel of mass 300 kg is centrally fitted, the system responds to an external disturbance and free vertical vibrations of amplitude 0.25 mm are set up. Calculate the frequency, the maximum velocity and the maximum acceleration for this vibration.

[41.1 Hz; 0.064 5 m/s; 16.67 m/s^2]

16. A thin uniform rod AB, of mass 1 kg and length 0.6 m, carries a concentrated mass of 2.5 kg at B. The rod is hinged at A and is maintained in a horizontal position by a vertical spring of stiffness 1.8 kN/m attached at its mid-point. Find the frequency of oscillation in the vertical plane.   [2.01 Hz]

17. A bar has a mass of 5 kg and is pivoted at one end. The radius of gyration of the bar about the pivot is 0.6 m. The bar is supported in the horizontal position by a vertical spring of stiffness 500 N/m and mass 3 kg which is attached to the bar at a point 0.4 m from the pivot. Determine the periodic time of oscillation in the vertical plane.   [0.98 s]

18. A balloon, ascending vertically with constant acceleration from rest on the ground, reaches a height of 360 m in one minute. It carries a pendulum clock which keeps correct time on the ground. Show that it will gain about 36.7 seconds per hour.

19. A connecting rod is swung as a compound pendulum about a knife-edge, first at its small end when the time for 100 complete swings is 99 s and then at its large end, when the time for 100 swings is 85 s. The distance between the two points of suspension measured on the rod is 300 mm. Find the position of its centre of gravity and its radius of gyration about its centre of gravity.

[95.7 mm from large end; 89.5 mm]

20. A wheel is mounted on a knife-edge on the inside surface of its rim at a distance of 400 mm from its centre of gravity. It is found to make 100 complete vibrations in 2 min 40 s. Calculate its radius of gyration about an axis through the centre of gravity. [307 mm]

21. In order to determine the radius of gyration of a wheel, it is swung from a knife-edge as a compound pendulum. When the knife-edge is 900 mm from the centre of gravity, it is found that the wheel makes twice as many swings in a given time as it does when the knife-edge is 125 mm from the centre of gravity. What is the radius of gyration of the wheel? [988.5 mm]

22. A torsional pendulum consists of a wire 0.5 m long, 10 mm diameter, fixed at its upper end and attached at its lower end to a disc of moment of inertia 0.06 kg m^2. The modulus of rigidity of the wire is 44 GN/m^2. Find the frequency of torsional oscillation of the disc.

If the maximum displacement to one side of the rest position is 5°, find the maximum angular velocity and acceleration of the disc.

[6.04 Hz; 3.31 rad/s; 125.5 rad/s^2]

23. A solid metal cylinder 450 mm diameter is suspended with its axis vertical by means of a wire coaxial with the cylinder and rigidly attached to it. The stiffness of the wire is 22 N m per radian of twist. Find the necessary mass of the cylinder so that when it is given a small angular displacement about its axis, it will make 40 vibrations per minute. [49.5 kg]

24. One end of a shaft, 0.9m long and 25 mm diameter, is fixed and a coaxial cylinder, 150 mm long and 100 mm diameter, is attached at the other end. Both the shaft and cylinder are made of steel of density 7.8 Mg/m^2. Calculate the frequency of torsional oscillation of the system, allowing for the inertia of the shaft.
$G = 85$ GN/m^2. [84.6 Hz]

25. A uniform shaft is rigidly fixed at both ends and a rotor is attached to the shaft at some intermediate point. Show that the frequency of torsional oscillation is a minimum when the rotor is attached at the mid-point.

26. A rotor of mass 34 kg and radius of gyration 1.15 m is fixed to one end of a shaft and another, of mass 16 kg and radius of gyration 1.4 m, is fixed to the other end. The shaft is 0.45 m long and 45 mm diameter. Calculate the frequency of torsional vibration if $G = 80$ GN/m^2. [9.9 Hz]

27. The moving parts of a radial engine have a total moment of inertia of 0.8 kg m^2 and are concentrated in the plane of the single crank pin. The engine is connected to a propeller of moment of inertia 15 kg m^2 by a hollow shaft of length 250 mm and outer and inner diameters 75 and 32 mm respectively. The stiffness of the crank alone is 2.5 MN m/rad. Determine the frequency of torsional vibration of the system if $G = 80$ GN/m^2. [152 Hz]

28. Two rotors A and B are fixed to the ends of a shaft of length 530 mm and the node is to be at a section C distant 330 mm from A. Rotor A has a mass of 40 kg and radius of gyration 140 mm while rotor B has a mass of 18 kg and radius of gyration 160 mm. If the diameter of part AC is 45 mm, find the diameter of part CB and the frequency of torsional oscillation. $G = 80 \text{ GN/m}^2$.

[34.7 mm; 56.1 Hz]

19 Transverse vibrations and whirling speeds

19.1 Light beam with single load

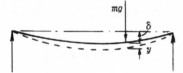

Figure 19.1

The beam shown in Figure 19.1 may be considered in the same way as a mass attached to the end of a spring, Section 18.3. If the mass m is displaced a distance y from the equilibrium position, the restoring force is given by Sy, where S is the stiffness of the beam, i.e., the force required at the load point per unit deflection at that point. This is of the form $S = kEI/l^3$ where k is a constant depending on the loading and method of support (see Chapter 5).

When released, the acceleration of the mass is given by

$$Sy = ma$$

Thus the acceleration is proportional to the displacement of the load, so that the motion is simple harmonic. The frequency is then given by

$$n = \frac{1}{2\pi}\sqrt{\left(\frac{a}{y}\right)} = \frac{1}{2\pi}\sqrt{\left(\frac{S}{m}\right)} \quad \text{(see Section 18.3)}$$

If the static deflection at the load point is δ, then $\delta = \dfrac{mg}{S}$

or

$$\frac{S}{m} = \frac{g}{\delta}$$

Hence

$$n = \frac{1}{2\pi}\sqrt{\left(\frac{g}{\delta}\right)}$$

If δ is measured in metres, $g = 9.81$ m/s^2

so that

$$n \approx \frac{1}{2\sqrt{\delta}} \text{ Hz} \tag{19.1}$$

19.2 Uniformly distributed load

Exact analysis of this case is beyond the scope of this book but a very close approximation may be obtained by assuming that the vibrating beam is of the same shape as the static deflection curve, i.e., that the amplitude Y at

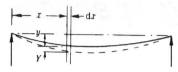

Figure 19.2

any point is proportional to the static deflection y at that point

or $\qquad\qquad Y = cy$ where c is a constant.

If the mass of the beam is m per unit length then, considering an element of length dx, Figure 19.2, the additional load required to deflect the beam a further distance Y

$$= mg\, dx \times \frac{Y}{y}$$

Work done by this load in moving through the distance Y

$$= \frac{1}{2}\left(mg\, dx \times \frac{Y}{y}\right) \times Y$$

$$= \tfrac{1}{2} mg\, c^2 y\, dx$$

Therefore, for the whole beam,

$$\text{work done} = \tfrac{1}{2} mg\, c^2 \int_0^l y\, dx \tag{19.2}$$

This represents the strain energy stored in the beam at maximum displacement, which must equal the kinetic energy of the beam as it passes through the mid-position.

At the mid-point, velocity of element $= \omega Y$ from equation (18.2)

Hence, K.E. of element in mid-position $= \tfrac{1}{2} m\, dx (\omega Y)^2$

$$= \tfrac{1}{2} m\omega^2 c^2 y^2\, dx$$

Therefore total K.E. of beam $= \tfrac{1}{2} m\omega^2 c^2 \int_0^l y^2\, dx \tag{19.3}$

Equations (19.2) and (19.3) give $\omega^2 = g\dfrac{\displaystyle\int_0^l y\, dx}{\displaystyle\int_0^l y^2\, dx}$

so that $\qquad\qquad n = \dfrac{\omega}{2\pi} = \dfrac{1}{2\pi}\sqrt{\left(g\dfrac{\displaystyle\int_0^l y\, dx}{\displaystyle\int_0^l y^2\, dx}\right)}$

If y is measured in metres,

$$n \approx \frac{1}{2}\sqrt{\left(\frac{\displaystyle\int_0^l y\, dx}{\displaystyle\int_0^l y^2\, dx}\right)} \tag{19.4}$$

For a simply supported beam, taking the origin at one end,

$$y = \frac{mg}{24EI} \left(l^3 x - 2lx^3 + x^4 \right)$$

Note: It is only *magnitudes* of deflections which are relevant to the subsequent analysis.

This gives

$$n = 4.935 \sqrt{\left(\frac{EI}{mgl^4} \right)}$$

$$= \frac{0.564}{\sqrt{\Delta}} \tag{19.5}$$

where Δ is the maximum static deflection of the beam.

For a cantilever, taking the origin at the fixed end,

$$y = \frac{mg}{24EI} \left(6l^2 x^2 - 4lx^3 + x^4 \right)$$

Note: It is only *magnitudes* of deflections which are relevant to the subsequent analysis.

This gives

$$n = 1.755 \sqrt{\left(\frac{EI}{mgl^4} \right)}$$

$$= \frac{0.624}{\sqrt{\Delta}} \tag{19.6}$$

This is known as the energy method. The error resulting from the initial assumption is extremely small and quite accurate results may also be obtained by assuming the shape to be a sine wave or parabola.

19.3 Several loads

(a) Energy method

Assuming that the vibrating beam is similar in shape to the static deflection curve, the frequency may be obtained in the same way as for a distributed load by equating the strain energy at maximum displacement to the kinetic energy in the mid-position. This leads to the equation

$$n = \frac{1}{2} \sqrt{\left(\frac{\Sigma my}{\Sigma my^2} \right)} \text{ Hz} \tag{19.7}$$

The deflection y under each load must be calculated with all the loads *acting together*.

Any distributed mass may be allowed for by adding to each concentrated mass the mass of the beam between the centres of the sections into which the loads divide the beam.

A good approximation can again be obtained by assuming the shape of the vibrating beam to be a sine wave or parabola instead of the static deflection curve.

(b) Dunkerley's method

This approximate method relates the frequency with all the loads acting together with those of each of the loads when acting alone. If n_1, n_2, etc., are the frequencies of the concentrated masses m_1, m_2, etc., when each acts alone on the beam and n_0 is the frequency of the beam due to its own distributed mass, then

$$\frac{1}{n^2} = \frac{1}{n_1^2} + \frac{1}{n_2^2} + \ldots + \frac{1}{n_0^2}$$

But $n_1 = \dfrac{1}{2\sqrt{\delta_1}}$, $n_2 = \dfrac{1}{2\sqrt{\delta_2}}$, etc., and $n_0 = \dfrac{0.564}{\sqrt{\Delta}}$ for a simply supported beam, so that

$$n = \frac{1}{\sqrt{(4\delta_1 + 4\delta_2 + \ldots + 3.144\Delta)}}$$

$$= \frac{1}{2\sqrt{\left(\delta_1 + \delta_2 + \ldots + \dfrac{\Delta}{1.27}\right)}} \tag{19.8}$$

This is an empirical relation and gives results which are less accurate than those given by the energy method.

19.4 Whirling speed of shafts

Figure 19.3 shows a shaft carrying a rotor of mass m, the c.g. of which has an eccentricity e relative to the shaft axis O. When the shaft rotates at a rate ω a centrifugal force $m\omega^2(y+e)$ will act upon the rotor, causing a deflection y relative to the static deflection δ. The restoring force due to the shaft stiffness S is Sy, so that, for equilibrium,

$$m\omega^2(y+e) = Sy$$

from which

$$y = \frac{e}{\dfrac{S}{m\omega^2} - 1} \tag{19.9}$$

When $\dfrac{S}{m\omega^2} = 1$, y will become infinite and the shaft is said to *whirl*.

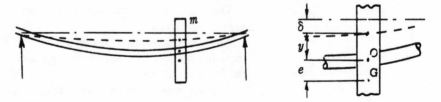

Figure 19.3

The whirling, or critical, speed ω_c is therefore given by

$$\omega_c^2 = \frac{S}{m} = \frac{g}{\delta}$$

so that the whirling speed

$$n_c = \frac{\omega_c}{2\pi} = \frac{1}{2\pi}\sqrt{\left(\frac{g}{\delta}\right)} \text{ rev/s}$$

$$\approx \frac{1}{2\sqrt{\delta}} \text{ where } \delta \text{ is in metres} \tag{19.10}$$

Thus the whirling speed is identical with the frequency of transverse vibration of the same shaft, carrying the same load and this applies equally to other systems of distributed and multiple loads.

At speeds other than the critical speed, the deflection of the shaft relative to the static deflection position may be expressed in the form

$$y = \frac{e}{\dfrac{\omega_c^2}{\omega^2} - 1} = \frac{\omega^2 e}{\omega_c^2 - \omega^2} \tag{19.11}$$

Equation (19.11) shows that when $\omega < \omega_c$, e and y are of the same sign, i.e. G is to the outside of O but when $\omega > \omega_c$, e and y are of opposite sign, so that G lies between O and the static deflection curve. As the speed increases, $y \to -e$ so that eventually G lies on the static deflection curve.

Figure 19.4 shows the numerical relation between y and ω. In practice, internal damping in the shaft material will prevent the deflection from becoming infinite when $\omega = \omega_c$ but the speed should be kept well away from the critical speed to prevent damage to the shaft.

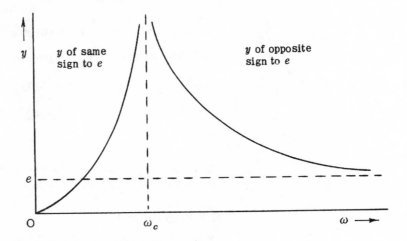

Figure 19.4

Worked examples

1. *A steel strip 10 mm wide and 0.8 mm thick rests on supports 200 mm apart and a mass of 0.15 kg is fixed to the strip at mid-span. Find the frequency of transverse vibration.*

If the greatest bending stress in the strip is 100 MN/m², calculate the amplitude of movement of the mass and also the least force on the supports. Neglect the mass of the strip. E = 200 GN/m².

$$\text{Static deflection under load} = \frac{Wl^3}{48EI} \quad \text{from equation (5.15)}$$

$$= \frac{0.15 \times 9.81 \times 0.2^3}{48 \times 200 \times 10^9 \times \dfrac{10 \times 0.8^3}{12 \times 10^{12}}}$$

$$= 0.002\,875 \text{ m}$$

$$\therefore n = \frac{1}{2\sqrt{0.002\,875}} \quad \text{from equation (19.1)}$$

$$= \underline{9.32 \text{ Hz}}$$

$$\text{Greatest force on strip} = \text{weight} + \text{maximum inertia force}$$

$$= mg + m\omega^2 Y \quad \text{where } Y \text{ is the amplitude}$$

$$= m\left[g + (2\pi n)^2 Y\right]$$

$$= 0.15\left[9.81 + (2\pi \times 9.32)^2 Y\right]$$

$$= 0.15(9.81 + 3\,430Y)$$

$$\sigma_{\max} = \frac{M}{Z} \quad \text{from equation (3.2)}$$

$$= \frac{Wl}{4} \Big/ \frac{bd^2}{6} = \frac{3Wl}{2bd^2}$$

$$\therefore 100 \times 10^6 = \frac{3 \times 0.15(9.81 + 3\,430Y) \times 0.2}{2 \times 10 \times 0.8^2 \times 10^{-9}}$$

from which
$$Y = 0.001\,29 \text{ m} \quad \text{or} \quad \underline{1.29 \text{ mm}}$$

$$\text{Least force on each support} = \tfrac{1}{2}\left(mg - m\omega^2 Y\right)$$

$$= \frac{0.15}{2}(9.81 - 3\,430 \times 0.001\,29)$$

$$= \underline{0.404 \text{ N}}$$

2. *A simply supported beam of length 3a carries two loads W situated at distances a and 2a from one end. Calculate the frequency of transverse vibration.*

(a) Energy method

Referring to Figure 19.5, the deflections under the loads are obtained by Macaulay's method (Section 5.5). Taking the origin at the left-hand end,

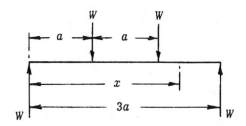

Figure 19.5

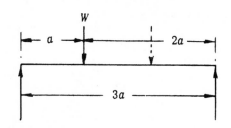

Figure 19.6

$$EI\frac{d^2 y}{dx^2} = Wx - W[x - a] - W[x - 2a]$$

$$\therefore EI\frac{dy}{dx} = \frac{W}{2}x^2 - \frac{W}{2}[x - a]^2 - \frac{W}{2}[x - 2a]^2 + A$$

$$\therefore EIy = \frac{W}{6}x^3 - \frac{W}{6}[x - a]^3 - \frac{W}{6}[x - 2a]^3 + Ax + B$$

When $x = 0$, $y = 0$, $\therefore B = 0$

When $x = 3a$, $y = 0$, $\therefore A = -Wa^2$

Therefore, when $x = a$, $EIy = \frac{Wa^3}{6} - Wa^3$

$$\therefore y = -\frac{5Wa^3}{6EI}$$

Note: It is only the *magnitude* of the deflection which is relevant to the subsequent analysis.

Since the beam is symmetrical, the deflection under the second load is identical.

Hence $$\Sigma my = 2 \times \frac{W}{g}\left(\frac{5Wa^3}{6EI}\right)$$

and $$\Sigma my^2 = 2 \times \frac{W}{g}\left(\frac{5Wa^3}{6EI}\right)^2$$

$$\therefore n = \frac{1}{2}\sqrt{\left(\frac{\Sigma my}{\Sigma my^2}\right)} = \frac{1}{2}\sqrt{\left(\frac{6EI}{5Wa^3}\right)}   \text{ from equation (19.7)}$$

$$= \underline{0.548\sqrt{\left(\frac{EI}{Wa^3}\right)}}$$

(b) Dunkerley's method

The numerical value of the deflection under each load assumed to be acting alone, Figure 19.6, is given by

$$\delta = \frac{Wa^2(2a)^2}{3EI(3a)} = \frac{4Wa^3}{9EI}   \text{ from equation (5.25)}$$

$$\therefore n = \frac{1}{2\sqrt{(\delta_1 + \delta_2)}} \quad \text{from equation (19.8)}$$

$$= \frac{1}{2\sqrt{\left(2 \times \dfrac{4Wa^3}{9EI}\right)}}$$

$$= 0.531\sqrt{\left(\frac{EI}{Wa^3}\right)}$$

3. *A shaft of diameter 40 mm is supported in flexible bearings 0.6 m apart. It carries a rotor of mass 100 kg at its centre, the c.g. of the rotor being 0.01 mm eccentric to the shaft axis.*

(a) Determine the critical speed of rotation, allowing for the mass of the shaft which has a density of 7.8 Mg/m³.

(b) Find the speed range over which the maximum deflection of the shaft relative to its static position will exceed 0.5 mm.
$E = 200 \ GN/m^2$.

The arrangement is shown in Figure 19.7.

Static deflection due to mass of rotor

$$= \frac{Wl^3}{48EI} \quad \text{from equation (5.15)}$$

$$= \frac{100 \times 9.81 \times 0.6^3}{48 \times 200 \times 10^9 \times \frac{\pi}{64} \times 0.04^4}$$

$$= 175.6 \times 10^{-6} \ \text{m}$$

Static deflection due to mass of shaft

$$= \frac{5wl^4}{384EI} \quad \text{from equation (5.17)}$$

$$= \frac{5}{384} \times \frac{\frac{\pi}{4} \times 0.04^2 \times 7.8 \times 10^3 \times 9.81 \times 0.6^4}{200 \times 10^9 \times \frac{\pi}{64} \times 0.04^4}$$

$$= 6.45 \times 10^{-6} \ \text{m}$$

$$\therefore n = \frac{1}{2\sqrt{\left(\delta + \dfrac{\Delta}{1.27}\right)}} \quad \text{from equation (19.8)}$$

$$= \frac{1}{2\sqrt{\left(175.6 \times 10^{-6} + \dfrac{6.45 \times 10^{-6}}{1.27}\right)}}$$

$$= 37.1 \ \text{rev/s}$$

From equation (19.11), $\quad y = \dfrac{e}{\dfrac{\omega_c^2}{\omega^2} - 1} = \dfrac{e}{\dfrac{n_c^2}{n^2} - 1}$

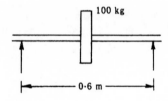

Figure 19.7

Since y may be of the same or opposite sign to e, depending on whether ω is less than or greater than ω_c,

$$0.5 = \pm \frac{0.01}{\dfrac{37.1^2}{n^2} - 1}$$

from which $n = \underline{36.8 \text{ and } 37.5 \text{ rev/s}}$

Further problems

4. Determine the frequency of transverse vibration of an *I*-section beam 6 m long simply supported at its ends which carries a uniformly distributed load of 750 kg/m. The second moment of area of the cross-section is 140×10^{-6} m^4 and $E = 200$ GN/m^2.   [8.46 Hz]

5. Find the frequency of transverse vibration of a cantilever turbine blade of uniform section 125 mm long, having a mass of 2 kg/m and second moment of area 2 700 mm^4. $E = 200$ GN/m^2.   [590 Hz]

6. A beam of I-section has a span of 3 m and is supported at the ends. The mass of the beam is 200 kg/m and the second moment of area of the section is 16×10^{-6} m^4. Two equal loads of one tonne are carried at points 1 m from each support.

Find the natural frequency of transverse vibration of the system, if $E = 200$ GN/m^2.   [8.8 Hz]

7. A shaft 50 mm diameter and 0.8 m long is simply supported at its ends and carries three loads each of 36 kg, one at the centre and one 0.2 m from each end. Calculate the frequency of transverse vibration, given that the deflection under the central load is $\dfrac{19 \, mgl^3}{384 \, EI}$ and under each end load it is $\dfrac{9 \, mgl^3}{256 \, EI}$. $E = 200$ GN/m^2.   [45.5 Hz]

8. A shaft is simply supported on bearings 3 m apart and carries five equal concentrated loads equally spaced with the end loads 0.3 m from each bearing. If the maximum static deflection is 2.5 mm, estimate the frequency of transverse vibration of the shaft when the static deflection is assumed to be (*a*) a sine wave, (*b*) a parabola.   [11.37 Hz; 11.30 Hz]

9. A beam of mass 18 kg/m is simply supported on a span of 3.6 m and carries a body of mass 250 kg at the centre. The second moment of area of the cross-section is 8×10^{-6} m^4 and $E = 200$ GN/m^2. Find the frequency of transverse vibrations.   [12.2 Hz]

10. A beam, simply supported on a span of 6 m, has a mass of 52 kg/m. The second moment of area of the cross-section is 120×10^{-6} m^4 and $E = 200$ GN/m^2. A load of 3 t is carried to a point 2.4 m from one support.

Find the frequency of transverse vibration of the beam, neglecting the mass of the beam. Then find the approximate frequency if the mass of the beam is taken into account.   [7.01 Hz; 6.99 Hz]

11. A light shaft, 20 mm diameter, is supported in flexible bearings 0.8 m apart and carries a rotor of mass 20 kg at its centre. Determine the whirling speed, given that $E = 200$ GN/m^2.

Determine also the speed range over which the deflection will exceed 2 mm if the eccentricity of the centre of gravity of the rotor from its geometrical axis is 0.5 mm.   [13.67 rev/s; 12.23 to 15.78 rev/s]

12. A light shaft carries a rotor in which the centre of gravity is 0.5 mm from the axis of rotation. If the whirling speed is 750 rev/min, find the speed range over which the deflection of the rotor, relative to the static position, will exceed 1.25 mm.   [633.8 to 968.2 rev/min]

13. A light shaft carrying a rotor is to be run at $1\frac{1}{2}$ times its critical speed. If the centrifugal deflection is not to exceed 0.25 mm, find the greatest permissible displacement of the centre of gravity from the axis of rotation.   [0.139 mm]

14. A light shaft carries a single disc whose centre of gravity is at a distance e from the shaft axis. The whirling speed is 3 600 rev/min and at a speed of 3 240 rev/min, the centre of gravity revolves in a circle of radius 3.8 mm. Calculate the distance e.   [0.722 mm]

20 Damped and forced vibrations

20.1 Introduction

All vibrations are damped, either externally by a dash-pot or eddy current damper, or by internal hysteresis forces within the system. The damping forces or torques are assumed to be proportional to the velocity of the vibration.

Damping will cause the vibration to die away unless sustained by an externally applied harmonic force or torque, which may be applied directly to the body or through its support. Examples of such a disturbance are an out-of-balance rotor or the reciprocating masses in an engine.

Immediately on starting, the motion of a body will be a combination of its free vibration and that due to the periodic disturbance. When the free vibration has died away, the steady-state vibration is of the same frequency as the disturbance.

The analysis of damped and forced vibrations necessitates the use of differential equations of motion, the acceleration of the mass being due to elastic forces, damping forces and/or disturbing forces.

20.2 Differential equation of free vibration

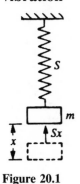

Figure 20.1

The motion of a freely vibrating body, Figure 20.1, is given by the equation

$$ma = F$$

i.e.

$$m\frac{d^2x}{dt^2} = -Sx$$

The negative sign arises because the restoring force is opposite in direction to the displacement.

This may be written as

$$\frac{d^2x}{dt^2} + \omega^2 x = 0$$

where

$$\omega^2 = \frac{S}{m}$$

The solution (see Appendix B) is

$$x = A \cos \omega t + B \sin \omega t \quad \text{where } A \text{ and } B \text{ are constants}$$

If the motion starts when $x = X$, corresponding with the point B of Figure 18.1, then $x = X$ when $t = 0$ and $\dfrac{dx}{dt} = 0$ when $t = 0$.

Thus $\qquad\qquad\qquad\qquad\qquad A = X \quad \text{and} \quad B = 0$

so that $\qquad\qquad\qquad\qquad\qquad x = X \cos \omega t$

This is an oscillatory motion of period $\dfrac{2\pi}{\omega}$, so that

$$T = 2\pi \sqrt{\left(\frac{m}{s}\right)}$$

The corresponding differential equation for angular motion is

$$I \frac{d^2\theta}{dt^2} = -q\theta$$

or $\qquad\qquad \dfrac{d^2\theta}{dt^2} + \omega^2\theta = 0 \quad \text{where } \omega^2 = \dfrac{q}{I}$

20.3 Damped vibrations

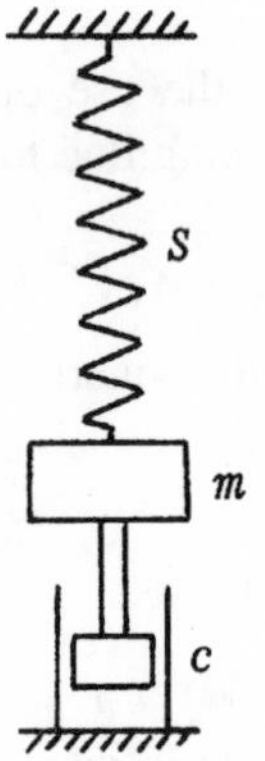

Figure 20.2

Figure 20.2 shows an elastic system represented by a mass m supported by a spring of stiffness S, the motion being opposed by a damper which exerts a damping force c per unit velocity; c is called the *damping coefficient*. The equation of motion is given by

$$m \frac{d^2x}{dt^2} = -Sx - c\frac{dx}{dt}$$

The negative signs arise because the restoring force is opposite in direction to the displacement and the damping force is opposite in direction to the velocity.

This may be written as

$$\frac{d^2x}{dt^2} + 2\mu\frac{dx}{dt} + \omega^2 x = 0 \qquad\qquad (20.1)$$

where $\qquad\qquad 2\mu = \dfrac{c}{m} \quad \text{and} \quad \omega^2 = \dfrac{S}{m}.$

If μ is equal to or greater than ω, the mass, when disturbed, will slowly return to its equilibrium position without oscillation but if μ is less than ω, the mass will overshoot its equilibrium position and then oscillate with decreasing amplitude until it eventually comes to rest.

The solution for this case (see Appendix B) is

$$x = e^{-\mu t}\{A \cos \sqrt{(\omega^2 - \mu^2)}t + B \sin \sqrt{(\omega^2 - \mu^2)}t\} \qquad (20.2)$$

The periodic time

$$T = \frac{2\pi}{\sqrt{(\omega^2 - \mu^2)}}$$ (20.3)

and the frequency

$$n = \frac{\sqrt{(\omega^2 - \mu^2)}}{2\pi}$$ (20.4)

Figure 20.3 illustrates the motion represented by equation (20.2)

If $x = X_1$ when $t = 0$

then $x = X_2$ when $t = T = \dfrac{2\pi}{\sqrt{(\omega^2 - \mu^2)}}$

$$\therefore \quad \frac{X_1}{X_2} = \frac{e^0(A\cos 0 + B\sin 0)}{e^{-\mu T}(A\cos 2\pi + B\sin 2\pi)} = e^{\mu T}$$ (20.5)

so that $\ln \dfrac{X_1}{X_2} = \mu T = \dfrac{2\pi\mu}{\sqrt{(\omega^2 - \mu^2)}}$ (20.6)

The term $\dfrac{2\pi\mu}{\sqrt{(\omega^2 - \mu^2)}}$ is known as the *logarithmic decrement*.

For angular vibrations, in which a rotor of moment of inertia I is supported by a shaft of torsional stiffness q and is subjected to a damping torque c per unit angular velocity, the equation of motion is

$$I\frac{d^2\theta}{dt^2} = -q\theta - c\frac{d\theta}{dt}$$

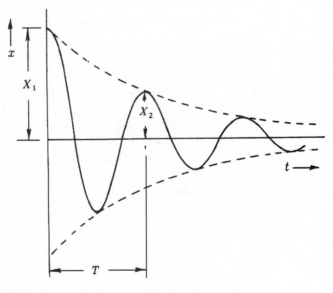

Figure 20.3

which can be written

$$\frac{d^2\theta}{dt^2} + 2\mu\frac{d\theta}{dt} + \omega^2\theta = 0 \tag{20.7}$$

where

$$2\mu = \frac{c}{I} \quad \text{and} \quad \omega^2 = \frac{q}{I}$$

The solution is similar to that for linear vibrations.

20.4 Forced vibrations

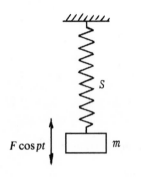

Figure 20.4

Figure 20.4 shows a system represented by a mass m supported by a spring of stiffness S, the mass being subjected to a harmonic disturbing force $F \cos pt$. The equation of motion is given by

$$m\frac{d^2x}{dt^2} = -Sx + F\cos pt$$

or

$$\frac{d^2x}{dt^2} + \omega^2 x = \frac{F}{m}\cos pt \text{ where } \omega^2 = \frac{S}{m} \tag{20.8}$$

The solution (see Appendix B) is

$$x = A\cos\omega t + B\sin\omega t + \frac{F\cos pt}{m(\omega^2 - p^2)} \tag{20.9}$$

The first two terms (i.e. the complementary function) represent the free vibration of the body, which dies out, leaving

$$x = \frac{F\cos pt}{m(\omega^2 - p^2)}$$

(i.e. the particular integral) to represent the steady-state or sustained vibration. This is a harmonic motion of frequency $\frac{p}{2\pi}$ Hz and amplitude

$$X = \frac{F}{m(\omega^2 - p^2)} \tag{20.10}$$

If p is less than ω, the body oscillates in phase with the disturbing force but if p is greater than ω, the body oscillates 180° out of phase with the disturbing force and the amplitude then becomes $F/(m(p^2 - \omega^2))$.

When $p = \omega$, the amplitude becomes infinite and *resonance* occurs. The frequency at resonance is known as the *critical* frequency of the system.

The force transmitted to the support is the sum of the weight of the body and the dynamic force due to the vibration.

Hence,   maximum force on support $= W + SX$

$$= mg + \frac{SF}{m(\omega^2 - p^2)} \tag{20.11}$$

For the corresponding angular case of a rotor of moment of inertia I supported by a shaft of torsional stiffness q and subjected to a harmonic torque $\tau \cos pt$, the equation of motion is

$$I\frac{d^2\theta}{dt^2} = -q\theta + \tau \cos pt$$

or $\qquad \dfrac{d^2\theta}{dt^2} + \omega^2\theta = \dfrac{\tau}{I}\cos pt \quad \text{where} \quad \omega^2 = \dfrac{q}{I} \qquad (20.12)$

20.5 Forced damped vibrations

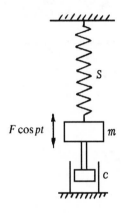

Figure 20.5

Figure 20.5 shows the combination of forcing and damping considered in the preceding articles.

This leads to the equation of motion

$$m\frac{d^2x}{dt^2} = -Sx - c\frac{dx}{dt} + F\cos pt$$

or $\qquad \dfrac{d^2x}{dt^2} + 2\mu\dfrac{dx}{dt} + \omega^2 x = \dfrac{F}{m}\cos pt \qquad (20.13)$

The complementary function is

$$x = e^{-\mu t}\{A\cos\sqrt{(\omega^2 - \mu^2)}t + B\sin\sqrt{(\omega^2 - \mu^2)}t\}$$

as in equation (20.2).

This motion dies away, leaving the steady-state vibration represented by the particular integral (see Appendix B), which is

$$x = \frac{F\cos(pt - \alpha)}{m\sqrt{(4\mu^2 p^2 + (\omega^2 - p^2)^2)}} \qquad (20.14)$$

where $\qquad \alpha = \tan^{-1}\dfrac{2\mu p}{\omega^2 - p^2} \qquad (20.15)$

This sustained motion has a frequency $\dfrac{p}{2\pi}$ Hz and amplitude

$$X = \frac{F}{m\sqrt{(4\mu^2 p^2 + (\omega^2 - p^2)^2)}} \qquad (20.16)$$

lagging behind the disturbing force by the *phase angle* α.

When $p = \omega$, the amplitude is not infinite, as in undamped vibrations, but is equal to $\dfrac{F}{2\mu m p}$ and the maximum amplitude does not occur when $p = \omega$ but when $\dfrac{dX}{dp} = 0$.

If the amplitude of the disturbing force F is constant, regardless of the frequency of application p, the amplitude of the mass varies in the forms shown in Figure 20.6, depending upon the ratio μ/ω. In most practical cases, however, such as out-of-balance rotors or piston inertia, the magnitude of F is proportional to p^2 and this leads to curves of the form shown in Figure 20.7. For both types of disturbing force, the effect of damping is small except in the region of resonance, i.e. when $p = \omega$.

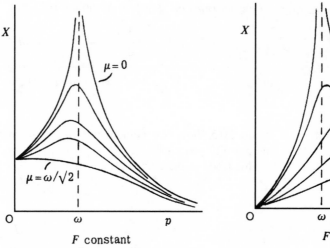

Figure 20.6   Figure 20.7

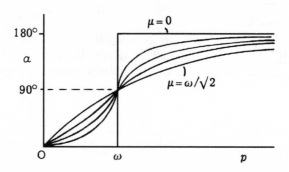

Figure 20.8

The variation of phase angle α with frequency p is shown in Figure 20.8. For forced damped angular vibrations, the corresponding equation of motion is

$$I\frac{\mathrm{d}^2\theta}{\mathrm{d}t^2} = -q\theta - c\frac{\mathrm{d}\theta}{\mathrm{d}t} + \tau\cos pt$$

or

$$\frac{\mathrm{d}^2\theta}{\mathrm{d}t^2} + 2\mu\frac{\mathrm{d}\theta}{\mathrm{d}t} + \omega^2\theta = \frac{\tau}{I}\cos pt \qquad (20.17)$$

20.6 Periodic movement of support

Vibrations may be forced upon a system by the periodic movement of the support instead of by the application of a harmonic force to the mass. In the undamped system shown in Figure 20.9, let the movement of the support be represented by the equation

$$y = Y\cos pt$$

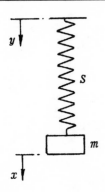

Figure 20.9

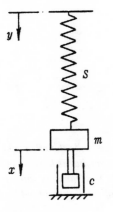

Figure 20.10

Then change of spring length $= x - y$

Therefore restoring force $\quad = S(x - y)$

The equation of motion is therefore

$$m\frac{\mathrm{d}^2x}{\mathrm{d}t^2} = -S(x - y)$$

or $\qquad \dfrac{\mathrm{d}^2x}{\mathrm{d}t^2} + \omega^2 x = \omega^2 y \quad \text{where} \quad \omega^2 = \dfrac{S}{m}$

$$= \omega^2 Y \cos pt \qquad (20.18)$$

The solution is similar to that for equation (20.8), giving

$$x = \frac{\omega^2 Y \cos pt}{\omega^2 - p^2} \qquad (20.19)$$

For the damped system shown in Figure 20.10, the equation of motion is

$$\frac{\mathrm{d}^2x}{\mathrm{d}t^2} + 2\mu\frac{\mathrm{d}x}{\mathrm{d}t} + \omega^2 x = \omega^2 Y \cos pt \qquad (20.20)$$

The solution is similar to that for equation (20.13), giving

$$x = \frac{\omega^2 Y \cos(pt - \alpha)}{\sqrt{(4\mu^2 p^2 + (\omega^2 - p^2)^2)}} \qquad (20.21)$$

Similar equations may be obtained for angular motion.

Amplitude curves for different degrees of damping are similar to those shown in Figure 20.6.

Worked examples

1. *A machine of mass 100 kg is supported on springs which deflect 20 mm under the load. The machine vibrates in a vertical plane and a dash-pot is fitted to reduce the amplitude of free vibration to one quarter of its initial value in two complete oscillations.*

Calculate the damping coefficient and compare the frequencies of the damped and undamped vibrations of the system.

$$T = 2\pi\sqrt{\left(\frac{g}{\delta}\right)} = \frac{2\pi}{\omega} \quad \text{from Sections 18.2 and 18.3}$$

$$\therefore \omega^2 = \frac{g}{\delta}$$

$$= \frac{9.81}{0.020} \quad = 490.5 \ (\text{rad/s})^2$$

$$\frac{X_1}{X_2} = e^{\mu T} \quad \text{from equation (20.5)}$$

Similarly $\quad \dfrac{X_2}{X_3} = e^{\mu T}$

$$\therefore \frac{X_1}{X_3} = \frac{X_1}{X_2} \times \frac{X_2}{X_3} = e^{2\mu T} = 4$$

$$\therefore \mu T = \frac{2\pi\mu}{\sqrt{(\omega^2 - \mu^2)}} = \ln 2$$

$$\therefore 2\pi\mu = 0.693\,1\sqrt{(490.5 - \mu^2)}$$

$$\therefore \mu = 2.43 = \frac{c}{2m} \quad \text{from equation (20.1)}$$

$$\therefore c = 2.43 \times 2 \times 100 = \underline{486 \text{ Ns/m}}$$

$$\text{Frequency of damped vibrations} = \frac{\sqrt{(\omega^2 - \mu^2)}}{2\pi} \quad \text{from equation (20.4)}$$

$$= \frac{\sqrt{(490.5 - 2.43^2)}}{2\pi}$$

$$= \underline{3.505 \text{ Hz}}$$

$$\text{Frequency of undamped vibrations} = \frac{\omega}{2\pi}$$

$$= \frac{\sqrt{(490.5)}}{2\pi}$$

$$= \underline{3.525 \text{ Hz}}$$

2. *A uniform bar of mass m and length l is hinged at one end while the other end is carried by a spring of stiffness S so that in the rest position, the bar is horizontal. Half-way along the bar, a dash-pot is attached which produces a damping force c per unit velocity. Obtain an expression for the periodic time of oscillation about the hinge.*

Referring to Figure 20.11, $I_A = \dfrac{ml^2}{3}$

Let the free end be depressed a small distance x and released.

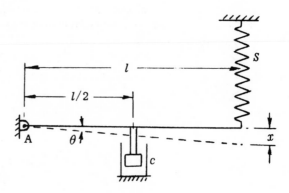

Figure 20.11

$$\text{Restoring force in spring} = Sx$$

$$\therefore \text{ moment about A} = Sx \times l$$

$$\text{Velocity of dash-pot} = \frac{1}{2}\frac{dx}{dt}$$

$$\therefore \text{ damping force} = \frac{c}{2}\frac{dx}{dt}$$

$$\therefore \text{ moment about A} = \frac{c}{2}\frac{dx}{dt} \times \frac{l}{2} = \frac{cl}{4}\frac{dx}{dt}$$

The equation of motion is therefore

$$I\frac{d^2\theta}{dt^2} = -Slx - \frac{cl}{4}\frac{dx}{dt}$$

i.e.
$$\frac{ml^2}{3} \times \frac{1}{l}\frac{d^2x}{dt^2} + \frac{cl}{4}\frac{dx}{dt} + Slx = 0 \quad \text{since} \quad x = l\theta \quad \text{and} \quad \frac{d^2x}{dt^2} = l\frac{d^2\theta}{dt^2}$$

i.e.
$$\frac{d^2x}{dt^2} + 2\mu\frac{dx}{dt} + \omega^2x = 0 \quad \text{where} \quad 2\mu = \frac{3c}{4m} \quad \text{and} \quad \omega^2 = \frac{3S}{m}$$

$$\therefore T = \frac{2\pi}{\sqrt{(\omega^2 - \mu^2)}} \quad \text{from equation (20.3)}$$

$$= \frac{2\pi}{\sqrt{\left(\frac{3S}{m} - \left(\frac{3c}{8m}\right)^2\right)}}$$

$$= \frac{16\pi m}{\sqrt{(192\ Sm - 9c^2)}}$$

3. *A light helical spring carries a mass of 5 kg at its lower end and the upper end is attached to a pin which moves in a vertical path with S.H.M., with a total stroke of 40 mm. When the frequency of oscillation is 200 cycles/min, the total movement of the mass is 30 mm. Find (a) the stiffness of the spring, (b) the maximum spring force, (c) the natural frequency of the system.*

From equation (20.19), $x = \dfrac{\omega^2 Y \cos pt}{\omega^2 - p^2}$

From Figure 20.6, it is evident that the amplitude of the mass can only be less than that of the disturbance if the frequency is above the critical, i.e. if $p > \omega$. Thus, for this condition, the mass is oscillating 180° out of phase with the disturbance and the amplitude is given by

$$X = \frac{\omega^2 Y}{p^2 - \omega^2} \quad \text{(as in Section 20.4)}$$

i.e.
$$15 = \frac{\omega^2 \times 20}{\left(200 \times \dfrac{2\pi}{60}\right)^2 - \omega^2}$$

from which $\qquad \omega^2 = 188(\text{rad/s})^2$

Hence stiffness $\qquad S = \omega^2 m$

$$= 188 \times 5 = \underline{940\ \text{N/m}}$$

As the mass and disturbance are 180° out of phase,

$$\text{maximum extension of spring} = \frac{0.04 + 0.03}{2} = 0.035\ \text{m}$$

$$\therefore\ \text{maximum spring force} = 0.035 \times 940 + 5 \times 9.81$$

$$= \underline{82\ \text{N}}$$

$$\text{Frequency } n = \frac{\omega}{2\pi} = \frac{\sqrt{188}}{2\pi} = \underline{2.18\ \text{Hz}}$$

4. *A machine, fixed to the floor of a workshop, produces a static deflection of 2 mm immediately under the machine. When the machine is working, there is an unbalanced mass which produces a vertical alternating force whose frequency is equal to the speed of the machine shaft. When the speed is 240 rev/min, the amplitude of the forced vibration of the floor is 1.25 mm. If the floor is assumed to be elastic and damping is neglected, what will be the amplitude of the forced vibration when the speed is 480 rev/min?*

At what speed will resonance occur?

$$\omega^2 = \frac{g}{\delta} = \frac{9.81}{0.002} = 4\,905(\text{rad/s})^2 \quad \text{as in Example 1}$$

From equation (20.10), $\quad X = \dfrac{F}{m(\omega^2 - p^2)}$

The inertia force on the unbalanced mass will be proportional to the square of the machine speed,

i.e. $\qquad\qquad\qquad\qquad F = kp^2$

When the machine speed is 240 rev/min, $p = 240 \times \dfrac{2\pi}{60} = 8\pi$ rad/s

When the machine speed is 480 rev/min, $p = 16\pi$ rad/s

Therefore, at 240 rev/min, $\qquad\qquad 1.25 = \dfrac{k(8\pi)^2}{m[4\,905 - (8\pi)^2]}$ $\qquad$ (1)

At 480 rev/min, $\qquad\qquad\qquad X = \dfrac{k(16\pi)^2}{m[4\,905 - (16\pi)^2]}$ $\qquad$ (2)

Therefore, from equations (1) and (2)

$$\frac{X}{1.25} = \frac{4\,905 - 64\pi^2}{64\pi^2} \times \frac{256\pi^2}{4\,905 - 256\pi^2}$$

from which $\qquad\qquad X = \underline{8.98\ \text{mm}}$

At resonance, $$p = \omega = \sqrt{4\,905} = 70 \text{ rad/s}$$

$$\therefore \text{ speed} = 70 \times \frac{60}{2\pi} = 669 \text{ rev/min}$$

5. *A machine, mounted on elastic supports, is free to vibrate vertically. The machine has a mass of 40 kg and rotor out-of-balance effects are equivalent to 2 kg at 150 mm radius. Resonance occurs when the machine is run at 621 rev/min, the amplitude of vibration at this speed being 45 mm. Determine the amplitude when running at 500 rev/min and find the angular position of the out-of-balance mass when the machine is at its highest position during vibration.*

Since the amplitude at resonance is not infinite, the vibration is damped and from equation (20.16),

$$X = \frac{F}{m\sqrt{(4\mu^2 p^2 + (\omega^2 - p^2)^2)}}$$

where $$F = mp^2 r = 2p^2 \times 0.15 = 0.3 p^2$$

At resonance, $$p = \omega = 621 \times \frac{2\pi}{60} = 65 \text{ rad/s}$$

and $$X = \frac{F}{2\mu m p}$$

i.e. $$0.045 = \frac{0.3 \times 65^2}{40 \times 2\mu \times 65}$$

$$\therefore 2\mu = 10.84 \text{ rad/s}$$

At 500 rev/min, $$p = 500 \times \frac{2\pi}{60} = 52.36 \text{ rad/s}$$

$$\therefore X = \frac{0.3 \times 52.36^2}{40\sqrt{(10.84^2 \times 52.36^2 + (65^2 - 52.36^2)^2)}}$$

$$= 0.013 \text{ m} \quad \text{or} \quad \underline{13 \text{ mm}}$$

From equation (20.15), the phase angle,

$$\alpha = \tan^{-1} \frac{2\mu p}{\omega^2 - p^2}$$

$$= \tan^{-1} \frac{10.84 \times 52.36}{65^2 - 52.36^2}$$

$$= \underline{21°}$$

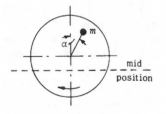

Figure 20.12

Figure 20.12 shows the rotor and the position of the out-of-balance mass when the machine is in the highest position.

6. *A rotor attached to the lower end of a vertical rod is observed to make one oscillation in $\frac{1}{2}$ s and the amplitude of the second oscillation is half that of the first. If the top of the rod is now compelled to make angular oscillations of period 2 s and amplitude 5°, find the amplitude of the rotor oscillations.*

As in equation (20.5),

$$\frac{\phi_1}{\phi_2} = e^{\mu T}$$

i.e.

$$2 = e^{\mu \times 1/2}$$

from which

$$\mu = 1.386 \text{ rad/s}$$

From equation (20.3),

$$T = \frac{2\pi}{\sqrt{(\omega^2 - \mu^2)}}$$

$$\therefore \frac{1}{2} = \frac{2\pi}{\sqrt{(\omega^2 - 1.386^2)}}$$

from which

$$\omega^2 = 159.1 \text{ (rad/s)}^2$$

If the angular amplitude of the periodic disturbance is γ, then, from equation (20.21), the amplitude of the rotor is given by

$$\phi = \frac{\omega^2 \gamma}{\sqrt{(4\mu^2 p^2 + (\omega^2 - p^2))}}$$

If the periodic time is 2 s,

$$p = \frac{2\pi}{2} = \pi \, rad/s$$

$$\therefore \phi = \frac{159.1 \times 5°}{\sqrt{(4 \times 1.386^2 \pi^2 + (159.1 - \pi^2)^2)}}$$

$$= \underline{5.33°}$$

Further problems

7. A mass suspended from a spring is subjected to damping proportional to the velocity. The frequency of damped vibrations is 1.5 Hz and the amplitude decreases to 20% of its initial value in one complete vibration. Determine the frequency of free undamped vibrations of the system.   [1.55 Hz]

8. A mass of 14 kg is suspended from a spring of stiffness 8.4 kN/m and due to damping, the amplitude of the vibration diminishes to 1/10 of its original value in two complete vibrations. Find the frequency of vibration and the value of the damping coefficient.   [3.83 Hz; 123.6 Ns/m]

9. A mass of 2.4 kg suspended from a spring is pulled downwards and released. The subsequent motion is controlled by a viscous damper such that the ratio of the first downward displacement to the third is 4 : 1 and five vibrations are completed in 4 s. Find the stiffness of the spring and the damping coefficient.
    [150 N/m; 4.16 Ns/m]

10. A disc of moment of inertia 0.6 kg m² is attached to one end of a shaft of stiffness 5 N m/rad, the other end of the shaft being fixed. Find the frequency of vibration of the disc when it is subjected to a damping torque of 0.3 N m s/rad and also the ratio of successive amplitudes on the same side of the equilibrium position.
    [0.458 Hz; 1.726]

11. A flywheel of mass 10 kg makes rotational oscillations under the control of a torsion spring of stiffness 4 N m/rad. Calculate the radius of gyration of the flywheel if the periodic time of oscillation is 2.5 s.

When a viscous damper is fitted to the system, the ratio of successive amplitudes on the same side is 0.1. Find the periodic time of the damped oscillation.

[0.252 m; 2.665 s]

12. A mass of 35 kg is supported by a spring of stiffness 25 kN/m and is acted upon by a disturbing force of amplitude 40 N and frequency 5 Hz. Find the amplitude of forced vibration.   [4.185 mm]

13. A mass of 100 kg is suspended from a spring of stiffness 4.5 kN/m. The upper end of the spring is given S.H.M. in a vertical direction by means of a crank 6 mm long. Determine the total vertical movement of the mass and also its maximum velocity when the crank is driven at 20 rad/s.   [1.52 mm; 15.2 mm/s]

14. An undamped vibrating system is excited by a sinusoidal force of constant amplitude but variable frequency. The amplitude of vibration is 25 mm at 10 Hz and this decreases continuously to 2.5 mm at 20 Hz. Determine the critical frequency of the system and the static deflection of the load.   [8.165 Hz; 3.73 mm]

15. An undamped vibrating system, excited by a sinusoidal force of constant magnitude but variable frequency, has a total travel of 100 mm at 3 Hz and 50 mm at 9 Hz and the critical frequency lies between these two rates. Determine the natural frequency of the system and the magnitude of the vibrating mass if the stiffness of the suspension is 7 kN/m.   [5.7 Hz; 5.38 kg]

16. An engine rests on an elastic foundation which deflects 0.85 mm under the dead load. Find the frequency of free vertical vibration.

If the engine has a mass of 1.25 t and when running at 450 rev/min there is an out-of-balance force of this frequency and amplitude 2.4 kN, find the amplitude of the forced vibration. Find also the maximum force exerted on the foundation.

[17.15 Hz; 0.206 mm; 15.22 kN]

17. A mass of 2 kg vibrating on a spring of stiffness 15 kN/m is subjected to a damping force of 7 Ns/m. Find the amplitude of the forced vibration of the mass when it is acted upon by a periodic disturbing force of 25 cos 100t N.

[4.95 mm]

18. A machine of mass $\frac{1}{2}$ t stands on a floor which deflects 6 mm under the load and the floor exerts a damping force of 6 kN per m/s. When the machine runs at 500 rev/min, the amplitude of vibration of the floor under the machine is 5 mm. Calculate the maximum value of the disturbing force within the machine.

What would be the amplitude at resonance?   [3.18 kN; 7.82 mm]

19. A mass of 100 kg is suspended from a spring of stiffness 10 kN/m and the motion of the mass is damped by a frictional resistance proportional to the velocity and of value 220 N s/m.

If the top of the spring vibrates vertically with an amplitude of 2.5 mm and frequency 20 Hz, find the amplitude of forced vibration of the mass.

[0.015 9 mm]

20. An engine of mass 240 kg is mounted on a spring support. When the engine is not running, free vertical vibrations are subject to damping forces which reduce the amplitude by 20% during each complete oscillation, the frequency of these oscillations being 9 Hz.

When the engine is running at 480 rev/min, there is an out-of-balance harmonic force of amplitude 200 N. Find the amplitude of steady-state vibration at this speed.

[1.183 mm]

21. An unbalanced engine of mass 200 kg is mounted on an elastic support and an increase of speed from 600 rev/min to 900 rev/min trebles the amplitude of the forced vibration. Find the damping coefficient required for a damping device which will reduce the amplitude of vibration at 600 rev/min by 10%.   [30.4 kN s/m]

21 Thermodynamic properties of substances

21.1 Properties of substances

A number of properties such as pressure, temperature, volume and energy are necessary to describe the *state* of a substance. Some of these properties depend on the mass of substance considered, e.g., volume and energy, and are called *extensive* properties whereas others, of which pressure and temperature are examples, are independent of mass and are termed *intensive*. Extensive properties may be divided by the mass with which they are associated and then are expressed specifically. Thus specific volume is volume/mass $\left(v = \dfrac{V}{m} \right)$.

Table 21.1 shows the properties of interest in thermodynamics and is used to define enthalpy and specific heat capacity as functions of other properties.

Property values are obtained by experiment and then tabulated or plotted. For most substances, any two independent properties such as pressure and temperature are sufficient to describe a state; all other data can then be obtained from tables.

Mathematical equations, called *equations of state*, may be formulated to replace the use of tables. These equations are too complex for everyday use but are useful for computer storage of data. A very simple equation which gives a good approximation to the behaviour of real gases may be used in elementary work; it is called the *ideal gas equation* and enables gas processes to be analysed by differentiation and integration.

21.2 Properties of vapours

The vapour in most common use is steam. The sequence of events which occurs when unit mass of water is heated at constant pressure may be used to describe the construction of property tables.

The apparatus is shown in Figure 21.1 As the *compressed liquid* is heated, the temperature T rises and the specific volume v increases until, at

Table 21.1

Property	Symbol	Unit	Notes
Pressure	p	N/m², bar, Pa	1 bar = 10^5 N/m² $= 10^5$ Pa
Thermodynamic or absolute temperature	T	K	
Volume	V	m³	
Specific volume	v	m³/kg	$v \equiv 1/\rho$
Density	ρ	kg/m³	$\rho \equiv 1/v$
Internal energy	U	kJ	
Specific internal energy	u	kJ/kg	
Enthalpy	H	kJ	$H = U + pV$
Specific enthalpy	h	kJ/kg	$h = u + pv$
Entropy	S	kJ/K	
Specific enthalpy	s	kJ/kg K	
Dryness fraction	x		
Specific heat capacity at constant pressure	c_p	kJ/kg K	$c_p \equiv \dfrac{\mathrm{d}h}{\mathrm{d}T}$
Specific heat capacity at constant volume	c_v	kJ/kg K	$c_v \equiv \dfrac{\mathrm{d}u}{\mathrm{d}T}$
Ratio of specific heat capacities	γ		$\gamma = \dfrac{c_p}{c_v}$
Specific gas constant	R	kJ/kg K	$R = c_p - c_v$
Molar or universal gas constant	R_o	kJ/kg mol K	$R_o = 8.3143$
Relative molecular mass	M		$MR = R_0$

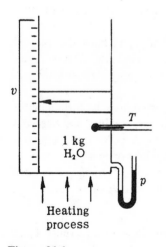

Figure 21.1

the *saturation temperature*, the liquid starts to evaporate and change from the *liquid phase* to the *vapour phase*. During the change of phase, T remains constant but v continues to increase rapidly. At the commencement of evaporation, the liquid is termed *saturated*. Between these two clearly defined states, there is a mixture of liquid and vapour, called a *wet vapour*. After the evaporation process has been completed, further heating results in a rise in T and an increase in v as the vapour becomes *superheated*. The difference between the saturation temperature and the temperature of the superheated vapour is called the *degree of superheat*.

The relation between T and v is shown in Figure 21.2, which also shows the effect of changing the pressure at which the process is conducted. By repeating the experiment at a series of different pressures, a complete picture of the behaviour of water is obtained, which shows that as the pressure is increased, the transition stage between liquid and vapour becomes progressively shorter until the critical temperature is reached. States above the critical temperature are *gaseous*.

On the pressure–specific volume diagram, Figure 21.3, point A is the *critical point*. The locus AB of saturated liquid states is called the *saturated liquid line* and points on it are given the suffix f; the locus AC of dry

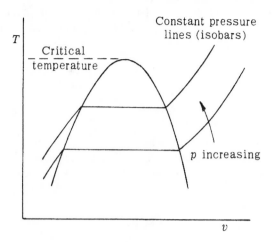

Figure 21.2

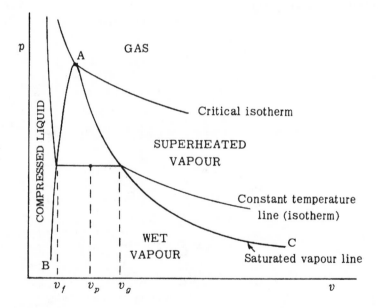

Figure 21.3

saturated vapour states is called the *saturated vapour line* and points on it are given the suffix *g*.

Between states *f* and *g*, where pressure and temperature are both constant and not independent, the water is partially vaporized and the properties are defined by the dryness fraction, *x*, given by

$$x = \frac{\text{mass of dry saturated vapour}}{\text{total mass of liquid and vapour}}$$

The mixture at P is made up of x parts of dry saturated vapour and $(1 - x)$ parts of saturated liquid. Thus the specific volume at P is given by

$$v_p = (1 - x)v_f + xv_g$$
$$= v_f + x(v_g - v_f) = v_f + xv_{fg}$$

from which

$$x = \frac{v_p - v_f}{v_g - v_f} = \frac{v_p - v_f}{v_{fg}} \qquad (21.1)$$

where $v_g - v_f$ is written v_{fg}.

Other properties at P are obtained in a similar manner. In particular, u and h which will be frequently required, are given by

$$u_p = u_f + xu_{fg} \quad \text{where} \quad u_{fg} = u_g - u_f \qquad (21.2)$$

and

$$h_p = h_f + xh_{fg} \quad \text{where} \quad h_{fg} = h_g - h_f \qquad (21.3)$$

21.3 Property tabulations and charts

Tables are normally available which give data for saturated liquid and saturated vapour, listing, for a given saturation pressure *or* temperature, some or all of the quantities $v_f, v_g, u_f, u_g, h_f, h_{fg}, h_g, s_f, s_{fg}$ and s_g. The tables

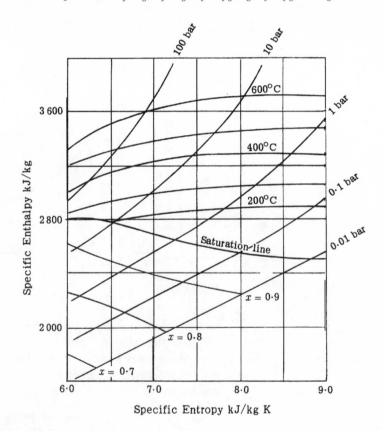

Figure 21.4

enable u, h or s to be obtained for saturated liquid, wet vapour and dry saturated vapour and they may also be used for compressed liquid by assuming that u, h and s are functions of temperature only.

Data for superheated vapour is given in a separate table listing some or all values of u, h, v (or ρ) and s for an entry of pressure *and* temperature.

Tables 21.2 and 21.3 (Example 1) show typical extracts from steam tables; linear interpolation may be used to determine intermediate values.

One chart is readily available which shows h as ordinate against s (specific entropy) as abscissa. It is called a Mollier chart and has contours enabling p, T or x to be used to fix a point on the chart from which h may be read off in either superheated, dry saturated or wet states. Figure 21.4 shows a Mollier chart for steam and Figures 21.5 and 21.6 show how steam processes will appear on both the Mollier and pressure–volume charts.

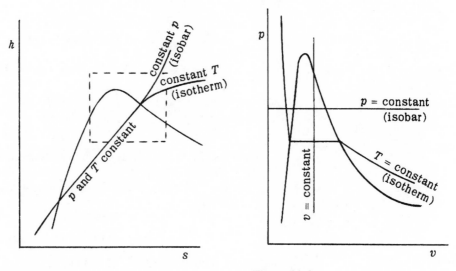

Figure 21.5   Figure 21.6

21.4 Properties of ideal and perfect gases

An *ideal gas* is described by the equation of state

$$pv = RT \tag{21.4}$$

where p is the pressure, v is the specific volume, T is the thermodynamic temperature and R is the specific gas constant.

This ideal gas equation incorporates the gas laws of Boyle and Charles and may also be written $\dfrac{pv}{T} = R$ so that, for a change of state from 1 to 2,

$$\frac{p_1 v_1}{T_1} = \frac{p_2 v_2}{T_2} \tag{21.5}$$

Since $v = \dfrac{V}{m}$, the ideal gas equation may also be written

$$pV = mRT \tag{21.6}$$

The value of R is obtained from the relation

$$R = \frac{R_0}{M}$$

where R_0 is the molar gas constant 8.3143 kJ/kg mol K and M is the relative molecular mass of the gas.

A *perfect gas* is an ideal gas with constant specific heat capacity. The specific heat capacities at constant volume and constant pressure are defined in Section 21.1 by the relations $c_v = \dfrac{du}{dT}$ and $c_p = \dfrac{dh}{dT}$ respectively. In this special case where c_v and c_p are constants, these relations may be integrated to give

$$\int_1^2 du = c_v \int_1^2 dT$$

i.e.

$$u_2 - u_1 = c_v(T_2 - T_1) \tag{21.7}$$

and

$$\int_1^2 dh = c_p \int_1^2 dT$$

i.e.

$$h_2 - h_1 = c_p(T_2 - T_1) \tag{21.8}$$

Equations (21.7) and (21.8) are frequently required for use in the energy equation, Chapter 22.

In Section 21.1, the specific enthalpy, h, is defined by

$$h = u + pv$$

Thus

$$h_2 - h_1 = (u_2 - u_1) + (p_2v_2 - p_1v_1)$$

and so, from equation (21.4), (21.7) and (21.8),

$$c_p(T_2 - T_1) = c_v(T_2 - T_1) + R(T_2 - T_1)$$

from which

$$R = c_p - c_v \tag{21.9}$$

21.5 Thermodynamic temperature

Charles's law states that the volume of a perfect gas is proportional to its absolute temperature, so that the temperature T used in the gas laws must be the absolute, or *thermodynamic*, temperature. The unit is the kelvin (K) and the thermodynamic temperature is obtained by adding 273 to the Celsius temperature.

Worked examples

1. *Using the extracts from steam tables given in Tables 21.2 and 21.3, determine:*
(a) the specific volume of steam of dryness fraction 0.8 at a pressure of 21 bar;
(b) the specific enthalpy of steam of dryness fraction 0.7 at a pressure of 22 bar;

(c) the specific enthalpy of compressed liquid at 25 bar and 214.8°C;
(d) the specific internal energy of saturated liquid at 23 bar;
(e) the specific volume of dry saturated steam at 21 bar;
(f) the specific enthalpy of steam at 15 bar and 300°C;
(g) the specific volume of steam at 16 bar and 400°C;
(h) the specific enthalpy of steam at 20 bar and 325°C;
(j) the state of steam at 25 bar and 223.9°C;
(k) the state of steam at 25 bar and 275°C.

Table 21.2. Saturated water and steam

p (bar)	t_{sat} (°C)	v_f (m³/kg)	v_g (m³/kg)	u_f (kJ/kg)	u_g (kJ/kg)	h_f (kJ/kg)	h_g (kJ/kg)	s_f (kJ/kg K)	s_g (kJ/kg K)
21	214.8	0.001 18	0.095 1	918	2 600	920	2 800	2.470	6.320
23	219.5	0.001 19	0.086 8	939	2 601	942	2 801	2.513	6.285
25	223.9	0.001 20	0.080 0	959	2 602	962	2 802	2.554	6.254

Table 21.3. Superheated steam

sat values	p (bar)		t(°C)				
			200	250	300	350	400
0.131 7	15	v m³/kg	0.132 4	0.152 0	0.170 0	0.186 5	0.202 9
2 792	t_{sat} =	h kJ/kg	2 796	2 925	3 039	3 149	3 256
6.444	198.9°C	s kJ/kg K	6.450	6.710	6.921	7.104	7.269
0.080 0	25	v m³/kg	–	0.087 0	0.098 9	0.109 7	0.120 0
2 802	t_{sat} =	h kJ/kg	–	2 881	3 010	3 128	3 239
6.257	223.9°C	s kJ/kg K	–	6.408	6.645	6.844	7.020

(a) $v = v_f + xv_g = 0.001\,18 + 0.8(0.095\,1 - 0.001\,18) = \underline{0.076\,3 \text{ m}^3/\text{kg}}$

(b) At 22 bar, $h_f = \frac{1}{2}(920 + 942) = 931$ kJ/kg

and $\qquad h_g = \frac{1}{2}(2\,800 + 2\,801) = 2\,800.5$ kJ/kg

$\therefore \quad h = h_f + xh_{fg} = 931 + 0.7(2\,800.5 - 931) = \underline{2\,240 \text{ kJ/kg}}$

(c) The enthalpy is a function of temperature alone and thus the value of h_f at 214.8°C is the required value, i.e., $h_f = \underline{920 \text{ kJ/kg}}$

The pressure of 25 bar is not required to determine this value.

(d) At 23 bar, $u_f = \underline{939 \text{ kJ/kg}}$

(e) At 21 bar, $v_g = \underline{0.095\,1 \text{ m}^3/\text{kg}}$

(f) At 15 bar and 300°C, $h = \underline{303\,9 \text{ kJ/kg}}$

(g) $v = 0.202\,9 - \frac{1}{10}(0.202\,9 - 0.120\,0) = \underline{0.194\,6 \text{ m}^3/\text{kg}}$

(h) At 20 bar and 300°C, $h = \frac{1}{2}(3\,039 + 3\,010) = 3\,024.5$ kJ/kg

At 20 bar and 350°C, $h = \frac{1}{2}(3\,149 + 3\,128) = 3\,138.5$ kJ/kg

∴ at 20 bar and 325°C, $h = \frac{1}{2}(3\,024.5 + 3\,138.5) = \underline{3\,081.5 \text{ kJ/kg}}$

(j) This cannot be determined as pressure and temperature are not independent and so a further property is required. Suppose that $h = 2\,300$ kJ/kg. Then, since $h < h_g$, the steam is wet and the dryness fraction is given by

$$x = \frac{h - h_f}{h_{fg}} = \frac{2\,300 - 962}{2\,802 - 962} = 0.727$$

(k) Since $t > t_{\text{sat}}$, the steam is superheated and the

$$\text{degree of superheat} = 275 - 223.9 = \underline{51.1 \text{ degrees}}$$

2. *A perfect gas has a relative molecular mass of 2 and the specific heat at constant pressure is 14.41 kJ/kg K. Determine the values of γ, R and c_v.*
$R_0 = 8.314\,3 \ kJ/kg \ mol \ K.$

$$R = \frac{R_0}{M} = \frac{8.314\,3}{2} = \underline{4.157 \text{ kJ/kg K}}$$

$$c_v = c_p - R = 14.41 - 4.157 = \underline{10.253 \text{ kJ/kg K}}$$

$$\gamma = \frac{c_p}{c_v} = \frac{14.41}{10.253} = \underline{1.4}$$

3. *What is the specific enthalpy change when oxygen, which may be considered a perfect gas of relative molecular mass 32, is heated from 20°C to 50°C? For oxygen, $\gamma = 1.4$. $R_0 = 8.314\,3$ kJ/kg mol K.*

From equation (21.9), $R = c_p - c_v$

$$\therefore \frac{R}{c_p} = \frac{c_p - c_v}{c_p} = \frac{\gamma - 1}{\gamma}$$

$$\therefore c_p = \frac{\gamma R}{\gamma - 1} = \frac{\gamma R_0}{(\gamma - 1)M}$$

$$= \frac{1.4 \times 8.314\,3}{0.4 \times 32} = 0.91 \text{ kJ/kg K}$$

$$h_2 - h_1 = c_p(T_2 - T_1) \quad \text{from equation (21.8)}$$

$$= 0.91([50 + 273] - [20 + 273])$$

$$= \underline{27.3 \text{ kJ/kg}}$$

4. *1 m^3 of nitrogen at a pressure of 15 kN/m^2 and a temperature of 115°C is heated to a temperature of 200°C. What is the internal energy change? For nitrogen, which may be treated as a perfect gas, $c_v = 0.743$ kJ/kg K and $R = 0.30$ kJ/kg K.*

From equation (21.7), $u_2 - u_1 = c_v(T_2 - T_1)$

$$= 0.743([200 + 273] + [115 + 273])$$

$$= 63.2 \text{ kJ/kg}$$

From equation (21.6), $pV = mRT$

$$\therefore m = \frac{15 \times 1}{0.3(115 + 273)} = 0.128\,8 \text{ kg}$$

Therefore change in internal energy,

$$U_2 - U_1 = 0.128\,8 \times 63.2 = \underline{8.14 \text{ kJ}}$$

Further problems

5. Using tables 21.2 and 21.3 determine the following data:
(i) What is the specific enthalpy of steam at 24 bar, dryness fraction 0.9?
(ii) What is the specific enthalpy of compressed liquid at 219.5°C, 30 bar?
(iii) What is the specific internal energy of steam at 15 bar, 250°C?
(iv) What is the specific enthalpy of steam at 22 bar, 375°C?
(v) What is the dryness fraction of steam at 21 bar, specific volume 0.06 m³/kg?
(vi) Steam at 25 bar has 76 degrees of superheat, what is the specific internal energy?
[2 617 kJ/kg; 942 kJ/kg; 2 697 kJ/kg; 3 189 kJ/kg; 0.629; 2 763 kJ/kg]

6. What mass of perfect gas occupies a volume of 2.1 m³ at a pressure of 5 bar and a temperature of 21°C? The gas has a relative molecular mass of 28.
[12 kg]

7. A perfect gas has a specific gas constant 0.287 kJ/kg K and $c_p = 1.005$ kJ/kg K. Determine c_v , γ and the relative molecular mass.
[0.718 kJ/kg K; 1.4; 29]

8. What is the enthalpy change of 2.2 kg of oxygen of relative molecular mass 32 when it is cooled from 312 K to 2°C? Oxygen may be treated as a perfect gas with $\gamma = 1.4$.
[73.8 kJ]

9. What is the specific internal energy change of air (which may be considered a perfect gas with $c_p = 1.005$ kJ/kg K and $R = 0.287$ kJ/kg K) when heated through 400°C?
[287.2 kJ/kg]

10. What is the specific enthalpy of
(i) dry saturated Refrigerant 12 at −5°C?
(ii) Refrigerant 12 at 0.423 MN/m² and 30°C?
(iii) Refrigerant 12 at 0.183 MN/m² and 0°C?
(iv) Ammonia at 5°C with dryness fraction 0.8?
(v) Ammonia at 0.19 MN/m² and 50°C?
(vi) dry saturated Refrigerant R134a at −15°C?
(vii) saturated liquid Refrigerant R134a at 30°C?

Note: Refrigerant 12 is also known as Freon 12
[185.4 kJ/kg; 205.2 kJ/kg; 190.2 kJ/kg; 120 1 kJ/kg; 1 579.4 kJ/kg; 389.18 kJ/kg; 242.0 kJ/kg]

22 Basic engineering thermodynamics

22.1 Introduction

Engineering thermodynamics is concerned with the transfer of energy in the working substance of a machine. Two kinds of systems are involved: (*a*) the *closed* or *non-flow* system in which a constant mass is enclosed by a boundary, which separates the system from the surroundings; (*b*) the *open* or *flow* system, through which there is a steady mass flow. Figure 22.1 shows an example of a closed system, where air is being compressed in a cylinder and Figure 22.2 shows an example of an open system, where steam is passing through a turbine. In the latter case, the boundary is replaced by a fixed framework, called the *control surface*, through which the working substance passes.

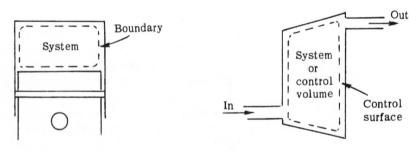

Figure 22.1   Figure 22.2

22.2 The first law of thermodynamics

The main source of energy in the world is heat transfer obtained by the combustion of fuel. It is therefore necessary, when applying the law of conservation of energy, to consider both heat and work transfers.

If a system undergoes a process in which a quantity of heat Q is transferred to the system and a quantity of work W is transferred *from* the system, the gain in energy is, from the law of conservation of energy, $Q - W$,

i.e. $$Q - W = E_{\text{final}} - E_{\text{initial}} = \Delta E \qquad (22.1)$$

where Δ denotes 'change in' and E represents energy.

This is known as the *first law of thermodynamics*. Heat transfer *to* a system is regarded as positive and work transfer *from* a system is regarded as positive.

If a system is insulated so that there is no heat transfer, it is called *adiabatic* and processes associated with it are adiabatic processes.

22.3 The non-flow energy equation

In thermodynamics, three forms of energy are involved: kinetic energy, potential energy and internal energy (U) due to the state of the system. In a closed system having a constant mass, there is no change in kinetic or potential energy so that the first law (equation 22.1) becomes the *non-flow energy equation*,

$$Q - W = \Delta U \tag{22.2}$$

Dividing equation (22.2) by the mass m of the system,

$$\frac{Q}{m} - \frac{W}{m} = \Delta \frac{U}{m}$$

or
$$q - w = \Delta u \tag{22.3}$$

where q is the specific heat transfer, w is the specific work transfer and u is the specific internal energy.

22.4 Reversibility

In the application of equation (22.3), the specific internal energy change Δu resulting from a change from one state to another is obtained from tables, as described in Chapter 21. The specific heat transfer q is difficult to measure or calculate accurately and is often the unknown quantity in an application.

The specific work transfer w may be obtained by measurement on an actual machine but in design calculations, it is necessary to predict the work transfer from the concept of an ideal reversible process in which there are no losses due to friction or other causes. In such an ideal case, the process is reversible since it is always possible to change the direction of the process and restore both the system and surroundings to their original state.

Reversible processes may be expressed by mathematical relations, from which the theoretical work transfer may be calculated. The work transfer in a real machine may be estimated by calculating the work in an equivalent reversible process and multiplying or dividing by a factor called the process efficiency (Section 23.6).

22.5 Displacement work transfer

In a non-flow system, such as a piston compressing a gas in a cylinder, Figure 22.3, work is done by displacing the boundary against a force (the pressure of the gas multiplied by the piston area). If an element of boundary of area dA is displaced a distance dl against a pressure p, the reversible work transfer,

$$dW = p\, dA\, dl = p\, dV$$

where dV is the change of volume.

For a complete system, $$W = \int p\,dV$$

or, dividing both sides by the mass of gas, m,

$$w = \int p\,dv \tag{22.4}$$

For the reversible process 1–2 of Figure 22.4, the work transfer $w = \int_1^2 p\,dv$, which is represented by the shaded area under the process line.

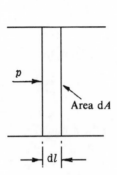

Figure 22.3

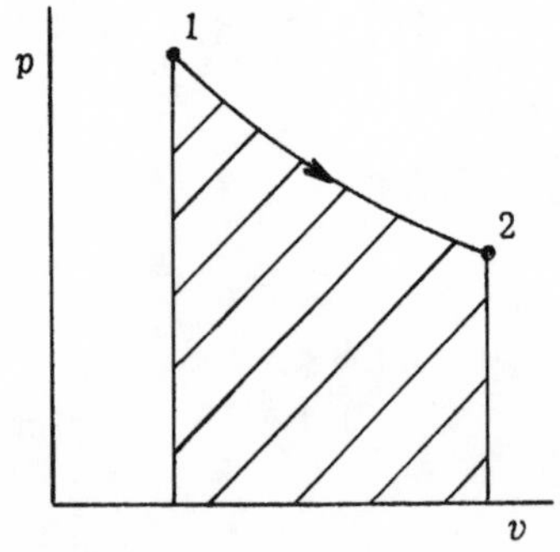

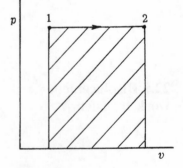

Figure 22.4

22.6 Reversible non-flow processes

There are five processes for which equation (22.4) needs to be evaluated.

(a) The reversible constant volume process

For this process, $v = $ constant, Figure 22.5

$$\therefore w = \int_1^2 p\,dv = 0 \tag{22.5}$$

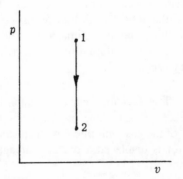

Figure 22.5

Figure 22.6

(b) The reversible constant pressure process

For this process, p = constant, Figure 22.6,

$$\therefore w = \int_1^2 p\,dv = p(v_2 - v_1) \tag{22.6}$$

(c) The reversible hyperbolic process

For this process, pv = constant, Figure 22.7. Let the constant be c.

Then

$$p = \frac{c}{v}$$

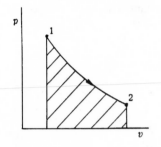

Figure 22.7

$$\therefore \qquad w = c \int_1^2 \frac{dv}{v} = c \ln \frac{v_2}{v_1}$$

Since $c = p_1v_1 = p_2v_2$, this may be expressed in the forms

$$\left.\begin{array}{l} w = p_1v_1 \ln \dfrac{v_2}{v_1} = p_2v_2 \ln \dfrac{v_2}{v_1} \\[2ex] w = p_1v_1 \ln \dfrac{p_1}{p_2} = p_2v_2 \ln \dfrac{p_1}{p_2} \end{array}\right\} \tag{22.7}$$

or

(d) The reversible polytropic process

For this process, pv^n = constant, Figure 22.8, where the index n has a numerical value usually between 1 and 2. Let the constant be c.

Then

$$p = \frac{c}{v^n}$$

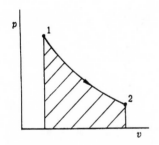

Figure 22.8

$$\therefore w = c \int_1^2 \frac{dv}{v^n}$$

$$= c \left[\frac{v^{1-n}}{1-n} \right]_1^2$$

Substituting $c = p_2v_2^n$ for limit 2 and $c = p_1v_1^n$ for limit 1,

$$w = \frac{p_2v_2^n \cdot v_2^{1-n} - p_1v_1^n \cdot v_1^{1-n}}{1-n}$$

$$= \frac{p_1v_1 - p_2v_2}{n-1} \tag{22.8}$$

(e) The reversible adiabatic process

For this process, pv^k = constant. This is similar to the preceding case, except that k has a value such that the heat transfer during the process is zero. By a similar analysis to case (d),

$$w = \frac{p_1 v_1 - p_2 v_2}{k - 1} \tag{22.9}$$

22.7 Special relations for perfect gases

(a) The reversible isothermal process

In an isothermal process, the temperature remains constant, so that the ideal gas equation $pv = RT$ becomes $pv = $ constant. This is the case of the hyperbolic process considered in Section 22.6(c) so that

$$w = p_1 v_1 \ln \frac{v_2}{v_1}$$

Since $p_1 v_1 = RT$ this may be written

$$w = RT \ln \frac{v_2}{v_1} = RT \ln \frac{p_1}{p_2} \tag{22.10}$$

(b) The reversible adiabatic process

From Section 22.6(e), $w = \dfrac{p_1 v_1 - p_2 v_2}{k - 1}$

and, for a perfect gas $\Delta u = c_v(T_2 - T_1)$ from equation (22.3). Hence, from the non-flow energy equation,

$$q - w = \Delta u,$$

$$0 - \frac{p_1 v_1 - p_2 v_2}{k - 1} = c_v(T_2 - T_1)$$

i.e.

$$\frac{R(T_2 - T_1)}{k - 1} = c_v(T_2 - T_1)$$

from which

$$k = \frac{R}{c_v} + 1$$

But $R = c_p - c_v$ and $\dfrac{c_p}{c_v} = \gamma$

$$\therefore k = \frac{c_p - c_v}{c_v} + 1 = \frac{c_p}{c_v} = \gamma$$

Thus, for a perfect gas, the reversible adiabatic process law is $pv^\gamma = c$ and combining $p_1 v_1^\gamma = p_2 v_2^\gamma$ with $\dfrac{p_1 v_1}{T_1} = \dfrac{p_2 v_2}{T_2}$ gives

$$\frac{v_2}{v_1} = \left(\frac{p_1}{p_2}\right)^{1/\gamma} = \frac{p_1}{p_2} \cdot \frac{T_2}{T_1}$$

from which
$$\frac{T_1}{T_2} = \left(\frac{p_1}{p_2}\right)^{\frac{\gamma-1}{\gamma}} \qquad (22.11)$$

or
$$\frac{p_1}{p_2} = \left(\frac{v_2}{v_1}\right)^{\gamma} = \frac{v_2}{v_1} \cdot \frac{T_1}{T_2}$$

from which
$$\frac{T_1}{T_2} = \left(\frac{v_2}{v_1}\right)^{\gamma-1} \qquad (22.12)$$

(c) The reversible polytropic process

In the general case of $pv^n = c$, equations (22.11) and (22.12) become

$$\frac{T_1}{T_2} = \left(\frac{p_1}{p_2}\right)^{\frac{n-1}{n}} \qquad (22.13)$$

and
$$\frac{T_1}{T_2} = \left(\frac{v_2}{v_1}\right)^{n-1} \qquad (22.14)$$

22.8 Applications of non-flow energy equation

The non-flow energy equation may be used for ideal processes in reciprocating machines such as spark ignition engines, diesel engines and steam motors. Assumptions may be made about a suitable process law and the displacement work transfer is then evaluated from $w = \int p\,dv$. the value of Δu is obtained from tables for vapours and from the equation $\Delta u = c_v(T_2 - T_1)$ for perfect gases.

(a) Reversible constant volume process

From equation (22.5), $w = 0$

Hence
$$q = u_2 - u_1 \qquad (22.15)$$

For a perfect gas,
$$q = c_v(T_2 - T_1) \qquad (22.16)$$

(b) Reversible constant pressure process

From equation (22.6), $w = p(v_2 - v_1)$

Hence
$$q - p(v_2 - v_1) = u_2 - u_1$$

or
$$q = (u_2 + pv_2) - (u_1 + pv_1)$$
$$= h_2 - h_1 \qquad (22.17)$$

For a perfect gas, $\Delta h = c_p(T_2 - T_1)$ from equation (21.8)

$$\therefore q = c_p(T_2 - T_1) \qquad (22.18)$$

(c) Reversible hyperbolic process

From equation (22.7), $w = p_1 v_1 \ln \dfrac{v_2}{v_1}$

Hence $q - p_1 v_1 \ln \dfrac{v_2}{v_1} = u_2 - u_1$ ⠀⠀⠀⠀⠀(22.19)

For a perfect gas, for which T is constant,

$$u_2 - u_1 = c_v(T_2 - T_1) = 0$$

$$\left. \begin{aligned} \therefore\; q = w &= p_1 v_1 \ln \frac{v_2}{v_1} \\ &= RT \ln \frac{v_2}{v_1} \end{aligned} \right\} \qquad (22.20)$$

(d) Reversible polytropic process

From equation (22.8), $w = \dfrac{p_1 v_1 - p_2 v_2}{n - 1}$

Hence $q - \dfrac{p_1 v_1 - p_2 v_2}{n - 1} = u_2 - u_1$ ⠀⠀⠀⠀⠀(22.21)

For a perfect gas,

$$q - \frac{R(T_2 - T_1)}{1 - n} = c_v(T_2 - T_1)$$

$$\therefore\; q = \left(\frac{R}{1 - n} + c_v \right)(T_2 - T_1)$$

$$= \left(\frac{R}{1 - n} + \frac{R}{\gamma - 1} \right)(T_2 - T_1)$$

$$= \frac{\gamma - n}{\gamma - 1} \cdot \frac{R}{1 - n}(T_2 - T_1)$$

$$= \frac{\gamma - n}{\gamma - 1} \cdot w \qquad (22.22)$$

(e) Reversible adiabatic process

From equation (22.9), $w = \dfrac{p_1 v_1 - p_2 v_2}{k - 1}$ and $q = 0$.

Hence $0 - \dfrac{p_1 v_1 - p_2 v_2}{k - 1} = u_2 - u_1$

i.e. $\dfrac{p_2 v_2 - p_1 v_1}{k - 1} = u_2 - u_1$ ⠀⠀⠀⠀⠀(22.23)

For a perfect gas, $k = \gamma$

$$\therefore \frac{p_2 v_2 - p_1 v_1}{\gamma - 1} = c_v(T_2 - T_1) \qquad (22.24)$$

Table 22.1. Summary of reversible non-flow processes for vapours

Name	Law	Displacement work $\int_1^2 p \, dv$	Δu	$q = w + \Delta u$
Constant volume process	$v = c$	0	$u_2 - u_1$	$q = (u_2 - u_1)$
Constant pressure process	$p = c$	$w = p(v_2 - v_1)$	$u_2 - u_1$	$q = p(v_2 - v_1) + (u_2 - u_1)$ or $q = h_2 - h_1$
Hyperbolic process	$pv = c$	$w = p_1 v_1 \ln\left(\dfrac{v_2}{v_1}\right)$	$u_2 - u_1$	$q = p_1 v_1 \ln\left(\dfrac{v_2}{v_1}\right) + (u_2 - u_1)$
Polytropic process	$pv^n = c$	$w = \dfrac{p_1 v_1 - p_2 v_2}{n - 1}$	$u_2 - u_1$	$q = \dfrac{p_1 v_1 - p_2 v_2}{n - 1} + (u_2 - u_1)$ or $q = \dfrac{(h_1 - h_2) + n(u_2 - u_1)}{n - 1}$
Adiabatic process	$pv^k = c$	$w = \dfrac{p_1 v_1 - p_2 v_2}{k - 1}$	$u_2 - u_1$	0

Table 22.2. Summary of reversible non-flow processes for gases

Name	Law	Displacement work $\int_1^2 p \, dv$	Δu	$q = w + \Delta u$
Constant volume process	$v = c$	0	$C_v(T_2 - T_1)$	$q = C_v(T_2 - T_1)$
Constant pressure process	$p = c$	$w = R(T_2 - T_1)$	$C_v(T_2 - T_1)$	$q = C_p(T_2 - T_1)$
Constant temp. (isothermal) process	$pv = c$ $T = c$	$w = RT_1 \ln\left(\dfrac{v_2}{v_1}\right)$	0	$q = RT_1 \ln\left(\dfrac{v_2}{v_1}\right)$
Polytropic process	$pv^n = c$	$w = \dfrac{R(T_1 - T_2)}{n - 1}$	$C_v(T_2 - T_1)$	$q = \left(\dfrac{\gamma - n}{\gamma - 1}\right) \dfrac{R(T_1 - T_2)}{(n - 1)}$
Adiabatic process	$pv^\gamma = c$	$w = \dfrac{R(T_1 - T_2)}{\gamma - 1}$	$C_v(T_2 - T_1)$	0

22.9 The steady flow energy equation

Figure 22.9 shows a system in which the control surface represents the fixed framework through which the fluid flows at a constant rate $\dot{m}$.

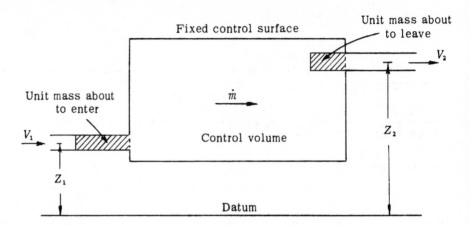

Figure 22.9

Let

p_1 = inlet pressure
v_1 = specific volume at inlet
V_1 = velocity at inlet
Z_1 = elevation of inlet
u_1 = specific internal energy at inlet

p_2 = outlet pressure
v_2 = specific volume at outlet
V_2 = velocity at outlet
Z_2 = elevation of outlet
u_2 = specific internal energy at outlet

Displacement work done as unit mass enters control volume = $-p_1v_1$
Displacement work done as unit mass leaves control volume = p_2v_2
Hence displacement work transfer rate, or power, to maintain the flow rate
$\dot{m} = \dot{m}(p_2v_2 - p_1v_1)$

If the heat transfer rate to the system is $\dot{Q}$ whilst the *useful* work transfer rate *from* the system is $\dot{W}_x$, then

the total work transfer rate, $\dot{W} = \dot{W}_x + \dot{m}(p_2v_2 - p_1v_1)$

Thus equation (22.1), written in terms of energy rates, becomes

$$\dot{Q} - \dot{W} = \dot{m}\Delta e$$

i.e. $$\dot{Q} - \{\dot{W}_x + \dot{m}(p_2v_2 - p_1v_1)\} = \dot{m}\Delta e$$

In flow systems, the change in energy includes kinetic and potential energies, as well as internal energy, hence

$$\dot{m}\Delta e = \dot{m}(u_2 - u_1) + \dot{m}\left(\frac{V_2^2}{2} - \frac{V_1^2}{2}\right) + \dot{m}(gZ_2 - gZ_1)$$

Thus

$$\dot{Q} - \{\dot{W}_x + \dot{m}(p_2v_2 - p_1v_1)\} = \dot{m}(u_2 - u_1) + \dot{m}\left(\frac{V_2^2}{2} - \frac{V_1^2}{2}\right)$$

$$+ \dot{m}(gZ_2 - gZ_1)$$

But $\qquad u_1 + p_1v_1 = h_1 \quad$ and $\quad u_2 + p_2v_2 = h_2,$

where h is the specific enthalpy, hence

$$\dot{Q} - \dot{W}_x = \dot{m}\Delta\left(h + \frac{V^2}{2} + gZ\right) \qquad (22.25)$$

where Δ denotes 'change in'.

Equation (22.25) is the steady flow energy equation, which relates the useful power $\dot{W}_x$ to the other energy change rates.

Dividing equation (22.25) by $\dot{m}$, putting $\dfrac{\dot{Q}}{\dot{m}} = q$ (the specific heat transfer) and $\dfrac{\dot{W}_x}{\dot{m}} = w_x$ (the specific useful work transfer), the specific steady flow energy equation is

$$q - w_x = \Delta\left(h + \frac{V^2}{2} + gZ\right) \qquad (22.26)$$

Since w_x is the specific *useful* work transfer it is not represented by $\int p\,dv$, which relates only to displacement work transfer in a non-flow process.

22.10 The continuity equation

In a steady flow process, the mass flow rate at inlet is equal to the mass flow rate at outlet. Thus, if the flow areas at inlet and outlet are A_1 and A_2 respectively, then

$$\dot{m} = \rho_1 A_1 V_1 = \rho_2 A_2 V_2 \qquad (22.27)$$

or $\qquad \dfrac{A_1 V_1}{v_1} = \dfrac{A_2 V_2}{v_2} \quad$ since $v = \dfrac{1}{\rho} \qquad (22.28)$

22.11 Applications of steady flow energy equation

In many cases the changes in kinetic and potential energy may be neglected and equation (22.26) then simplifies to

$$q - w_x = \Delta h \qquad (22.29)$$

This is similar to equation (22.3) but w_x represents only the useful work done.

Values of Δh for vapours are obtained from tables and for perfect gases, they are given by $\Delta h = c_p(T_2 - T_1)$, from equation (21.8).

The reversible process laws $(v = c, \; p = c, \; pv = c, \; pv^n = c, \; pv^\gamma = c)$ may be required to deduce unknown values of pressure, temperature, etc.

(a) Turbines and rotary compressors, Figure 22.10.

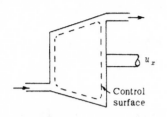

These machines are considered to be adiabatic since the mass flow rate is large enough to make the specific heat transfer negligible compared with the large specific work transfer. If the changes in kinetic and potential energy are also assumed negligible, then

$$-w_x = \Delta h = h_2 - h_1 \tag{22.30}$$

Figure 22.10

(b) Steam generators and boilers, Figure 22.11.

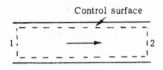

These are workless devices and if kinetic and potential energy changes are assumed negligible, then

$$q = \Delta h = h_2 - h_1 \tag{22.31}$$

Figure 22.11

(c) Heat exchangers, Figure 22.12.

A perfect heat exchanger is both workless and externally adiabatic since all the energy leaving one stream enters the other stream. If changes in kinetic and potential energy are assumed negligible,

then for stream A, $\dot{Q} = \dot{m}_A(h_2 - h_1)$

and for stream B, $\dot{Q} = \dot{m}_B(h_4 - h_3)$

Thus $-\dot{m}_A(h_2 - h_1) = \dot{m}_B(h_4 - h_3)$ (22.32)

(d) Nozzles, Figure 22.13.

Nozzles are workless devices which increase the kinetic energy of a stream of fluid. They are usually assumed adiabatic and if changes in

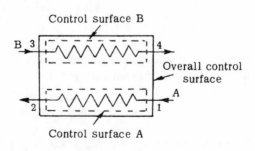

Figure 22.12

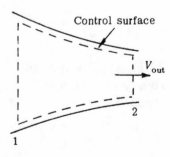

Figure 22.13

potential energy are negligible, then

$$0 = \Delta h + \Delta \frac{V^2}{2}$$

If the inlet velocity is very small compared with the exit velocity,

$$V_{\text{out}} = \sqrt{(-2\Delta h)} = \sqrt{(2(h_1 - h_2))} \tag{22.33}$$

(e) Throttle valves, Figure 22.14.

A valve is a workless, adiabatic expansion device in which changes in kinetic and potential energy are assumed negligible, so that $\Delta h = 0$,

i.e. $$h_2 = h_1 \tag{22.34}$$

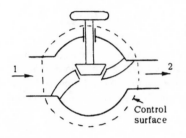

Figure 22.14

Table 22.3. Summary for flow devices

Device	$q - w_x = \Delta(h + \frac{V^2}{2} + gz)$
Heat exchanger, steam generator or boiler	$q = \Delta h$ if $\Delta\left(\frac{V^2}{2} + gz\right)$ negligible
Adiabatic turbine or rotary compressor	$-w_x = \Delta h$ if $\Delta\left(\frac{V^2}{2} + gz\right)$ negligible
Adiabatic nozzle or diffuser	$0 = \Delta\left(h + \frac{V^2}{2}\right)$ if Δgz negligible
Adiabatic throttle valve	$0 = \Delta h$ if $\Delta\left(\frac{V^2}{2} + gz\right)$ negligible

Worked examples

1. *A piston and cylinder mechanism contains 2 kg of a perfect gas. The gas expands reversibly and isothermally from a pressure of 10 bar and a temperature of 327°C to a pressure of 1.8 bar. Calculate the work transfer, the heat transfer and the change in specific enthalpy of the gas.*
$R = 0.3 \ kJ/kg \ K$ and $\gamma = 1.4$.

For a reversible isothermal process for a perfect gas,

$$pv = c$$

and the specific work transfer in this non-flow process is given by

$$w = RT \ln \frac{p_1}{p_2} \qquad \text{from equation (22.10)}$$

$$\therefore \quad \text{for 2 kg of gas, } W = 2 \times 0.3 \times (327 + 273) \ln \frac{10}{1.8}$$

$$= \underline{615 \;\; \text{kJ}}$$

Since this is positive it represents work output *from* the system.

For a non-flow process, $Q - W = \Delta U = mc_v(T_2 - T_1)$ but for an isothermal process, $T_2 = T_1$ so that $\Delta U = 0$.

Hence
$$Q = W = \underline{615 \text{ kJ}}$$

Since this is positive, it represents heat transfer to the system.

The specific enthalpy change, $\Delta h = c_p(T_2 - T_1)$ but since $T_2 = T_1$

$$\Delta h = \underline{0}$$

2. *The gas expanding in the combustion space of a reciprocating engine has an initial pressure of 50 bar and an initial temperature of 1 623 °C. The initial volume is 50 000 mm³ and the gas expands through a volume ratio of 20 according to the law $pV^{1.25} = constant$. Calculate the work transfer and heat transfer in the expansion process.*

$R = 270 \; J/kg \; K \; and \; c_v = 800 \; J/kg \; K.$

For this polytropic non-flow process,

$$w = \frac{p_1 v_1 - p_2 v_2}{n - 1} \quad \text{from equation (22.8)}$$

$$\therefore \; W = mw = \frac{p_1 V_1 - p_2 V_2}{n - 1}$$

But
$$V_2 = 20V_1$$

and
$$p_2 = p_1 \left(\frac{V_1}{V_2}\right)^{1.25} = \frac{p_1}{20^{1.25}} = \frac{50}{20^{1.25}} = 1.18 \text{ bar}$$

Hence
$$W = \frac{50\,000 \times 10^{-9}(50 - 20 \times 1.18) \times 10^5}{1.25 - 1}$$

$$= \underline{527 \text{ J}} \; (\textit{from the system, since positive})$$

From equations (22.2) and (21.7)

$$Q - W = \Delta U = mc_v(T_2 - T_1)$$

$$m = \frac{p_1 V_1}{RT_1} \quad \text{from equation (21.6)}$$

$$= \frac{50 \times 10^5 \times 50\,000 \times 10^{-9}}{270(1\,623 + 273)} = 0.49 \times 10^{-3} \text{ kg}$$

and
$$\frac{T_1}{T_2} = \left(\frac{v_2}{v_1}\right)^{n-1} \quad \text{from equation (22.14)}$$

$$\therefore \; T_2 = \frac{1\,623 + 273}{20^{0.25}} = 896 \text{ K}$$

Hence $\qquad Q - 527 = 0.49 \times 10^{-3} \times 800(896 - 1896)$

from which $\qquad Q = \underline{135\ J}$ (*to* the system, since positive)

3. *A reciprocating steam engine cylinder contains 2 kg of steam at a pressure of 30 bar and a temperature of 300°C. The steam expands reversibly according to the law* $pv^{1.2} = constant$ *until the pressure is 2 bar. Calculate the final state of the steam, the work transfer and the heat transfer in the process.*

From superheated steam tables at 30 bar and 300°C,

$$v = 0.081\,2\ \text{m}^3/\text{kg} \quad \text{and} \quad u = 2\,751\ \text{kJ/kg}$$

Final specific volume of steam,

$$v_2 = v_1 \left(\frac{p_1}{p_2} \right)^{\frac{1}{1.2}}$$

$$= 0.081\,2 \left(\frac{30}{2} \right)^{\frac{1}{1.2}} = 0.77\ \text{m}^3/\text{kg}$$

From the tables at 2 bar, $v_g = 0.885\,6\ \text{m}^3/\text{kg}$, so that the final state of the steam is wet.

The dryness fraction, $x = \dfrac{v_p - v_f}{v_{fg}} \qquad$ from equation (21.1)

and from the tables at 2 bar,

$$v_f = 0.001\,06\ \text{m}^3/\text{kg}$$

$$\therefore\ v_{fg} = 0.885\,6 - 0.001\,06 = 0.884\,5\ \text{m}^3/\text{kg}$$

$$\therefore\ x = \frac{0.77 - 0.001\,06}{0.884\,5} = \underline{0.868}$$

The work transfer, $\qquad W = m \left(\dfrac{p_1 v_1 - p_2 v_2}{n - 1} \right). \qquad$ from equation (22.8)

$$= 2 \left(\frac{30 \times 10^5 \times 0.081\,2 - 2 \times 10^5 \times 0.77}{1.2 - 1} \right)\ \text{J}$$

$$= \underline{900\ \text{kJ}} \quad (\textit{from} \text{ the system, since positive})$$

The heat transfer, $\qquad Q = W + \Delta U$

$$= 900 + 2(u_2 - 2\,751)$$

From the tables at 2 bar, $\quad u_g = 2\,530\ \text{kJ/kg} \quad \text{and} \quad u_f = 505\ \text{kJ/kg}$

so that $\qquad u_{fg} = 2\,025\ \text{kJ/kg}$

$$\therefore\ u_2 = u_g + x u_{fg}$$

$$= 505 + 0.868 \times 2\,025 = 2\,263\ \text{kJ/kg}$$

$$\therefore\ Q = 900 + 2(2\,263 - 2\,751)$$

$$= \underline{-76\ \text{kJ}} \quad (\textit{from} \text{ the system, since negative})$$

4. *Steam at a pressure of 60 bar and a temperature of 500°C enters an adiabatic turbine with a velocity of 20 m/s and expands to a pressure of 0.5 bar. At exit, the steam has a velocity of 200 m/s and a dryness fraction of 0.98. The turbine is required to develop 1 MW. Determine the mass flow rate of steam required and the cross-sectional area of flow at the turbine exit.*

From the steam chart, the specific enthalpy at entry, $h_1 = 3\,421$ kJ/kg and at exit, $h_2 = 2\,599$ kJ/kg.

This data may also be obtained from steam tables; h_1 is read directly from the superheat table and h_2 is obtained from the relation

$$h_2 = h_f + xh_{fg}$$

At 0.5 bar, $h_f = 340$ kJ/kg and $h_{fg} = 2\,305$ kJ/kg

Hence $h_2 = 340 + 0.98 \times 2\,305 = 2\,599$ kJ/kg

From equation (22.25), the steady flow energy equation is

$$\dot{Q} - \dot{W}_x = \dot{m}\Delta\left(h + \frac{V^2}{2} + gZ\right)$$

But $\dot{Q} = 0$ and ΔgZ is assumed to be negligible, so that

$$-\dot{W}_x = \dot{m}\left\{(h_2 - h_1) + \left(\frac{V_2^2}{2} - \frac{V_1^2}{2}\right)\right\}$$

i.e. $-10^3 = \dot{m}\left\{2\,599 - 3\,421 + \dfrac{1}{2 \times 10^3}(200^2 - 20^2)\right\}$ kW

from which $\dot{m} = \underline{1.25 \text{ kg/s}}$

At exit, the specific volume,

$$v = (1 - x)v_f + xv_g$$

From tables, $v_f = 0.001$ m^3/kg   and $v_g = 3.239$ m^3/kg

$$\therefore\ v = (1 - 0.98) \times 0.001 + 0.98 \times 3.239$$

$$= 3.175 \text{ m}^3/\text{kg}$$

From equation (22.27), $\dot{m} = \dfrac{AV}{v}$

$$\therefore\ A = \frac{1.25 \times 3.175}{200}$$

$$= \underline{0.019\,8 \text{ m}^2}$$

5. *Air, which may be considered a perfect gas, enters an adiabatic nozzle with negligible velocity. The entry pressure is 6 bar and the exit pressure is 1 bar; the entry temperature is 760 K. The flow throughout the nozzle is reversible and the mass flow rate is 2 kg/s. Calculate the exit velocity and the exit diameter. R = 0.287 kJ/kg K and γ = 1.4.*

The nozzle is shown in Figure 22.15.

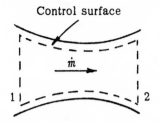

Control surface

$\dot{m}$

1   2

Figure 22.15

For this process, as flow is reversible and adiabatic,

$$T_2 = T_1 \left(\frac{p_2}{p_1} \right)^{\frac{\gamma - 1}{\gamma}} \quad \text{from equation (22.11)}$$

$$= 760 \left(\frac{1}{6} \right)^{\frac{0.4}{1.4}} = 455 \text{ K}$$

From equation (22.33), $V_2 = \sqrt{(2(h_1 - h_2))}$ assuming ΔgZ is negligible, and from equation (21.8), $h_1 - h_2 = c_p(T_1 - T_2)$

$$c_p - c_v = R \quad \text{and} \quad \gamma = \frac{c_p}{c_v}$$

$$\therefore c_p = \frac{\gamma R}{\gamma - 1}$$

$$= \frac{1.4 \times 0.287}{1.4 - 1} = 1.005 \text{ kJ/kg K}$$

$$\therefore V_2 = \sqrt{(2 \times 1.005 \times 10^3 (760 - 455))}$$

$$= \underline{781 \text{ m/s}}$$

From equations (22.27) and (21.4),

$$\dot{m} = \frac{AV}{v} \quad \text{and} \quad v = \frac{RT}{p}$$

$$\therefore A = \frac{\dot{m}RT}{pV}$$

$$= \frac{2 \times 287 \times 455}{1 \times 10^5 \times 781} = 0.003\,34 \text{ m}^2$$

$$\therefore d = \sqrt{\left(\frac{4 \times 0.003\,34}{\pi} \right)} = \underline{0.065 \text{ m}}$$

6. *A turbine passes 12 000 kg/h of steam to an adiabatic condenser at a pressure of 0.05 bar with dryness fraction 0.82. The condensate leaves the condenser at a temperature of 28°C. The condenser is cooled by water which enters at a temperature of 15°C and leaves at a temperature of 22°C. Calculate the mass flow rate of cooling water required if all changes in kinetic and potential energy may be neglected.*

The system is shown in Figure 22.16.

Applying the steady flow energy equation to the steam,

$$\dot{Q} - \dot{W}_x = \dot{m} \left(\Delta h + \Delta \frac{V^2}{2} + \Delta gZ \right)$$

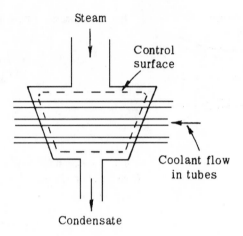

Figure 22.16

Since $\qquad \dot{W}_x = 0, \qquad \Delta \dfrac{V^2}{2} = 0 \quad$ and $\quad \Delta gZ = 0,$

the heat transfer rate from the steam is given by

$$\dot{Q} = \dot{m}(h_2 - h_1)$$

From tables at 0.05 bar,

$$h_f = 138 \text{ kJ/kg}, \qquad h_{fg} = 2\,423 \text{ kJ/kg}$$

and the saturation temperature is 32.9°C. Thus the condensate is liquid below the saturation temperature. It is normally assumed that specific enthalpy is a function of temperature for the liquid state, so that, at 28°C, $h_f = 117.3$ kJ/kg.

Thus $\qquad \dot{Q} = \dot{m}\{h_2 - (h_f + xh_{fg})\}$

$$= 12\,000\{117.3 - (138 + 0.82 \times 2\,423)\}$$

$$= -24 \times 10^6 \text{ kJ/h}$$

This heat transfer from the steam is numerically equal to the heat transfer to the cooling water.

Applying the steady flow energy equation to the cooling water,

$$\dot{Q} = \dot{m}\Delta h$$

From tables at 15°C, $h_f = 62.9$ kJ/kg and at 22°C, $h_f = 92.2$ kJ/kg so that mass flow rate of cooling water $= \dfrac{\dot{Q}}{\Delta h}$

$$= \frac{24 \times 10^6}{92.2 - 62.9}$$

$$= \underline{822 \times 10^3 \text{ kg/h}}$$

Further problems

7. A reciprocating compressor delivers 0.1 kg/s of air at a pressure of 12 bar. The air enters the compressor at a pressure of 1 bar and a temperature of 15°C.

Calculate the delivery temperature of the air, the work transfer rate and the heat transfer rate in the compression process for:

(i) reversible polytropic compression, $pv^{1.2}$ = constant;
(ii) reversible adiabatic compression;
(iii) reversible isothermal compression.

Air may be treated as a perfect gas for which the specific heat capacity at constant pressure, c_p = 1.005 kJ/kg K and γ = 1.4.

[436 K, -21.2 kW, -10.6 kW; 586 K, -21.4 kW, 0; 288 K, -20.5 kW, -20.5 kW]

8. A reciprocating internal combustion engine has a clearance volume of 0.0001 m³ and a compression ratio (volume ratio) of 10. The pressure and temperature of the combustion gases when the piston is at top dead centre are 4000 kN/m² and 1800°C respectively. Assuming that the expansion process follows a law $pv^{1.3}$ = constant, calculate the work transfer in this process and the temperature of the gases at the end of the process.

The combustion gases may be considered to be perfect gases with specific heat capacity at constant pressure c_p = 1.15 kJ/kg K and γ = 1.33.

[0.663 kJ; 1040 K]

9. A reciprocating steam motor is supplied with dry saturated steam at a pressure of 16 bar. The stroke of the motor is 0.8 m and the bore is 0.3 m. The clearance volume is negligible. The steam enters the cylinder at constant pressure for $\frac{1}{4}$ of the stroke and then expands reversibly according to a law pV = constant. Calculate the pressure of the steam and the state of the steam (dryness fraction if wet, temperature if superheated) at the end of the expansion process. What is the work transfer and the heat transfer in the process?

Note: If Thermodynamic Tables in S.I. (Metric) Units, R. W. Haywood, Cambridge University Press, 1968, are used to solve this problem then an interesting exercise in interpolation is obtained.   [4 bar; 169°C; 31.4 kJ; 31.46 kJ]

10. A piston and cylinder mechanism has its piston fixed so that the volume contained is 0.0025 m³. The mechanism is filled with wet steam at a pressure of 2 bar. The steam is heated until it reaches the critical point. The piston is released and the steam expands adiabatically to a pressure of 2 bar and a volume of 0.5 m³. Calculate the mass of steam in the mechanism and the dryness fraction of the steam after expansion and the work transfer in the expansion process.

[0.788 kg; 0.715; 48 kJ]

11. A mass of gas occupying 0.08 m³ at 6 kN/m² and 80°C is expanded reversibly in a non-flow process according to a law $pV^{1.2}$ = constant. The pressure at the end of expansion is 0.7 kN/m². The gas is then heated at constant pressure to the original temperature. The specific heat capacities at constant pressure and constant volume are 1.00 and 0.74 kJ/kg K respectively.

Determine (*a*) the work transfer in the expansion process;
          (*b*) the heat transfer in the expansion process;
          (*c*) the volume at the end of the heating process;
          (*d*) the change in internal energy during the heating process.

[0.73 kJ; 0.32 kJ; 0.684 m³; 0.41 kJ]

12. A vertical cylinder 150 mm in diameter contains water to a depth of 20 mm at a temperature of 151.8°C. A frictionless piston rests on the water and a force of 10.5 kN is applied to the piston. Atmospheric pressure is 105 kN/m². The cylinder is heated and 600 kJ are transferred to the H_2O. Calculate the dryness fraction of the steam formed in the cylinder.   [0.87]

13. An ideal centrifugal air compressor takes in air at 1 bar, 15°C and compresses it reversibly and adiabatically to a pressure of 4 bar. Calculate the delivery temperature of the gas. If kinetic energy and potential energy changes are negligible, calculate the specific work transfer in the compression process.

Air may be assumed to be a perfect gas with specific heat capacity at constant pressure $c_p = 1.005$ kJ/kg K and $\gamma = 1.4$.   [428.5 K; −140.9 kJ/kg]

14. A reversible adiabatic air turbine drives a small generator which requires a power of 2 kW. The air supply for the turbine is provided by a reservoir and the pressure and temperature at turbine entry may be considered constant at 9 bar, 20°C respectively. The velocity of the air at inlet to the turbine is small and may be neglected but at exit the velocity is 55 m/s. The exit pressure is 1.2 bar. Calculate the air temperature at exit from the turbine and the mass flow rate of air, stating any assumptions made.

Air may be considered a perfect gas for which the specific heat capacity at constant pressure $c_p = 1.005$ kJ/kg K and $\gamma = 1.4$.

[165 K; 0.015 7 kg/s, assumes $\Delta gZ = 0$]

15. Combustion gases enter an adiabatic gas turbine at a pressure of 420 kN/m² and a temperature of 900 K. The gases leave the turbine at a temperature of 680 K with a velocity of 70 m/s. The turbine delivers 7 500 kW when 30 kg/s of combustion gases are used. Calculate the velocity of the gases entering the turbine, ignoring changes in potential energy.

If the turbine had been reversible and the exit pressure 143 kN/m², what would have been the temperature of the gases at the end of the expansion process and the required mass flow rate of gases? (The velocities may be assumed to be unaltered.)

The combustion gases may be assumed to be perfect gases with specific heat capacity at constant pressure $c_p = 1.14$ kJ/kg K and $\gamma = 1.4$.

[53.8 m/s; 661 K; 27.6 kg/s]

16. A one pass steam generator consists of a heated tube. Water enters the tube at 11 bar, 20°C and leaves as steam at 10 bar, 400°C. The tube is 50 mm in diameter and the mass flow rate of steam is 1 100 kg/h. Calculate the heat transfer rate in the steam generator.   [0.97 MW]

17. Steam at a pressure of 10 bar and a temperature of 350°C enters a nozzle with a velocity of 150 m/s. The steam expands reversibly and adiabatically in the nozzle to a pressure of 1 bar and a dryness fraction of 0.99. Calculate the velocity of the steam at exit from the nozzle and the exit area of the nozzle for a mass flow rate of 0.6 kg/s.   [1 018 m/s; 988 mm²]

23 The second law of thermodynamics

23.1 Introduction

When a process or series of processes returns a working substance to its initial state, a *cycle* has been performed. In a cycle, the change in any property must be zero and the first law becomes

$$\sum q = \sum w \qquad (23.1)$$

Where $\sum$ denotes 'the sum round the cycle'. On a property diagram, the processes representing a cycle must form a closed path such as in Figure 23.6.

The primary object of a power engineer is to design a machine which will accept a heat transfer and give a work transfer. Figure 23.1 shows a machine in which the working substance accepts heat transfer Q_1 from a high temperature energy reservoir and gives a work transfer W. In this ideal case $W = Q_1$ but this is impossible to attain since heat is always transferred to a second energy reservoir at a lower temperature, as shown in Figure 23.2.

In this case, $W = Q_1 - Q_2$ and the existence of the second heat transfer leads to the second law of thermodynamics: 'It is impossible to construct a machine which will give continuous positive work transfer whilst exchanging heat with a single reservoir'.

The machine shown in Figure 23.2 is known as a heat-engine and the measure of its success is the comparison of W with Q_1.

This is expressed by the thermal efficiency,

$$\eta_{th} = \frac{\text{work transfer from engine}}{\text{heat transfer to engine}} = \frac{W}{Q_1} = \frac{Q_1 - Q_2}{Q_1} = 1 - \frac{Q_2}{Q_1} \qquad (23.2)$$

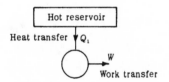

Figure 23.1

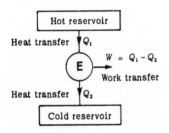

Figure 23.2

23.2 The ideal heat engine and the reversed heat engine

In an ideal heat engine, all processes are reversible and there are no losses; such an engine is the most efficient possible between the chosen reservoirs. This is known as *Carnot's principle* and the efficiency of an ideal heat engine is called the *Carnot efficiency*.

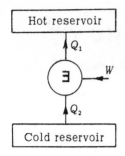

Figure 23.3

Carnot's principle may be demonstrated by considering the existence of a reversed heat engine, which is a device which accepts work transfer in order to transfer heat from a cold reservoir to a hot reservoir. This is shown in Figure 23.3 and is called a *refrigerator*. An ideal reversed heat engine will perform reversible processes and will be the exact 'reverse' to the ideal heat engine.

Let a non-ideal heat engine drive an ideal reversed heat engine and assume that the non-ideal engine has a greater thermal efficiency than that of an ideal engine. The combination is shown in Figure 23.4; the work transfer from the non-ideal engine is W and the work transfer required by the reversed engine is W_{rev}. There is therefore an excess of work $(W - W_{rev})$ available and omitting the redundant hot reservoir, which receives exactly the same amount of energy as it delivers, a device is obtained which produces work whilst exchanging heat with a single reservoir, Figure 23.5. Such a device is impossible according to the second law of the thermodynamics and it is therefore not possible for the non-ideal engine to be more efficient than the ideal reversible engine.

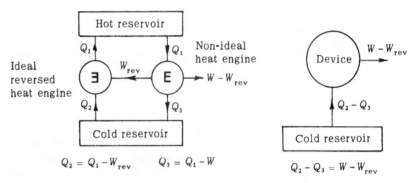

Figure 23.4   **Figure 23.5**

23.3 The ideal heat engine cycle

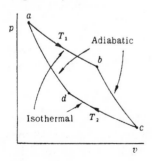

Figure 23.6

The ideal reversible heat engine cycle must consist of two reversible isothermal heat transfer processes, one in which Q_1 is transferred from the hot reservoir at temperature T_1 and one in which Q_2 is transferred to the cold reservoir at temperature T_2. To complete the cycle, two processes without heat transfer, i.e. adiabatic processes, are required. Such a cycle is called a *Carnot cycle*.

Figure 23.6 shows the Carnot cycle for a perfect gas on a pressure–specific volume diagram. The processes ab and cd are isothermal $(pv = c)$ and processes bc and da are adiabatic $(pv^\gamma = c)$.

Let there be a mass m of gas in the processes. Then, from equation (22.20), heat transfer to cycle in ab,

$$Q_1 = mRT_1 \ln \frac{v_b}{v_a}$$

and heat transfer from cycle in cd,

$$Q_2 = mRT_2 \ln \frac{v_c}{v_d}$$

From equation (22.12)
$$\frac{T_1}{T_2} = \left(\frac{v_d}{v_a}\right)^{\gamma-1} = \left(\frac{v_c}{v_b}\right)^{\gamma-1}$$

so that
$$\frac{v_b}{v_a} = \frac{v_c}{v_d}$$

Thus, thermal efficiency,
$$\eta_{th} = \frac{Q_1 - Q_2}{Q_1}$$

$$= \frac{mRT_1 \ln \dfrac{v_b}{v_a} - mRT_2 \ln \dfrac{v_b}{v_a}}{mRT_1 \ln \dfrac{v_b}{v_a}}$$

$$= \frac{T_1 - T_2}{T_1} = \frac{T_{max} - T_{min}}{T_{max}} \quad (23.3)$$

The efficiency of an ideal heat engine is therefore independent of the gas used and depends only upon temperatures. For this efficiency to be a maximum, T_1 should be as high as possible and T_2 as low as possible. This is valid for all power producing cycles, whether ideal or not, and there is constant effort to find materials able to withstand higher temperatures.

The same efficiency may be obtained for vapour processes.

Although the Carnot cycle is the ideal heat engine cycle, giving the maximum possible efficiency between two reservoirs, there are considerable practical difficulties in devising machinery to perform the cycle and it is not used except as a measure of the ultimate efficiency possible. Instead, constant pressure and constant volume cycles are considered for which there are realistic practical counterparts but the digression from the Carnot cycle means that these cycles will have an efficiency which is less than that dictated by the second law. Such cycles are considered in Chapters 24, 25 and 26.

23.4 Entropy

Entropy may be defined as a property which remains constant during a reversible adiabatic process. The symbol for entropy is S and specific entropy (S/m) is denoted by s. In a reversible process which is not adiabatic, there will be an entropy change and the specific heat transfer q is given by the relation

$$q = \int_1^2 T \, ds \quad (23.4)$$

If a reversible process path is drawn on a property diagram with axes T and s, Figure 23.7, then equation (23.4) shows that the specific heat

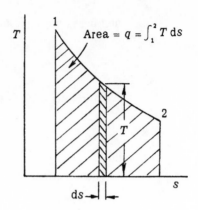

Figure 23.7

transfer is represented by the area under the process path. For a reversible isothermal process, the $T - s$ diagram will be a horizontal line and for a reversible adiabatic process, it will be a vertical line.

Equation (23.4) is *not* applicable to *irreversible* processes.

23.5 Entropy changes in adiabatic processes

Many machines are approximately adiabatic and so the process is of practical interest. In a reversible process, the entropy is constant and $q = \int_1^2 T ds = 0$, but in an actual irreversible process, there is no continuous relation between T and s and $q \neq \int_1^2 T ds$. No direct deduction may be made about the entropy change in such a process but it may be determined by considering the change in *enthalpy*.

This is most easily seen in the case of a perfect gas, for which enthalpy is a function of temperature ($\Delta h = c_p \Delta T$). The friction in an irreversible process causes the temperature of the gas to be higher than it would have been in a frictionless (reversible) process. In the $T - s$ diagrams showing adiabatic compression and expansion processes, Figure 23.8, the

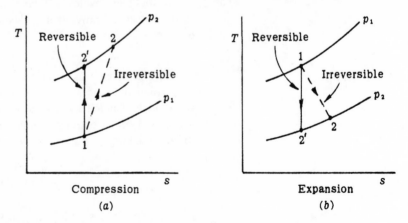

Figure 23.8

vertical lines 1–2′ represent the reversible processes for which the entropy is constant and the sloping lines 1–2 show the irreversible process where $T_2 > T_2'$. It is clearly seen that the entropy has increased in both cases.

For a vapour, the enthalpy at the end of the irreversible adiabatic process is higher than that at the end of the reversible process and Figure 23.9 shows that the entropy increases in the irreversible process 1–2. If the end state 2 is in the wet vapour region, Figure 23.9(a), the temperature will not alter, but the *dryness fraction* is increased. If the end state 2 is in the superheat region, Figure 23.9(b), the vapour *temperature* will be higher in the irreversible process.

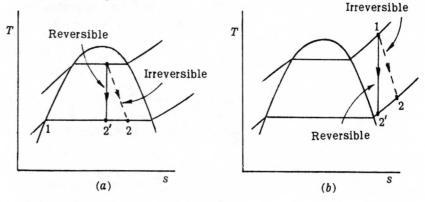

Figure 23.9

23.6 Isentropic efficiency

The entropy change in an irreversible adiabatic process leads to *process efficiency*. The ideal constant entropy process is termed *isentropic* and the ratio of the specific work transfer in the ideal process to that in the actual process is called the *isentropic efficiency*. Since the majority of adiabatic machines are flow processes, isentropic efficiency is usually expressed in terms of the useful work w_x.

For the compression process, Figure 23.8(a), $\qquad \eta_{\text{isen}} = \dfrac{w_{x_{12'}}}{w_{x_{12}}}$ (23.5)

and for the expansion process, Figure 23.8(b), $\qquad \eta_{\text{isen}} = \dfrac{w_{x_{12}}}{w_{x_{12'}}}$

(23.6)

If changes in kinetic and potential energy are negligible, equation (22.29) may be used to rewrite these expressions in terms of specific enthalpy change and for a perfect gas, enthalpy change may be expressed by temperature change, from equation (21.8).

Thus, for compression processes,

$$\eta_{\text{isen}} = \frac{h_{2'} - h_1}{h_2 - h_1} \quad \text{which becomes} \quad \frac{T_{2'} - T_1}{T_2 - T_1} \quad \text{for a perfect gas}$$

and for expansion processes,

$$\eta_{\text{isen}} = \frac{h_1 - h_2}{h_1 - h_{2'}} \quad \text{which becomes} \quad \frac{T_1 - T_2}{T_1 - T_{2'}} \quad \text{for a perfect gas}$$

The physical interpretation of this efficiency is that an irreversible compression process requires more work than an ideal process and an irreversible expansion process gives less work than an ideal process.

In the special case of an adiabatic nozzle expansion process, when the object is to produce kinetic energy rather than work, the kinetic energy produced is less in the actual process than in the ideal process. Isentropic efficiency for a nozzle is expressed by

$$\eta_{\text{isen}} = \frac{\text{actual exit K.E.}}{\text{ideal exit K.E.}}$$

and for a nozzle with negligible inlet velocity and no change in potential energy, the exit K.E. is $(h_1 - h_2)$, from equation (22.33). The process is again represented by the expansion in Figure 23.8(a) and the nozzle efficiency,

$$\eta_{\text{isen}} = \frac{h_1 - h_2}{h_1 - h_{2'}} \quad \text{which becomes} \quad \frac{T_1 - T_2}{T_1 - T_{2'}} \quad \text{for a perfect gas}$$

The expression for efficiency is identical for the expansion whether it is performed in a turbine or a nozzle.

23.7 Processes on the $T - s$ diagram

Figure 23.10 shows the Carnot cycle and heat transfers involved, represented on a $T - s$ diagram. Gas processes are represented on $T - s$ diagram in Figure 23.11 and Figure 23.12 shows the vapour processes.

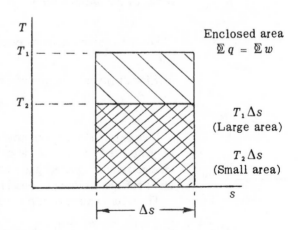

Figure 23.10

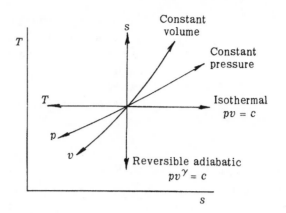

Figure 23.11

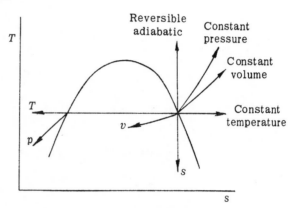

Figure 23.12

23.8 Heat transfer across a finite temperature difference

All heat transfers require a temperature difference between the systems exchanging energy and the greater the temperature difference, the greater the heat transfer rate.

Consider the adiabatic system shown in Figure 23.13 with a small heat transfer rate $\delta \dot{Q}$ from a body A at temperature T_a to a body B at temperature $T_b (T_a > T_b)$. Let $\delta \dot{Q}$ be small enough to leave T_a and T_b unaltered for a short time.

$$\text{Rate of change of entropy of A} = -\frac{\delta \dot{Q}}{T_a}$$

and

$$\text{rate of change of entropy of B} = +\frac{\delta \dot{Q}}{T_b}$$

$$\therefore \text{ rate of change of entropy of system, } \Delta \dot{S} = \frac{\delta \dot{Q}}{T_b} - \frac{\delta \dot{Q}}{T_a}$$

$$= \frac{\delta \dot{Q}(T_a - T_b)}{T_a T_b} \quad (23.7)$$

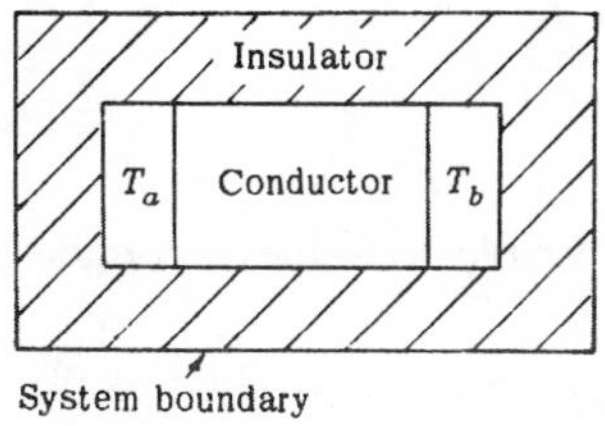

Figure 23.13

Since $T_a > T_b$, equation (23.7) shows that $\Delta \dot{S}$ is positive and the entropy of an adiabatic system is increasing. It may therefore be deduced that *a heat transfer process is irreversible*. Thus any heat transfer across a finite temperature difference is irreversible, which means that even if friction could be eliminated as a source of irreversibility, the second law of thermodynamics states that *any heat transfer which occurs will make the process less than perfect.*

23.9 Entropy changes for vapours and perfect gases

(a) **Vapours**

Specific entropy is one of the properties given in vapour data tables and changes in entropy are obtained by subtraction of extracted values.

(b) **Perfect gases**

The first law, $q - w = \Delta u$, may be written, for a reversible process,

$$\int T \, ds - \int p \, dv = \int du = \int c_v \, dT \tag{23.8}$$

from which

$$ds = \frac{p \, dv}{T} + c_v \frac{dT}{T}$$

From the ideal gas equation,

$$\frac{p}{T} = \frac{R}{v}$$

so that

$$ds = R \frac{dv}{v} + c_v \frac{dT}{T}$$

Thus, by integration,

$$\Delta s = R \ln \frac{v_2}{v_1} + c_v \ln \frac{T_2}{T_1} \tag{23.9}$$

Since $\dfrac{v_2}{v_1} = \dfrac{p_1}{p_2} \cdot \dfrac{T_2}{T_1}$, then

$$\Delta s = -R \ln \frac{p_2}{p_1} + R \ln \frac{T_2}{T_1} + c_v \ln \frac{T_2}{T_1}$$

$$= c_p \ln \frac{T_2}{T_1} - R \ln \frac{p_2}{p_1} \tag{23.10}$$

or since $\dfrac{T_2}{T_1} = \dfrac{p_2}{p_1} \cdot \dfrac{v_2}{v_1}$, then

$$\Delta s = R \ln \frac{v_2}{v_1} + c_v \ln \frac{p_2}{p_1} + c_v \ln \frac{v_2}{v_1}$$

$$= c_p \ln \frac{v_2}{v_1} + c_v \ln \frac{p_2}{p_1} \tag{23.11}$$

23.10 Summary of useful second law information

(a) Heat engines must reject heat. It is not possible to achieve complete conversion of heat to work.

(b) The measure of success of conversion of heat to work is called thermal efficiency, which has a maximum possible value $(T_{max} - T_{min})/T_{max}$.

(c) Thermal efficiency is improved by increasing T_{max} and decreasing T_{min}.

(d) Most losses are caused by friction or heat transfer across a large temperature difference.

(e) Losses can be quantified by a process efficiency, the most useful being the isentropic efficiency of an adiabatic process.

Worked examples

1. *Steam flows along a horizontal duct. At one point in the duct, the pressure of the steam is 1 bar and the temperature is 425°C. At a second point, some distance from the first, the pressure is 1.5 bar and the temperature is 500°C. Assuming the flow to be adiabatic, determine whether the flow is accelerating or decelerating.*

Since the flow is adiabatic, the specific entropy will remain constant if the flow is reversible but will increase if the flow is irreversible.

From tables   for point 1   ($p = 1$ bar, $T = 425°$C), $s_1 = 8.616$ kJ/kg K

and   for point 2   ($p = 1.5$ bar, $T = 500°$C), $s_2 = 8.646$ kJ/kg K

Since $s_2 > s_1$, the flow is from point 1 to point 2.
Applying the steady flow equation,

$$q - w_x = \Delta h + \Delta \frac{V^2}{2} + \Delta gZ$$

But $q = 0$, $w_x = 0$ and $\Delta gZ = 0$ so that

$$\Delta h + \Delta \frac{V^2}{2} = 0$$

From tables at point 1, $h_1 = 3\,330.5$ kJ/kg

and at point 2, $h_2 = 3\,488$ kJ/kg

Thus $$(3\,488 - 3\,330.5) + \frac{V_2^2 - V_1^2}{2} = 0$$

Therefore $$V_2^2 - V_1^2 = -303 \text{ kJ/kg}$$

so that $V_1 > V_2$, i.e. the flow is *decelerating*.

2. *Determine the maximum possible efficiency in the following cases, assuming that the minimum temperature that can be used in any power producing plant is approximately 20°C (fixed by ambient atmospheric conditions):*
(a) *a steam power station in which the maximum steam temperature is 600°C;*
(b) *a gas turbine plant in which the maximum gas temperature is 1100 K;*
(c) *a reciprocating internal combustion engine in which the maximum temperature is 3000 K.*
Comment on the reality of these efficiencies when applied to actual cases.

The maximum possible efficiency is the Carnot efficiency, equation (23.3),

i.e. $$\eta_{th} = \frac{T_1 - T_2}{T_1}$$

where T_1 and T_2 are the maximum and minimum cycle temperatures.

(a) Steam plant:

$$\eta_{th} = \frac{(600 + 273) - (20 + 273)}{(600 + 273)} = 0.664$$

(b) Gas turbine plant:

$$\eta_{th} = \frac{1\,100 - (20 + 273)}{1\,100} = 0.733$$

(c) Reciprocating engine:

$$\eta_{th} = \frac{3\,000 - (20 + 273)}{3\,000} = 0.905$$

None of these efficiencies can be achieved in actual plants. Practical values of efficiencies are:

steam plant	0.4
gas turbine plant	0.25
reciprocating engine	0.35

The steam plant approaches close to the ideal because the condensing process takes place at a temperature close to 20°C at a pressure below atmospheric. Gas turbines and reciprocating engines exhaust at approximately atmospheric pressure with temperatures greatly in excess of 20°C. The reciprocating engine maximum temperature is only a transient value which is not representative of the maximum temperature of the whole working substance.

There are also many other factors which affect the actual efficiencies achieved.

3. *An adiabatic steam turbine expands steam from a pressure of 60 bar and a temperature of 500°C to a pressure of 0.04 bar. The isentropic efficiency of the turbine is 0.82 and changes in kinetic and potential energy may be neglected.*

Determine the state of the steam at exit from the turbine and the specific work transfer.

Figure 23.14 represents the expansion process on the Mollier chart.
From steam tables or chart at 60 bar and 500°C,

$$h_1 = 3\,421 \text{ kJ/kg}$$

and

$$s_1 = 6.879 \text{ kJ/kg K}$$

On the chart, a vertical line is drawn from 1 to 2' representing the ideal reversible process and at 0.04 bar, $h_{2'} = 2\,070$ kJ/kg

Alternatively, from tables, $s_{2'} = s_1$ and $x_{2'}$ is found using $s_f = 0.422$ kJ/kg K and $s_{fg} = 8.051$ kJ/kg K.

Then

$$6.879 = 0.422 + x_{2'} \times 8.051$$

from which

$$x_{2'} = 0.802$$

Thus

$$h_{2'} = h_f + x h_{fg}$$

$$= 121 + 0.802 \times 2\,433 = 2\,070 \text{ kJ/kg}$$

From equation (23.6), $\eta_{isen} = \dfrac{W_{x12}}{W_{x12'}}$ which becomes

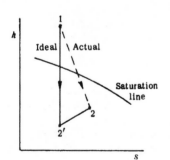

Figure 23.14

$$\eta_{\text{isen}} = \frac{h_1 - h_2}{h_1 - h_{2'}} \text{ since } \Delta\frac{V^2}{2} \text{ and } \Delta gZ \text{ are zero .}$$

i.e.
$$0.82 = \frac{3\,421 - h_2}{3\,421 - 2070}$$

from which
$$h_2 = 2\,317 \text{ kJ/kg}$$

By inspection of the chart, $x_2 = 0.905$ or from tables at 0.04 bar,

$$h_2 = h_f + xh_{fg}$$

i.e.
$$2\,317 = 121 + x_2 \times 2\,433$$

from which
$$x_2 = \underline{0.905}$$

Applying the steady flow energy equation,

$$q - w_x = \Delta h + \Delta\frac{V^2}{2} + \Delta gZ$$

But $q = 0$, $\Delta\dfrac{V^2}{2} = 0$ and $\Delta gZ = 0$, so that

$$-w_x = 2\,317 - 3\,421$$

$$w_x = \underline{1\,104 \text{ kJ/kg}}$$

4. *A rotary air compressor takes in air (which may be treated as a perfect gas) at a pressure of 1 bar and a temperature of 20°C and compresses it adiabatically to a pressure of 6 bar. The isentropic efficiency of the process is 0.85 and changes in kinetic and potential energy may be neglected. Calculate the specific entropy change of the air. Take $R = 0.287$ kJ/kg K and $c_p = 1.006$ kJ/kg K.*

Figure 23.15 represents the process on a $T - s$ diagram.

For the ideal process,
$$\frac{T_2}{T_1} = \left(\frac{p_2}{p_1}\right)^{\frac{\gamma-1}{\gamma}}$$

From the relations $\dfrac{c_p}{c_v} = \gamma$ and $c_p - c_v = R$

$$\frac{\gamma - 1}{\gamma} = \frac{R}{c_p}$$

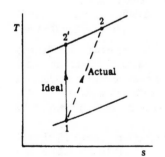

Figure 23.15

Hence
$$T_{2'} = (20 + 273)\left(\frac{6}{1}\right)^{\frac{0.287}{1.006}} = 488 \text{ K}$$

From equation (23.5), $\eta_{\text{isen}} = \dfrac{w_{x_{12'}}}{w_{x_{12}}}$ which reduces to

$$\eta_{\text{isen}} = \frac{T_{2'} - T_1}{T_2 - T_1} \text{ for a perfect gas when } \Delta\frac{V^2}{2} \text{ and}$$

$$\Delta gZ \text{ are negligible,}$$

i.e.

$$0.85 = \frac{488 - 293}{T_2 - 293}$$

from which

$$T_2 = 522 \ K$$

The specific entropy change is given by

$$\Delta s = c_p \ln \frac{T_2}{T_1} - R \ln \frac{p_2}{p_1} \quad \text{from equation (23.10)}$$

$$= 1.006 \ln \frac{522}{293} - 0.287 \ln \frac{6}{1}$$

$$= \underline{0.074 \ \text{kJ/kg K}}$$

This is positive, as expected, in an irreversible adiabatic process.

Further problems

5. At some point in a horizontal adiabatic duct the pressure and temperature of the air flowing through it are found to be 1.8 bar, 25°C respectively. At a second point the pressure and temperature are found to be 1.75 bar, 28°C respectively. In which direction is the flow?

For air, the specific heat capacity at constant pressure, $c_p = 1.005$ kJ/kg K and $\gamma = 1.4$.   [from 1.75 bar to 1.8 bar]

6. The isentropic efficiency of an adiabatic gas turbine is 0.88. The gases enter the turbine at a temperature of 940 K and a pressure of 500 kN/m² and expand through a pressure ratio of 4.8. When changes in kinetic and potential energy are neglected, calculate the specific work transfer from the turbine and the exit area required for a power output of 8 000 kW with an exit velocity of 75 m/s.

For the gas, the specific heat capacity at constant pressure, $c_p = 1.15$ kJ/kg K and $\gamma = 1.33$.   [306.5 kJ/kg; 0.64 m²]

7. The propelling nozzle of an aircraft jet engine has an isentropic efficiency of 0.9. The gases enter the adiabatic nozzle at a pressure of 170 kN/m² and a temperature of 850 K with negligible velocity. The gases leave the nozzle at a pressure of 60 kN/m². Calculate the gas velocity at exit from the nozzle.

If the engine propels an aircraft at 200 m/s, calculate the specific thrust of the engine using the momentum equation.

For the gas, the specific heat capacity at constant pressure, $c_p = 1.15$ kJ/kg K and $\gamma = 1.33$.   [635 m/s; 4.35 Ns/kg]

8. A steam turbine receives steam from a steam generator at a pressure of 50 bar and a temperature of 400°C. The condenser pressure is 0.05 bar. The steam mass flow rate is 2.9 kg/s and the power output is 2 800 kW. If changes in kinetic and potential energy are negligible, calculate the isentropic efficiency of the turbine and the state of the steam at exit from the turbine.   [83%; $x = 0.864$]

9. A perfect gas, for which the specific heat capacity at constant volume, $c_v = 700$ J/kg K and $\gamma = 1.4$, undergoes a reversible non-flow process in which the pressure and temperature change from 15 bar, 760°C to 2 bar, 25°C. Calculate the specific entropy change of the gas in the process.

If the process may be represented by a straight line on the $T - s$ chart, calculate the specific work transfer in the process.   [−0.653 kJ/kg K; 80 kJ/kg]

10. (*a*) Saturated mercury vapour enters a turbine at a pressure of 4 bar. It leaves at 1 bar. The isentropic efficiency of expansion is 0.85. Calculate the specific work transfer from the turbine.

(*b*) The mercury is condensed at constant pressure by water entering at 15 bar, 20°C. The mercury leaves as saturated liquid at 1 bar and the H_2O as superheated steam at 15 bar and 250°C. Calculate the mass flow rate of H_2O per unit mass flow rate of mercury.   [31.4 kJ/kg; 0.096]

11. Steam flows steadily with low velocity through an adiabatic machine. The mass flow rate is 8 kg/min and at the boundaries of the machine the state of the steam is: Boundary X, 90 bar, 700°C; Boundary Y, 30 bar, 500°C.

(*a*) Is entry at X or Y?

(*b*) What work transfer rate is associated with the machine?

(*c*) What sort of machine is it likely to be?

(*d*) What is the isentropic efficiency of the machine?

[X; 55.7 kW; turbine; 0.986]

24 Steam plant

24.1 Cycle performance parameters

Engineers use a number of parameters to determine the performance of a cycle; some of these are defined below, and will be used in Chapters 24, 25 and 26.

Specific work (w) is the net work transfer per unit mass from the cycle in kJ/kg.

Thermal efficiency (η_{th}) is the ratio $\dfrac{\text{specific work transfer}}{\text{heat transfer to cycle}}$

$$= \frac{\text{heat transfer to cycle} - \text{heat transfer from cycle}}{\text{heat transfer to cycle}}$$

Specific consumption is the mass flow rate of working substance required to produce unit power output,

i.e. $\quad$ specific consumption $= \dfrac{3\,600}{\text{specific work}} (\text{kg/kW h})$

Mean effective pressure – for non-flow cycles, the m.e.p. is the average pressure which, acting over the same swept volume, gives the same work transfer,

i.e. $\quad$ m.e.p. $= \dfrac{\text{specific work} \times \text{mass of substance}}{\text{swept volume}} = \dfrac{\int p \, dV}{\text{swept volume}}$

The m.e.p. for a cycle is shown in Figure 24.1.

It is important to realize that ideal cycle parameters do not allow for losses and other efficiencies may also be used: for steam plant we need

$$\text{Boiler efficiency} = \frac{\dot{m}_{H_2O}(h_3 - h_2)}{\dot{m}_{fuel} \times \text{calorific value}}$$

$$\text{Overall efficiency of a steam plant} = \frac{\text{shaft power from plant}}{\dot{m}_{fuel} \times \text{calorific value}}$$

24.2 The Rankine cycle

Steam plant has a long history but is now used mainly for electricity generation. The ideal cycle for steam plant is known as the Rankine cycle, which is a constant pressure flow process cycle. A simple steam plant is shown in Figure 24.2 and the processes are set out in Table 24.1. The steady flow energy equation is used to analyse each process, changes in kinetic and potential energy being neglected.

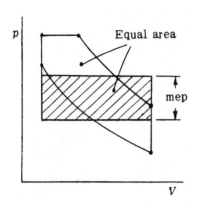

Figure 24.1

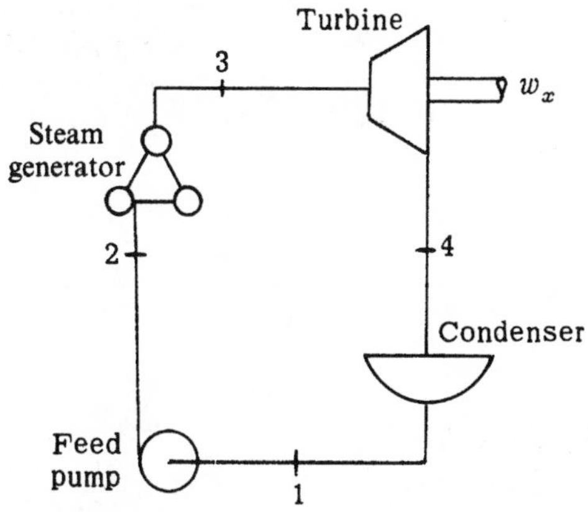

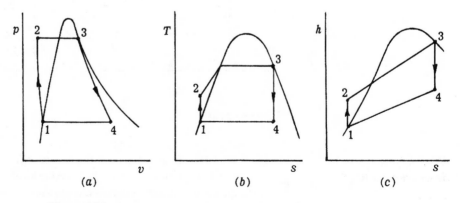

Figure 24.2

Table 24.1. Rankine cycle processes for the plant shown in Figure 24.2

Process	Description	$q - w_x = \Delta h$
1–2	Reversible, adiabatic compression of saturated water in a *feed pump*	$-w_x = \Delta h = h_2 - h_1$
2–3	Reversible, constant pressure heat transfer to cycle in a *steam generator*	$q = \Delta h = h_3 - h_2$
3–4	Reversible, adiabatic expansion of steam in a *turbine*	$-w_x = \Delta h = h_4 - h_3$ or $w_x = h_3 - h_4$
4–1	Reversible, constant pressure heat transfer from cycle in a *condenser*	$q = \Delta h = h_1 - h_4$

Figure 24.3

These processes are shown on $p - v$, $T - s$ and $h - s$ diagrams in Figures 24.3 (a), (b) and (c).

The specific work of the cycle,

$$w = (h_3 - h_4) - (h_2 - h_1) \qquad (24.1)$$

The thermal efficiency of the cycle,

$$\eta_{th} = \frac{(h_3 - h_4) - (h_2 - h_1)}{h_3 - h_2} \qquad (24.2)$$

$$= \frac{(h_3 - h_4) - (h_2 - h_1)}{(h_3 - h_1) - (h_2 - h_1)} \qquad (24.3)$$

The feed pump work, $h_2 - h_1$, is usually very small compared with the turbine work, $h_3 - h_4$, and the boiler heat transfer, $h_3 - h_2$ and so may be neglected.

Thus
$$\eta_{th} = \frac{h_3 - h_2}{h_3 - h_1}$$

and
$$w = h_3 - h_4 \qquad (24.4)$$

If superheated steam is produced in a steam generator, the cycle will appear in different form on the property diagrams, Figures 24.4 (a) and (b), but equations (24.1), (24.2), (24.3) and (24.4) are still valid.

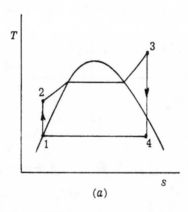 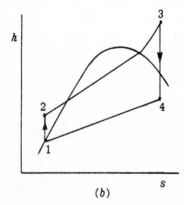

(a) (b)

Figure 24.4

24.3 Cycle temperatures

The use of superheated steam will improve efficiency since it increases the maximum cycle temperature to the metallurgical limit of the materials used.

Efficiency can also be improved by lowering the minimum cycle temperature but this is fixed by the ambient cooling water temperature used in the condensers. This means that steam is usually condensed at about 27°C at the corresponding saturation pressure of 3.6 kN/m² which is well below atmospheric pressure.

24.4 Steam plant cycles with isentropic efficiency

The effect of isentropic efficiency in steam plant is only considered for the expansion process since, as stated in Section 24.2, the feed pump work is so small as to be negligible.

Figure 24.5 shows the actual expansion process 3–4, compared with the ideal process 3–4′. As defined in Section 23.6, the isentropic efficiency.

$$\eta_{\text{isen}} = \frac{w_{x34}}{w_{x34'}} = \frac{h_3 - h_4}{h_3 - h_{4'}} \tag{24.5}$$

and this shows that the turbine work output is reduced. Since the heat transfer to the cycle is unaltered, the plant cycle efficiency will be less than the ideal Rankine cycle efficiency,

$$\eta_{\text{cycle}} = \frac{h_3 - h_4}{h_3 - h_2} \tag{24.6}$$

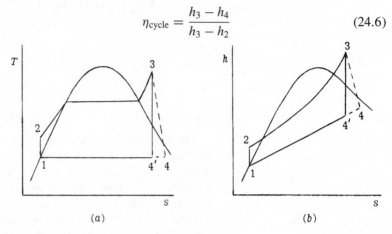

(a)   (b)

Figure 24.5

24.5 Methods of improving steam plant cycle efficiency

As stated in Section 24.3, the use of superheat up to the metallurgical limit of the materials used will improve cycle efficiency. When this limit is reached, the steam generator pressure may also be increased, giving further improvement in efficiency at the expense of wetter steam in the expansion stage, Figure 24.6.

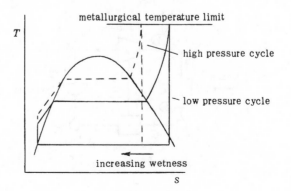

Figure 24.6

This can be overcome by reheating the steam part way through the expansion process but this will require the use of two turbines. The layout of the plant and the resulting $T - s$ diagram for the cycle are shown in Figure 24.7 (*a*) and (*b*).

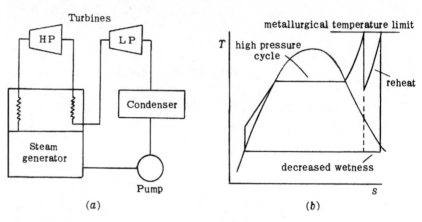

(*a*)   (*b*)

Figure 24.7

The least efficient part of the heat transfer in the steam generator is the heating of the cold feed water because this takes place at temperatures well below the cycle maximum temperature and often with large temperature differences (Section 23.8). To overcome this problem, *feed water heaters* are used to raise cycle efficiency.

Figure 24.8(*a*) shows the plant layout and Figure 24.8(*b*) shows the cycle. For a mass flow of 1 kg/s entering the turbine, a small amount m kg/s (0.1 to 0.2 kg/s) is bled off part way through the expansion and passed to the feed water heater. The remaining steam, $(1-m)$ kg/s, is fully expanded, condensed and pumped to the feed heater where it mixes with the bled mass. The amount of steam bled *ideally* just condenses and raises the feed water temperature to the saturation temperature corresponding to the bleed pressure.

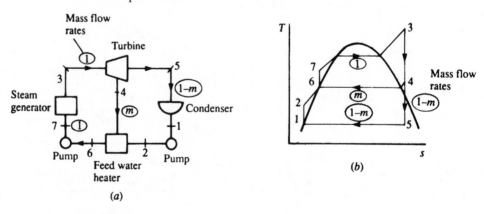

(*a*)

Figure 24.8

The heat transfer to the steam generator is now reduced and occurs at a higher temperature. However, the work produced by the turbine is reduced as the full mass is not completely expanded. The new higher efficiency, neglecting feed pump work, is given by

$$\eta = \frac{1(h_3 - h_4) + (1 - m)(h_4 - h_5)}{1(h_3 - h_6)} \qquad (24.7)$$

The bled mass m is found by applying the steady flow energy equation to the workless adiabatic feed heater.

$$q - w_x = \Delta h$$

$$\therefore \Delta h = 0 \quad \text{i.e.} \quad h_{\text{in}} = h_{\text{out}}$$

Thus $mh_4 + (1 - m)h_2 = h_6$

But $h_2 = h_1$ since feed pump work is negligible,

so that
$$m = \frac{h_6 - h_1}{h_4 - h_1} \qquad (24.8)$$

In modern power stations, several stages of feed heating are used to raise the feed water temperature step-by-step towards the steam generator saturation temperature.

24.6 Combined heat and power plant

A *combined heat and power plant* uses the energy in steam plant condenser cooling water for district heating or process heating in factories. To raise the temperature of the coolant to a useful value, the pressure in the condenser is often raised to the saturation pressure corresponding to the process temperature, Figure 24.9.

The efficiency of such a plant is given by

$$\eta = \frac{\text{total energy output in work and heat}}{\text{energy supplied in fuel}} \qquad (24.9)$$

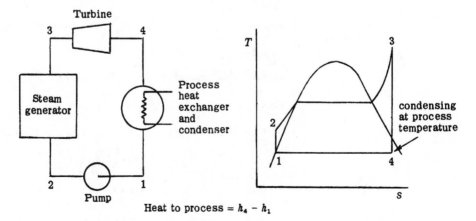

Figure 24.9

This plant is also called a *back pressure turbine plant*. Another form of steam plant to supply both heat and power uses a *pass-out turbine*. In this plant some steam is bled at a pressure suited to other uses, the remaining steam in the turbine continues to expand to the normal condensing pressure.

Worked examples

1. *(a) An ideal steam plant uses the Rankine cycle; the boiler pressure is 12 bar and the condenser pressure is 0.04 bar. Calculate the thermal efficiency of the plant, neglecting the feed pump work.*

(b) If the boiler delivers superheated steam at a pressure of 12 bar and a temperature of 400°C, what is the new thermal efficiency?

(c) Comment on the results obtained.

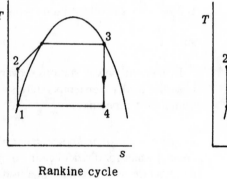
Rankine cycle

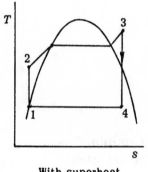
With superheat

Figure 24.10

Figure 24.10 shows the cycles involved. In both cases the thermal efficiency is given by equation (24.3),

i.e.
$$\eta_{th} = \frac{(h_3 - h_4) - (h_2 - h_1)}{(h_3 - h_1) - (h_2 - h_1)}$$

which reduces to $\quad \eta_{th} = \dfrac{h_3 - h_4}{h_3 - h_1} \quad$ when the feed pump work, $h_2 - h_1$,

is neglected.

(a) From chart and tables, $h_1 = 121$ kJ/kg, $h_3 = 2\,784$ kJ/kg and $h_4 = 1\,962$ kJ/kg.

Hence
$$\eta_{th} = \frac{2\,784 - 1\,962}{2\,784 - 121} = 0.308$$

(b) From chart and tables, $h_1 = 121$ kJ/kg, $h_3 = 3\,260$ kJ/kg and $h_4 = 2\,225$ kJ/kg.

Hence
$$\eta_{th} = \frac{3\,260 - 2\,225}{3\,260 - 121} = 0.329$$

(c) It can be seen that the use of superheat has improved the cycle efficiency and if the superheat is increased further, the efficiency will increase further. Steam plant uses superheat temperatures up to the metallurgical limit of the materials used in the superheater tubes of the steam generator.

It will also be seen that the dryness fraction of the steam at the end of the expansion process is greater than that with saturated steam, i.e., the expanded steam contains fewer water droplets which erode turbine blades. The use of superheated steam is essential to the operation of modern steam turbines.

2. *The steam plant of Example 1(b) is fitted with a feed heater, bleeding steam from the turbine at 1.3 bar. The amount of steam bled is just enough to raise the feed water temperature to the saturation temperature at 1.3 bar.*

Determine the mass bled per kg of steam in the steam generator and the cycle efficiency.

Figure 24.8(*b*) shows the cycle.

From steam tables or chart, $h_4 = 2\,728$ kJ/kg

From steam tables, $h_6 = h_f$ at 1.3 bar $= 449.2$ kJ/kg

From Example 1(*b*), $h_3 = 3\,260$ kJ/kg, $h_5 = 2\,225$ kJ/kg, $h_1 = 121$ kJ/kg

$$\text{Thus mass bled, } m = \frac{h_6 - h_1}{h_4 - h_1} \text{ from equation (24.8)}$$

$$= \frac{449.2 - 121}{2\,728 - 121} = \underline{0.126 \text{ kg}}$$

$$\text{Cycle efficiency, } \eta = \frac{1(h_3 - h_4) + (1 - m)(h_4 - h_5)}{1(h_3 - h_6)} \text{ from equation (24.7)}$$

$$= \frac{1(3\,260 - 2\,728) + (1 - 0.126)(2\,728 - 2\,225)}{1(3\,260 - 449.2)}$$

$$= 0.346 \quad \text{or} \quad \underline{34.6\%}$$

Note: The addition of feed heating has clearly improved the cycle efficiency.

3. *In a steam plant (steam generator, high pressure turbine, low pressure turbine, condenser and feed pump), the steam generator exit state is 4 MN/m², 600°C. The high pressure turbine expands the steam, to 0.6 MN/m² with isentropic efficiency 0.8. The steam is then reheated at constant pressure to 600°C and expanded in the low pressure turbine to 0.004 MN/m² with isentropic efficiency 0.8. The steam is condensed without subcooling.*

Determine the cycle efficiency, neglecting feed pump work.

Figure 24.11 shows the cycle. From chart and tables,

$$h_1 = h_2 = 121.4 \text{ kJ/kg}$$

$$h_3 = 3\,675 \text{ kJ/kg}$$

$$h_{4'} = 3\,050 \text{ kJ/kg}$$

$$h_5 = 3\,700 \text{ kJ/kg}$$

$$h_{6'} = 2\,495 \text{ kJ/kg}$$

The plant layout is shown in Figure 24.7(*a*).

For the high pressure turbine,

$$\eta_{\text{isen}} = \frac{h_3 - h_4}{h_3 - h_{4'}} \text{ from equation (24.5)}$$

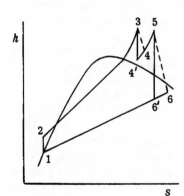

Figure 24.11

i.e $$0.8 = \frac{3\,675 - h_4}{3\,675 - 3\,050}$$

from which $$h_4 = 3\,175 \text{ kJ/kg}$$

Similarly, for the low pressure turbine,

$$\eta_{isen} = \frac{h_5 - h_6}{h_5 - h_6}$$

i.e. $$0.8 = \frac{3\,700 - h_6}{3\,700 - 2\,495}$$

from which $$h_6 = 2\,736 \text{ kJ/kg}$$

$$\text{Specific work output} = (h_3 - h_4) + (h_5 - h_6)$$

$$\text{Specific heat transfer to cycle} = (h_3 - h_2) + (h_5 - h_4)$$

$$\therefore \text{ cycle efficiency} = \frac{(h_3 - h_4) + (h_5 - h_6)}{(h_3 - h_2) + (h_5 - h_4)}$$

i.e. $$\eta = \frac{(3\,675 - 3\,175) + (3\,700 - 2\,736)}{(3\,675 - 121.4) + (3\,700 - 3\,175)}$$

$$= 0.359 \quad \text{or} \quad \underline{35.9\%}$$

Further problems

4. An ideal steam plant uses the Rankine cycle. The steam generator pressure is 20 bar and the condenser pressure is 0.04 bar. Neglecting the feed pump work, determine the thermal efficiency and the specific work transfer from the cycle. Determine also the dryness fraction of the steam at exit from the turbine and the Carnot efficiency of the cycle.

[33.4%; 896 kJ/kg; 0.734; 37.7%]

5. (i) In order to improve the performance of the steam plant of question 4 the pressure at which the steam is generated is increased to 40 bar. Calculate the new cycle efficiency and the specific work transfer and the dryness fraction at turbine exit.

(ii) To improve the performance of the cycle further and to avoid too low a dryness fraction at turbine exit the steam is superheated in the steam generator to a temperature of 700°C. Calculate the new cycle efficiency and specific work transfer and the dryness fraction at turbine exit.

(iii) To improve the performance of the cycle further, a feed heater is fitted, bleeding steam from the turbine at 3.6 bar. The amount of steam bled is just enough to raise the feed water temperature to the saturation temperature at 3.6 bar. Determine the new cycle thermal efficiency and specific work transfer.

[(i) 36.3%; 975 kJ/kg; 0.7, (ii) 42.5%; 1 610 kJ/kg; 0.893, (iii) 44.73%; 1 483.2 kJ/kg]

6. A steam plant is to be designed using the Rankine cycle with superheat. The steam enters the turbine through a pipe 0.6 m in diameter and expands to 0.05 bar. In order to produce 12 MW, a mass flow rate of 32 000 kg/h is to be used and the dryness fraction at turbine exit must not be less than 0.9. Determine the steam generator pressure, the steam temperature, the cycle efficiency and the velocity of the steam in the inlet pipe. Feed pump work may be neglected and the change in kinetic energy across the turbine is negligible. (The Mollier chart may be used in this problem.)   [23 bar; 590°C; 38.3%; 5.56 m/s]

7. A steam plant uses steam delivered from the steam generator at 6 MN/m^2, 700°C and expanded in the turbine to 0.005 MN/m^2 with isentropic efficiency 0.85. Assuming the condensate is saturated liquid and feed pump work is negligible, determine the cycle thermal efficiency and the steam mass flow rate for an output of 3 MW.   [36.8%, 2.17 kg/s]

8. A steam plant uses two turbines with reheating between them. The isentropic efficiency of both turbines is 0.8. The steam leaves the steam generator at 6 MN/m^2, 550°C and expands to 0.5 MN/m^2. Reheat at this pressure raises the temperature to 500°C before expansion in the second turbine to 0.004 MN/m^2. The steam flow rate is 4.8 kg/s. Determine the power output and the thermal efficiency of the plant.
[6.94 MW, 36.6%]

25 Gas turbine plant

25.1 Introduction

Gas turbine plant is used mainly for aircraft propulsion but is also used in ships, locomotives, automobiles, pumping and power generation. Aircraft application is not dealt with here since a knowledge of gas dynamics in intakes and nozzles is required.

25.2 The Joule cycle

The ideal gas turbine plant cycle is known as the Joule cycle, which is a constant pressure flow process cycle. A typical arrangement is shown in Figure 25.1 and the processes are set out in Table 25.1. The steady flow energy equation is used to analyse each process, changes in kinetic and potential energy being neglected.

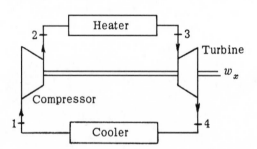

Figure 25.1

Table 25.1. Joule cycle processes for the plant shown in Figure 25.1.

Process	Description	$q - w_x = \Delta h = c_p \Delta T$ for gases
1–2	Reversible, adiabatic compression in a *rotary compressor*	$-w_x = h_2 - h_1$ $= c_p(T_2 - T_1)$
2–3	Reversible, constant pressure heat transfer to cycle in a *heater*	$q = h_3 - h_2$ $= c_p(T_3 - T_2)$
3–4	Reversible, adiabatic expansion in a *turbine*	$-w_x = h_4 - h_3$ $w_x = c_p(T_3 - T_4)$
4–1	Reversible, constant pressure heat transfer from cycle in a *cooler*	$q = h_1 - h_4$ $= c_p(T_1 - T_4)$

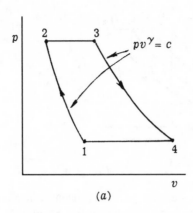

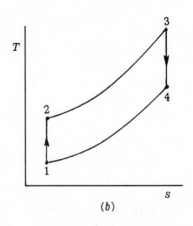

(a)   (b)

Figure 25.2

These processes are shown on $p - v$ and $T - s$ diagrams in Figures 25.2(a) and (b).

The specific work of this cycle,

$$w = c_p(T_3 - T_4) - c_p(T_2 - T_1) \qquad (25.1)$$

The thermal efficiency of the cycle,

$$\eta_{\text{th}} = \frac{c_p(T_3 - T_4) - c_p(T_2 - T_1)}{c_p(T_3 - T_2)}$$

$$= 1 - \frac{T_4 - T_1}{T_3 - T_2}$$

From equation (22.11), $\dfrac{T_3}{T_4} = \dfrac{T_2}{T_1} = \left(\dfrac{p_2}{p_1}\right)^{\frac{\gamma-1}{\gamma}}$

$$= r_p^{\frac{\gamma-1}{\gamma}} \quad \text{where } r_p \text{ is the cycle pressure ratio}$$

Thus $$\eta_{\text{th}} = 1 - \frac{1}{r_p^{\frac{\gamma-1}{\gamma}}} \qquad (25.2)$$

25.3 Cycle temperatures

Equation (25.2) shows that as the pressure ratio increases, the cycle efficiency increases. There is, however, a practical limit to the way in which it may be interpreted.

There is a metallurgical limit to the maximum cycle temperature, T_3, and if this is set as shown in Figure 25.3, it will be seen that many Joule cycles can be drawn with increasing pressure ratio ($A > B > C$) and increasing efficiency. However, it is found that the work output from the cycle increases to a peak at a particular pressure ratio and this must be considered an important design consideration since at the maximum value of r_p, the work becomes zero, Figure 25.4.

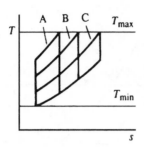

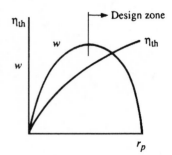

Figure 25.3   Figure 25.4

25.4 Gas turbine plant cycle with isentropic efficiency

In a gas turbine plant, it is usual to use a combustion chamber (instead of the heater in the Joule cycle. Section 25.1) and to burn the fuel directly in the air after compression. It is then necessary to discard the gases after expansion in the turbine without cooling and to take a fresh supply of cool air. This makes for a compact plant of high power/weight ratio; the layout is shown in Figure 25.5.

The effect of isentropic efficiency applies to both compression and expansion processes, Figure 25.6. For the compression process,

$$\eta_{isen} = \frac{w_{x_{12'}}}{w_{x_{12}}} = \frac{h_{2'} - h_1}{h_2 - h_1} = \frac{T_{2'} - T_1}{T_2 - T_1} \quad (25.3)$$

and for the expansion process,

$$\eta_{isen} = \frac{w_{x_{34}}}{w_{x_{34'}}} = \frac{h_3 - h_4}{h_3 - h_{4'}} = \frac{T_3 - T_4}{T_3 - T_{4'}} \quad (25.4)$$

The turbine work output is reduced and the compressor work requirement is increased so that the net work from the plant, and hence the plant efficiency, is less than that for the ideal Joule cycle.

It should be noted that since the working substance changes from air in the compressor to combustion products in the turbine, the values of the specific heat and γ will be different for each process.

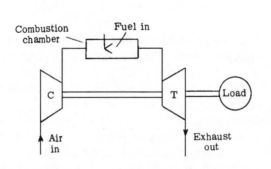

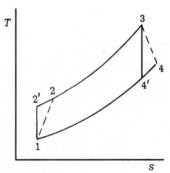

Figure 25.5   Figure 25.6

Thus the specific work,

$$w = c_{p_{34}}(T_3 - T_4) - c_{p_{12}}(T_2 - T_1) \tag{25.5}$$

and the efficiency of the cycle is given by

$$\eta_{\text{cycle}} = \frac{c_{p_{34}}(T_3 - T_4) - c_{p_{12}}(T_2 - T_1)}{c_{p_{23}}(T_3 - T_2)} \tag{25.6}$$

25.5 Methods of improving gas turbine plant efficiency

The cycle maximum temperature T_3 should be as high as the metallurgical limit of the materials used. It is then possible to determine analytically the pressure ratio required for maximum efficiency *or* maximum output. For aircraft use, the lightest plant is chosen but for industrial applications such as power and pumping stations, and for use in ships, trains and automobiles, where volume and weight considerations are not so vital, it is usual to fit a heat exchanger. Figure 25.7(a) shows such a plant, in which a heat exchanger is placed between the hot exhaust gas leaving the turbine and the cooler air leaving the compressor. This saves fuel since part of the heating process is achieved with energy which is otherwise wasted. This leads to an improvement in plant efficiency.

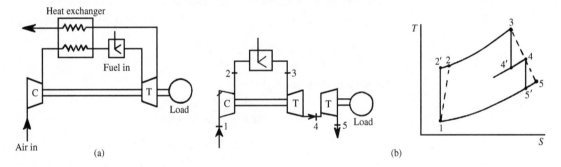

Figure 25.7

25.6 The *free* power turbine

For some purposes it may be useful to have power delivered by a gas turbine *mechanically* separate from the compressor. Effectively the compressor–turbine unit acts as a hot gas generator for the *free* gas turbine which may be used for traction in a vehicle or a ship or any other device needing separate speed control. Problems are solved by calculating the value of T_4 and hence the intermediate pressure. The *free turbine* power may then be calculated. (Figure 25.7(b)).

Worked Examples

1. *An ideal gas turbine plant uses a Joule cycle with a perfect gas for which* $c_p = 5.193$ *kJ/kg K and* $\gamma = 1.67$. *The gas commences the compression process at a temperature of* $200\,°C$ *and the maximum cycle temperature is* $1\,200\,°C$. *The cycle pressure ratio is 10.*

Calculate the specific heat transfer to the cycle, the specific work transfer in the expansion process and the thermal efficiency of the cycle. Changes in kinetic and potential energy may be neglected.

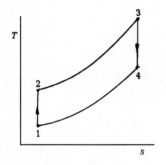

Figure 25.8

The cycle is shown in Figure 25.8.

For the reversible adiabatic compression process,

$$\frac{T_2}{T_1} = \left(\frac{p_2}{p_1}\right)^{\frac{\gamma-1}{\gamma}}$$

$$\therefore T_2 = 473 \times 10^{0.4}$$

$$= 1\,188 \text{ K}$$

For the reversible adiabatic expansion process,

$$\frac{T_3}{T_4} = \left(\frac{p_3}{p_4}\right)^{\frac{\gamma-1}{\gamma}}$$

$$\therefore T_4 = \frac{1\,473}{10^{0.4}} = 587 \text{ K}$$

The specific heat transfer to the cycle in process 2–3 is given by

$$q = \Delta h = c_p(T_3 - T_2) \quad \text{from the steady flow energy equation}$$

$$= 5.193(1\,473 - 1\,188) = \underline{1\,478 \text{ kJ/kg}}$$

The specific work transfer in the expansion process 3–4 is given by

$$-w_x = \Delta h = c_p(T_4 - T_3) \quad \text{from the steady flow energy equation}$$

$$= 5.193(587 - 1\,473)$$

$$\therefore w_x = \underline{4\,610 \text{ kJ/kg}}$$

The thermal efficiency is given by

$$\eta_{th} = \frac{\text{net work transfer}}{\text{heat transfer to cycle}}$$

$$= \frac{c_p(T_3 - T_4) - c_p(T_2 - T_1)}{c_p(T_3 - T_2)}$$

$$= \frac{(1\,473 - 587) - (1\,188 - 473)}{1\,473 - 1\,188} = \underline{0.608}$$

Alternatively, $\quad \eta_{th} = 1 - \dfrac{1}{r_p^{\frac{\gamma-1}{\gamma}}} \quad$ from equation (25.2)

$$= 1 - \frac{1}{10^{0.4}} = \underline{0.608}$$

Note: This problem represents a closed cycle gas turbine using a special gas with a high value of specific heat to raise specific output. This could also be achieved by pressurizing the circuit.

2. *A gas turbine plant takes in air at 100 kN/m², 16°C and compresses it to 500 kN/m² with isentropic efficiency 0.84. The air enters a combustion chamber and the combustion products leave at 490 kN/m², 600°C. The turbine expands the gases to 100 kN/m² with isentropic efficiency 0.88. The plant produces 1500 kW.*

Determine the air mass flow rate, neglecting the fuel mass flow rate. If the calorific value of the fuel (Section 29.3) is 45000 kJ/kg and the combustion efficiency is 95%, what is the air–fuel ratio?

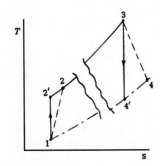

Figure 25.9

For air, $\quad c_p = 1.0 \ kJ/kg \ K \quad and \quad \gamma = 1.4$

For gases, $c_p = 1.15 \ kJ/kg \ K \quad and \quad \gamma = 1.3$

Figure 25.9 shows the cycle.

For the compressor,

$$\frac{T_{2'}}{T_1} = \left(\frac{p_2}{p_1}\right)^{\frac{\gamma-1}{\gamma}} \qquad \text{from equation (22.11)}$$

$$\therefore T_{2'} = 289 \left(\frac{500}{100}\right)^{\frac{0.4}{1.4}}$$

$$= 458 \ K$$

From equation (25.3), $\qquad \eta_{\text{isen}} = \dfrac{T_{2'} - T_1}{T_2 - T_1}$

i.e. $\qquad 0.84 = \dfrac{458 - 289}{T_2 - 289}$

from which $\qquad T_2 = 490.2 \ K$

For the turbine, $\qquad \dfrac{T_3}{T_{4'}} = \left(\dfrac{p_3}{p_4}\right)^{\frac{\gamma-1}{\gamma}} \qquad$ from equation (22.11)

i.e. $\qquad \dfrac{873}{T_{4'}} = \left(\dfrac{490}{100}\right)^{\frac{0.3}{1.3}}$

from which $\qquad T_{4'} = 605 \ K$

From equation (25.4) $\qquad \eta_{\text{isen}} = \dfrac{T_3 - T_4}{T_3 - T_{4'}}$

i.e. $\qquad 0.88 = \dfrac{873 - T_4}{873 - 605}$

from which $\qquad T_4 = 637.2 \ K$

Let the mass flow rate for the air be $\dot{m}$; since the mass flow rate for the fuel is to be neglected, this is also the mass flow rate for the gas.

$$\text{Net power} = \text{turbine power} - \text{compressor power}$$

i.e. $\qquad \dot{W}_x = \dot{m} c_{p_{\text{gas}}}(T_3 - T_4) - \dot{m} c_{p_{\text{air}}}(T_2 - T_1)$

i.e. $\qquad 1\,500 = \dot{m}\{1.15(873 - 637.2) - 1.0(490.2 - 289)\}$

from which $\qquad \underline{\dot{m} = 21.44 \ \text{kg/s}}$

In the combustion chamber, the energy released by burning the fuel raises the temperature from T_2 to T_3. Hence, if the flow rate of fuel is $\dot{m}_{\text{fuel}}$,

$$\text{combustion efficiency} = \frac{\dot{m} c_{p_{\text{gas}}}(T_3 - T_2)}{\dot{m}_{\text{fuel}} \times \text{calorific value}}$$

i.e. $\qquad 0.95 = \dfrac{\dot{m} \times 1.15(873 - 490.2)}{\dot{m}_{\text{fuel}} \times 45\,000}$

from which   air–fuel ratio, $\dfrac{\dot{m}}{\dot{m}_{\text{fuel}}} = \underline{97.1}$

Note: This air–fuel ratio is higher than that for a diesel or petrol engine to avoid *sustained* high combustion temperatures which would burn out the combustion chamber.

3. *In the gas turbine plant shown in Figure 25.10, air enters the compressor at 100 kN/m², 290 K and is compressed with isentropic efficiency 0.84 to 500 kN/m². The air then passes through a heat exchanger in which the temperature is raised to 620 K. The gases enter the turbine at 920 K and expand with isentropic efficiency 0.88 to 100 kN/m².*

Determine the overall efficiency of the plant if the combustion efficiency is 95%.

For air, $\gamma = 1.4$ and $c_p = 1.0 \, kJ/kg$

For gas, $\gamma = 1.3$ and $c_p = 1.15 \, kJ/kg$

The plant and cycle are shown in Figure 25.10.

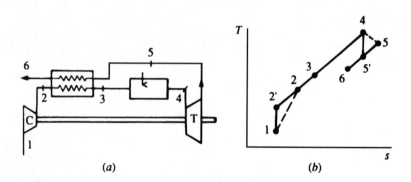

(a)   (b)

Figure 25.10

For the compressor, $\dfrac{T_{2'}}{T_1} = \left(\dfrac{p_2}{p_1}\right)^{\frac{\gamma-1}{\gamma}}$ from equation (22.11)

$$\therefore T_2' = 290 \left(\frac{500}{100}\right)^{\frac{0.4}{1.4}} = 459.3 \text{ K}$$

From equation (25.3), $\eta_{\text{isen}} = \dfrac{T_{2'} - T_1}{T_2 - T_1}$

i.e. $0.84 = \dfrac{459.3 - 290}{T_2 - 290}$

from which $T_2 = 491.5 \text{ K}$

For the turbine, $\dfrac{T_4}{T_{5'}} = \left(\dfrac{p_4}{p_5}\right)^{\frac{\gamma-1}{\gamma}}$ from equation (22.11)

$$\therefore T_5 = \frac{920}{\left(\dfrac{500}{100}\right)^{\frac{0.3}{1.3}}} = 634.6 \text{ K}$$

From equation (25.4),

$$\eta_{\text{isen}} = \frac{T_4 - T_5}{T_4 - T_{5'}}$$

i.e.

$$0.88 = \frac{920 - T_5}{920 - 634.6}$$

from which

$$T_5 = 668.8 \text{ K}$$

Specific work from cycle $= c_{p_{\text{gas}}}(T_4 - T_5) - c_{p_{\text{air}}}(T_2 - T_1)$ from equation (25.5)

$$= 1.15(920 - 668.8) - 1.0(491.5 - 290)$$

$$= 87.4 \text{ kJ/kg}$$

Specific heat transfer to cycle from fuel $= c_{p_{\text{gas}}}(T_4 - T_3)$

$$= 1.15(920 - 620)$$

$$= 345 \text{ kJ/kg}$$

Specific energy supplied in fuel $= \dfrac{345}{\eta_{\text{combustion}}}$

$$= \frac{345}{0.95} = 363.2 \text{ kJ/kg}$$

Overall efficiency of plant $= \dfrac{\text{specific work transfer from cycle}}{\text{specific energy supplied in fuel}}$

$$= \frac{87.4}{363.2}$$

$$= 0.241 \text{ or } \underline{24.1\%}$$

Note: Without the heat exchanger, the overall efficiency of the plant would be 16.9%.

Further problems

4. A gas turbine is to be designed to operate a Joule cycle using air as a working substance. A thermal efficiency of 40% is required with a specific work output of 200 kJ/kg. The minimum cycle temperature is 15°C. Determine the cycle pressure ratio and the maximum cycle temperature that is required.

[5.98; 977 K]

5. A Joule cycle using air as a working substance has a pressure ratio of 10 to 1. The air enters the compressor at 100 kN/m², 16°C and leaves the heater at a temperature of 1 000°C. Calculate the specific work from the cycle, the specific heat transfer to the cycle and the cycle thermal efficiency. For air $c_p = 1.0$ kJ/kg K and $\gamma = 1.4$.   [344.5 kJ/kg; 715 kJ/kg; 48.2%]

6. A Joule cycle with isentropic efficiency effects has a pressure ratio of 10 to 1. The air enters the compressor at 100 kN/m², 16°C and leaves the heater at 1 000°C. The isentropic efficiency of compression and expansion processes is 0.82. Calculate the specific work from the cycle and the thermal efficiency of the cycle. For air $c_p = 1.0$ kJ/kg K and $\gamma = 1.4$.   [175.2 kJ/kg; 26.7%]

Compare the answers of questions 5 and 6.

7. Air enters the compressor of a gas turbine plant at 101 kN/m^2, 17°C and is compressed to 808 kN/m^2 with isentropic efficiency 0.8. The air is heated in a combustion chamber to 1 200 K. The gases expand in the turbine with isentropic efficiency 0.84 to 101 kN/m^2. Determine the plant efficiency and the power output with an air mass flow rate of 30 kg/s. The mass flow rate of fuel may be neglected. The values of c_p and γ are 1.0 kJ/kg K and 1.4 for air, and 1.15 kJ/kg K and 1.3 for gases.   [20.8%, 4.42 MW]

8. A gas turbine plant produces 40 kW with an air flow rate of 0.5 kg/s. The compressor pressure ratio is 4 and inlet conditions are 101 kN/m^2, 17°C. The compressor and turbine isentropic efficiencies are both 0.82. The pressure at turbine entry is 5% less than at compressor exit. The mass flow rate of fuel may be neglected. The turbine exit pressure is 101 kN/m^2. Determine the maximum cycle temperature and the plant efficiency. The values of c_p and γ are 1.0 kJ/kg K and 1.4 for air, and 1.15 kJ/kg K and 1.3 for gases.   (Ans.: 1 160 K, 10%)

9. Equation (25.1), $w = c_p(T_3 - T_4) - c_p(T_2 - T_1)$, gives the specific work from a Joule cycle. Using the reversible adiabatic relationship to eliminate T_2 and T_4, determine the pressure ratio for maximum work output.   [$r_p = (T_3/T_1)^{\gamma/2(\gamma-1)}$]

10. Air enters the compressor of a gas turbine plant at 101 kN/m^2, 17°C and is compressed to 808 kN/m^2 with isentropic efficiency 0.8. The air enters a heat exchanger where the temperature is raised by the exhaust gases to 724 K (Figure 25.7) before entering the combustion chamber which it leaves at 1 200 K. The gases expand in the turbine to 101 kN/m^2 with isentropic efficiency 0.84. The mass flow rate of fuel may be neglected. Determine the plant efficiency and the power output with an air mass flow rate of 30 kg/s. The values of c_p and γ are 1.0 kJ/kg K and 1.4 for air, and 1.15 kJ/kg K and 1.3 for gases.   [27%; 4.42 MW]

Compare the answers of questions 7 and 10.

11. A gas turbine of the type shown in Figure 25.7(b) has a separate power turbine to drive the load. The compressor takes in 10 kg/s of air at 0.1 MN/m^2, 160°C and compresses it to 0.7 MN/m^2. The turbine driving the compressor has an isentropic efficiency of 85% and the isentropic efficiency of the compressor is 80%. The maximum cycle temperature is 700°C and there are no losses in the combustion chamber. The separate power turbine expands the gases to atmospheric pressure with isentropic efficiency 85%. Assume that the mass flow rate of gases is equal to the mass flow rate of air.

For air $c_p = 1.005$ kJ/kg K and $\gamma = 1.4$; for combustion and expansions $c_p = 1.15$ kJ/kg K and $\gamma = 1.33$.

Referring to Figure 25.7(b), calculate

(i) the temperature at compressor exit (T_2)
(ii) the temperature at high pressure turbine exit (T_4)
(iii) the pressure ratio of the high pressure turbine (p_3/p_4)
(iv) the pressure ratio of the low pressure turbine (p_4/p_5)
(v) the exit temperature from the low pressure turbine (T_5)
(vi) the power output of the plant ($\dot{W}_x$)
(vii) the overall efficiency of the plant (η_0).

Note. This is quite a difficult problem and the multiple answers demanded provide guidance to the order of working out the solution.

[557.6 K; 738.3 K; 3.79; 1.85; 648.8 K; 1 028.4 kW; 21.5%]

26 Reciprocating internal combustion engines

26.1 The Otto, diesel and dual cycles

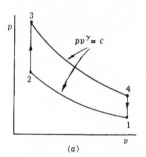

(a)

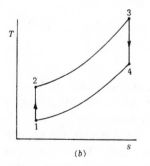

(b)

Figure 26.1

Spark ignition engines using petrol as a fuel and compression ignition engines using diesel oil as a fuel are the most common power plants in use today.

The ideal cycles for these engines are based on the Otto cycle, which is a constant volume, non-flow process cycle taking place in a piston-and-cylinder mechanism. The processes, set out in Table 26.1, are analysed using the non-flow energy equation; these are shown on $p - v$ and $T - s$ diagrams in Figure 26.1 (*a*) and (*b*).

Table 26.1. Otto cycle processes

Process	Description (Otto cycle)	$q - w = \Delta u = c_v \Delta T$ for gases
1–2	Reversible, adiabatic compression	$-w = (u_2 - u_1)$ $= c_v(T_2 - T_1)$
2–3	Reversible, constant volume heat transfer to cycle	$q = (u_3 - u_2)$ $= c_v(T_3 - T_2)$
3–4	Reversible, adiabatic expansion	$-w = (u_4 - u_3)$ $w = c_v(T_3 - T_4)$
4–1	Reversible, constant volume heat transfer from cycle	$q = (u_1 - u_4)$ $= c_v(T_1 - T_4)$

The specific work for the cycle,

$$w = c_v(T_3 - T_4) - c_v(T_2 - T_1) \tag{26.1}$$

The thermal efficiency of the cycle.

$$\eta_{th} = \frac{c_v(T_3 - T_4) - c_v(T_2 - T_1)}{c_v(T_3 - T_2)}$$

$$= 1 - \frac{T_4 - T_1}{T_3 - T_2}$$

From equation (22.12),

$$\frac{T_3}{T_4} = \frac{T_2}{T_1} = \left(\frac{v_1}{v_2}\right)^{\gamma-1} = r_v^{\gamma-1}$$

where r_v is the cycle compression ratio

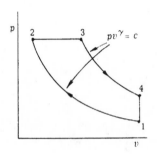

Figure 26.2

Thus
$$\eta_{th} = 1 - \frac{1}{r_v^{\gamma-1}} \qquad (26.2)$$

In the diesel cycle shown in Figure 26.2, the heat transfer process to the cycle is at constant pressure and in the dual cycle, Figure 26.3, it is partly at constant volume and partly at constant pressure.

In each case, the thermal efficiency may be obtained by the application of the non-flow energy equation.

For the diesel cycle,

$$\eta_{th} = \frac{c_p(T_3 - T_2) - c_v(T_4 - T_1)}{c_p(T_3 - T_2)}$$

which becomes

$$\eta_{th} = 1 - \frac{1}{r_v^{\gamma-1}} \left[\frac{r_c^{\gamma} - 1}{\gamma(r_c - 1)}\right] \qquad (26.3)$$

where r_c is the cut-off ratio v_3/v_2.

For the dual cycle,

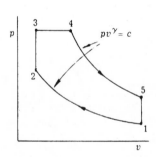

Figure 26.3

$$\eta_{th} = \frac{c_p(T_4 - T_3) + c_v(T_3 - T_2) - c_v(T_5 - T_1)}{c_p(T_4 - T_3) + c_v(T_3 - T_2)}$$

which becomes

$$\eta_{th} = 1 - \frac{1}{r_v^{\gamma-1}} \left[\frac{r_p r_c^{\gamma} - 1}{(r_p - 1) + \gamma r_p(r_c - 1)}\right] \qquad (26.4)$$

where r_c is the cut-off ratio, v_4/v_3 and r_p is the pressure ratio, p_3/p_2.

In both cases, r_v is the compression ratio and in both cases the efficiency is less than that of the Otto cycle *with the same compression ratio*.

The diesel cycle is no longer relevant to modern compression ignition engines but the dual cycle is considered to give a good representation of the compression ignition combustion process. The Otto cycle is considered to be representative of the spark ignition combustion process.

26.2 Reciprocating engine cycles

In the reciprocating engine, the fuel is burned directly in air after compression. It is then necessary to discard the gases after expansion without cooling and to take in a fresh supply of cool air. It is not easy to separate the processes since they all take place in the same variable volume space.

To investigate the cycle, a pressure–volume graph is obtained for the real engine by using an *indicator*. This practical diagram has rounded corners, as shown in Figure 26.4, unlike the ideal cases shown in Figures 26.1, 26.2 and 26.3 for the Otto, diesel and dual cycles. The area enclosed on the indicator diagram represents cycle work and the maximum cycle pressure may also be measured, for comparison with the ideal cycle.

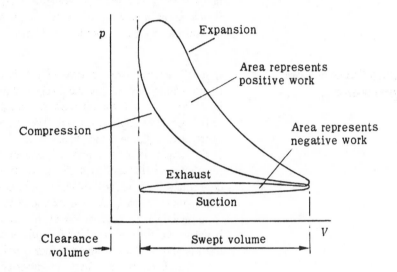

Figure 26.4

It is usual to evaluate the *mean effective pressure* (m e p) of both cycles for comparison and then evaluate the power from equation (26.7).

Net work per cycle = positive work area − negative work area

$$\text{Actual cycle m e p} = \frac{\text{net work per cycle}}{\text{swept volume}} \tag{26.5}$$

$$\text{Ideal cycle m e p} = \frac{\text{ideal work per cycle}}{\text{swept volume}} \tag{26.6}$$

Cycle power output = m e p × swept volume × cycle frequency   (26.7)

See also Chapter 30 for performance tests.

In modern high-speed engines, it is essential to use electronic transducers to obtain pressure–crank angle $(p - \theta)$ diagrams, which can be stored in a data logger or displayed on an oscilloscope. If required, this stored data can

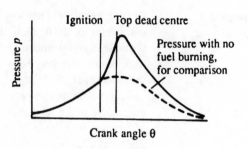

Figure 26.5

be converted to a pressure–volume diagram by a computer program but it is not essential to do this, for such a program could evaluate mean effective pressure as a direct calculation. Figure 26.5 shows such a diagram.

26.3 Combustion processes

It would appear from equations (26.2) and (26.4) that the efficiency of a reciprocating engine can be improved by a high compression ratio but in practice, the compression ratio is limited by the combustion process.

In the petrol engine, air and fuel are premixed and ignited by an electric spark and a flame front spreads across the combustion chamber. If the design and mixture are correct, there are no problems but if the compression ratio exceeds about 9 to 1, the mixture tends to explode prematurely. It is also found that the fuel will not ignite unless the air–fuel ratio lies between 10 to 1 and 20 to 1. A ratio of about 15 to 1 is ideal (see Chapter 29).

In the compression ignition engine, this explosion problem does not occur because the fuel is supplied in a different manner. The compression ratio used is around 20 to 1 and the fuel is injected into the air when the temperature achieved by the large compression ratio is high enough to cause the fuel to ignite without a spark. However, because the fuel is in liquid droplet form and the air–fuel ratio still needs to be correct (about 15 to 1), the burning droplets move rapidly through the air in the cylinder seeking oxygen to maintain a correct *local* air–fuel ratio. Overall, only enough fuel is injected to satisfy the demand of the engine so that, although burning is sustained by the *locally* correct air–fuel ratio, the *overall* ratio varies from around 80 to 1 at idling to 18 to 1 at maximum power. If more fuel than this is supplied, it issues from the exhaust as black smoke.

This method of combustion, combined with a high compression ratio, enables the compression ignition engine to achieve a higher efficiency in practice than the spark ignition engine.

26.4 Total energy plant

Power plant coolants and exhausts contain a great deal of energy and there is an increasing tendency to design plant to utilize this energy. An *exhaust-*

gas-heated boiler plant uses the exhaust gases from a diesel or gas turbine plant, which are at a relatively high temperature, to produce hot water or steam. Figure 26.6 shows such a scheme.

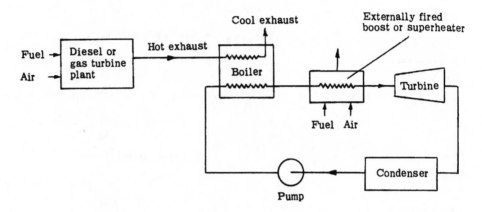

Figure 26.6

The efficiency of these types of plant is given by

$$\eta_{plant} = \frac{\text{total energy obtained from plant}}{\text{energy supplied in fuel}} \tag{26.8}$$

This efficiency will be much greater than that of the simple steam, diesel or gas turbine plant, in which the condenser or exhaust energy is wasted.

26.5 Turbocharging and intercooling

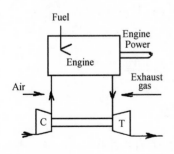

Figure 26.7

Exhaust gas turbocharging is a means of improving the power output of an engine by increasing the mass of air forced into the cylinders. It is particularly well suited to compression ignition engines. Exhaust gas is passed from the engine at a pressure greater than atmospheric to a gas turbine which in turn drives a compressor to supply the extra air to the cylinders of the engine (Figure 26.7). The necessary extra fuel is provided by the injection system. In large marine engines and truck engines, two-stage compression with intercooling between the stages may be used. (Supercharging of an engine normally refers to compressors driven by the engine.)

Worked examples

1. *A diesel cycle using air as the working substance has a compression ratio of 18 to 1. The lowest cycle temperature is 15°C and at this point the pressure is 1.1 bar. The maximum cycle temperature is 1 800 K. Calculate the cut-off ratio, the maximum cycle pressure, the specific heat transfer from the cycle and the thermal efficiency.*

Air may be treated as a perfect gas, for which $c_v = 0.718$ kJ/kg K and $c_p = 1.006$ kJ/kg K.

The cycle is shown in Figure 26.8.

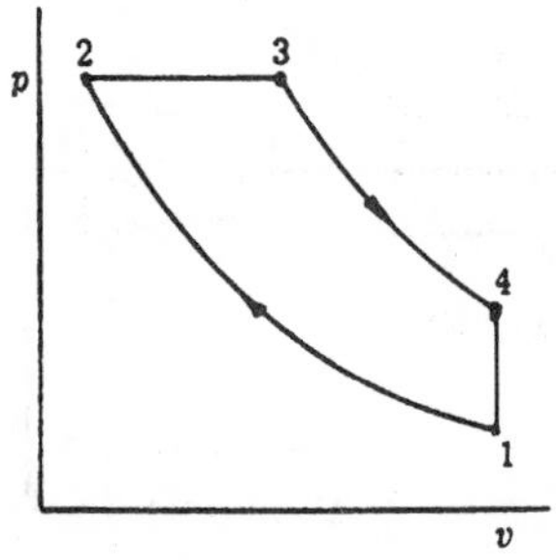

$$\gamma = \frac{1.006}{0.718} = 1.4$$

The maximum cycle pressure p_2 is obtained in the reversible adiabatic process 1–2.

$$\frac{p_2}{p_1} = \left(\frac{v_1}{v_2}\right)^{\gamma}$$

$$\therefore\ p_2 = 1.1 \times 18^{1.4}$$

$$= \underline{62.7\ bar}$$

Figure 26.8

Also

$$\frac{T_2}{T_1} = \left(\frac{v_1}{v_2}\right)^{\gamma-1}$$

$$\therefore\ T_2 = (273 + 15) \times 18^{0.4} = 918\ K$$

In the process 2–3,

$$\frac{p_2 v_2}{T_2} = \frac{p_3 v_3}{T_3}$$

But $p_2 = p_3$ so that the cut-off ratio,

$$\frac{v_3}{v_2} = \frac{T_3}{T_2} = \frac{1\,800}{918} = \underline{1.96}$$

This cut-off ratio represents cut-off at 5.7% of the stroke. In the reversible adiabatic expansion process,

$$\frac{T_4}{T_3} = \left(\frac{v_3}{v_4}\right)^{\gamma-1}$$

$$\therefore\ T_4 = 1\,800 \times \left(\frac{1.96}{18}\right)^{0.4} = 741\ K$$

Heat transfer from cycle $= c_v(T_4 - T_1)$

$$= 0.718(741 - 288) = \underline{326\ kJ/kg}$$

The thermal efficiency is given by

$$\eta_{\text{th}} = \frac{c_p(T_3 - T_2) - c_v(T_4 - T_1)}{c_p(T_3 - T_2)}$$

$$= \frac{1.006(1\,800 - 918) - 0.718(741 - 288)}{1.006(1\,800 - 918)}$$

$$= \underline{0.632}$$

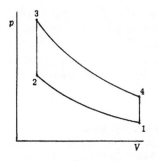

Figure 26.9

2. *The area of an indicator diagram of a single cylinder petrol engine is 15 cm².
The spring constant for the pressure scale is 1 mm = 21 kN/m² and the length of the
diagram is 50 mm. The compression ratio of the engine is 8. Compare the mean effective pressure of the engine with that of the Otto cycle with the same compression ratio,
using air as a working substance. The Otto cycle maximum temperature is 2 000 K
and minimum pressure and temperature are 100 kN/m² and 300 K, respectively.*
   For air, $c_v = 0.72$ kJ/kg K, $\gamma = 1.4$ and $R = 0.286$ kJ/kg K.

For the indicator diagram,

$$\text{m e p} = \text{mean height} \times \text{pressure scale}$$

$$= \frac{15 \times 10^2}{50} \times 21$$

$$= 630 \text{ kN/m}^2$$

Figure 26.9 shows the Otto cycle.

Net work $= m\{c_v(T_3 - T_4) - c_v(T_2 - T_1)\}$ where m is the mass of air.

Swept volume $= V_1 - V_2$

$$\therefore \text{m e p} = \frac{\text{work done}}{\text{swept volume}} \qquad \text{from equation (26.6)}$$

$$= \frac{mc_v\{(T_3 - T_4) - (T_2 - T_1)\}}{V_1 - V_2}$$

$$m = \frac{p_1 V_1}{RT_1} = \frac{100}{0.286 \times 300} V_1 = 1.165 V_1$$

And from equation (22.12),

$$\frac{T_2}{T_1} = \frac{T_3}{T_4} = \left(\frac{V_1}{V_2}\right)^{\gamma - 1} = 8^{0.4} = 2.297$$

$$\therefore T_2 = 2.297 T_1 \quad \text{or} \quad T_2 - T_1 = 1.297 T_1$$

and $$T_3 = 2.297 T_4 \quad \text{or} \quad T_3 - T_4 = 1.297 T_3$$

Hence $$\text{m e p} = \frac{1.165 V_1 \times 0.72 \times 1.297(T_3 - T_1)}{V_1 - \dfrac{V_1}{8}}$$

$$= \frac{0.165 \times 0.72 \times 1.297(2\,000 - 300)}{0.875}$$

$$= 707 \text{ kN/m}^2$$

$$\therefore \text{ratio } \frac{\text{actual m e p}}{\text{Otto m e p}} = \frac{630}{707} = \underline{0.89}$$

Further problems

In the problems below air may be treated as a perfect gas having a specific heat capacity at constant volume $c_v = 0.718$ kJ/kg K and a ratio of specific heat capacities $\gamma = 1.4$.

3. An ideal diesel cycle using air as a working substance has a compression ratio of 16 to 1. The maximum cycle temperature is 1 700°C and the minimum cycle temperature is 15°C. Calculate:

(i) the specific heat transfer to the cycle;
(ii) the specific work transfer in the adiabatic expansion process;
(iii) the thermal efficiency of the cycle. What proportion of the Carnot efficiency does this represent?   [1 105.7 kJ/kg; 770 kJ/kg; 60.1%, 70.3%]

4. An ideal Otto cycle using air as a working substance has a compression ratio of 16 to 1. Compare the thermal efficiency of this cycle with that of the Diesel cycle in question 3. The maximum cycle temperature in the Otto cycle is (i) 1 600°C, (ii) 1 700°C, (iii) 1 800°C. Determine the specific work transfer from the cycle in each case with a minimum cycle temperature of 15°C. Comment on the effect of maximum cycle temperature on efficiency and output.
[67%; 481 kJ/kg; 529 kJ/kg; 578 kJ/kg]

5. Derive an expression for the thermal efficiency of the Otto cycle in terms of the compression ratio of the cycle.

For such a cycle using air with a compression ratio of 7, the temperature at the beginning of compression is 50°C and the pressure is 1.02 bar. The cycle maximum temperature is 2 000°C. Determine the mean effective pressure of the cycle.
[7.77 bar]

6. An ideal dual cycle using air as a working substance has a compression ratio of 16 to 1. The heat transfer to the cycle at constant volume is equal to the heat transfer to the cycle at constant pressure. The maximum cycle temperature is 1 700°C and the minimum cycle temperature is 15°C. Calculate the thermal efficiency of the cycle and compare it with the values obtained in questions 3 and 4. Evaluate the mean effective pressure of the cycle if the minimum cycle pressure is 1 bar.
[65.8%; 7.87 bar]

7. An ideal air engine uses the Otto cycle with compression ratio 9.2. Determine the mean effective pressure if the minimum cycle temperature and pressure are 20°C, 100 kN/m^2 and the maximum cycle temperature is 2 000 K.

When a petrol engine is built, the actual m e p is 78% of the ideal. Determine the power output of a 1 500 cc engine for a speed of 2 000 cycles/min. $c_v = 0.72$ kJ/kg K, $\gamma = 1.4$ and $R = 0.286$ kJ/kg K.   [726 kN/m^2, 28.4 kW]

8. A diesel engine plant with an efficiency of 35% produces 2 000 kW and has 2 800 kW of energy in the exhaust gas at high temperature. The exhaust gas enters a heat exchanger and produces steam at 3 MN/m^2, 300°C. The gases leave the heat exchanger having transferred 80% of their energy to the steam. The steam expands in a turbine to 0.004 MN/m^2 with isentropic efficiency 0.8. Determine the efficiency of the *total energy* plant.   [46.5%]

27 Refrigerators and heat pumps

27.1 Introduction

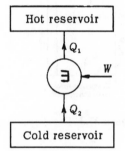

Figure 27.1

In Section 23.2, the concept of the ideal reversed heat engine was introduced; this machine transfers heat from a cold reservoir to a hot reservoir and requires a work input, Figure 27.1. If the object of the machine is the heat transfer from the cold reservoir, it is called a *refrigerator* but if the object is the heat transfer to the hot reservoir, it is called a *heat pump*.

For the refrigerator, the *coefficient of performance* is given by

$$\text{cop}_{\text{ref}} = \frac{\text{heat transfer from cold reservoir}}{\text{work transfer to cycle}}$$

$$= \frac{Q_2}{W} \tag{27.1}$$

For the heat pump, the corresponding coefficient of performance,

$$\text{cop}_{\text{hp}} = \frac{\text{heat transfer to hot reservoir}}{\text{work transfer to cycle}}$$

$$= \frac{Q_1}{W} \tag{27.2}$$

The ideal reversed heat engine cycle is the reversed Carnot cycle and if an analysis similar to that of Section 23.3 is made, it will be seen that

$$\text{cop}_{\text{ref}} = \frac{T_2}{T_1 - T_2} \tag{27.3}$$

and

$$\text{cop}_{\text{hp}} = \frac{T_1}{T_1 - T_2} \tag{27.4}$$

where T_1 is the temperature of the hot reservoir and T_2 is the temperature of the cold reservoir.

These expressions give the maximum possible coefficients of performance for a reversed heat engine working between reservoir temperatures

T_1 and T_2 but, as for the Carnot cycle, Section 23.3, the reversed Carnot cycle is not a practical proposition. There are various practical refrigeration and heat pump cycles, the most common being the vapour compression cycle, but their coefficients of performance are much lower than that of the ideal reversed Carnot cycle.

27.2 Vapour compression cycle

There are several designs of refrigeration plant which use different working processes but the most common type uses a vapour compression cycle and achieves its heat transfer by evaporation and condensation at constant pressure. In these processes of *phase change*, the temperature is also constant at the *saturation* value (see Chapter 21).

The ideal vapour compression cycle is a constant pressure flow process cycle. A typical arrangement is shown in Figure 27.2 and Figure 27.3 shows a domestic refrigerator.

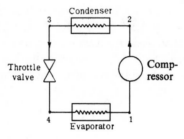

Figure 27.2

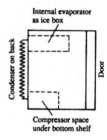

Figure 27.3

The processes are set out in Table 27.1 and analysed by the steady flow energy equation, changes in kinetic and potential energy being neglected.

The processes forming the cycle are shown on $p - v$, $T - s$ and $p - h$ diagrams in Figures 27.4(a), (b) and (c) (refrigeration engineers use $p - h$ diagrams). It can be seen that the cycle is similar to a reversed Rankine cycle, Section 24.2, except that the expansion of wet vapour is more convenient in a throttle valve than in any other expansion device.

Table 27.1. Processes for the ideal vapour compression cycle

Process	Description	$q - w_x = \Delta h$
1–2	Reversible, adiabatic compression of a superheated vapour in a *compressor*	$-w_x = h_2 - h_1$
2–3	Reversible constant pressure heat transfer from the cycle in a *condenser*	$q = h_3 - h_2$
3–4	*Irreversible* expansion of a vapour in a *throttle valve*	$h_3 = h_4$
4–1	Reversible constant pressure heat transfer to the cycle in an *evaporator*	$q = h_1 - h_4$

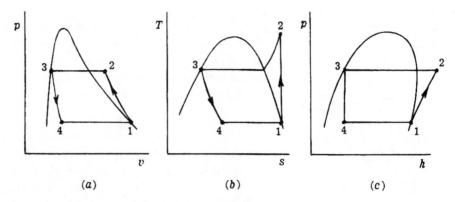

Figure 27.4

If the plant is designed as a refrigerator or freezer, the condenser temperature T_3 must be greater than the coolant temperature in order to reject heat and the lowest cycle temperature T_4 must be less than the cold space temperature in order to receive heat. Conversely, for a heat pump, Section 27.4, the condenser temperature T_3 must be greater than the hot space temperature and the lowest cycle temperature T_4 must be less than the source temperature (sea, river, etc.).

For a refrigerator or freezer, the cooling capacity, called the *refrigeration effect*, is $h_1 - h_4$ and

$$\text{cop}_{\text{ref}} = \frac{h_1 - h_4}{h_2 - h_1} \tag{27.5}$$

For a heat pump, the heating capacity, called the *specific output* or *heating effect*, is $h_2 - h_3$ and

$$\text{cop}_{\text{hp}} = \frac{h_2 - h_3}{h_2 - h_1} \tag{27.6}$$

27.3 Refrigerants

Working substances for refrigerators, freezers and heat pumps need to have freezing points lower than the lowest cycle temperature T_4, which could be $-50°C$ in a freezer plant, and critical points higher than the condenser temperature T_3. Various substances have been used including Freon 12, ammonia, methyl chloride and carbon dioxide. Properties of refrigerants are available in tables and on charts. It is now thought that Freon 12 has an adverse effect on the ozone layer of the atmosphere and it is no longer used in new plant. A different substance, R 134a, is now available and new property tables are available for student use. Examples 1, 3, and 4 allow for this change.

27.4 Applications of heat pump plant

The idea of using free low-temperature energy for heating purposes is attractive and many schemes are proposed for heat pumps but each requires careful costing to ensure its benefits. Equation (27.4) shows

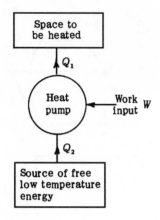

Figure 27.5

that the ideal coefficient of performance of a reversed Carnot cycle is given by $\dfrac{T_{max}}{T_{max} - T_{min}}$ but practical plant coefficients of performance are considerably less.

Consider the scheme shown in Figure 27.5. Suppose that the source temperature is $10°C$ and the delivery temperature is $50°C$.

$$\text{Ideal cop} = \frac{323}{40} = 8.08$$

This looks very attractive, but in practice the overall coefficient of performance might be about 3. This means that for a 1 kW input, the heat pump will take 2 kW from the free source and deliver 3 kW to the space being heated. The cost is therefore a 1 kW input, which will be supplied to an electric motor, using electricity generated at about $33\frac{1}{3}\%$ efficiency, burning 3 kW of fuel. The cost of this fuel, plus generating and distribution costs, will appear in the electricity bill, which could then exceed the cost of buying the fossil fuel (oil, gas or coal) and burning it in the heated space in an efficient space heater.

To estimate the likely success of heat pump schemes, the following points should be noted:

1. The free source temperature should be as high as possible.
2. The temperature rise should be as low as possible.
3. The capital cost of the scheme should be as low as possible.
4. The waste energy to be recovered should be adequate to justify the scheme and be available when required (e.g. heating is needed in winter rather than in summer).

Thus complex heat pump schemes attempting large rises in temperature from low temperature sources are unlikely to recoup their capital costs by savings in running costs in a commercially acceptable period of time.

Figure 27.6 shows two successful schemes in which both the cooling and heating are essential processes. The heat pump uses the waste energy from the cooling to assist in the heating and the saving in running costs

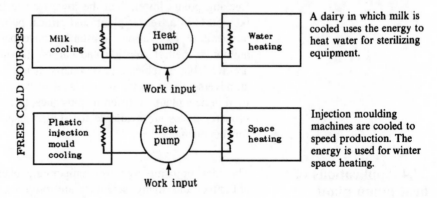

Figure 27.6

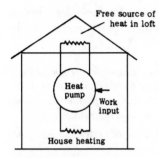

Figure 27.7

27.5 Refrigeration plant with isentropic efficiency

is obvious. Since the heat pump is similar in design to the refrigerator normally used for the cooling, the extra capital cost is small and any schemes involving the use of waste heat from essential cooling to provide necessary heating are likely to be successful.

In applications where the cooling is not essential, it is less likely that a heat pump would prove economic. Figure 27.7 shows a home-heating scheme which would probably be cheaper to run than full price electricity but could be dearer to run than solid fuel heating. The installation costs would be very important in determining the cost effectiveness of such a scheme.

In practical plant, slight superheating at entry to the compressor avoids lubrication problems associated with wet vapour in the compressor. Subcooling at condenser exit is also beneficial, since it increases the cooling capacity of the plant.

The effect of isentropic efficiency is to increase the work needed in the compression process and thus to reduce the coefficient of performance. Since the temperature at the compressor exit is higher, the amount of subcooling that can be achieved in the condenser, and hence the refrigeration effect, will be reduced by isentropic efficiency. There will also be pressure losses due to friction in the condenser and evaporator, which will further reduce the coefficient of performance. Figure 27.8 shows the plant cycle.

The isentropic efficiency of the compression process is given by

$$\eta_{\text{isen}} = \frac{h_{2'} - h_1}{h_2 - h_1} \tag{27.7}$$

and the plant cycle coefficient of performance

$$\text{cop}_{\text{ref}} = \frac{h_1 - h_4}{h_2 - h_1} \tag{27.8}$$

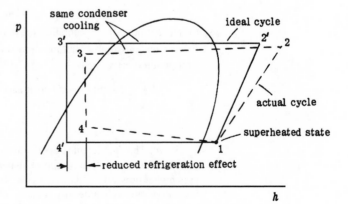

Figure 27.8

27.6 Heat pump plant with isentropic efficiency

Worked examples

The effects of isentropic efficiency will be similar to those described in Section 27.5 for the refrigerator. The coefficient of performance will be reduced (see also Section 27.4).

1. *An ideal vapour compression refrigerator has an evaporator temperature of $-15°C$ and a condenser temperature of $30°C$. Compare the condenser pressures, evaporator pressures, refrigeration effects and coefficients of performance when using (a) Freon 12, (b) ammonia and (c) R 134a as the working substance.*

The cycle is shown in Figure 27.4(*c*) and the coefficient of performance is given by equation (27.5).

(*a*) From charts or tables, $h_1 = 181$ kJ/kg, $h_2 = 205.8$ kJ/kg,

$$h_3 = h_4 = 64.6 \text{ kJ/kg}.$$

Condenser pressure = saturation pressure at $30°C = 0.745$ MN/m^2

Evaporator pressure = saturation pressure at $-15°C = 0.183$ MN/m^2

Refrigeration effect = $h_1 - h_4 = 116.4$ kJ/kg

$$\text{Coefficient of performance} = \frac{h_1 - h_4}{h_2 - h_1}$$

$$= \frac{181 - 64.6}{205.8 - 181} = \underline{4.69}$$

(*b*) From charts or tables, $h_1 = 1\,426$ kJ/kg, $h_2 = 1\,659.9$ kJ/kg,

$$h_3 = h_4 = 323.1 \text{ kJ/kg}.$$

Condenser pressure = saturation pressure at $30°C = 1.167$ MN/m^2

Evaporator pressure = saturation pressure at $-15°C = 0.236$ MN/m^2

Refrigeration effect = $h_1 - h_4 = 1\,102.9$ kJ/kg

$$\text{Coefficient of performance} = \frac{h_1 - h_4}{h_2 - h_1}$$

$$= \frac{1\,426 - 323.1}{1\,659.9 - 1\,426} = \underline{4.71}$$

(*c*) From charts or tables, $h_1 = 389.18$ kJ/kg, $h_2 = 422$ kJ/kg

$$h_3 = h_4 = 242 \text{ kJ/kg}.$$

Condenser pressure = saturation pressure at $30°C = 0.767$ MN/m^2

Evaporator pressure = saturation pressure at $-15°C = 0.164$ MN/m^2

Refrigeration effect = $h_1 - h_4 = 147.18$ kJ/kg

$$\text{Coefficient of performance} = \frac{h_1 - h_4}{h_2 - h_1}$$

$$= \frac{389.18 - 242}{422 - 389.18} = \underline{4.484}$$

Comment: The coefficients of performance are approximately the same and thus the power requirements will be approximately the same for a given task. The ammonia uses higher pressures and has a higher refrigeration effect and so will need a smaller mass flow rate. When density is taken into account, it is found that the volume flow rate of Freon is 1.7 times than that of ammonia and hence the compressor will need to be larger or to rotate faster.

The refrigeration effect with R 134a is larger than that of Freon 12 but the coefficient of performance is the lowest of the three substances.

2. *A vapour compression heat pump using Freon 12 has an evaporator exit pressure of 0.261 MN/m² at 0°C with compressor exit pressure of 1.219 MN/m² at 75°C. The Freon leaves the condenser at 40°C. Determine the compression isentropic efficiency, the heating effect and the cycle coefficient of performance. What is the coefficient of performance of the ideal cycle working between the same condenser and evaporator pressures?*

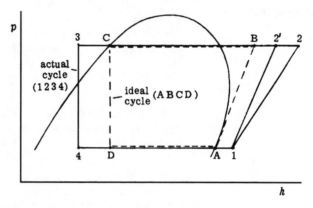

Figure 27.9

The cycle is shown in Fig. 27.9.

The evaporation temperature is −5°C, so that the vapour enters the compressor with 5 K of superheat.

The condensation temperature is 50°C, so that the vapour leaves the compressor with 25 K of superheat and leaves the condenser with 10 K of subcooling.

Process 1–2 is the actual process and process 1–2′ is the ideal isentropic process. From charts and tables,

$$h_1 = 188.6 \text{ kJ/kg}, h_2 = 226.5 \text{ kJ/kg}, h_{2'} = 218.5 \text{ kJ/kg}, h_3 = h_4 = 74.6 \text{ kJ/kg}$$

From equation (27.7), $\eta_{\text{isen}} = \dfrac{h_{2'} - h_1}{h_2 - h_1}$

$$= \frac{218.5 - 188.6}{226.5 - 188.6} = \underline{0.79}$$

Heating effect $= h_2 - h_3$

$$= 226.5 - 74.6 = \underline{151.9 \text{ kJ/kg}}$$

From equation (27.6), $\text{cop} = \dfrac{h_2 - h_3}{h_2 - h_1}$

$$= \frac{226.5 - 74.6}{226.5 - 188.6} = \underline{4.01}$$

For the ideal cycle with no superheat at evaporator exit, no subcooling at condenser exit and with isentropic compression (as shown by ABCD in Figure 27.9),

$$h_A = 185.4 \text{ kJ/kg}, h_B = 213 \text{ kJ/kg}, h_C = h_D = 84.9 \text{ kJ/kg}$$

$$\text{Ideal cop} = \frac{h_B - h_C}{h_B - h_A}$$

$$= \frac{213 - 84.9}{213 - 185.4} = \underline{4.64}$$

Note: The heating effect is reduced with no subcooling to 121.8 kJ/kg. Subcooling is beneficial in heat pump and refrigeration cycles.

Further problems

3. An ideal vapour compression heat pump has an evaporator temperature of −5°C and a condenser temperature of 50°C. Compare the condenser pressures, the evaporator pressures, the specific outputs and the coefficients of performance when using (*a*) Freon 12, (*b*) ammonia and (*c*) R 134a, as the working substance.

[1.219 MN/m^2, 0.261 MN/m^2, 128 kJ/kg, 4.65;
2.033 MN/m^2, 0.355 MN/m^2, 1,281 kJ/kg, 4.85;
1.31 MN/m^2, 0.234 MN/m^2, 164 kJ/kg, 4.1]

4. An ideal vapour compression refrigerator chills 10 kg/s of water from 25°C to 5°C. The plant uses ammonia with an evaporator temperature of −5°C and a condenser temperature of 35°C. Determine the power required by the refrigerator. If the working substances were (*a*) Freon 12 and (*b*) R 134a, would more or less power be required? [146.8 kW; 4 kW more; 18 kW more]

5. An ideal vapour compression heat pump uses Freon 12 and has a lake as a heat source. The lake minimum temperature is 4°C. The output from the heat pump is at 50°C. Determine the volume flow rate of Freon 12 at entry to the compressor for a heating load of 10 kW, and the coefficient of performance.

[0.23 m^3/min; 5.37]

6. An ideal vapour compression freezing plant has an evaporator temperature of −40°C and a condenser temperature of 30°C. The plant uses Freon 12 and freezes 1.6 t/h of fish. The reduction in specific enthalpy of the fish is 420 kJ/kg. Determine the coefficient of performance, the power input to the plant and the mass flow rate of Freon 12. [2.43; 76.82 kW; 1.78 kg/s]

7. A vapour compression refrigeration plant using ammonia has an evaporator temperature of −5°C and a condenser temperature of 30°C. The gas enters the compressor with 25 degrees of superheat and leaves the condenser with 10 degrees of subcooling. The isentropic efficiency of the compression is 0.8. Determine the refrigeration effect and the cycle coefficient of performance. If the cooling load is 20 kW, what must be the mass flow rate of ammonia?

[1 224.3 kJ/kg; 5.29, 0.016 kg/s]

8. A heat pump using Freon 12 takes energy from the sea at an evaporator temperature of 0°C and delivers it to an hotel at a condenser pressure of 1.219 MN/m^2. The Freon is 20 degrees superheated at compressor entry and 20 degrees subcooled at condenser exit. The compression isentropic efficiency is 0.85. The heating load is 30 kW.

(*a*) Determine the coefficient of performance of the plant cycle.

(*b*) If the price of electricity is 5 p per kWh, determine the cost of running continuously for 180 days and compare this with burning solid fuel for the same period. Solid fuel costs £65 per tonne and burns with 70% efficiency, with calorific value 32 000 kJ/kg. [5.3; £1 223; £131 cheaper]

28 Reciprocating air compressors

28.1 Introduction

There are various air compressor designs, which may be divided into three classes: (*a*) reciprocating compressors using a piston and cylinder mechanism, (*b*) positive displacement rotary compressors, of which the Roots blower is an example, and (*c*) rotary compressors with axial or radial flow, in which the pressure rise is achieved by diffusing kinetic energy given to the air by the power input. The latter type is used in gas turbine plants.

This chapter deals with reciprocating compressors characterized by low flow rates but which can achieve high output pressures in a simple manner; these are very common in industry. The two parameters of interest to the engineer are the air flow rate, usually given as the *free air delivery* (i.e. the volume flow rate at specified atmospheric conditions) and the power input needed to drive the compressor.

28.2 Air compression cycle

The working cycle is shown in Figure 28.1. An important feature is the clearance volume, where some of the compressed air is trapped when delivery ceases and then re-expands. This causes the *induced volume* to be less than the *swept volume* and the *volumetric efficiency* of the compressor is defined as

$$\eta_{\text{vol}} = \frac{\text{induced volume}}{\text{swept volume}} \tag{28.1}$$

The induced and delivered volumes each contain the same mass of air and it is this amount which is important.

The ideal air compression process for minimum work input is isothermal but real processes are better represented by polytropic compression and expansion which allow for temperature changes of the air. These are minimised by water cooling. An isothermal efficiency is sometimes used to compare the actual work needed for compression with the minimum isothermal work.

The processes used are set out in Table 28.1 and the displacement work transfer is evaluated for each process.

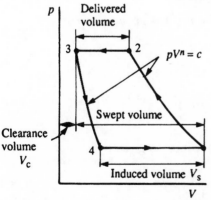

Figure 28.1

Table 28.1. Processes for the air compression cycle

Process	Description	Displacement work transfer $= \int_1^2 p\,dV$
1–2	Reversible, polytropic compression, $pV^n = c$	$W = \dfrac{(p_1 V_1 - p_2 V_2)}{n - 1}$
2–3	Delivery of compressed air at constant pressure and temperature	$W = p_2(V_3 - V_2)$
3–4	Reversible, polytropic expansion, $pV^n = c$	$W = \dfrac{(p_3 V_3 - p_4 V_4)}{n - 1}$
4–1	Induction of fresh air at constant pressure and temperature	$W = p_1(V_1 - V_4)$

It is assumed that the index n in the expansion and compression processes is the same ($1 < n < 1.4$).

The total work in the 'cycle' can be obtained by addition,

i.e.
$$W = \frac{p_1 V_1 - p_2 V_2}{n - 1} + p_2(V_3 - V_2) + \frac{p_3 V_3 - p_4 V_4}{n - 1} + p_1(V_1 - V_4)$$

$$(28.2)$$

It is convenient to rearrange this expression in terms of the important parameters which include the mass of air induced and delivered (i.e. excluding the trapped air which never escapes); and the pressure ratio of the compressor.

Let m be the mass of air delivered per cycle, m_c be the trapped clearance mass and v_1, v_2, v_3 and v_4 be the specific volumes at points 1, 2, 3 and 4 respectively.

Then $V_1 = (m + m_c)v_1$, $V_2 = (m + m_c)v_2$, $V_3 = m_c v_3$ and $V_4 = m_c v_4$

Thus
$$W = \frac{p_1(m + m_c)v_1 - p_2(m + m_c)v_2}{n - 1} + p_2(m_c v_3 - (m + m_c)v_2)$$

$$+ \frac{p_3 m_c v_3 - p_4 m_c v_4}{n - 1} + p_1((m + m_c)v_1 - m_c v_4)$$

But $p_1 = p_4$ and $p_2 = p_3$

Also $v_1 = v_4$ and $v_3 = v_2$,   so that

$$W = m\left(\frac{n}{n - 1}\right)(p_1 v_1 - p_2 v_2)$$

or, for a mass flow rate $\dot{m}$, the power input is given by

$$\dot{W} = \dot{m}\left(\frac{n}{n - 1}\right)(p_1 v_1 - p_2 v_2) \qquad (28.3)$$

Since air is a perfect gas, $p_1 v_1 = RT_1$ and $p_2 v_2 = RT_2$

Hence
$$\dot{W} = \dot{m}\left(\frac{n}{n-1}\right)R(T_1 - T_2)$$

$$= \dot{m}\left(\frac{n}{n-1}\right)RT_1\left(1 - \frac{T_2}{T_1}\right) \tag{28.4}$$

For a polytropic process,

$$\frac{T_2}{T_1} = \left(\frac{p_2}{p_1}\right)^{\frac{n-1}{n}} = r_p^{\frac{n-1}{n}} \qquad \text{from equation (22.13)}$$

where r_p is the pressure ratio $\dfrac{p_2}{p_1} = \dfrac{p_3}{p_4}$

Thus power input,

$$\dot{W} = \dot{m}\left(\frac{n}{n-1}\right)RT_1\left(1 - r_p^{\frac{n-1}{n}}\right) \tag{28.5}$$

The mass flow rate may be determined from the required volume flow rate $\dot{V}$ of air at the specified pressure p and T,

i.e.
$$\dot{m} = \frac{p\dot{V}}{RT}$$

28.3 Volumetric efficiency

Using the definition of volumetric efficiency in equation (28.1) and referring to Figure 28.1,

$$\eta_{\text{vol}} = \frac{V_1 - V_4}{V_s} = \frac{(V_s + V_c) - V_c r_p^{1/n}}{V_s}$$

$$= 1 - \frac{V_c}{V_s}(r_p^{1/n} - 1) \tag{28.6}$$

referred to inlet conditions.

Volumetric efficiency is often expressed in terms of the free air delivery rather than the induced volume, to enable comparisons between compressors to be made.

Let the free air delivery be V_a at pressure and temperature p_a and T_a.

Then
$$\frac{(V_1 - V_4)p_1}{T_1} = \frac{V_a p_a}{T_a} \qquad \text{from equation (21.5)}$$

Thus
$$\eta_{\text{vol}} = \frac{V_1 - V_4}{V_s} = \frac{V_a}{V_s} \cdot \frac{p_a}{p_1} \cdot \frac{T_1}{T_a}$$

Therefore, from equation (28.6),

$$\eta_{\text{vol}} = \left\{1 - \frac{V_c}{V_s}(r_p^{1/n} - 1)\right\}\frac{p_1}{p_a} \cdot \frac{T_a}{T_1} \tag{28.7}$$

referred to free air conditions.

28.4 Multistage compressors

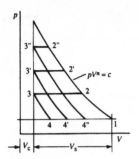

Figure 28.2

As the required pressure ratio increases, it may be seen from Figure 28.2 that the re-expanded trapped air occupies more volume at the inlet pressure. This means that the amount of air induced and delivered falls with increasing pressure ratio and this results in a reduced volumetric efficiency.

If the volumetric efficiency falls below about 65%, the compressor becomes uneconomic and in order to achieve high pressure ratios with adequate flow, it is necessary to use multi-stage compressors. In such machines, intercooling is used between the stages to reduce the temperature and volume of the air and hence the amount of work needed in the following stage, Figure 28.3.

Perfect intercooling reduces the air at entry to the second and subsequent stages to the initial temperature T_1 and Figure 28.4 shows the processes in a two-stage air compressor.

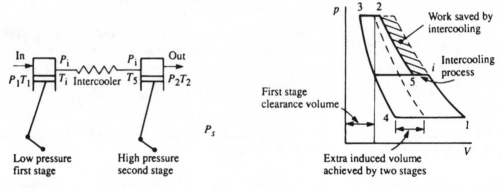

Figure 28.3

Figure 28.4

The total power required,

$$\dot{W} = \dot{m} \left(\frac{n}{n-1} \right) \left\{ RT_1 \left[1 - \left(\frac{p_i}{p_1} \right)^{\frac{n-1}{n}} \right] + RT_5 \left[1 - \left(\frac{p_2}{p_i} \right)^{\frac{n-1}{n}} \right] \right\}$$

(28.8)

With perfect intercooling, $T_5 = T_1$, so that the power required,

$$\dot{W} = \dot{m} \left(\frac{n}{n-1} \right) RT_1 \left\{ \left[1 - \left(\frac{p_i}{p_1} \right)^{\frac{n-1}{n}} \right] + \left[1 - \left(\frac{p_2}{p_i} \right)^{\frac{n-1}{n}} \right] \right\}$$

For minimum work required with perfect intercooling, the ideal intermediate pressure is obtained by differentiating with respect to p_i and equating to zero.

Thus

$$\frac{d\dot{W}}{dp_i} = - \left(\frac{n-1}{n} \right) p_i^{\left(\frac{n-1}{n} \right) - 1} + \left(\frac{n-1}{n} \right) p_i^{-\left(\frac{n-1}{n} \right) - 1} = 0$$

from which $p_i^2 = p_1 p_2$

Hence $\dfrac{p_i}{p_1} = \dfrac{p_2}{p_i}$ (28.9)

Thus the minimum work with perfect intercooling occurs when the pressure ratios are equal, or

$$r_{p_{\text{stage}}} = \sqrt{(r_{p_{\text{overall}}})}$$

For N stages, it may be shown that with perfect intercooling, the minimum work is required if

$$r_{p_{\text{stage}}} = \sqrt[N]{(r_{p_{\text{overall}}})}$$ (28.10)

These conditions will result in equal work per stage, so that

$$\dot{W} = N\dot{m}\left(\frac{n}{n-1}\right)RT_1(1 - r_{p_{\text{stage}}}^{(n-1)/n})$$ (28.11)

However, if intercooling is not perfect, equation (28.8) must be used to determine the total power in the case of a two-stage compressor.

Since the air entering the first stage of a multi-stage compressor must pass through succeeding stages, it is the volumetric efficiency of the first stage which determines the free air delivery of the compressor and as the pressure ratio of this stage has been reduced, the compressor volumetric efficiency will be improved by the use of stages.

28.5 Practical air compression

The preceding work concerns ideal compressors. In practice the pressure in the cylinder must be less than atmospheric to allow air to enter the cylinder and the pressure in the receiver must be less than the delivery pressure of the compressor to allow the air to enter the receiver. Both these needs involve extra work. There is also a safety aspect in that air that has been compressed is hot and will hold moisture which will condense in the receiver as the air cools. Similarly the lubrication of the compressor will allow oil vapour to pass to the receiver so that there is a danger of explosion or fire requiring regular draining of these substances. In circumstances where purity is vital, compressors without oil lubrication are available and desiccants are used to remove the water vapour.

Worked examples

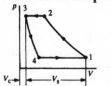

Figure 28.5

1. *A single-stage reciprocating compressor takes in air at a pressure of 96 kN/m² and a temperature of 305 K. The air is compressed to a pressure of 725 kN/m² and delivered to a reservoir. The clearance volume of the compressor is 5 per cent of the swept volume and both the compression and expansion processes may be represented by a reversible relation of the form $pV^{1.3} = $ constant. Determine the compressor volumetric efficiency referred to atmospheric conditions of 101.3 kN/m² and 292 K and the power input for a mass flow rate of 0.1 kg/s. For air $R = 0.287$ kJ/kg K.*

The cycle is shown in Figure 28.5.

From equation (28.7),

$$\eta_{vol} = \left\{ 1 - \frac{V_c}{V_s}(r_p^{1/n} - 1) \right\} \frac{p_1}{p_a} \cdot \frac{T_a}{T_1}$$

$$= \left\{ 1 - 0.05 \left[\left(\frac{725}{96} \right)^{1/1.3} - 1 \right] \right\} \frac{96}{101.3} \cdot \frac{292}{305}$$

$$= 0.738 \quad \text{or} \quad \underline{73.8\%}$$

From equation (28.5),

$$\dot{W} = \dot{m} \left(\frac{n}{n-1} \right) RT_1(1 - r_p^{(n-1)/n})$$

$$= 0.1 \left(\frac{1.3}{0.3} \right) \times 0.287 \times 305 \left[1 - \left(\frac{725}{96} \right)^{\frac{0.3}{1.3}} \right]$$

$$= \underline{-22.55 \text{ kW}}$$

2. *In an ideal four-stage reciprocating air compressor, the inlet pressure is 96 kN/m² and the inlet temperature is 300 K. The air is delivered at a pressure of 27.6 MN/m². The compressor is designed for the minimum power requirement and has perfect intercooling. The reversible compression and expansion processes both conform to the relation $pV^{1.2} = C$. Calculate the interstage pressures, the air delivery temperature and the power input for a free air delivery of 2 m³/s measured at 101 kN/m², 16°C. The compressor mechanical efficiency is 75%. For air R = 0.287 kJ/kg K.*

The stage pressure ratio with perfect intercooling is obtained from equation (28.10),

i.e. $$r_{p_{stage}} = \sqrt[4]{\left(\frac{27.6 \times 10^6}{96 \times 10^3} \right)} = 4.117$$

Thus the interstage pressure are

$$96 \times 4.117 = 395.3 \text{ kN/m}^2,$$

$$395.3 \times 4.117 = 1\,627.8 \text{ kN/m}^2$$

$$1\,627.8 \times 4.117 = 6\,702.7 \text{ kN/m}^2$$

From equation (22.13), since at entry to each stage the air is cooled back to 300 K, the air delivery temperature,

$$T_{del} = T_1(r_{p_{stage}})^{\frac{n-1}{n}}$$

$$= 300 \times 4.117^{\frac{0.2}{1.2}} = 379.8 \text{ K}$$

Mass flow rate of free air

$$= \frac{p\dot{V}}{RT} = \frac{101 \times 2}{0.287 \times 289} = 2.435 \text{ kg/s}$$

Hence, from equation (28.11),

$$\dot{W} = N\dot{m}\left(\frac{n}{n-1}\right)RT_1(1 - r_{p_{\text{stage}}}^{(n-1)/n}) \times \frac{1}{\eta_{\text{mech}}}$$

$$= 4 \times 2.435\left(\frac{1.2}{0.2}\right) \times 0.287 \times 300(1 - 4.117^{\frac{0.2}{1.2}}) \times \frac{1}{0.75}$$

$$= -1\,784.5 \text{ kW}$$

Further problems

For the problems below take $R = 0.287$ kJ/kg K.

3. A single-stage reciprocating compressor takes in 0.4 m^3/s of air at 96 kN/m^2, 25°C and delivers it at 350 kN/m^2. The clearance volume is 5 per cent of the swept volume. Determine the power required if both compression and expansion follow a law $pV^{1.25}$ = constant and the volumetric efficiency of the compressor at inlet conditions. [56.7 kW; 90.9%]

4. A single-stage reciprocating air compressor operates at 1 200 rev/min. The bore and stroke of the compressor are each 80 mm. The clearance volume is 6% of the swept volume and the volumetric efficiency referred to inlet conditions of 96 kN/m^2, 30°C is 0.82. What is the delivery pressure and the power required if compression and expansion both follow a law $pV^{1.3}$ = constant and the mechanical efficiency is 75%? [583 kN/m^2, 1.87 kW]

5. A single-stage reciprocating air compressor has a clearance volume of 6% of the swept volume. Air enters at 98 kN/m^2, 30°C and is compressed to 490 kN/m^2 according to a law $pV^{1.2}$ = constant. The air flow rate is 1.2 kg/s. Assuming that expansion follows the same process law as compression, determine the power input to the compressor, the free air delivery and the volumetric efficiency referred to ambient conditions of 101 kN/m^2, 16°C. [192.6 kW; 1.033 2 m^3/s; 76.9%]

6. A three-stage reciprocating air compressor delivers 0.1 kg/s of air at 7 500 kN/m^2. The inlet conditions are 98 kN/m^2, 30°C. Assuming that intercooling is perfect and that all compressions and expansions follow a law $pV^{1.25}$ = constant, find the interstage pressures and the power input to the compressor. [416 kN/m^2; 1 766.5 kN/m^2; 43.74 kW]

7. A two-stage reciprocating air compressor is designed for minimum power and has a first stage with clearance volume 5% of the swept volume. The swept volume is 0.12 m^3 and the speed of rotation is 240 rev/min. Ambient air at 100 kN/m^2, 16°C is taken in, compressed and delivered at 3 MN/m^2. Assuming all compressions and re-expansions follow the law $pV^{1.2}$ = constant, and that the compressor has perfect intercooling, determine the air mass flow rate and the power required from the driving motor if the mechanical efficiency of the compressor is 80%. [0.489 kg/s; 199.2 kW]

8. A two-stage reciprocating compressor fitted with an intercooler takes in air at 80 kN/m^2, 20°C and delivers it at 1.7 MN/m^2. Ambient conditions are 101 kN/m^2, 16°C. The low pressure delivery pressure is 0.4 MN/m^2 and the pressure and temperature at entry to the high pressure cylinder are 0.38 MN/m^2, 30°C. The first stage clearance volume is 5% of the swept volume and all compression and expansion processes follow a law $pV^{1.22}$ = constant. Determine the volumetric efficiency referred to ambient conditions and the power input for a flow rate of 0.4 kg/s with a mechanical efficiency of 80%. [67.4%; 153.3 kW]

29 Combustion

29.1 Introduction

When a fuel is burned in air, energy is made available by the chemical changes which occur in the combustion process. Ideally this energy may be (*a*) continuously transferred during the process so that the reaction is isothermal and the final temperature of the products is the same as the initial temperature of the reactants, or (*b*) retained in the products, which will reach a high temperature in an adiabatic reaction.

29.2 Measurement of energy release

When a solid or liquid fuel is burned, the energy release is measured in an isothermal, constant volume reaction performed in a special combustion chamber known as a *bomb*. The fuel is burned in oxygen and the process is constrained to be almost isothermal by placing the bomb in a water-filled calorimeter. The temperature of the whole apparatus is allowed to rise by an amount large enough to be measured accurately but small enough to be considered isothermal.

The bomb and calorimeter are shown in Figure 29.1.

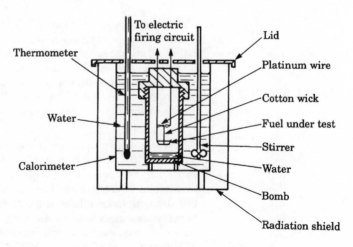

Figure 29.1

The heat transfer to the water,

$$Q = m_e c \Delta T \qquad (29.1)$$

where m_e is the water equivalent of the apparatus, c is the specific heat capacity of the water and ΔT is the small temperature rise.

If m_f is the mass of fuel consumed, then the energy release per unit mass is $\dfrac{m_e c \Delta T}{m_f}$.

The energy release when a gaseous fuel is burned is measured in an isothermal, constant pressure reaction. The metered stream of gas is burned with air and the energy released is transferred to the cooling water flowing in tubes around which the combustion products flow. The temperature of the products at exit from the calorimeter is equal to the temperature of the reactants at entry.

The gas calorimeter is shown in Figure 29.2.

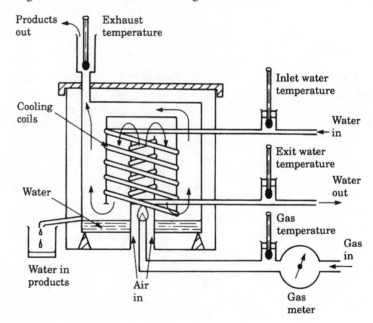

Figure 29.2

The heat transfer rate to the cooling water,

$$\dot{Q} = \dot{m}_w c \Delta T \qquad (29.2)$$

where $\dot{m}_w$ is the cooling water mass flow rate, c is the specific heat capacity of water and ΔT is the temperature rise of the water.

If $\dot{V}_f$ is the volume flow rate of gas, then the energy release per unit volume is $\dfrac{\dot{m}_w c \Delta T}{\dot{V}_f}$.

29.3 Calorific value

The calorific value of a fuel is defined as the amount of energy released per unit mass or volume of fuel when completely burned under specified conditions.

If a hydrocarbon fuel is burned, the products of the reaction will contain H_2O. If the H_2O is in the liquid phase, the measurement gives the *higher calorific value* (H.C.V.) but if the H_2O is in the vapour phase, the *lower calorific value* (L.C.V.) is obtained.

The relation between the two calorific values is as follows:

For the bomb experiment, H.C.V. = L.C.V. + mass of $H_2O \times u_{fg}$.

For the gas experiment,   H.C.V. = L.C.V. + mass of $H_2O \times h_{fg}$.

In practical combustion processes, not all the energy may be released due to incomplete combustion. In such a case, the combustion efficiency is defined by

$$\eta_{comb} = \frac{\text{actual energy release}}{\text{calorific value of fuel}} \qquad (29.3)$$

Combustion efficiencies are usually greater than 90 per cent.

29.4 Chemistry of combustion

The products of a combustion reaction may be determined from the chemical equation of the combustion process and it is also possible to calculate the amount of oxygen or air required to burn a given fuel. Most of the fuels of interest to engineers are hydrocarbon compounds or mixtures; these contain hydrogen, carbon and some impurities and form carbon dioxide, carbon monoxide and steam when burned.

For solid or liquid fuels, a mass analysis is often used and for gases, a volumetric analysis may be used but a chemical formula may be used for either. Chemical equations can only be manipulated in chemical units and it is necessary to express mass or volume in chemical terms before solving.

The *relative atomic mass* of an element gives its mass on a scale in which carbon is the basic unit with a numerical value of 12. The *relative molecular mass* of an element or compound is the sum of the atomic masses of the atoms which comprise the molecule.

One kg mole of a substance contains a mass in kg numerically equal to the relative molecular mass so that a mass analysis may be converted to a chemical or molar analysis by dividing the mass of each substance by its own relative molecular mass. For example, 100 kg of a fuel which has a mass analysis of 84 per cent carbon and 16 per cent hydrogen will contain 84/12 kg moles of carbon and 16/2 kg moles of hydrogen. Chemically this may be written $7C + 8H_2$.

Avogadro's law states that equal volumes of gases at the same pressure and temperature contain equal numbers of moles; thus a volumetric analysis is identical with a chemical or molar analysis. For example, 100 kg moles of gas of volumetric analysis 80 per cent hydrogen and 20 per cent carbon monoxide contains 80 kg moles of hydrogen and 20 kg moles of carbon monoxide. Chemically this may be written $80H_2 + 20CO$.

Chemical equations are formed by writing the reactants on one side of the equation and the products on the other. The equation is balanced by conserving the number of atoms of each element on either side. Thus $80H_2 + 20CO + 50O_2 = 80H_2O + 20CO_2$ is a balanced equation; there are 160 atoms of hydrogen, 120 atoms of oxygen and 20 atoms of carbon on either side.

Substance		Relative atomic mass		Relative molecular mass
Hydrogen	H	1	H_2	2
Oxygen	O	16	O_2	32
Carbon	C	12	C	12
Nitrogen	N	14	N_2	28
Sulphur	S	32	S	32
Carbon monoxide	–	–	CO	28
Carbon dioxide	–	–	CO_2	44
Steam	–	–	H_2O	18
Sulphur dioxide	–	–	SO_2	64

29.5 Combustion in air

In engineering plant, combustion takes place in air, which is usually considered to be a perfect gas of relative molecular mass 29. It contains 21 per cent oxygen and 79 per cent nitrogen by volume or 23.3 per cent oxygen and 76.7 per cent nitrogen by mass.

When chemical equations are formed, both the oxygen and the nitrogen should be included since the nitrogen will affect the pressure and temperature of the products even though it is inert in the reaction.

The principal objects of a chemical approach are:

(a) to determine the chemically correct, or *stoichiometric*, air–fuel ratio for a particular fuel to burn completely so that there is no oxygen in the products, i.e. the minimum air requirement;

(b) to obtain the products of combustion in either

  (i) the stoichiometric reaction, or

  (ii) a reaction with excess air, the excess being defined by

$$\frac{x - x_{stoic}}{x_{stoic}} \times 100 \text{ per cent, where } x \text{ is the air–fuel ratio;}$$

(c) to determine the air–fuel ratio used in a reaction in which the analysis of the products of combustion are known.

29.6 Analysis of products of combustion

Since the products are gaseous, the analysis is by volume. An analysis which excludes the H_2O content is termed *dry* whereas one which includes the H_2O is termed *wet*. In practical combustion processes, the percentage of carbon dioxide may be continuously metered to give an indication of the success of the combustion or alternatively the gas may be completely analysed.

The complete analysis may be made with an *Orsat apparatus* in which a sample of gas is passed through a series of chemicals, each of which absorbs one of the constituents of the gas. After each absorption, the contraction in volume of the gas measured at constant pressure and temperature gives a measure of the volumetric contribution of the absorbed gas. The Orsat apparatus yields a dry analysis since the measurements are made in a water-sealed burette and so no H_2O determination can be made (Figure 29.3).

Once the Orsat analysis is known, the air–fuel ratio used may be calculated by the method shown in examples 3 and 4.

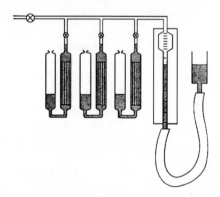

Figure 29.3

It is possible to use electronic transducers to determine the CO, CO_2 and O_2 content of exhaust gases rather than the Orsat apparatus. This enables continuous computer control and display of the process and combustion efficiency.

29.7 Pollution

Combustion control to avoid pollution is an important consideration for internal combustion engines, which are tested at intervals to ensure that exhaust gases meet legal requirements. Combustion processes in multi-cylinder vehicle engines which do not run at constant speed, but vary with traffic conditions, are difficult to control. Progress has been made by *mapping* various parameters which are stored electronically in the emission control system. When combined with the output from a λ-meter ($\lambda = 1$ is the stoichiometric air–fuel ratio, see Section 29.5) monitoring the exhaust gas oxygen content, which should be zero with complete combustion of a stoichiometric mixture, a signal can be sent to the fuel injection system to control the amount of fuel to be injected. This method, accompanied by a catalytic exhaust converter, reduces pollution.

This is not an ideal system because the information is out of date by the time it is used since three or more cylinders will have fired since the last measurement was made in the cylinder about to fire. Improvements are continuous, particularly in mapped information updating, but control will always be difficult.

Worked examples

1. *A gaseous fuel consists of 50% methane (CH_4), 25% hydrogen, 12% carbon monoxide, 3% oxygen, 6% carbon dioxide and 4% nitrogen by volume. Calculate the stoichiometric air–fuel ratio by volume and the wet volumetric and mass analysis of the products of combustion.*

Consider 100 moles of fuel and 100 X moles of air in a stoichiometric reaction. The reaction equation is

$$50CH_4 + 25H_2 + 12CO + 3O_2 + 6CO_2 + 4N_2 + 21XO_2 + 79XN_2$$

$$= aCO_2 + bH_2O + cN_2$$

Since the atoms on either side of the equation must be balanced, then,

for carbon C, $\qquad\qquad\qquad\qquad 50 + 12 + 6 = a \qquad\qquad$ (1)

for hydrogen H_2, $\qquad\qquad\qquad\qquad 100 + 25 = b \qquad\qquad$ (2)

for oxygen O_2, $\qquad\qquad \dfrac{12}{2} + 3 + 6 + 21X = a + \dfrac{b}{2} \qquad\qquad$ (3)

and, for nitrogen N_2, $\qquad\qquad\qquad 4 + 79X = c \qquad\qquad$ (4)

From equations (1), (2) and (3),

$$15 + 21X = (50 + 12 + 6) + \left(\frac{100 + 25}{2}\right)$$

from which $\qquad\qquad\qquad\qquad X = \underline{5.5}$

Since 100 X moles of air combined with 100 moles of fuel, then X represents the stoichiometric air–fuel ratio by volume.

Product analysis

From equation (4), $\qquad c = 4 + 79 \times 5.5 = 438$

Product	Moles	Volumetric analysis (%)	Mol. mass	Mass	Mass analysis (%)
CO_2	68	$\dfrac{68}{631} \times 100 = 10.8$	44	$68 \times 44 = 2\,992$	$\dfrac{2\,992}{17\,506} \times 100 = 17.1$
H_2O	125	$\dfrac{125}{631} \times 100 = 19.8$	18	$125 \times 18 = 2\,250$	$\dfrac{2\,250}{17\,506} \times 100 = 12.9$
N_2	438	$\dfrac{438}{631} \times 100 = 69.4$	28	$438 \times 28 = 12\,264$	$\dfrac{12\,264}{17\,506} \times 100 = 70$
Total	631	100		17 506	100

2. (a) *A fuel for a petrol engine consists of 86% carbon and 14% hydrogen by mass. Calculate the stoichiometric air–fuel ratio by mass.*

(b) *If the fuel is burnt with 10% excess air, calculate the dry volumetric analysis of the products of combustion.*

(a) Consider 100 kg of fuel and 100 X moles of air in the stoichiometric reaction. The fuel will contain $\frac{86}{12}$ moles of carbon and $\frac{14}{2}$ moles of hydrogen.

Thus $\qquad \dfrac{86}{12}C + \dfrac{14}{2}H_2 + 21XO_2 + 79XN_2 = aCO_2 + bH_2O + cN_2$

Then, for balance:

Carbon (C), $\qquad\qquad\qquad\qquad \dfrac{86}{12} = a \qquad\qquad$ (1)

Hydrogen (H_2), $\qquad\qquad\qquad\qquad \dfrac{14}{2} = b \qquad\qquad$ (2)

Oxygen (O_2),

$$21X = a + \frac{b}{2} \qquad (3)$$

Therefore

$$21X = \frac{86}{12} + \frac{14}{4}$$

from which

$$X = 0.507$$

Thus 100×0.507 moles of air combine with 100 kg of fuel. The relative molecular mass of air is 29 (Section 29.5) so that stoichiometric air–fuel ratio

$$= \frac{100 \times 0.507 \times 29}{100} = \underline{14.71}$$

(b) With 10% excess air,

$$\frac{86}{12}C + \frac{14}{2}H_2 + 21 \times 0.507 \times \frac{110}{100}O_2 + 79 \times 0.507 \times \frac{110}{100}N_2$$

$$= aCO_2 + bH_2O + cN_2 + dO_2$$

The values of a and b are unchanged since there is no more fuel to burn but c is different.

For balance:

Oxygen (O_2),

$$21 \times 0.507 \times \frac{110}{100} = \frac{86}{12} + \frac{14}{4} + e$$

Nitrogen (N_2),

$$79 \times 0.507 \times \frac{110}{100} = d$$

Hence $d = 44.1$ and $e = 1.04$

Dry analysis of products

Product	Moles	Volumetric analysis (%)
CO_2	7.17	$(7.17/52.31) \times 100 = 13.7$
O_2	1.04	$(1.04/52.31) \times 100 = 1.9$
N_2	44.1	$(44.1/52.31) \times 100 = 84.4$
Total	52.31	100

3. *The Orsat analysis of the products of combustion of a hydrocarbon fuel in air in a diesel engine is CO_2 10%, O_2 8%, CO 0%. Calculate the percentage excess air used in the combustion and the mass analysis of the fuel.*

The Orsat figures give the dry volumetric analysis of the products; the remainder of the exhaust is assumed to be nitrogen. Consider 100 moles of dry products and let $100\,X$ moles of air be used. If there are a moles of carbon and b moles of hydrogen in the reaction, then

$$a\,C + bH_2 + 21X\,O_2 + 79X\,N_2 = 10\,CO_2 + 8O_2 + 82N_2 + cH_2O$$

The H_2O is included as an unknown since the Orsat test makes no determination of this product.

For balance:

Carbon (C),

$$a = 10 \qquad (1)$$

Hydrogen (H_2),

$$b = c \qquad (2)$$

Oxygen (O_2),

$$21X = 10 + 8 + \frac{c}{2} \qquad (3)$$

Nitrogen (N_2), $\qquad\qquad\qquad\qquad 79X = 82 \qquad\qquad\qquad$ (4)

Thus $X = 1.038$, $\quad a = 10$, $\quad b = 7.6$ and $\quad c = 7.6$.

$$\text{Therefore air–fuel ratio} = \frac{100 \times 1.038 \times 29}{10 \times 12 + 7.6 \times 2} = \underline{22.3}$$

In a stoichiometric reaction, the carbon and hydrogen burn to form the same amount of CO_2 and H_2O but with no excess oxygen and less nitrogen. Thus the stoichiometric air–fuel ratio is obtained from

$$10C + 7.6H_2 + 21XO_2 + 79XN_2 = 10CO_2 + 7.6H_2O + dN_2$$

For oxygen balance, $\qquad\qquad\qquad 21X = 10 + \frac{7.6}{2} = 13.8$

$$\therefore \quad X = 0.657$$

Therefore stoichiometric air–fuel ratio

$$= \frac{100 \times 0.657 \times 29}{10 \times 12 + 7.6 \times 2} = 14.1$$

$$\text{Therefore percentage excess air} = \frac{x - x_{\text{stoic}}}{x_{\text{stoic}}} \times 100 = \frac{22.3 - 14.1}{14.1} \times 100$$

$$= 58.2$$

Mass analysis of fuel

$$\text{Mass of carbon} = 10 \times 12 = 120 \text{ units}$$

$$\text{Mass of hydrogen} = 7.6 \times 2 = 15.2 \text{ units}$$

$$\text{Therefore percentage carbon} = \frac{120}{135.2} \times 100 = \underline{88.8}$$

$$\text{and percentage hydrogen} = \frac{15.2}{135.2} \times 100 = \underline{11.2}$$

4. *The fuel used in a reciprocating engine has a mass analysis 88.8% carbon and 11.2% hydrogen. The Orsat analysis of the products is CO_2 10%, O_2 8% and CO 0%. Calculate the air–fuel ratio used.*

It may be seen that this example uses the data of the previous example. If the Orsat analysis *and* fuel composition are known, the calculation of the air–fuel ratio is simple.

The combustion equation for $100a$ kg of fuel and $100\,X$ moles of air giving 100 moles of dry products is

$$\frac{88.8}{12}a + \frac{11.2}{2}a + 21XO_2 + 79XN_2 = 10CO_2 + 8O_2 + 82N_2 + bH_2O$$

The carbon atom balance gives $\qquad \dfrac{88.8}{12}a = 10 \quad \text{i.e.} \quad a = \dfrac{12 \times 10}{88.8}$

The nitrogen atom balance gives $\qquad 79X = 82 \quad \text{i.e.} \quad X = \dfrac{82}{79}$

$$\text{Air–fuel ratio} = \frac{100X \times 29}{100a} = \frac{\dfrac{82}{79} \times 29}{\dfrac{12 \times 10}{88.8}} = \underline{22.3}$$

This is the same as before and inspection shows that

$$\text{air–fuel ratio} = \frac{29}{79 \times 12} \times \frac{\%N_2 \text{ in Orsat analysis}}{\%CO_2 \text{ in Orsat analysis}} \times \text{mass } \%C \text{ in fuel}$$

For a particular fuel, the last term is a constant and the air–fuel ratio

$$= \text{constant} \times \frac{\%N_2 \text{ in Orsat analysis}}{\%CO_2 \text{ in Orsat analysis}}$$

Further problems

Relative atomic and molecular masses given in the table of Section 29.4 may be used in the questions below. Air may be assumed to be 21% oxygen and 79% nitrogen by volume or 23.3% oxygen and 76.7% nitrogen by mass.

5. A coal gas consists of 54% hydrogen, 9% carbon monoxide, 25% methane (CH_4), 1% oxygen, 3% carbon dioxide and 8% nitrogen by volume. The gas is burned with 40% excess air. Determine the air–fuel ratio by volume and the dry volumetric analysis of the exhaust gas.   [5.36; 7.5% CO_2; 86% N_2; 6.5% O_2]

6. A liquid fuel contains 83% carbon and 17% hydrogen by mass. It is burned at an air–fuel ratio (by mass) of 18 to 1 in a motor car engine. Determine the mass analysis of the products of combustion.

[16% CO_2; 8% H_2O; 72.8% N_2; 3.2% O_2]

7. An Orsat apparatus gave the following results when used to analyse the exhaust gases from the combustion of a hydrocarbon fuel in air. The fuel contains 81% carbon by mass. CO_2 9%, O_2 8%, CO 0%. Determine the air–fuel ratio used.

[22.8]

8. A 4-stroke, 4 cylinder petrol engine uses a hydrocarbon fuel. A dry volumetric analysis of the products gave: 6.34% CO_2, 11.26% CO, 82.4% N_2. There was no oxygen or unburnt fuel in the exhaust. Determine the mass percentages of carbon and hydrogen in the fuel and the air–fuel ratio by mass used.

[84% C; 16% H_2, 12]

9. A dry analysis of combustion products gave the following results: CO_2 11%; O_2 7%, CO 0%. The hydrocarbon fuel used was burned in air. Determine the mass analysis of the fuel and the air–fuel ratio used.   [89.7% C; 10.3% H_2; 20.4]

10. The lower calorific value of a fuel used in a gas turbine combustion chamber is 43 000 kJ/kg. The fuel reacts with and heats 35 kg/s of air which enters the chamber at 450 K. The products of combustion leave the chamber at 1 150 K and the mean specific heat capacity at constant pressure in the process is 1.15 kJ/kg K. The fuel has a mass analysis 82% carbon, 18% hydrogen. Determine the mass flow rate of fuel and the volumetric analysis of the products of combustion.

[0.668 kg/s; 3.7% CO_2; 4.8% H_2O; 14.4% O_2; 77.1% N_2]

11. A liquid fuel for use in a spark ignition engine has a mass analysis 70% octane (C_8H_{18}) and 30% benzene (C_6H_6). Determine the stoichiometric air–fuel ratio by mass and the percentage of oxygen by volume in the exhaust when burned with 10 per cent excess air.   [14.52; 1.8%]

12. A fuel gas contains the following components by volume: H_2 23%, CO 12%, CO_2 16%, CH_4 5%, N_2 44%. When completely burned in air it is found that the percentage of CO_2 in the exhaust is identical to that in the fuel. Calculate the air–fuel ratio (by volume). Does this ratio represent excess air or a stoichiometric ratio?   [1.24; excess]

30 Plant performance

30.1 Efficiency

It was seen in Chapters 24, 25, 26 and 27 that for applications to real plant cycles allowance must be made for the irreversibilities of real processes. In particular, process efficiency of expansion or compression, imperfections of combustion and pressure losses due to friction in flow will combine to make the efficiency of real plant considerably less than that of ideal cycles. Additionally, the work transfer calculated for the working substance is transmitted to the output shaft by mechanical devices which are subject to friction. To allow for the friction losses we define the mechanical efficiency of a work-producing engine as

$$\eta_{mech} = \frac{\text{external work transfer at output shaft}}{\text{internal work transfer associated with working substance}}$$

For a work-absorbing machine (such as a compressor) this ratio must be inverted.

External work transfer rate, or power, is sometimes known as *brake* power since it is measured with a brake (dynamometer) and internal work transfer rate is sometimes known as *indicated* power since, for a reciprocating machine, it can be measured with a mechanical or electronic indicator.

Thus

$$\eta_{mech} = \frac{\text{brake power}}{\text{indicated power}} = \frac{\text{external power}}{\text{internal power}}$$

for a work producing motor, engine or plant, and

$$\eta_{mech} = \frac{\text{indicated power}}{\text{brake power}} = \frac{\text{internal power}}{\text{external power}}$$

for a work absorbing compressor.

The overall efficiency of an engine or plant is given by

$$\eta_{overall} = \frac{\text{brake or external power}}{\text{rate of energy supply in fuel}} \tag{30.1}$$

This will depend on the mechanical efficiency, combustion efficiency and cycle efficiency.

The rate of energy supply in equation (30.1) may be considered in a different way leading to a parameter called *specific fuel consumption* (sfc) normally applied to internal combustion engines. Specific fuel consumption is the mass flow rate of fuel used by an engine to produce unit power and is quoted in kg/kWh. As a practical piece of information calculated from measurements made when testing an engine on a 'brake' (dynamometer) to determine power output, the term is usually recorded as brake specific fuel consumption (bsfc)

$$\text{bsfc} = \frac{\dot{m}_{\text{fuel}}}{\text{brake power}}$$

Ultimately it is the overall efficiency which concerns the engineer and this, together with an energy balance drawn up from the steady flow equation, is the usual way of presenting plant performance.

30.2 Energy balance

Consider the plant shown in Figure 30.1.

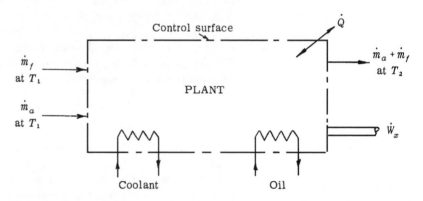

Figure 30.1

Let $\dot{m}_f$ = mass flow rate of fuel

$\dot{m}_a$ = mass flow rate of air

$\dot{m}_c$ = mass flow rate of coolant

$\dot{m}_o$ = mass flow rate of oil

T_1 = air and fuel inlet temperature

T_2 = exhaust gas temperature

ΔT_c = coolant temperature rise

ΔT_o = oil temperature rise

CV = calorific value of fuel

c_{pg} = specific heat capacity at constant pressure of exhaust gas

c_{pa} = specific heat capacity at constant pressure of air

c_{pf} = specific heat capacity at constant pressure of fuel

c_{po} = specific heat capacity at constant pressure of oil

c_{pc} = specific heat capacity at constant pressure of coolant

$\dot{W}_x$ = external power

$\dot{Q}$ = possible heat transfers

T_0 = datum temperature for enthalpy and calorific value

The control surface is carefully drawn round the plant to include those effects considered relevant. The steady flow energy equation is then written in a form to allow for the various flows entering and leaving. Thus

$$\dot{Q} - \dot{W}_x = \sum_{\text{out}} \dot{m}\left(h + \frac{V^2}{2} + gZ\right) - \sum_{\text{in}} \dot{m}\left(h + \frac{V^2}{2} + gZ\right)$$

If changes in kinetic and potential energy are considered negligible, then

$$\dot{Q} - \dot{W}_x = (\dot{m}_f + \dot{m}_a)c_{pg}(T_2 - T_0) - \dot{m}_a c_{pa}(T_1 - T_0) - \dot{m}_f c_{pf}(T_1 - T_0)$$
$$- \dot{m}_f CV + \dot{m}_c c_{pc}\Delta T_c + \dot{m}_o c_{po}\Delta T_0 \tag{30.2}$$

$$\text{The overall efficiency, } \eta_{\text{overall}} = \frac{\dot{W}_x}{\dot{m}_f CV} \tag{30.3}$$

Equation (30.3) is often represented by a tabular energy balance to give a clearer presentation.

Energy balance at T_0:

Rate of energy supply			*Rate of energy consumption*		
	(kW)	(%)		(kW)	(%)
By fuel enthalpy			By brake power		
By air enthalpy			By exhaust gas enthalpy		
By combustion			By transfer to coolant		
			By transfer to oil		
TOTAL		100	TOTAL		$100 - a$
			Random energy transfer found by difference		a

Worked examples

1. *In a trial on a single cylinder compression ignition engine, the following data were obtained:*

Indicated power from indicator output	6.1 kW
Fuel mass flow rate	0.0004 kg/s
Air mass flow rate	0.0105 kg/s
Fuel and air inlet temperature	15°C (288 K)
Engine cooling water inlet temperature	15°C
Engine cooling water outlet temperature	28°C
Engine cooling water mass flow rate	425 kg/h
Brake load	79 N
Brake speed	1 600 rev/min

The exhaust gases are cooled in a calorimeter by a mass flow rate of 47 kg/h of water entering at 15°C and leaving at 57.5°C. The gases leave the calorimeter at 30°C.
Other data:

Lower calorific value of fuel	42 500 kJ/kg
Brake power = 3.72 × brake load × brake speed × 10^{-5} kW	
Specific heat capacity of water	4.18 kJ/kg
Specific heat capacity of exhaust gas	1.15 kJ/kg

Draw up an energy balance for the engine based on a temperature of 288 K and determine the mechanical and overall efficiencies of the engine.

The plant is shown diagrammatically in Figure 30.2.

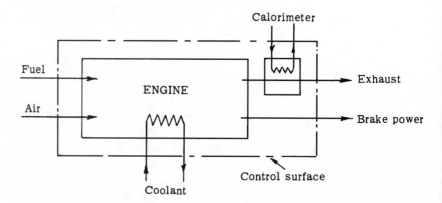

Figure 30.2

Applying the steady flow energy equation, allowing for the various flows:

$$\dot{Q} - \dot{W}_x = (\dot{m}_f + \dot{m}_a)c_{pg}(T_2 - T_0) - \dot{m}_a c_{pa}(T_1 - T_0) - \dot{m}_f c_{pf}(T_1 - T_0)$$
$$- \dot{m}_f CV + \dot{m}_c c_{pc}\Delta T_c + \dot{m}_{cal}c_{p_{cal}}\Delta T_{cal}$$

where $\dot{m}_{cal}$ = mass flow rate of calorimeter cooling water

$c_{p_{cal}}$ = specific heat capacity at constant pressure of calorimeter cooling water

ΔT_{cal} = temperature rise of calorimeter cooling water

The remaining symbols are defined in Section 30.2.

Since the air and fuel enter at 288 K and the balance is drawn up at 288 K, $T_1 = T_0$ so that

$$\dot{Q} - \dot{W}_x = (\dot{m}_f + \dot{m}_a)c_{pg}(T_2 - T_0) - \dot{m}_f CV + \dot{m}_c c_{pc}\Delta T_c + \dot{m}_{cal}c_{p_{cal}}\Delta T_{cal} \quad (1)$$

Rate of energy supplied by fuel $= 0.0004 \times 42\,500 = 17.0$ kW

Brake power, $\dot{W}_x = \dfrac{3.72 \times 79 \times 1\,600}{10^5} = 4.71$ kW

Exhaust gas enthalpy rate above 288 K $= (0.0004 + 0.0105) \times 1.15 \times (30 - 15)$

$$= 0.188 \text{ kW}$$

Energy transfer rate to exhaust calorimeter $= \dfrac{47}{3\,600} \times 4.18 \times (57.5 - 15)$

$$= 2.32 \text{ kW}$$

Energy transfer rate to engine coolant $= \dfrac{425}{3\,600} \times 4.18 \times (28 - 15)$

$$= 6.42 \text{ kW}$$

Friction power loss $=$ indicated power–brake power $= 6.1 - 4.71$

$$= 1.39 \text{ kW}$$

Hence, in equation (1),

$$17.0 - 4.71 = 0.188 + 2.32 + 6.42 + \text{unaccounted losses}$$

The friction loss is not included in this equation since it is already included in the energy transferred to the engine coolant.

Thus unaccounted losses $= 3.36$ kW

These quantities are set out in tabular form below:

Rate of energy supply			Rate of energy consumption		
	(kW)	(%)		(kW)	(%)
By fuel	17.00	100	By friction	1.39	8.2
			By brake power	4.71	27.7
			By exhaust gas:		
			in gas	0.188	
					14.8
			in calorimeter	2.32	
			By engine coolant	6.42	37.8
Total	17.00	100	Total	13.64	80.3
			Unaccounted loss by difference	3.36	19.7
				17.00	100

$$\text{Mechanical efficiency} = \frac{\text{brake power}}{\text{indicated power}} = \frac{4.71}{6.1} \times 100$$

$$= \underline{77.3\%}$$

$$\text{Overall efficiency} = \frac{\text{brake power}}{\text{rate of energy supply in fuel}} = \frac{4.71}{17.00} \times 100$$

$$= \underline{27.7\%}$$

Further problems

2. In a trial on a vapour compression refrigeration plant, 500 kg of ice at 0°C is produced in 1 h from water at 20°C. The power input to the compressor is measured by the brake load on the electric motor drive and at a speed of 1 420 rev/min, the load is 146 N on a brake arm 0.9 m long. The electric motor has an efficiency of 97%. The compressor is water-cooled and 0.12 kg/s of water enters at 20°C and leaves at 29.7°C. The condenser of the refrigerator is cooled with 1.82 kg/s of water, entering at 20°C and leaving at 29.5°C.

Draw up an energy balance for the plant and determine the coefficient of performance.

[In: water 14.9%, ice 59.3%, electricity 25.8%; Out: compressor coolant 6.2%, condenser coolant 92.6%, electrical loss 0.8%, unaccounted 0.4%; cop, 2.24]

3. A steam plant consists of a steam generator, turbine, condenser and feed pump. Draw up an energy balance at 17°C from the data below and determine the overall efficiency of the plant.

Steam generator:
    Fuel mass flow rate   0.25 kg/s
    Air mass flow rate   5.02 kg/s
    Lower calorific value of fuel   42 000 kJ/kg
Turbine:
    Brake power output   2 473 kW
Condenser:
    Mass flow rate of cooling water   53 kg/s
    Temperature of cooling water at entry   17°C
    Temperature of cooling water at exit   46°C
Feed pump:
    Electric motor drive, power input 11 kW
Other data:
    Specific heat capacity at constant pressure of water   4.18 kJ/kg K
    Specific heat capacity at constant pressure of exhaust gas   1.21 kJ/kg K
    Temperature of fuel and air at inlet   17°C
    Temperature of exhaust gases   211°C
    Other auxiliary power inputs (oil pumps, etc.)   13 kW

[In: fuel 99.77%, other 0.23%; Out: turbine 23.5%, condenser 61.1%, exhaust 11.8%, unaccounted 3.6%; 23.5%]

4. In a trial on a four cylinder motor car engine the following data were obtained. Determine the mechanical efficiency, the overall efficiency, the b s f c and draw up an energy balance at 19°C.

    Brake power   75 kW
    Friction power   17 kW

Fuel mass flow rate   0.008 kg/s
Air mass flow rate   0.144 kg/s
Fuel and air inlet temperature   19°C
Engine cooling water inlet temperature   17°C
Engine cooling water outlet temperature   39°C
Engine cooling water mass flow rate   1.9 kg/s
Exhaust gas temperature   327°C
Lower calorific value of fuel   42 000 kJ/kg
Specific heat capacity at constant pressure of exhaust gas   1.23 kJ/kg K
Specific heat capacity at constant pressure of water   4.18 kJ/kg K

[81.5%; 22.3%; 0.384 kg/kWh; In: fuel 100%; Out: power 22.3%,
exhaust 17.2%, coolant 52.0%, unaccounted 8.5%]

5. The following data were obtained in a trial on a steam generator. Draw up an energy balance at 15°C based on the higher calorific value of the fuel, 45 000 kJ/kg, and determine the efficiency of the steam generator defined by the ratio energy transfer to steam to energy supply in fuel.

Pressure of steam generated   20 bar
Temperature of steam generated   350°C
Temperature of feed water supply   50°C
Mass flow rate of fuel   340 kg/h
Mass flow rate of feed water   4 000 kg/h
Mass analysis of fuel 84% carbon, 16% hydrogen

Air and fuel are supplied at 15°C and the exhaust gases leave at 190°C.

The steam generator is supplied with 10% excess air and it may be assumed that the fuel is completely burned. The specific heat capacity at constant pressure of the exhaust gases is 1.23 kJ/kg K.

Air may be assumed to be 21% oxygen and 79% nitrogen by volume, and to have a molecular weight of 29.

[In: fuel 4 250 kW, feed 163 kW; Out: steam 3 418 kW, exhaust 360 kW,
unaccounted 635 kW; 76.5%]

31 Heat transfer

31.1 Introduction

Heat transfer is the transfer of energy due to temperature difference. There are three different modes of heat transfer.

Conduction

Energy is transferred by molecular collision and is most important in solids, where molecular density is high.

Convection

Energy is transferred by motion of the molecules of a fluid. The energy transfer in the thin layer immediately adjacent to a solid boundary is by conduction and this results in a temperature difference between that layer and the body of the fluid. This temperature difference causes density gradients which result in motion within the fluid.

If the motion is solely caused by density gradients, the convection is termed *free* but if the motion is enhanced by a pump or blower it is termed *forced*.

Radiation

Energy is transferred by electromagnetic waves emitted by a body (solid or fluid) due to its temperature. No contact is necessary between the emitter and receiver and radiation never ceases. It is not a significant mode of heat transfer at low temperatures.

All three modes of heat transfer may occur concurrently but it is often possible to ignore one in comparison with the other two, e.g. heat transfer in a solid is by conduction alone, conduction is negligible in gases and is often negligible in fluids, radiation is often insignificant below 500°C.

Problems can be broadly divided into two groups:
(a) those involving energy transfer from one fluid to another fluid, separated from the first by a solid boundary in which radiation is negligible. This entails convection from the fluid to the solid and conduction through the solid. Examples are heat exchangers and walls of houses.

(b) those involving heat transfer by radiation combined with conduction and convection. Examples are boiler furnaces and combustion chambers.

31.2 Convection from a fluid to a solid boundary

The heat transfer rate $\dot{Q}$ is given by

$$\dot{Q} = hA\theta \qquad (31.1)$$

where A is the heat transfer area,
$\quad\quad\theta$ is the temperature difference between the surface and fluid,
and $\quad h$ is the surface heat transfer coefficient.

The value of h is dependent on whether the flow is laminar or turbulent in character, whether the convection is free or forced and the properties of the fluid involved. It may be determined by mathematical analysis, analogies and dimensional analysis methods.

31.3 Electrical analogy

If equation (31.1) is written

$$\dot{Q} = \frac{\theta}{\dfrac{1}{hA}}$$

and compared with Ohm's law,

$$I = \frac{V}{R},$$

it will be seen that the heat transfer rate $\dot{Q}$ is analogous with current I and temperature difference θ is analogous with potential difference V. It therefore follows that $\dfrac{1}{hA}$ is analogous with electrical resistance R and problems involving series of thermal resistances may be solved by applying the corresponding rules for series of electrical resistances.

31.4 Conduction through a solid

The heat transfer rate is given by Fourier's equation

$$\dot{Q} = -kA\frac{dT}{dx} \qquad (31.2)$$

where k is the thermal conductivity of the solid, A is the heat transfer area, T is the temperature and x is distance. The negative sign arises because positive heat transfer occurs in the direction of negative temperature gradient.
Common cases are as follows:

(a) Steady conduction through a single layer plane wall, Figure 31.1

$$\dot{Q} = -kA\frac{dT}{dx}$$

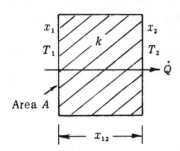

Figure 31.1

$$\therefore \int_1^2 \dot{Q}\, dx = - \int_1^2 kA\, dT$$

i.e.
$$\dot{Q}(x_2 - x_1) = kA(T_1 - T_2)$$

Putting $x_2 - x_1 = x_{12}$, the thickness of the layer, then

$$\dot{Q} = \frac{kA(T_1 - T_2)}{x_{12}} \tag{31.3}$$

Equation (31.3) may be written in terms of the electrical analogy

$$\dot{Q} = \frac{T_1 - T_2}{\dfrac{x_{12}}{kA}}$$

from which it may be seen that the thermal resistance of the wall is $\dfrac{x_{12}}{kA}$.

(b) Steady conduction through a multi-layer plane wall, Figure 31.2

The heat transfer through each layer must be the same,

i.e.
$$\dot{Q} = \frac{k_{12}A(T_1 - T_2)}{x_{12}} = \frac{k_{23}A(T_2 - T_3)}{x_{23}} = \frac{k_{34}A(T_3 - T_4)}{x_{34}} \tag{31.4}$$

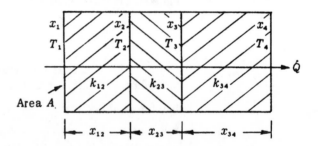

Figure 31.2

This simultaneous equation may be solved to give three unknown quantities usually $\dot{Q}$, T_2 and T_3. However, using the electrical analogy, the thermal resistances of the layers may be added in the same way as electrical resistances in series to give

$$\dot{Q} = \frac{T_1 - T_4}{\dfrac{x_{12}}{k_{12}A} + \dfrac{x_{23}}{k_{23}A} + \dfrac{x_{34}}{k_{34}A}} = \frac{T_1 - T_4}{\sum \dfrac{x}{kA}} \tag{31.5}$$

(c) Steady radial conduction through a single layer cylindrical pipe, Figure 31.3

Consider an elementary cylinder of length l, radius r and thickness dr. If the temperature difference across this cylinder is dT, then Fourier's equation

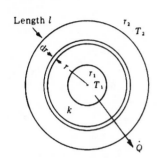

Figure 31.3

becomes

$$\dot{Q} = -2\pi r l k \, \frac{dT}{dr}$$

Therefore, for the whole cylinder,

$$\int_1^2 \dot{Q}\frac{dr}{r} = -\int_1^2 2\pi l k \, dT$$

from which

$$\dot{Q} = \frac{2\pi l k (T_1 - T_2)}{\ln \dfrac{r_2}{r_1}} \qquad (31.6)$$

This may be written in terms of the electrical analogy

$$\dot{Q} = \frac{T_1 - T_4}{\dfrac{\ln \dfrac{r_2}{r_1}}{2\pi l k}}$$

from which it may be seen that the thermal resistance of the wall is $\dfrac{\ln \dfrac{r_2}{r_1}}{2\pi l k}$.

(d) Steady radial conduction through a multi-layer cylindrical wall, Figure 31.4

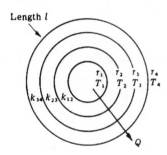

Figure 31.4

The heat transfer rate through each layer must be the same,

i.e. $\quad \dot{Q} = \dfrac{2\pi l k_{12}(T_1 - T_2)}{\ln \dfrac{r_2}{r_1}} = \dfrac{2\pi l k_{23}(T_2 - T_3)}{\ln \dfrac{r_3}{r_2}} = \dfrac{2\pi l k_{34}(T_3 - T_4)}{\ln \dfrac{r_4}{r_3}}$

$$(31.7)$$

As in case (b) the heat transfer rate may be obtained by dividing the overall temperature difference by the cumulative resistance,

i.e. $\quad \dot{Q} = \dfrac{T_1 - T_4}{\dfrac{\ln(r_2/r_1)}{2\pi l k_{12}} + \dfrac{\ln(r_3/r_2)}{2\pi l k_{23}} + \dfrac{\ln(r_4/r_3)}{2\pi l k_{34}}}$

$$= \dfrac{T_1 - T_4}{\sum \dfrac{\ln(r_{\text{outer}}/r_{\text{inner}})}{2\pi l k}} \qquad (31.8)$$

31.5 Steady heat transfer between fluids separated by a solid boundary

The heat transfer rate is determined by adding the thermal resistance at each fluid/solid interface to the thermal resistance of the dividing wall.

For a plane boundary, Figure 31.5,

$$\dot{Q} = \frac{T_1 - T_2}{\dfrac{1}{h_1 A} + \sum \dfrac{x}{kA} + \dfrac{1}{h_2 A}} \qquad (31.9)$$

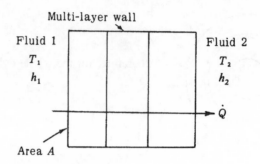

Figure 31.5

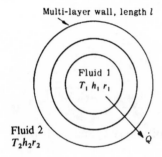

Figure 31.6

For a cylindrical boundary, Figure 31.6,

$$\dot{Q} = \frac{T_1 - T_2}{\dfrac{1}{h_1 A_1} + \sum \dfrac{\ln(r_{\text{outer}}/r_{\text{inner}})}{2\pi l k} + \dfrac{1}{h_2 A_2}}$$

$$= \frac{T_1 - T_2}{\dfrac{1}{2\pi r_1 l h_1} + \sum \dfrac{\ln(r_{\text{outer}}/r_{\text{inner}})}{2\pi l k} + \dfrac{1}{2\pi r_2 l h_2}} \tag{31.10}$$

Equations (31.9) and (31.10) enable the effect of each thermal resistance to be seen and these equations are convenient for investigating any changes which may be necessary in a practical application.

31.6 Overall heat transfer coefficient

The heat transfer rate may be written

$$\dot{Q} = UA\theta \tag{31.11}$$

where U is the overall heat transfer coefficient.

For a plane wall, comparison with equation (31.9) shows that

$$\frac{1}{U} = \frac{1}{h_1} + \sum \frac{x}{k} + \frac{1}{h_2} \tag{31.12}$$

Values of U are tabulated for some building materials.

For a cylindrical wall, the overall heat transfer coefficient is expressed in terms of the length of the pipe. Thus

$$\dot{Q} = U' l \theta \tag{31.13}$$

Comparison with equation (31.10) shows that

$$\frac{1}{U'} = \frac{1}{2\pi r_1 h_1} + \sum \frac{\ln(r_{\text{outer}}/r_{\text{inner}})}{2\pi k} + \frac{1}{2\pi r_2 h_2} \tag{31.14}$$

31.7 Heat transfer by radiation

When radiation effects are small, they may be ignored or allowed for with an estimated surface heat transfer coefficient for radiation. Thus for radiation effects alone,

$$\dot{Q} = h_r A\theta \tag{31.15}$$

and for a combination of convection and radiation,

$$\dot{Q} = (h + h_r)A\theta \tag{31.16}$$

When radiation effects are considerable, this approximation is invalid and the laws of radiation must be used:

(a) Stefan–Boltzmann law for ideal black body radiation which gives the emissive power $\dot{E}_b''$ in W/m^2

$$\dot{E}_b'' = \sigma T^4 \tag{31.17}$$

where T is the temperature in kelvins and σ is a constant of value 5.67×10^{-8} W/m^2 K^4.

(b) Planck's Law for the distribution of black body radiation by wavelength λ.

$$\dot{E}_{b_\lambda}'' = c_1/\lambda^5 \left(e^{c_2/\lambda T} - 1\right) \tag{31.18}$$

where $c_1 = 3.745 \times 10^{-6}$ Wm^2 and $c_2 = 1.438\,8 \times 10^{-2}$ mK.

(c) Wien's law for the wavelength of maximum emissive power

$$\lambda_{\max} T = 2.9 \times 10^{-3} \text{ mK} \tag{31.19}$$

Ideal black bodies absorb all incident radiation. Unfortunately real bodies are not *black* and act as selective emitters and absorbers leading to complex calculations. A partial solution may be obtained by defining a *grey* body with constant emissivity $\in$ and constant absorptivity α which are less than unity. Kirchoff's law states that for such a body $\in = \alpha$ so that emission and absorption are both reduced equally. Incident radiation may also be reflected or transmitted and further properties reflectivity ρ and transmissivity τ are defined such that $\rho + \tau + \alpha = 1$. Figure 31.7

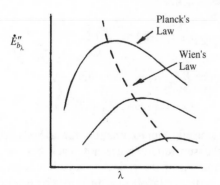
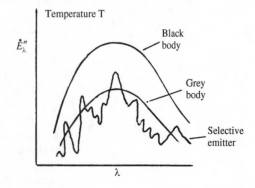

Figure 31.7

illustrates the laws of radiation. For simple calculations intervening media are considered transparent to radiation ($\tau = 1$). This is not true. It is also clear that if there are several receivers in one space there will be complex transfers between them since radiation that is not absorbed or transmitted will be reflected to other receivers. Calculations to solve radiation problems between non-enclosing surfaces will need information of how much a surface can *see* of other surfaces *in view*. Tables or charts are available for this purpose.

Only one type of radiation problem can be solved with a simple approach, that being the heat transfer to or from a grey body in black surroundings, the media between having $\tau = 1$ (Figure 31.8).

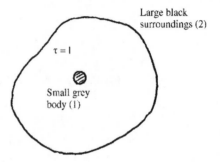

Figure 31.8

The heat transfer rate $\dot{Q}''_{21}$ is the difference between the energy absorbed by the grey body and the energy emitted by the grey body. Let K be the fraction of the energy emitted by the black surroundings that is incident on the grey body. Then

$$\dot{Q}_{21} = \alpha_1 K \dot{E}''_{b_2} - \dot{E}''_{g_1}$$

$$= \alpha_1 K A_2 \sigma T_2^4 - \epsilon_1 A_1 \sigma T_1^4$$

For a grey body $\alpha_1 = \epsilon_1$ hence

$$\dot{Q}_{21} = \epsilon_1 \sigma A_1 \left[K \frac{A_2}{A_1} T_2^4 - T_1^4 \right]$$

If $T_2 = T_1$, $\dot{Q} = 0$ hence $K A_2 / A_1 = 1$ and thus

$$\dot{Q}_{21} = \epsilon_1 \sigma A_1 (T_2^4 - T_1^4) \tag{31.20}$$

This result enables a useful approximation for a body which may be considered grey. If the surroundings are not black but are large enough so that a negligible amount of the reflected radiation is reincident on the grey body, the result is also valid. It may therefore be used for small grey bodies in large surroundings. For this case we can determine the necessary enhancement of the surface heat transfer coefficient to allow for radiation effects (equation (31.16)).

Then
$$\in_1 \sigma A_1(T_2^4 - T_1^4) = h_r(T_2 - T_1)$$

$$h_r = \in_1 \sigma(T_1 + T_2)(T_1^2 + T_2^2) \qquad (31.21)$$

31.8 Lagging

This section uses the theory of Section 31.5 to consider the specific problems involved in lagging. In general, lagging hot surfaces reduces heat losses; however, for small pipes up to about 12 mm in diameter the increase in surface area may increase the losses. For larger pipes and plane surfaces, lagging will reduce heat losses and it is possible to calculate an economic thickness of lagging based on installed insulation costs, fuel cost and an acceptable payback period.

Calculations to determine the economic thickness for lagging pipes may be made with suitable nomograms in fuel efficiency guides issued by the (UK) Energy Efficiency Office or by direct application of the appropriate equations in Section 31.6. The calculations are lengthy and several would be needed to produce a graph which should exhibit the features of Figure 31.9. A computer program for lagging would be an alternative way of finding an optimum solution. Insulation for buildings may be dealt with in a similar way but the variation in building structures complicates the calculations. The table below shows the orders of magnitude of the thermal conductivity of lagging materials but does not discuss their suitability for any particular application.

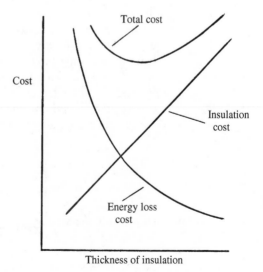

Figure 31.9

Cork	0.036 W/m K	Foamed in Urethane	0.026 W/m K
Glass Wool	0.037 W/m K	Rock Wool	0.040 W/m K
Glass fibre	0.035 W/m K	Kapok	0.035 W/m K
Magnesia (204°C)	0.080 W/m K	Felt (150°C)	0.095 W/m K

31.9 Boundary layers and convection heat transfer

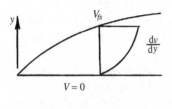

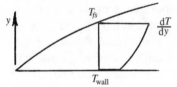

Figure 31.10

When a fluid flows over a solid boundary the fluid particles immediately adjacent to the boundary are at rest and there is a velocity gradient in the thin film of fluid close to the boundary as the velocity changes from zero to the free stream value V_{fs}. This thin film is called the boundary layer. Similar situations occur in a thin film known as the thermal boundary layer. At the boundary the fluid temperature is equal to the wall temperature and in the thermal boundary layer there is a temperature gradient as the fluid temperature changes to the free stream value T_{fs}. The thickness of these boundary layers is not equal (Figure 31.10).

In order to solve problems of convection heat transfer, equation (31.1) has to be used. This equation involves the surface heat transfer coefficient h which is not a simple parameter and it is important to be able to estimate a value for any particular situation. Typical values are:

Free convection in gases	0.5 to 600 W/m K
in liquids	50 to 2 000 W/m K
Forced convection in gases	10 to 700 W/m K
in liquids	100 to 10 000 W/m K

An understanding of boundary layers is required to make a reasonable estimate of surface heat transfer coefficients. Some mathematical analysis is possible but is not considered here.

Flow in a boundary layer has particular characteristics which may be illustrated with a flat plate discussion. The boundary layer starts at the leading edge and increases in thickness y as distance from the leading edge increases (Figure 31.11). The flow in zone A is smooth and is termed laminar. At some point in the flow the smoothness begins to change and zone B is a transition to zone C in which the flow is randomly disturbed by eddies and is termed turbulent. When applied to flow in a tube, the flow pattern is similar but the boundary layer grows from entry around the circumference and eventually meets to produce fully developed flow in

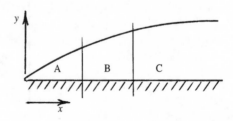

Figure 31.11

which the values of the free stream velocity and temperature are replaced by the bulk velocity V_b and the bulk temperature T_b (Figure 31.12). In simple calculations for turbulent flow, V_b and T_b could be measured with a flow meter and a thermometer, respectively, but in laminar flow a traverse method might be required to calculate

$$V_b = \int \frac{V\,\mathrm{d}A}{A} \quad \text{and} \quad T_b = \int \frac{TV\,\mathrm{d}A}{AV_b}$$

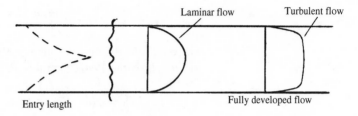

Figure 31.12

Experiment shows that the change from laminar to turbulent flow occurs at predictable conditions determined by the Reynolds number (Re) of the flow (see Section 37.4)

$$Re = \frac{\rho V l}{\mu}$$

where ρ is the fluid density, μ is the fluid viscosity, V is the free stream velocity (bulk velocity for tube flow) and l is some representative length dimension (distance from the leading edge for a plate and diameter for a tube).

For a plane surface with forced flow; if $Re < 500\,000$ flow is laminar and if $Re > 500\,000$ flow is turbulent.

For forced flow inside a tube; if $Re < 2\,300$ flow is laminar and if $Re > 6\,000$ flow is fully turbulent.

Transition zones and tube entry flows are special problems and are not considered here.

31.10 Forced convection

Mathematical analysis of the flow in boundary layers is complex and in order to determine surface heat transfer coefficients it is simpler to use an approach by dimensional analysis (see Chapter 38).

In forced convection in tubes the variables considered are:

h the surface heat transfer coefficient
V the bulk velocity of the fluid
d the tube diameter
μ the dynamic viscosity of the fluid

ρ the density of the fluid
c_p the specific heat of the fluid at constant pressure
k the thermal conductivity of the fluid.

The relation is formed by writing

$$h = \phi(V, d, \mu, k, \rho, c_p)$$

and by the methods of dimensional analysis it is found that;

$$\frac{hd}{k} = \phi \left[\left(\frac{\rho V d}{\mu} \right) \left(\frac{\mu c_p}{k} \right) \right] \tag{31.22}$$

where hd/k is the Nusselt number Nu, $\rho V d/\mu$ is the Reynolds number Re and $\mu c_p/k$ is the Prandtl number Pr so that we may write

$$Nu = \phi(Re, Pr) \text{ or } Nu = \text{ constant } Re^a Pr^b \tag{31.23}$$

The fluid properties in tube flow are normally taken at T_b the bulk temperature of the fluid unless there is a large temperature difference between the wall and the fluid when the values of μ and k are taken at the film temperature T_f

$$T_f = (T_b + T_{\text{wall}})/2 \tag{31.24}$$

Experimental data is used to determine the values of the indices and constants used in equation (31.23). There are many correlations for turbulent flow in a tube. An equation for normal use is the Dittus–Boelter relation

$$Nu = 0.023 \, Re^{0.8} Pr^n \tag{31.25}$$

The index n in this equation is 0.4 for heating the fluid and 0.3 for cooling. The range of validity for this relation is $Re > 6\,000$, $0.7 < Pr < 160$ and $l/d > 60$.

Other simple relations include:

	Laminar flow	Turbulent flow
Plane	$Nu = 0.664 \, Re^{0.5} \, Pr^{0.33}$	$Nu = 0.036 \, Re^{0.8} \, Pr^{0.33}$
Tube	$Nu = 3.66$ if d/L is small	$NU = 0.023 \, Re^{0.8} \, Pr^{0.4}$

An alternative approach to dimensional analysis for the determination of surface heat transfer coefficients is to use the Reynolds analogy between heat transfer and momentum transfer in fluid flow which results in:

$$\text{Stanton number } St = \frac{Nu}{Re \, Pr} = \frac{f}{2} \tag{31.26}$$

where f is the friction factor in fluid flow (Section 37.7). For turbulent flow in smooth tubes,

$$f = 0.079 \, Re^{-0.25} \tag{31.27}$$

Reynolds analogy was later improved by Colburn to

$$St \, Pr^{2/3} = f/2 \quad \text{for } 0.5 < Pr < 100 \tag{31.28}$$

31.11 Free convection

Free convection will occur when an object submerged in a fluid has a different temperature to that of the fluid. The temperature difference will cause a heat transfer resulting in a change in temperature and density of the fluid in contact with the submerged object. This fluid will rise or fall as a result of the density change. To allow for this the velocity parameter used in forced convection is replaced by the temperature difference θ and the product of the coefficient of cubical expansion of the fluid β with gravitational acceleration g.

Thus in free convection

$$h = \phi(l, \mu, k, \rho, c_p, (\beta g), \theta)$$

and dimensional analysis results in the relation

$$Nu = \phi(Pr, Gr) \tag{31.29}$$

where $Pr = \mu c_p/k$, the Prandtl number and $Gr = \rho^2(\beta g)\theta l^3/\mu^2$, the Grashof number.

For liquids the property β is found in tables and for ideal gases $\beta = 1/T$, where T is the *free stream temperature* of the fluid outside the boundary layer.

It is found that the onset of turbulence in free convection is determined by the value of the Rayleigh number, Ra:

$$Ra = Pr \, Gr = \frac{\rho^2 l^3 c_p (\beta g)}{\mu k} \tag{31.30}$$

Empirical correlations for free convection over *vertical* plane or cylindrical surfaces with uniform wall temperature are

$$\left. \begin{array}{ll} 10^4 < Ra < 10^9 \text{ Laminar flow} & Nu = 0.59 \, (Pr \cdot Gr)^{1/4} \\ 10^9 < Ra < 10^{13} \text{ Turbulent flow} & Nu = 0.13 \, (Pr \cdot Gr)^{1/3} \end{array} \right\} \tag{31.31}$$

The characteristic length for Gr is height and the resulting values of Nu are a mean over the whole surface. Fluid transport properties are found in tables and evaluated at T_{film}, the mean between T_{wall} and T_{ambient}.

For free convection in air *at moderate temperatures* and at atmospheric pressure, the relations of equations (31.31) may be simplified and are shown in equations (31.32) together with relations for free convection from horizontal cylinders.

Vertical plates and cylinders of height l	$h_{mean} = 1.42(\theta/l)^{1/4}$ $h_{mean} = 1.31(\theta)^{1/3}$	$10^4 < Ra < 10^9$ $10^9 < Ra < 10^{13}$	Laminar Turbulent
Horizontal cylinders of diameter d	$h_{mean} = 1.32(\theta/d)^{1/4}$ $h_{mean} = 1.24(\theta)^{1/3}$	$10^4 < Ra < 10^9$ $10^9 < Ra < 10^{12}$	Laminar Turbulent

$$(31.32)$$

31.12 Heat transfer with change of phase

Steam and refrigeration plant both use processes involving evaporation and condensation, requiring an understanding of phase change heat transfer. Evaporation is a more complex process to consider since it may involve both high pressure and high temperature, making safety an important factor in design.

Pool boiling

Consider a heating surface which is submerged in a liquid and gradually increase the surface temperature. Distinct patterns of heat transfer can be observed which depend on the temperature difference between heating surface and fluid (Figure 31.13(a)).

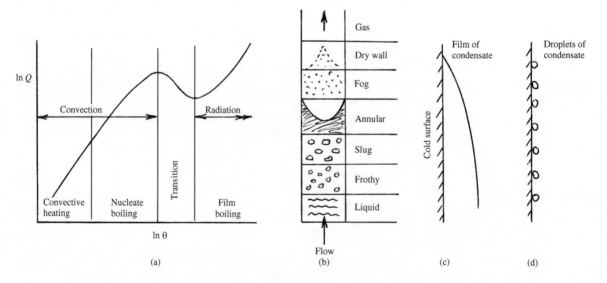

Figure 31.13

Convection heat transfer regimes

(*i*) Surface temperature less than saturation temperature; no boiling.

(*ii*) Surface temperature greater than saturation temperature; bubbles of vapour form and rise but condense as they rise.

(*iii*) Nucleate boiling; vapour bubbles reach the liquid surface and the heat transfer reaches a maximum until the bubbles coalesce on the heating surface insulating the surface causing the surface temperature to rise and possibly fail. This is termed *burnout*.

Radiation regime

Provided the heating surface has not failed, it will become completely *dry* and radiation will then be the dominant mode of heat transfer.

Similar patterns occur in forced flow in a vertical tube, the liquid flow becomes a vapour as it passes along the tube until at the *dry wall* state the heat transfer regime is that of a *gas* in a hot tube (Figure 31.13(b)). Horizontal tube flow will be similar but gravitational effects will occur as the liquid evaporates causing an asymmetric flow pattern which may cause surface failure. Analysis of evaporation processes is not included in this book.

Film condensation

When condensate forms a continuous liquid layer over a condensing surface it is termed film condensation and is relatively simple to analyse (Figure 31.13(c)).

Vapour may also condense as droplets and remain in droplet form to flow across a *non-wettable* condensing surface. This enhances the heat transfer rate but is complex to analyse and difficult to achieve in practice (Figure 31.13(d)).

The results of an analysis of *film* condensation for vertical and inclined plane surfaces and the outside of horizontal tubes is well documented. For a plate of length L inclined at an angle ϕ to the horizontal, the *mean* surface heat transfer coefficient is given by

$$h_{\text{mean}} = 0.943 \left[\frac{\rho_f(\rho_f - \rho_g)gh_{fg}k_f^3 \sin\phi}{\mu_f L(T_g - T_w)} \right]^{1/4} \tag{31.33}$$

A *vertical* plate equation is obtained with $\phi = 90$ degrees so that $\sin\phi = 1$.

For a horizontal tube of diameter D

$$h_{\text{mean}} = 0.725 \left[\frac{\rho_f(\rho_f - \rho_g)gh_{fg}k_f^3}{\mu_f D(T_g - T_w)} \right]^{1/4} \tag{31.34}$$

where ρ = density, k = thermal conductivity, μ = dynamic viscosity, T = temperature, g = gravitational acceleration. h_{fg} = specific enthalpy of the phase change vapour to liquid, and suffices f and g represent saturated liquid and vapour, respectively.

Properties for equations (31.33) and (31.34) should be evaluated at the film temperature $T_{\text{film}} = 0.5(T_{\text{wall}} + T_{\text{saturation}})$.

It should be noted that a vertical tube has a smaller condensation heat transfer coefficient than a horizontal tube which means that tubular condensers are normally arranged to have horizontal tubes in vertical tiers. In such an arrangement it is assumed that the condensate from an upper tube flows down on to the tube below in which case it is found that the heat transfer coefficient h_N for one tier of N tubes of diameter d is given by

$$h_N = \left[\frac{1}{N^{0.25}}\right] h_{\text{tube}} \tag{31.35}$$

where h_{tube} is found by use of equation (31.34).

31.13 Heat exchangers

A heat exchanger is used to transfer energy from a hot fluid to a cold fluid. Plane or tubular walls are used to keep the fluids separated and these walls may be finned to allow more area for heat transfer. The heat transfer requirements should be satisfied but there should also be an attempt to minimize pressure losses in the heat exchanger and the design must be such that cleaning is possible. Heat transfer coefficients are larger in forced, turbulent flow and most heat exchangers will be designed using this flow regime.

The most common form of heat exchanger is known as a recuperator and consists of a thin, good conducting, metal boundary separating a hot fluid from a cold fluid. When clean, a boundary of this type has negligible thermal resistance. Recuperators appear in various forms (Figure 31.14) and may be classified by construction:concentric tubes with parallel or counter flow, shell and tube, plate, crossflow and finned tube matrix.

There are also regenerative heat exchangers which recycle energy using heat transfer to improve plant efficiency. Examples include feed water heating in steam plant (Figure 24.8), combined heat and power plant (Figure 24.9) and gas turbine exhaust gas reheating (Figure 25.7). Another example of regeneration is to heat a matrix with hot waste gas and then cool the matrix with cold fresh gas. This may be achieved intermittently or continuously by use of a rotating matrix design known as a *thermal wheel*.

31.14 Finned surfaces

Figure 31.14 shows that crossflow designs such as a car radiator are made with a finned tube matrix to obtain more surface area on the air flow side where surface heat transfer coefficients are small compared with those inside the water-filled flat tubes. This gives a complex outer surface. Such surfaces have been analysed and the results are often presented graphically.

Simple shapes such as a pin fin or a rectangular fin can be analysed without great difficulty but tapered fins and annular fins need complex mathematics and are usually approached through graphs of fin efficiency (Figure 31.15).

$$\eta_{\text{fin}} = \frac{\text{Actual heat transfer from the fin}}{\text{Rate if the whole fin were at the wall temperature}}$$

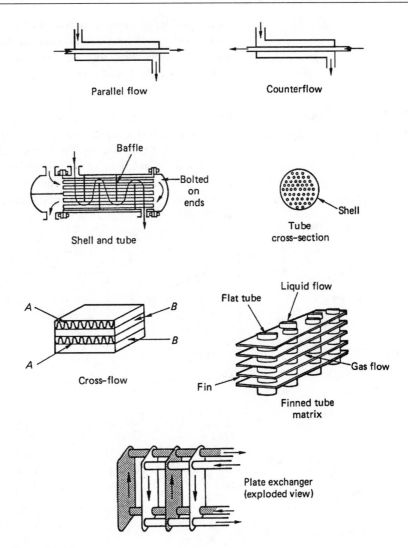

Parallel flow

Counterflow

Baffle

Bolted
on
ends

Shell and tube

Shell

Tube
cross-section

Cross-flow

Flat tube

Liquid flow

Fin

Gas flow

Finned tube
matrix

Plate exchanger
(exploded view)

Figure 31.14

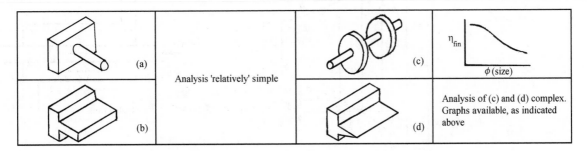

| (a) | Analysis 'relatively' simple | (c) | η_{fin} vs ϕ (size) |
| (b) | | (d) | Analysis of (c) and (d) complex. Graphs available, as indicated above |

Figure 31.15

Many problems may be solved by evaluating the denominator of the expression above and multiplying by the fin efficiency obtained from standard graphs. Fins that are less than 70% efficient are a poor design and a reasonable working target might be 85%.

31.15 Log mean temperature difference

Consideration of the purpose of a heat exchanger shows that one stream of fluid gets hotter as the other gets colder (unless one or both of the streams is changing phase). The flows exchanging heat may be in *parallel flow* (unidirectional) or *counter flow* but in either case the difference in temperature between the two streams will be changing (Figure 31.16). It is therefore necessary to find a representative temperature difference between the fluids.

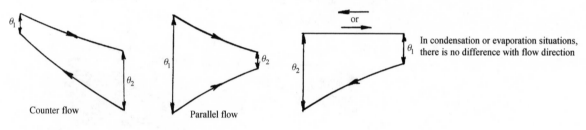

Figure 31.16

Consider the parallel flow case (Figure 31.17) and let U be the overall heat transfer coefficient (Section 31.6). At entry to the heat exchanger the temperature difference is θ_1 and at exit θ_2. The mass flow rates of flow of hot and cold streams are $\dot{m}_H$ and $\dot{m}_C$ having specific heat capacities c_{pH} and c_{pC}. For the element of area dA the temperature changes in hot and

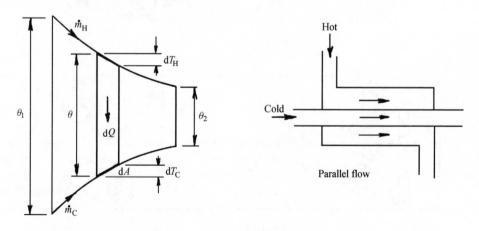

Figure 31.17

cold streams are dT_H and dT_C and the temperature difference between the streams at this point is θ.

Let the heat transfer at the element of area dA be $d\dot{Q}$.

Then
$$d\dot{Q} = U\,dA\theta \qquad \text{(i)}$$

Also

$$d\dot{Q} = -\dot{m}_H c_{p_H} \Delta T_H \qquad \text{(ii)}$$

$$= +\dot{m}_C c_{p_C} \Delta T_C \qquad \text{(iii)}$$

$$\theta = T_H - T_C \qquad \text{(iv)}$$

Hence

$$d\theta = dT_H - dT_C \qquad \text{(v)}$$

From (ii), (iii) and (v),

$$d\theta = -d\dot{Q}\left(\frac{1}{(\dot{m}_H C_{p_H})} + \frac{1}{\dot{m}_C c_{p_C}}\right) \qquad \text{(vi)}$$

From (ii) and (vi),

$$\frac{d\theta}{\theta} = -U\,dA\left(\frac{1}{\dot{m}_H c_{p_H}} + \frac{1}{\dot{m}_C c_{p_C}}\right) \qquad \text{(vii)}$$

Integrate (vii)

$$\ln\frac{\theta_2}{\theta_1} = -UA\left(\frac{1}{\dot{m}_H c_{p_H}} + \frac{1}{\dot{m}_C c_{p_C}}\right) \qquad \text{(viii)}$$

From (vi) by integration,

$$\theta_1 - \theta_2 = -\dot{Q}\left(\frac{1}{\dot{m}_H c_{p_H}} + \frac{1}{\dot{m}_C c_{p_C}}\right) \qquad \text{(ix)}$$

From (viii) and (ix)

$$\dot{Q} = UA\left[\frac{\theta_1 - \theta_2}{\ln(\theta_1/\theta_2)}\right] \qquad (31.36)$$

$[(\theta_1 - \theta_2)/\ln(\theta_1/\theta_2)]$ is the *log mean temperature difference* and is used in equations (31.11) and (31.13) (Section 31.6) to give $\dot{Q} = UA\theta_{\text{LMTD}}$ for a plane wall and $\dot{Q} = U'L\theta_{\text{LMTD}}$ for a tube. When θ_{LMTD} is calculated it is found that with the same mass flow rates and the same inlet temperatures counterflow gives a larger mean temperature difference which requires a smaller heat exchanger. Thus counterflow heat exchangers are preferred. In a case where one fluid is changing phase (evaporation or condensation) the value of θ_{LMTD} is not affected by the flow directions.

There are heat exchangers in which flow is neither counter nor parallel, examples include crossflow and multi-pass designs and a different analysis is required. The necessary calculations have been made for a range of designs to determine a multiplying factor F to be applied to equation (31.34).

$$\dot{Q} = UAF\theta_{\text{LMTD}} \text{ (or } \dot{Q}' = U'LF\theta_{\text{LMTD}})\tag{31.37}$$

There is an important convention for this application to crossflow and multi-pass heat exchangers in that θ_{LMTD} should be based on *counterflow*.

A simple example of this type of heat exchanger is shown in Figures 31.18(*a*) and (*b*). If baffles are added to the multi-pass design of Figure 31.18(*a*), it can be seen in Figure 31.18(*c*) that the flow has become much more complicated and the correction factor for this particular design is obtained from the chart shown in Figure 31.18(*d*). Charts are available for other designs such as crossflow, etc.

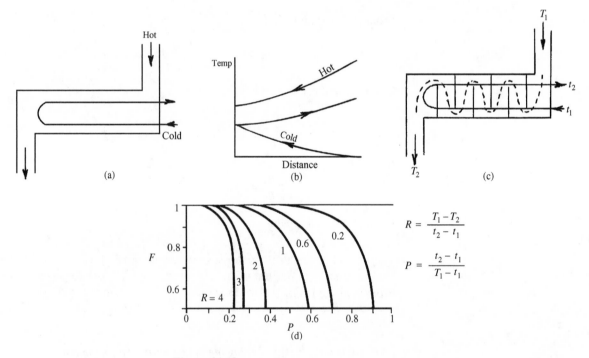

Figure 31.18

The charts assume that the heat exchanger is externally adiabatic. The heat transfer is

$$\dot{m}c_{p_{\text{tube}}}(t_2 - t_1) = \dot{m}c_{p_{\text{shell}}}(T_1 - T_2)$$

and a factor R is defined by

$$R = \frac{T_1 - T_2}{t_2 - t_1} = \frac{\dot{m}c_{p_{\text{tube}}}}{\dot{m}c_{p_{\text{shell}}}}$$

and used as a contour on the chart. A further factor P, plotted from the horizontal axis, is defined by

$$P = \frac{t_2 - t_1}{T_1 - t_1}$$

P enables the multiplying factor F to be obtained from the vertical axis.

Example 9 demonstrates this method for multi-pass or crossflow heat exchangers.

31.16 Heat exchanger fouling

Heat exchange surfaces become contaminated by scale and sludge. This is termed *fouling* and causes extra thermal resistance to heat transfer so that regular cleaning is required. Design must allow for this occurrence so that acceptable performance is still available at the time cleaning is deemed to be due. Fouling can be anticipated at the design stage by including an extra resistance in equation (31.10) as shown below:

$$\dot{Q} = \frac{T_1 - T_2}{\dfrac{1}{2\pi r_1 l h_1} + \sum \dfrac{\ln(r_{\text{outer}}/l_{\text{inner}})}{2\pi l K} + \dfrac{1}{2\pi r_2 l h_2} + R_{\text{fouling}}} \qquad (31.38)$$

Fouling resistance values are of the order 0.2 to 2 K/kW.

31.17 Heat exchanger efficiency

Heat exchanger performance is assessed by the effectiveness E:

$$E = \frac{\text{Actual heat transfer rate}}{\text{Maximum possible heat transfer rate}} \qquad (31.39)$$

The denominator in this expression would be the heat transfer rate when the minimum temperature difference at one 'end' of the exchanger falls to zero (Figure 31.19). This would require an infinite area for heat exchange. Heat exchanger effectiveness usually lies between 0.6 and 0.8 depending on the fouling state.

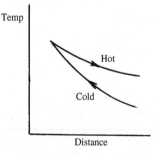

Figure 31.19

31.18 A simple approach to heat exchanger 'design'

If a heat transfer rate $\dot{Q}$ is required in a tubular heat exchanger, there are five equations available (Figure 31.20):

(i) $\dot{Q} = U'LF\theta_{\text{LMTD}}$ or $UAF\theta_{\text{LMTD}}$ ($F = 1$ if simple counter or parallel flow)

(ii) $\dot{Q} = \dot{m}_H C_{p_H}(T_1 - T_2)$

(iii) $\dot{Q} = \dot{m}_C C_{p_C}(T_3 - T_4)$

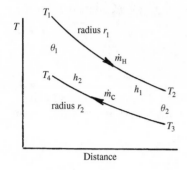

Figure 31.20

(iv) $\theta_{\text{LMTD}} = \dfrac{(\theta_1 - \theta_2)}{\ln(\theta_1/\theta_2)}$

(v) $\dfrac{1}{U'} = \dfrac{1}{2\pi r_1 h_1} + \dfrac{1}{2\pi r_2 h_2}$ or $\dfrac{1}{U} = \dfrac{1}{2\pi r_1 l h_1} + \dfrac{1}{2\pi r_2 l h_2}$

Equation (v) assumes the tube walls are thin (so that r_1 is equal to r_2) and offer negligible thermal resistance. If this is not so, equation (31.14) or (31.12) applies.

The value of the *inside* surface heat transfer coefficient may be calculated from the appropriate relation

$$Nu = \phi(Re, Pr)$$

Thus for a given tube diameter the tube length can be found provided the outside surface heat transfer coefficient is given since *this requires calculations beyond the scope of this book*. Similarly, no consideration is given to the effects of tiers of tubes (equation (31.35)) at this stage of design. If there are n tubes in the design, we can find the tube length by dividing the required heat transfer rate by the number of tubes to be used. In this case

$$\dot{m} = n \cdot \dot{m}_{\text{tube}} \quad \text{and} \quad \dot{Q} = n \cdot \dot{Q}_{\text{tube}} \tag{31.40}$$

31.19 Heat exchanger design criteria

There are many types of heat exchanger and several approaches to design. Large–scale computer programs are available. It is clear that there will be many variables and much trial and error will be needed to achieve an optimum solution. In any design the final arbiter may well be the costs which will involve three elements:

Capital costs   (size, shape, material)
Running costs   (pressure loss, maintenance)
Peripheral costs (coolant cost, instrumentation, etc.)

Worked examples

1. *An oven door consists of a layer of insulation sandwiched between a layer of steel on the inner face and a layer of wood on the outer face. The thickness of the steel is 20 mm and that of the wood 12 mm. The adhesive used to bond the wood to the insulation has a maximum safe working temperature of 93°C. The temperature of the inside surface of the door is 537°C and that of the outside 15.5°C.*

Determine the minimum permissible thickness of insulation, ignoring the thickness and resistance of the adhesive.

$k_{\text{steel}} = 41.5$ W/m K, $k_{\text{insulation}} = 0.19$ W/m K   and $k_{\text{wood}} = 1.9$ W/m K.

The door is shown in Figure 31.21. From equation (31.3), the heat transfer through each layer is given by

$$\dot{Q} = \frac{kA(T_1 - T_2)}{x_{12}} \quad \text{and this is the same for each layer.}$$

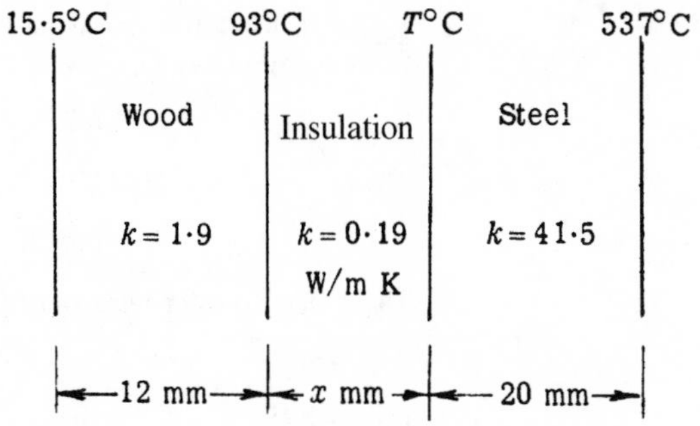

15·5°C　　　　93°C　　　　T°C　　　　537°C

Wood	Insulation	Steel
$k = 1.9$	$k = 0.19$ W/m K	$k = 41.5$

$\leftarrow$ 12 mm $\rightarrow$ | $\leftarrow x$ mm $\rightarrow$ | $\leftarrow$ 20 mm $\rightarrow$

Figure 31.21

For the wood,

$$\dot{Q} = \frac{1.9A(93 - 15.5)}{12 \times 10^{-3}} = 12.3 \times 10^3 A \qquad (1)$$

For the steel,

$$\dot{Q} = \frac{41.5A(537 - T)}{20 \times 10^{-3}} = 2.08(537 - T) \times 10^3 A \qquad (2)$$

Thus, from equations (1) and (2), $T = 531.1°C$.
For the insulation,

$$\dot{Q} = \frac{0.19A(531.1 - 93)}{x \times 10^{-3}} \qquad (3)$$

and hence, from equations (1) and (3),

$$x = \underline{6.78 \text{ mm}}$$

2. *A cast iron central heating pipe has an inner diameter of 100 mm and a wall thickness of 5 mm. The pipe feeds a radiator with water at a temperature of 85°C and the inner wall temperature of the pipe may be assumed equal to the water temperature. The thermal conductivity of cast iron is 52 W/m K and the outer surface heat transfer coefficient is 20 W/m² K.*

Calculate the surface temperature of the pipe and the heat transfer rate per metre of pipe in a room at 19°C.

The pipe is shown in Figure 31.22. From equation (31.6), the heat transfer rate by conduction through the pipe wall is given by

$$\dot{Q} = \frac{2\pi l k (T_1 - T_2)}{\ln \dfrac{r_2}{r_1}}$$

$$= \frac{2\pi \times 1 \times 52(85 - T_2)}{\ln \dfrac{55}{50}}$$

$$= 3\,425(85 - T_2)W$$

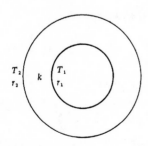

Figure 31.22

The heat transfer rate from the pipe surface in surroundings at temperature T is, from equation (31.1),

$$\dot{Q} = hA\theta \quad \text{where} \quad \theta = T_2 - T_1 \quad \text{and} \quad A = 2\pi r_2 l$$

Thus
$$\dot{Q} = 20 \times 2\pi \times 55 \times 10^{-3} \times 1 \times (T_2 - 19)$$

$$= 6.912(T_2 - 19) W$$

These heat transfer rates must be equal, so that

$$3\,425(85 - T_2) = 6.912(T_2 - 19)$$

from which
$$T_2 = \underline{84.87°C}$$

and
$$\dot{Q} = \underline{455.5 \ W/m}$$

Note: It will be seen that the temperature difference between the inside and outside surfaces of the pipe is very small and it is often assumed that with thin walls having a high thermal conductivity, the temperature change is negligible.

3. *A pipe of 50 mm bore and 25 mm wall thickness is made of material of thermal conductivity 52 W/m K and is lagged by cork 13 mm thick with thermal conductivity 0.04 W/m K.*

The pipe carries saturated steam at a pressure of 10 bar in surroundings at 18 °C. The surface heat transfer coefficient on the inside of the pipe is 15 W/m² K and on the outside of the cork, it is 10 W/m² K.

Calculate the heat transfer rate per unit length of pipe.

The pipe is shown in Figure 31.23. At 10 bar, the saturation temperature of the steam is 179.9°C.

The thermal resistance at the outside surface of the cork per metre length

$$= \frac{1}{2\pi rh} = \frac{1}{2\pi \times 63 \times 10^{-3} \times 10}$$

$$= 0.252 \ m \ K/W$$

The thermal resistance at the inside surface of the pipe per metre length

$$= \frac{1}{2\pi rh} = \frac{1}{2\pi \times 25 \times 10^{-3} \times 15}$$

$$= 0.425 \ m \ K/W$$

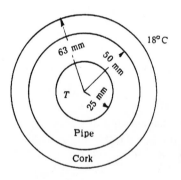

Figure 31.23

The thermal resistance of the pipe per metre length

$$= \frac{\ln \dfrac{r_2}{r_1}}{2\pi k} = \frac{\ln \dfrac{50}{25}}{2\pi \times 52} = 0.002\,12 \ m \ K/W$$

The thermal resistance of the cork per metre length

$$= \frac{\ln \dfrac{r_2}{r_1}}{2\pi k} = \frac{\ln \dfrac{63}{50}}{2\pi \times 0.04} = 0.918 \ m \ K/W$$

Thus the overall thermal resistance per metre length

$$= 0.252 + 0.425 + 0.002\,12 + 0.918$$

$$= 1.597 \ m \ K/W$$

Therefore heat transfer rate per metre length,

$$\dot{Q} = \frac{\text{overall temperature difference}}{\text{overall resistance per metre}}$$

$$= \frac{179.9 - 18}{1.597} = \underline{101.3 \text{ W}}$$

4. *An electric heater is placed in a large room. The heater surface temperature is 750°C and it has a surface area of 0.1 m². The heater is fitted with a reflector so that all the energy supplied is emitted to the room. The room surface temperature is 15°C and the total surface area is 50 m².*

Assuming that the room and heater may be considered as ideal black surfaces and that 1/500 of the radiation emitted by the room is incident on the heater, determine
(a) the heat transfer rate from the heater to the room;
(b) the radiation heat transfer coefficient from the fire to the room.

From equation (31.17), the energy radiation rate from the fire

$$= A\sigma T^4$$

$$= 0.1 \times 5.67 \times 10^{-8} \times (750 + 273)^4$$

$$= 6211 \text{ W}$$

Similarly, the energy radiation rate from the room

$$= 50 \times 5.67 \times 10^{-8} \times (15 + 273)^4$$

$$= 19504 \text{ W}$$

Of this quantity emitted by the room, 1/500 is incident on the fire, to that energy incident on fire $= \dfrac{19504}{500} = 39 \text{ W}$

Therefore heat transfer rate from fire to room,

$$\dot{Q} = 6211 - 39 = \underline{6172 \text{ W}}$$

From equation (31.16), heat transfer coefficient,

$$h_r = \frac{\dot{Q}}{A\theta}$$

$$= \frac{6172}{0.1(1023 - 288)} = \underline{83.97 \text{ W/m}^2\text{K}}$$

5. *A small rocket combustion chamber is being tested in a large laboratory with all surfaces at a temperature of 16°C. The surface area of the combustion chamber is 1.75 m² and the surface temperature is 600°C. The emissivity of the combustion chamber surface is 0.27. Calculate the radiation heat transfer rate and the surface heat transfer coefficient for radiation effects. The Stefan–Boltzmann constant $\sigma = 5.67 \times 10^{-8}$ W/m² K⁴.*

This problem can be treated as a small grey body in large surroundings.

From equation (31.20)

$$\dot{Q}_{21} = \epsilon_1\, \sigma A_1(T_2^4 - T_1^4)$$

where 2 is the grey body combustion chamber and 1 is the laboratory.

$$\dot{Q}_{21} = 0.27 \times (5.67 \times 10^{-8}) \times 1.75 \times (873^4 - 289^4) \times 10^{-3}\ \text{kW}$$

The radiation heat transfer rate to the laboratory $\dot{Q} = \underline{15.37\ \text{kW}}$.
The surface heat transfer coefficient for radiation is given by equation (31.21):

i.e. $$h = 0.27(5.67 \times 10^{-8}) \times (289 + 873) \times (289^2 + 873^2)$$

$$= \underline{15.04\ \text{W/m}^2\ \text{K}}.$$

6. (i) *Water flows at 2 m/s through a long pipe 30 mm in diameter. The water is being heated and the mean bulk temperature is 70°C. Estimate the surface heat transfer coefficient in turbulent flow between the pipe and the water using;*

(a) *the Dittus–Boelter equation $Nu = 0.023\ Re^{0.8}\ Pr^{0.4}$ for heating provided $Re > 6\,000$;*
(b) *Reynolds analogy $St = Nu/Re\ Pr = f/2$ where $f = 0.079\ Re^{-0.25}$;*
(c) *Reynolds–Colburn analogy $St.\ Pr^{2/3} = f/2$.*

(ii) *Calculate the mean surface heat transfer coefficient for dry saturated steam at 80°C condensing on the outside surface of a horizontal tube of diameter 30 mm through which water flows at a 'mean' temperature of 20°C.*

Assume the wall temperature to be equal to the water temperature of 20°C (see equation (31.34).

(i) (a) Dittus–Boelter
Check $Re > 6\,000$; at 70°C, $\rho = 980\ \text{kg/m}^3$, $\mu = 0.4 \times 10^{-3}\ \text{kg/ms}$, $Pr = 2.53$

$$Re = \frac{\rho V d}{\mu} = \frac{980 \times 2 \times 0.03}{0.4 \times 10^{-3}} = 147\,000$$

$$Nu = 0.023\ Re^{0.8} Pr^{0.4} = 0.023(147\,000)^{0.8}(2.53)^{0.4} = 453.89$$

At 70°C, k = 0.662 W/m K, hence

$$h = Nu \times (k/d) = \frac{453.89 \times 0.662 \times 10^{-3}}{0.03} = \underline{10\ \text{kW/m}^2\text{K}}$$

(b) Reynolds analogy

$$St = \frac{Nu}{Re \cdot Pr} = \frac{f}{2};\quad \text{for turbulent flow } f = 0.079\ Re$$

$$Re = 147\,000 \quad \text{and} \quad Pr = 2.53 \text{ so that } \frac{f}{2} = \frac{0.079(147\,000)^{-0.25}}{2} = 0.002$$

$$St = 0.002 = Nu/Re \cdot Pr$$

$$Nu = 0.002\ Re \cdot Pr = (0.002 \times 147\,000 \times 2.53) = 750.24$$

Hence

$$h = Nu \times k/d = \frac{(750.24 \times 0.662 \times 10^{-3})}{0.03} = \underline{16.56 \text{ kW/m}^2\text{K}}$$

(c) Reynolds–Colburn analogy

$St \cdot Pr^{2/3} = f/2$ hence we divide the Reynolds analogy result by $Pr^{2/3}$, i.e.

$$h = 16.56 \div (2.53)^{2/3} = \underline{8.91 \text{ kW/m}^2\text{K}}$$

Note that the Reynolds–Colburn analogy is closer to the Dittus–Boelter result than the simple Reynolds analogy (see also example 19).

(ii)

From equation (31.34) the condensing surface heat transfer coefficient is given by

$$h_{\text{mean}} = \left[\frac{\rho_f(\rho_f - \rho_g)g - h_{fg}k_f^3}{\mu_f D(T_s - T_w)} \right]^{1/4}$$

(see Section 31.12 for the symbol definitions).

From tables at the *film temperature* $(80 + 20)/2 = 50°C$,

$$\rho_f = 990 \text{ kg/m}^3, \quad \rho_g = 0.083 \text{ kg/m}^3, \quad h_{fg} = 2\,382.9 \text{ kJ/kg},$$

$$\mu_f = 0.544 \text{ g/sm}, \quad k_f = 0.643 \text{ W/m K}, \text{ hence from equation (31.34)}:$$

$$h_{mean} = 0.725 \left[\frac{990(990 - 0.083)9.81 \times 2\,382.9(0.643 \times 10^{-3})^3}{(0.544 \times 10^{-3})(30 \times 10^{-3})(80 - 20)} \right]^{1/4}$$

$$= \underline{6.43 \text{ kW/m}^2\text{K}}$$

7. *A counter-flow heat exchanger cools 0.4 kg/s of oil of specific heat 2.18 kJ/kg K from 80°C to 30°C with 0.3 kg/s of water of specific heat 4.2 kJ/kg K entering at 20°C. The overall heat transfer coefficient U for the heat exchanger is 1.3 kW/m² K. Determine the heat transfer area required.*

Heat transfer rate from oil, $\qquad \dot{Q} = \dot{m}c_p\Delta T$

$$= 0.4 \times 2.18 \times (80 - 30) = 43.6 \text{ kJ/s}$$

This is taken up by the water, which leaves at temperature T.

Hence $\qquad\qquad\qquad 0.3 \times 4.2(T - 20) = 43.6$

from which $\qquad\qquad\qquad T = 54.6°C$

Figure 31.16 shows the counter-flow arrangement, in which

$$\theta_1 = 80 - 54.6 = 25.4°C$$

and $\qquad\qquad\qquad \theta_2 = 30 - 20 = 10°C$

From equation (31.36),

$$\text{log mean temperature difference} = \frac{25.4 - 10}{\ln \dfrac{25.4}{10}} = 16.52°C$$

From equation (31.11), $\dot{Q} = UA\theta$

$$\therefore A = \frac{\dot{Q}}{U\theta} = \frac{4 \times 2.18 \times (80 - 30)}{1.3 \times 16.52} = \underline{2.03 \text{ m}^2}$$

Note: The arithmetic mean temperature difference is 17.8°C, which would underestimate the area required.

8. (a) *Water at a temperature of 77°C flows through a horizontal circular steel pipe of diameter 100 mm. The thermal resistance of the pipe wall is negligible. The ambient temperature is 10°C. Calculate the surface heat transfer coefficient for free convection.*

(b) *Such a pipe could transfer more energy if fins were used (Section 31.14). The effect of fins on the surface temperature will be small as the thermal conductivity of steel is large. Calculate the heat transfer rate if annular fins of twice the pipe diameter pitched at 100 fins/metre were fitted and the pipe had the same surface temperature and the same surface heat transfer coefficient but with 85 percent efficiency to allow for the fall in temperature to the fin perimeter (Figure 31.15 (c)).*

(a)

$$T = 77 + 273 = 350 \text{ K, hence } \beta = \frac{1}{350}.$$

$$Ra = Gr \cdot Pr = \frac{9.81}{350} \times \frac{0.1^3 (77 - 10)(1.009)^2}{(2.075)^2 \times 10^{-10}} \times 0.697 = 3.09 \times 10^6$$

This is a problem of air at a *moderate temperature* and the simplified relations of equation (31.32) apply.

Flow is laminar and $h_{\text{mean}} = 1.32 \left(\dfrac{\theta}{d}\right)^{1/4}$

$$= 1.32 \left[\frac{(77 - 10)}{0.1}\right]^{1/4}$$

$$= 6.715 \text{ W/m}^2 \text{ K}$$

This is a small heat transfer coefficient and the heat transfer will be small.

$$\dot{Q}' = 6.715 \times (\pi \times 0.1 \times 1)(77 - 10) = 141.3 \text{ W/m}$$

(b)

$$\dot{Q}' = 141.3 + h(A_{\text{fin}} \times 100 \times \theta \times \eta_{\text{fin}})$$

$$= 141.3 + 6.715(2 \times \pi/4(0.2^2 - 0.1^2)) \times 100 \times (77 - 10) \times 0.85$$

$$= 141.3 + 1\,802.3$$

Heat transfer rate with fins $= 1\,943.6$ W/m.

This example clearly shows why fins are important if heat transfer rates are low.

9. *A one shell pass, two tube pass heat exchanger of the design shown in Figure 31.18 (c) has water in the tube and engine oil in the shell. The oil enters* at $130°C$ and has to be cooled to $80°C$ before returning to the engine. Cooling water at $25°C$ is available at a flow rate of $5\,040$ kg/h and is be heated to $70°C$ for other uses. The overall heat transfer coefficient, U, is 310 W/m^2 K. (U could be determined if the surface heat transfer coefficient on the outside of the tube was calculated but this is beyond the scope of this book.) Calculate the heat transfer area needed.

From tables at $(70 + 25)/2$, $c_{p\text{water}} = 4.181$ kJ/kg K.

$$\dot{Q} = (mc_p\Delta T)_{\text{water}} = (5\,040/3\,600) \times 4.181(70 - 25) = 263 \text{ kW}$$

From the F chart of Figure 31.18(d),

$$R = (T_1 - T_2)/(t_2 - t_1) = (130 - 80)/(70 - 25) = 1.11; \text{contour on Figure 31.18}(c)$$

$$P = (t_2 - t_1)/(T_1 - t_1) = (70 - 25)/(130 - 25) = 0.429; \text{x axis on Figure 31.18}(c)$$

Hence $F = 0.88$ and $\dot{Q} = UAF\theta_{\text{LMTD}}$ where θ_{LMTD} is for *counterflow*

$$\theta_{\text{LMTD}} = \frac{(130 - 70) - (80 - 25)}{\ln[(130 - 70)/(80 - 25)]} = 58°C$$

Hence

$$A = \frac{\dot{Q}}{UF\theta_{\text{LMTD}}} = \frac{263}{310 \times 0.88 \times 58} = \underline{16.62 \text{ m}^2}$$

10. *A single tube counterflow heat exchanger is required to heat 1.2 kg/s of water from $10°C$ to $30°C$ for use in an office heating system. The office is adjacent to a workshop in which liquid at $35°C$ is available. The heat exchanger tube diameter is 30 mm and the walls are of negligible thickness offering no thermal resistance. The mean specific heat capacity of the liquid is 1.7 kJ/kg K and the mass flow rate is 3 kg/s. The heat transfer coefficient on the outside of the tube is 1.7 kW/m^2 K. Calculate the length of tube required to produce 2 kW of energy for heating. Figure 31.24 shows the temperature profile. Water properties at a mean temperature of $20°C$ are obtained from tables.*

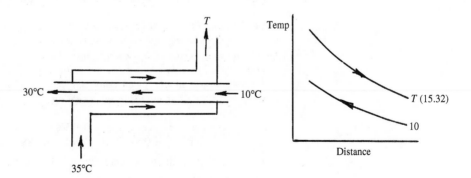

Figure 31.24

The heat transfer rate $\dot{Q}$ in the exchanger is

$$\dot{Q} = (\dot{m}C_p\Delta T)_{\text{water}} = (\dot{m}C_p\Delta T)_{\text{liquid}}$$

$$= 1.2 \times 4.182 \times (30 - 10) = 3 \times 1.7 \times (35 - T)$$

where T is the exit temperature of the hot fluid; $T = 15.32°C$.

Log mean temperature difference $= \dfrac{(\theta_1 - \theta_2)}{\ln(\theta_1/\theta_2)} = \dfrac{(5.32 - 5)}{\ln(5.32/5)} = 5.16 \text{ K}$

Now find the surface heat transfer coefficient on the inside (water side) of the tube. For this we need the Reynolds number $Re = \dfrac{\rho V d}{\mu}$.

$$\text{Water velocity } V = \frac{\dot{m}}{\rho A} = \frac{1.2 \times 4}{1\,000 \times \pi(0.03)^2} = 1.7 \text{ m/s}$$

$$Re = \frac{1\,000 \times 1.7 \times 0.03}{1.002 \times 10^{-3}} = 50\,898 \text{ (turbulent)}$$

Hence

$$Nu = 0.023\, Re^{0.8} Pr^{0.4} = 0.023 \times 50\,898^{0.8} \times 6.95^{0.4} = 285.9$$

But

$$Nu = (h \times d)/k \text{ so that } h = \frac{285.9 \times 0.603}{0.03 \times 10^3} = 5.75 \text{ kW/m}^2\text{K}$$

Since the tube walls are of negligible thickness, Section 31.18, equation (v) applies.

$$\frac{1}{U'} = \frac{1}{2\pi r h_1} + \frac{1}{2\pi r h_2} = \frac{1}{2\pi r}\left[\frac{1}{h_1} + \frac{1}{h_2}\right]$$

$$= \frac{1}{2\pi \times 0.015}\left[\frac{1}{1.7} + \frac{1}{5.75}\right]$$

i.e.

$$U' = 0.124\, \frac{\text{kW}}{\text{mK}}$$

Hence $\dot{Q} = U' L \theta_{\text{LMTD}} = 0.124 \times L \times 5.16$.
But $\dot{Q}$ must be 2 kW, hence the length of the tube $L = \underline{3.13 \text{ m.}}$

11. *A multitube condenser with 90 tubes is required to remove 1 050 kW using cooling water entering at 10°C and leaving at 20°C. The saturation temperature of the condensing substance is 30°C (Figure 31.25).*

The condenser tubes are 30 mm in diameter and the surface heat transfer coefficient on the outside (condensing) surface is 2 kW/m² K. Assuming that all the tubes achieve the same heat transfer, calculate the length of a tube.

For water at a mean temperature of 15°C; $\rho = 1\,000$ kg/m³, $Pr = 8.12$, $\mu = 1.152 \times 10^{-3}$ kg/ms, $k = 0.595$ W/m K, $c_p = 4.188$ kJ/kg K.

The heat transfer rate to the cooling water is 1 050 kW.

Thus $1\,050$ kW $= \dot{m} \times 4.188 \times (20 - 10)$, where $\dot{m}$ is the mass flow rate of coolant.

Thus $\dot{m} = 25.07$ kg/s.

The mass flow rate of coolant per tube $= 25.07/90 = 0.278$ kg/s.

The log mean temperature difference $\theta_{\text{LMTD}} = \dfrac{(20 - 10)}{\ln(20/10)} = 14.43 \text{ K}$

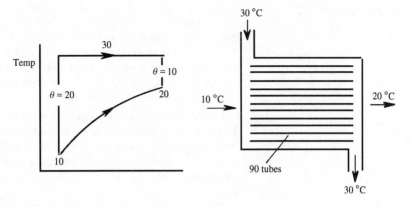

Figure 31.25

Heat transfer rate *per tube* $= (1\,050/90) = 11.67$ kW

For *one* tube, $Re = \rho Vd/\mu$ and $V = \dot{m}/A$, hence $Re = \dot{m}d/A$

But $A = \pi(d/2)^2$,   hence $Re = 4\dot{m}/(\pi d\mu)$

$$= (4 \times 0.278 \times 10^3)/(\pi \times 0.03 \times 1.152)$$

$$= 10\,242$$

(This is not an essential way of calculating Re which may also be found from $Re = \rho Vd/\mu$ but V will have to be calculated.)

$$Nu = 0.023Re^{0.8}Pr^{0.4} = 0.023(10\,242)^{0.8}(8.12)^{0.4}$$

$$= 0.023 \times 1\,615.5 \times 2.31$$

$$= 85.83 = hd/k = (h \times 0.03)/0.595$$

Thus $h = 1\,702.3$ W/m^2 K

The overall heat transfer coefficient per unit length for one tube is given by

$$\frac{1}{U'} = \frac{1}{2\pi 0.015}\left[\frac{1\,000}{1\,702.3} + \frac{1}{2}\right]$$

hence $U' = 0.086\,7$ kW/mK

The heat transfer per tube $= 11.67$ kW (see above) $= U'L\theta_{\text{LMTD}}$

$$\therefore 11.67 = 0.086\,7 \times L \times 14.43 \text{ (kW/mK)}$$

giving $\underline{L = 9.33 \text{ m}}$ for each tube.

Further problems

12. An off-peak electric heater in the form of a rectangular box is 1 m wide by 0.3 m deep by 0.8 m high. The surface of the heater, with the exception of the base and the bottom 0.1 m of all sides, is at a constant temperature of 80°C. The remaining parts of the surface are at the ambient temperature 17°C. Calculate the energy output of the heater over a 24 hour period and the rate of electrical energy input if power is supplied for 8 hours in every 24 hours. The surface heat transfer coefficient of the box is 6 W/m²K. [19.25 kWh; 2.4 kW]

13. The temperature of the inside surface of a pair of refrigerator doors is −15°C. The doors are made of an interior layer of plastic 5 mm thick, a layer of insulating foam 80 mm thick and a 1 mm layer of steel. Each door is 2.1 m high and 0.8 m wide. Calculate the heat transfer rate through the doors if the outside surface is at a temperature of 18°C. What is the temperature on the inside surface of the steel layer?

Thermal conductivity of plastic = 3.95 W/m K; thermal conductivity of insulating foam = 0.05 W/m K; thermal conductivity of steel = 26 W/m K.

[70 W; 17.992°C]

14. A steel pipe carries saturated steam at a pressure of 50 bar. The pipe has an inner diameter of 120 mm and walls 15 mm thick. The pipe is lagged with a layer of diatomaceous earth 20 mm thick, a layer of magnesia 40 mm thick and a layer of binding material 10 mm thick. Calculate the heat transfer rate per metre length from the pipe assuming that the inner wall of the pipe is at the steam temperature and the outer layer of the binding material is at the ambient temperature of 19°C.

Thermal conductivity of steel = 14.0 W/m K; thermal conductivity of diatomaceous earth = 0.036 W/mK; thermal conductivity of magnesia = 0.024 W/m K; thermal conductivity of binding material = 1.0 W/m K. [69.6 W/m].

15. Steam flows through a pipe 3 000 m long with an entry point 2 500 m below the exit. The steam enters the pipe at 60 bar, 500°C and leaves at 50 bar, 450°C. The steam mass flow rate is 11 500 kg/h. The pipe, which is 0.5 m inner diameter is insulated with material 0.75 m outer diameter and the mean surface temperature of the insulation is 8°C greater than the temperature of the surroundings. Calculate the surface heat transfer coefficient of the insulation. [4.53 W/m²K]

16. A furnace wall 0.25 m thick is built of brick having thermal conductivity 0.8 W/m K. The temperature of the surroundings is 20°C and of the furnace wall 1 100°C. The heat losses are too great and a layer of insulation 0.25 m thick is added to the outside with thermal conductivity 0.061 W/m K. The combined surface heat transfer coefficient for radiation and convection from the insulation is 7.2 W/m²K.

Calculate the temperature of the outside surface of the insulation and the heat transfer rate per unit area. Assuming that the surface heat transfer coefficient remained unchanged, what percentage saving in heat loss has been achieved by adding the insulation? [53°C; 238 W/m²; 90 %]

17. An electric fire element which may be considered an ideal black surface is 0.25 m long and 18 mm in diameter. Electrical energy is supplied to the element at a rate of 1 kW. Assuming that 95 % of the energy supplied is dissipated by radiation and that radiation from the surroundings to the element is negligible, calculate:

(i) the temperature of the element;
(ii) the radiation heat transfer coefficient in surroundings at 20°C;
(iii) the convection heat transfer coefficient in surroundings at 20°C.

The Stefan–Boltzmann constant is 5.67×10^{-8} W/m^{2}K^4.

[770°C; 89.7 W/m^2 K; 4.72 W/m^{2}K]

18. A cylindrical wood burning stove 0.6 m high and 0.3 m diameter is placed on a plinth at the centre of a large room. The surface temperature of the stove is 200°C and the emissivity of the surface is 0.6. The room is kept at a steady temperature of 17°C by the stove. Ignoring the heat transfer from the stove base and plinth, calculate:

(i) the heat transfer rate to the room by radiation;
(ii) the heat transfer rate to the room by convection if the surface heat transfer coefficient for convection is 20 W/m^2 K;
(iii) the enhanced surface heat transfer coefficient allowing for both convection and radiation effects.   [930 W; 2 328 W; 28 W/m^2 K]

19. (i) Water flows at 3 m/s through a long pipe 25 mm in diameter. The water is being *cooled* and the mean bulk temperature is 40°C. Estimate the surface heat transfer coefficient in turbulent flow between the water and the pipe using:

(a) the Dittus–Boelter equation $Nu = 0.023\,Re^{0.8}Pr^{0.3}$ for cooling provided $Re > 6\,000$;
(b) the Reynolds analogy $St = Nu/(Re \cdot Pr) = f/2$;
(c) the Reynolds–Colburn analogy $St\,Pr = f/2$.

[(a) 10.012 kW/m^{2}K; (b) 26.7 kW/m^{2}K; (c) 10.085 kW/m^{2}K]

(ii) Calculate the mean surface heat transfer coefficient for dry saturated steam at a temperature of 70°C condensing on the surface of a vertical plate of height 0.8 m having a surface temperature of 10°C.   [3.487 kW/m^{2}K]

(iii) If the plate is inclined at 50° to the horizontal, what would be the value of the mean surface heat transfer coefficient?   [3.262 kW/m^{2}K]

20. A heat exchanger in which 0.4 kg/s of water at a mean temperature of 20°C is used to cool oil at a mean temperature of 61°C, consists of a flat plate 0.21 m wide by 2 m long separating the two flows. The change in temperature of the water is 12 K and the oil temperature is to be reduced by 38°C.

Calculate the permissible oil flow rate and the overall heat transfer coefficient between the two streams. If the resistance of the plate is negligible and the surface heat transfer coefficient on the water side is 2.3 kW/m^{2}K, calculate the surface heat transfer coefficient on the oil side. (LMTD not required.)

Specific heat capacity at constant pressure of water, $c_p = 4\,178$ J/kg K;
Specific heat capacity at constant pressure of oil, $c_p = 2\,131$ J/kg K.

[0.247 kg/s; 1.16 kW/m^{2}K; 2.36 kW/m^{2}K]

21. Saturated steam at a pressure of 0.08 bar condenses to become saturated liquid in a thin condenser tube. The steam mass flow rate is 0.01 kg/s and the condenser cooling water is at a mean temperature of 20°C. The condenser tube is 10 m long and 40 mm inside diameter with a wall thickness of 2 mm.

Calculate the surface heat transfer coefficient between the tube and the steam if conduction resistance through the tube is negligible and the surface heat transfer coefficient between the water and the tube is 1.5 kW/m^{2}K. (LMTD not required.)

[1.92 kW/m^{2}K]

22. A heat exchanger cools oil from 180°C to 60°C with water which rises in temperature from 10°C to 50°C. What are the LMTD values in parallel and counter flow? What will be the ratio of area needed in parallel flow to that needed in counter flow?

Using the specific heat values of example 20, determine the mass flow rate of water that will cool 0.6 kg/s of oil and the minimum area required with an overall heat transfer coefficient of 1.98 kW/m^{2}K.   [56°; 83.7°; 1.48; 0.92 kg/s; 0.93 m^2]

23. A heater is required to keep the air temperature in a well insulated, small room at 20°C. A vertical, flat plate, electric heater is chosen. The heater is 0.5 m high and 1 m wide and has a surface temperature regulated to 52°C. The heater surface has 100 vertical fins equally spaced across the width of the heater and the fins have an efficiency of 85% (Figure 31.15(b)).

The back of the heater is well insulated and it may be assumed that no energy passes through that side of the heater which is against the wall.

The fins are 30 mm high and 3 mm wide. The heater has a remote sensor in the room placed so that the whole room reaches the required temperature. Calculate the heat transfer rate to the room when the heating is about to be cut off by the sensor. Assume the equations (31.32) apply.   [329.7 W]

24. A one shell pass, two tube pass heat exchanger of the type shown in Figure 31.18(c) heats 2.5 kg/s of water in the tube from 25°C to 85°C. The energy is obtained from hot oil cooling in the shell. The oil enters at 150°C and leaves at 72°C. The overall heat transfer coefficient is 270 W/m^{2}K.

Calculate the area needed to achieve this task.   [59.82 m^2]

25. Hot gas flows through a tubular heat exchanger at 0.275 kg/s. It is cooled by water from 380°C to 100°C. The water flow rate is 1 440 kg/h and the entry temperature of the water is 10°C. The specific heat capacity at constant pressure of the gases is 1.125 kJ/kg K and of the water is 4.19 kJ/kg K. The overall heat transfer coefficient is 250 W/m^{2}K. Calculate the area required in (a) counter flow and (b) parallel flow.   [1.92 m^2; 2.37 m^2]

26. Dry saturated steam at a pressure of 0.28 MN/m^2 is condensed on the outside of a tube of diameter 52 mm. Assume that the tube walls are thin and have no thermal resistance. The outside (condensing side) surface heat transfer coefficient is 18 kW/m^{2}K. Cooling water enters the tube at 25°C and leaves at 85°C. The condensate is a saturated liquid. Calculate the mass flow rate of condensate and the length of tube required.   [0.312 kg/s; 15.32 m]

27. A multitube heat exchanger has 20 thin walled tubes of 25 mm diameter contained in the shell. Air in the tubes flows at a *total* mass flow rate of 0.3 kg/s, entering at a temperature of 2°C and leaving at 27°C. The velocity of the air in the tubes is 4 m/s. The fluid in the shell is condensing at 40°C and the surface heat transfer coefficient on the outside of the tube is 0.3 kW/m^{2}K. Calculate the length of each tube.   [2.65 m]

28. A multitube counter flow heat exchanger cools 6 000 kg/h of hot oil which enters the shell at 120°C and leaves at 60°C. The specific heat capacity of the oil is 2.1 kJ/kg K. Twenty smooth tubes of 30 mm diameter are contained in the shell and water enters these tubes at 20°C. The total mass flow rate of water is 1.26 kg/s. The tube walls are thin and have negligible thermal resistance. The surface heat transfer coefficient on the outside of the tubes is 0.8 kW/m^{2}K. Calculate the length of each tube.   [6.16 m]

32 Pressure and pressure measurement

32.1 Introduction

A fluid is defined as a substance which cannot resist shear stress and thus offers no resistance to change of shape. Fluids may be divided into liquids and gases; a quantity of liquid occupies a fixed volume, has a free horizontal surface and is virtually incompressible, whereas a quantity of gas completely fills the enclosing vessel and is easily compressed.

32.2 Pressure exerted by a fluid

Since a fluid cannot resist shear stress, the pressure exerted by a fluid on any surface is always normal to that surface.

Consider a small triangular element ABC at a point in a fluid, Figure 32.1. If the area of the face AC is A, then the area of AB is $A \sin \theta$ and that of BC is $A \cos \theta$. If the pressures on AC, AB and BC are p_1, p_2 and p_3 respectively, then, for horizontal equilibrium,

$$p_1 A \times \sin \theta = p_2 \times A \sin \theta, \quad \text{i.e.} \ p_1 = p_2$$

and for vertical equilibrium

$$p_1 A \times \cos \theta = p_3 \times A \cos \theta, \quad \text{i.e.} \ p_1 = p_3$$

Thus the pressure at a point in a fluid is the same in all directions.

Figure 32.2 shows a column of fluid at rest, the pressure on the upper and lower surfaces being p_0 and p respectively. If the density of the fluid is ρ, then for vertical equilibrium,

$$pA = p_0 A + \rho g A h$$

i.e. $$p = p_0 + \rho g h \qquad (32.1)$$

p is the total, or *absolute*, pressure and if p_0 represents atmospheric pressure, then $p - p_0$ is the *gauge* pressure.

Equation (32.1) shows that the pressure increases with depth but this variation is ignored in the case of gases due to the low densities involved.

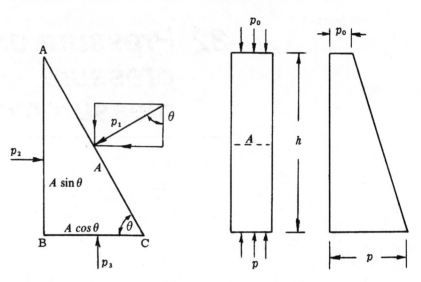

Figure 32.1   **Figure 32.2**

The only important exception is in the variation of pressure with altitude in the earth's atmosphere (Section 33.6)

In the case of liquid pressure on a surface, it is only the gauge pressure which is relevant since the atmospheric pressure also acts on the reverse side of the surface. For such applications, the effective pressure at a depth h is merely that required to balance the weight of the liquid above,

i.e. $$p = \rho gh \qquad (32.2)$$

32.3 Units of pressure and density

The basic unit of pressure is the N/m^2, which is termed the pascal (Pa). The pressure exerted by the atmosphere is $101\,325\ N/m^2$, so that the pascal represents a very small pressure. More practical units are kN/m^2, and the bar (1 bar $= 10^5\ N/m^2$). It will be seen that atmospheric pressure is approximately 1 bar and is usually taken as such unless otherwise stated.

The basic unit of density is the kg/m^3, the density of water being $10^3\ kg/m^3$ or $1\ t/m^3$. Since $1 m^3 = 10^3 l$, the density of water is also $1\ kg/l$.

Relative density is the ratio $\dfrac{\text{density of substance}}{\text{density of water}}$. Thus oil of relative density 0.8 has a density of $800\ kg/m^3$.

Equation (32.2) will give pressure in N/m^2 when ρ is in kg/m^3, g is in m/s^2 and h is in metres. It is often convenient to express pressure in terms of a height (or *head*) of liquid, given by $h = p/\rho g$. Thus an atmospheric pressure of $101.3\ kN/m^2$ is the pressure exerted by a column of water 10.34 m high or by a column of mercury (relative density 13.6) 760 mm high.

32.4 Archimedes principle

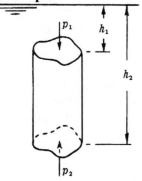

Figure 32.3

Figure 32.3 shows a solid body of cross-sectional area A immersed in a fluid. The horizontal forces on the body are in equilibrium but, in the vertical direction,

$$\text{force on upper surface} = p_1 A = \rho g h_1 A$$

$$\text{and force on lower surface} = p_2 A = \rho g h_2 A$$

$$\therefore \text{ resultant upthrust} = \rho g A(h_2 - h_1)$$

$$= \text{weight of displaced fluid}$$

This is known as *Archimedes principle*.

The upthrust acts through the centre of gravity of the displaced fluid.

32.5 Manometers

A manometer is a simple form of pressure gauge which indicates the pressure in a pipe or vessel by the displacement of a liquid column. The following are the most common types of manometer and for simplicity, it will be assumed that the liquid whose pressure is being measured is water.

(a) Piezometer tube, Figure 32.4

This is simply a glass tube inserted in the top of the pipe; the pressure then forces some water up the tube until equilibrium is attained.

The pressure is given by h m of water or $10^3 \times 9.81h$ N/m^2.

(b) U-tube, Figure 32.5

If the pressure in the pipe is too high to use a piezometer tube, a U-tube filled with a heavy liquid, such as mercury, is used. If the relative density of

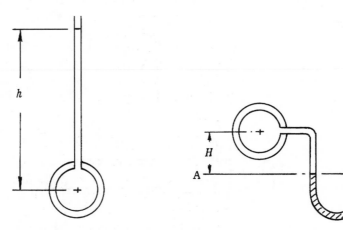

Figure 32.4   **Figure 32.5**

the liquid in S, then x m of the liquid exerts the same pressure as Sx m of water.

Hence, equating pressures on the two sides at level AA,

$Sx = H + h$ where h is the pressure in the pipe in metres of water.

i.e. $$h = Sx - H \qquad (32.3)$$

If a U-tube is used for measuring gas pressure, water or a lighter liquid may be used in the tube, depending upon the magnitude of the pressure being measured. In such an application the height H is irrelevant due to the low density of the gas.

(c) U-tubes for pressure difference

To measure small difference in pressure between two points in a pipe, an inverted U-tube is used, Figure 32.6.

The difference in pressure between points 1 and 2 is given directly by

$$h_1 - h_2 = h \qquad (32.4)$$

The sensitivity of such a manometer can be increased by filling the space above the water columns with a liquid lighter than water. If the relative density of this liquid is S, the difference of pressure is represented by x m of water, less Sx m of liquid.

i.e. $$h_1 - h_2 = x(1 - S) \qquad (32.5)$$

For the measurement of large pressure differences between two points, the U-tube is arranged below the pipe and is filled with mercury or other heavy liquid, Figure 32.7. If the specific gravity of the liquid is S then

$$h_1 - h_2 = (S - 1)x \qquad (32.6)$$

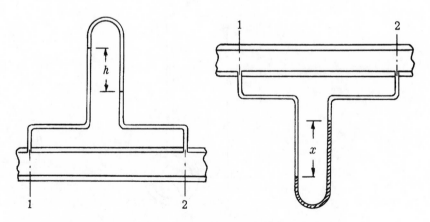

Figure 32.6   Figure 32.7

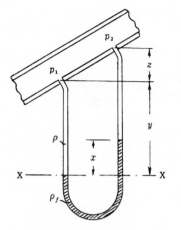

Figure 32.8

(d) General case of differential manometer

Let the density of the liquid in the pipe or vessel be ρ and that of the manometer liquid be ρ_f. If the difference between the levels of the measuring points is z and the difference in levels of manometer liquid is x, Figure 32.8, then, equating pressures at level XX,

$$p_1 + \rho g y = p_2 + \rho g(y + z - x) + \rho_f g x$$

i.e.
$$p_1 - p_2 = \rho g z + (\rho_f - \rho)g x$$

or
$$\frac{p_1 - p_2}{\rho g} = z + \left(\frac{\rho_f}{\rho} - 1\right) x \qquad (32.7)$$

If the ends of the manometer are at the same level and the liquid in the pipe is water, this reduces to equation (32.6).

This type of manometer may be used for a range of pressures by choosing the ratio ρ_f/ρ to give a reasonable value for x. It may also be modified by varying the cross-sectional area of the limbs (see Example 2). In the application shown in Figure 32.9, one limb has been made very large so that the liquid level remains virtually constant and only one limb reading need be taken. In Figure 32.10, the reading limb is inclined to increase the sensitivity.

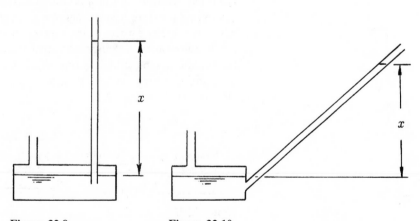

Figure 32.9   **Figure 32.10**

32.6 Bourdon gauge

For pressure measurements where high precision is not required, a Bourdon gauge is commonly used, Figure 32.11. The pressure is applied to a closed bent tube of elliptical cross-section which tends to straighten and this movement is used to move a needle round a dial through a toothed sector and pinion. The dial is normally calibrated to read zero at atmospheric pressure and hence indicates gauge pressure.

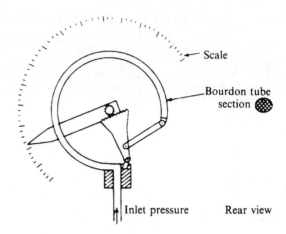

Figure 32.11

32.7 Pressure transducers

When pressure is changing, it is often necessary to record the pressure continuously over a period of time. A barograph is a simple device used to record atmospheric pressure but when changes are rapid, this simple bellows and mechanical linkage is inadequate and quick response devices are used which have an electrical output for recording on an oscilloscope, pen recorder or $u - v$ recorder or for feeding into a data logger, controller or microprocessor. A device which takes in a pressure signal and produces an electrical output signal proportional to the pressure is called a *transducer*.

In some pressure transducers, the fluid pressure acts on a diaphragm, to which is fitted a resistance strain gauge. Stretching of the diaphragm due to the pressure causes a change in the resistance of the strain gauge which, with the aid of a suitable circuit, produces an electric potential proportional to the pressure. The potential can then be applied to an oscilloscope, controller, etc., as shown in Fig 32.12.

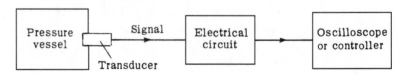

Figure 32.12

Other transducers use change in capacitance or impedance of the sensing device or make use of a piezoelectric crystal to produce a small electric signal under the action of the applied pressure.

32.8 Micromanometers

For the measurement of very small pressure differences, a micromanometer is used. Two liquids are used and the ends are enlarged to give a large miniscus movement for a small pressure difference.

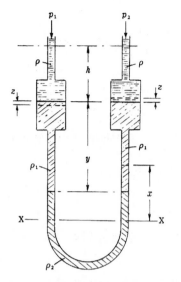

Figure 32.13

Figure 32.13 shows such a manometer. The density of the fluid under investigation is ρ and those of the manometer liquids are ρ_1 and ρ_2. The cross-sectional areas of the tube and reservoir are a and A respectively.

Let the applied pressures be p_1 and p_2, where $p_1 > p_2$. If this results in a movement z in the reservoir levels and a difference of levels in the tube x, then

$$a \times \frac{x}{2} = A \times z$$

or

$$z = \frac{ax}{2A} \qquad (32.8)$$

If the initial level of the common surface below the reservoir levels is y, then, equating pressures at the new common surface at level XX,

$$p_1 + \rho g(h + z) + \rho_1 g \left(y + \frac{x}{2} - z \right)$$

$$= p_2 + \rho g(h - z) + \rho_1 g \left(y - \frac{x}{2} + z \right) + \rho_2 g x$$

which simplifies to

$$p_1 - p_2 = g\{(\rho_2 - \rho_1)x + 2(\rho_1 - \rho)z\}$$

$$= gx \left\{ (\rho_2 - \rho_1) + (\rho_1 - \rho) \frac{a}{A} \right\} \quad (32.9)$$

substituting for z from equation (32.8).

If the ratio a/A is very small, this reduces to

$$p_1 - p_2 = gx(\rho_2 - \rho_1) \qquad (32.10)$$

Thus, if the difference of the manometer liquid densities, $\rho_2 - \rho_1$, is small, the reading x is large for a small applied pressure difference, $p_1 - p_2$.

To improve accuracy in reading this type of instrument, a graduated telescope may be used to view the meniscus and the whole instrument may be fitted with a micrometer screw to return the meniscus to its original position. The pressure difference is then obtained from the micrometer movement. Examples of these specialist instruments are the Chattock gauge and the Krell gauge, which can measure pressure differences as small as 0.02 N/m^2 (0.002 mm H$_2$O).

Worked examples

1. *A cylindrical buoy floats in sea water with its axis vertical so that it is two-thirds submerged. The buoy is 0.8 m in diameter and 2 m in height and is fabricated from iron plate 12 mm thick. Calculate the mass of iron chain securing the buoy. The density of iron is 7700 kg/m^3 and of sea water is 1030 kg/m^3.*

Figure 32.14 shows the buoy in equilibrium under the action of the weight W, the upthrust U and the chain force F.

The upthrust is equal to the weight of water displaced,

i.e.

$$U = \frac{\pi}{4} \times 0.8^2 \times \frac{2}{3} \times 2 \times 1030 \times 9.81 = 6750 \text{ N}$$

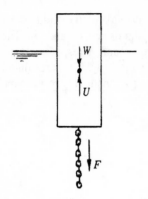

Figure 32.14

Weight of buoy,

$$W = \left(\pi \times 0.8 \times 2 + 2 \times \frac{\pi}{4} \times 0.8^2\right) \times \frac{12}{10^3} \times 7\,700 \times 9.81$$

$$= 5\,450 \text{ N}$$

$$\therefore F = U - W$$

$$= 6\,750 - 5\,450 = 1\,300 \text{ N}$$

The chain force is the difference between the weight of the chain and the upthrust on the chain. If the volume of the chain is V, then

$$1\,300 = 7\,700 \times 9.81 \times V - 1\,030 \times 9.81 \times V$$

from which $V = 0.0199 \text{ m}^3$

Therefore mass of chain $= 7\,700 \times 0.0199 = \underline{152 \text{ kg}}$

2. *A pressure gauge for the measurement of small pressures consists of a vertical glass U-tube with internal diameter 10 mm. The upper ends of the tube are enlarged to form reservoirs of diameter 60 mm. One limb contains water and is connected to the unknown pressure while the other contains paraffin and is open to the atmosphere. Calculate the gauge pressure applied to the water limb when the water-paraffin meniscus is displaced 10 mm.*
Density of paraffin $= 850 \text{ kg/m}^3$.

Let the initial level of the common surface be at XX, Figure 32.15, and let the heights of the paraffin and water columns above this level be h_p and h_w respectively.

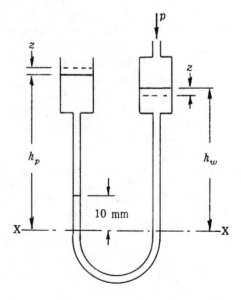

Figure 32.15

Equating pressures at XX:

$$\rho_p g h_p = \rho_w g h_w$$

$$\therefore h_w = \frac{850}{1\,000} h_p$$

$$= 0.85 h_p$$

When the common surface is displaced 10 mm,

$$z = 10 \times \frac{\frac{\pi}{4} \times 10^2}{\frac{\pi}{4} \times 60^2}$$

$$= \frac{10}{36} \text{ mm}$$

$$= 0.000\,278 \text{ m}$$

Equating gauge pressures at level XX when the pressure p is applied to the water limb,

$$\rho_p g(h_p + z - 0.01) + \rho_w g \times 0.01 = \rho_w g(h_w - z) + p$$

i.e. $850g(h_p + 0.000\,278 - 0.01) + 1\,000g \times 0.01 = 1\,000g(0.85h_p$

$$- 0.000\,278) + p$$

from which $p = \underline{19.77 \text{ N/m}^2}$

This is a pressure of approximately 0.000 2 atmosphere.

3. *An open vertical cylinder has a connection from the bottom to one arm of a vertical U-tube containing mercury, the other arm being open to atmosphere. When the cylinder is empty, the level of the mercury in each of the tubes is initially 2 m below a fixed point A on the cylinder. Water is now poured into the cylinder up to the level of A. Determine the difference of levels of mercury in the two tubes.*

The cylinder diameter is 24 mm and the diameter of the tube is 6 mm. If a frictionless plunger weighing 40 N is now inserted into the cylinder, determine the new difference in mercury levels, assuming no leakage.

Relative density of mercury = 13.6.

If the difference in levels of mercury is x, Figure 32.16 (a), then, equating pressures at level XX,

$$2 + \frac{x}{2} = 13.6x$$

$$\therefore x = \underline{0.152\,7 \text{ m}}$$

When the plunger is added, let the new difference in levels be y, Figure 32.16 (b). Then surface in cylinder falls a distance z such that

$$\frac{\pi}{4} \times 0.024^2 \times z = \frac{\pi}{4} \times 0.006^2 \times \left(\frac{y}{2} - \frac{x}{2} \right)$$

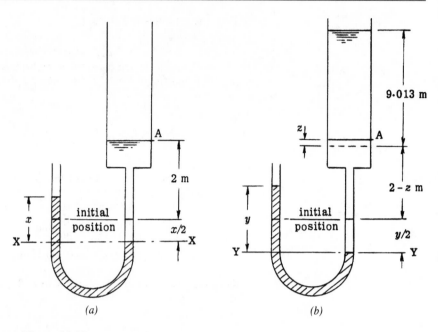

Figure 32.16

from which
$$z = \frac{y - 0.1527}{32}$$

The effect of the weight of the plunger is equivalent to adding further water of the same weight. Hence, if the additional depth of water is h,

$$\frac{\pi}{4} \times 0.024^2 \times h \times 10^3 \times 9.81 = 40$$

from which
$$h = 9.013 \text{ m}$$

Equating pressures at level YY,

$$9.013 + (2 - z) + \frac{y}{2} = 13.6y$$

i.e.
$$11.013 - \frac{y - 0.1527}{32} = 13.1y$$

from which
$$y = \underline{0.839 \text{ m}}$$

Note: If the small movement z is ignored, $y = \underline{0.841 \text{ m}}$

Further problems

4. A hollow float is in the form of a closed cylinder of diameter 140 mm and height 80 mm. It is made from thin sheet metal weighing 120 N/m². What force is necessary just to submerge the float completely in paraffin of relative density 0.8?
[1.75 N]

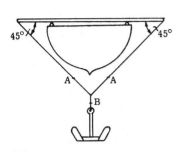

Figure 32.17

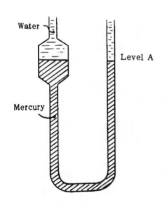

Water

Level A

Mercury

Figure 32.18

5. A barge has a mass of 100 t.

(a) What will be the immersed volume when floating in fresh water?

(b) If the cross-sectional area at the water-line is 80 m², what load would depress the barge 0.5 m, assuming the sides of the barge to be vertical?

(c) By how much will the loaded barge rise if it moves from fresh water to sea water, of relative density 1.025?   [100 m³; 39.24 kN; 0.042 7 m]

6. In order to lay out an anchor of mass 600 kg, it is suspended below a boat by a wire strop, as shown in Figure 32.17. The density of the anchor material is 7 000 kg/m³. Determine the tensions in the strop at points A and B. The density of sea water is 1 030 kg/m³.   [5.015 kN; 3.54 kN]

7. A small sunken vessel is to be raised by filling the hold with watertight drums of 5 m³ capacity. The drums each have a mass of 120 kg and the vessel commences to rise when 70 of the drums have been pumped out. The total number of drums in the hold is 100. Calculate the mass of the vessel, neglecting the volume of metal in the vessel and drums. The density of sea water is 1 025 kg/m³.   [346.5 t]

8. A mercury U-tube manometer is used to measure the difference in pressure across a flowmeter in a horizontal pipe, the water from the pipe being in contact with the mercury in the manometer. The meter determines the flow rate of water according to the relation $Q = 0.26\sqrt{h}$ where Q is the flow rate in m³/s and h is the difference in head across the meter in metres of water.

Calculate the manometer reading when the flow rate is 0.7 m³/s. The relative density of mercury is 13.6.   [0.573 m]

9. A differential pressure gauge for the measurement of small air pressures consists of a vertical glass U-tube having a uniform cross-section of internal diameter 6 mm. The upper ends of the tube are enlarged to a diameter of 48 mm. The gauge is filled with water and kerosene and the free surfaces of the liquids are in the enlarged ends. The air pressure to be measured is applied to the top of the limb containing water, the other limb being open to atmosphere. Calculate the gauge pressure of the air which will cause a displacement of the meniscus of 25 mm from the balance position. The relative density of kerosene is 0.8.   [55.95 N/m²]

10. Figure 32.18. shows a special manometer for use with a flowmeter which determines the flow rate in a horizontal pipe according to the relation $Q = 0.07\sqrt{h}$ where Q is the flow rate in m³/s and h is the difference in head across the meter in metres of water. By how much will level A move when the flow rate is 0.08 m³/s? The diameter of the larger limb is 3.8 times that of the smaller limb and the relative density of the mercury is 13.6.   [97.1 mm]

11. A manometer of the type shown in Figure 32.15 has a U-tube of 12 mm bore and enlarged ends of 60 mm diameter. The liquids used in the manometer are oil of density 890 kg/m³ and water of density 1 000 kg/m³. An air pressure 25 N/m² greater than atmospheric is applied to the water limb and the oil limb is left open to the atmosphere. Calculate the movement of the oil–water meniscus.

[13.73 mm]

12. A manometer of the type shown in Figure 32.13 has a U-tube of 12 mm bore and enlarged ends of 60 mm diameter. The liquids used in the manometer are oil of density 890 kg/m³ and water of density 1 000 kg/m³. A pressure difference of 25 kN/m² is applied by a gas of density 1.1 kg/m³ to the liquid in the enlarged ends. Calculate the reading of the manometer.   [17.4 mm]

33 Hydrostatics

33.1 Forces on immersed surfaces

Since the pressure in a liquid increases with the depth, the resultant force on an immersed surface will depend on the average pressure and will always act at a point below the centroid of the area.

33.2 Total force on a vertical surface

Figure 33.1 shows a vertical rectangular surface of breadth b extending to a depth h below the surface. The pressure distribution diagram is a triangle, of maximum ordinate $\rho g h$.

The total force F is the average ordinate of the triangle (which is the pressure at the centroid G), multiplied by the area bh.

i.e. $$F = \tfrac{1}{2}\rho g \times bh = \tfrac{1}{2}\rho g b h^2 \qquad (33.1)$$

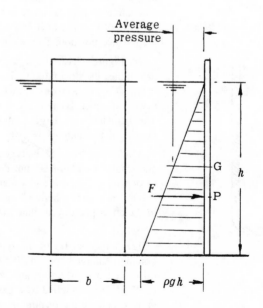

Figure 33.1

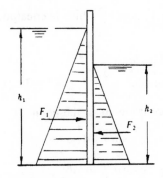

Figure 33.2

This force acts through the centroid of the triangle, which is at a depth $2h/3$ below the surface. The point of application of the force, F, is called the *centre of pressure*.

If a vertical surface has water on both sides, such as a lock gate, Figure 33.2, then

$$F_1 = \tfrac{1}{2}\rho g b h_1^2 \quad \text{acting at } 2h_1/3 \text{ below the surface}$$

and

$$F_2 = \tfrac{1}{2}\rho g b h_2^2 \quad \text{acting at } 2h_2/3 \text{ below the surface}$$

The resultant force, $F = F_1 - F_2$ and the height h of its point of application from the base is obtained by taking moments about the base,

i.e.

$$Fh = F_1\frac{h_1}{3} - F_2\frac{h_2}{3}$$

As $h_2 \to h_1$, $F \to 0$ and $h \to \dfrac{h_1}{3}$.

33.3 Total force on any plane surface

Figure 33.3 shows a plane surface of area A immersed in a liquid, the surface being inclined at an angle θ to the horizontal.

Pressure on elementary strip at distance x from O

$$= \rho g\, x \sin\theta$$

$$\therefore \text{ force on strip} = \rho g\, x \sin\theta \times b\, dx$$

$$\therefore \text{ force on surface} = \rho g \sin\theta \int xb\, dx$$

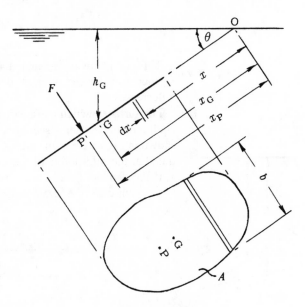

Figure 33.3

But $\int xb\,dx$ is the first moment of the area A about O, which is equal to Ax_G.

Thus
$$F = \rho g\, Ax_G \sin\theta = A(\rho g\, h_G)$$
$$= \text{area} \times \text{pressure at centroid} \qquad (33.2)$$

To find the line of action of F, moments are taken about the point O.

$$\text{Moment of force on strip about O} = \rho g\, x \sin\theta\, b\,dx \times x$$

$$\therefore \text{ total moment about O} = \rho g \sin\theta \int bx^2\,dx$$

But $\int bx^2\,dx$ is the second moment of the area about the point O, I_O

i.e.
$$\text{total moment about O} = \rho g \sin\theta I_O$$

Also
$$\text{moment of } F \text{ about O} = Fx_p = \rho g Ax_G \sin\theta \cdot x_p$$

$$\therefore x_p = \frac{\rho g \sin\theta I_O}{\rho g \sin\theta Ax_G} = \frac{I_O}{Ax_G}$$

i.e.

$$\text{distance of centre of pressure from O} = \frac{\text{second moment of area about O}}{\text{first moment of area about O}}$$

$$(33.3)$$

By the theorem of parallel axes (see Appendix A),
$$I_O = I_G + Ax_G^2 = A(k_G^2 + x_G^2)$$
$$\therefore x_p = \frac{A(k_G^2 + x_G^2)}{Ax_G} = x_G + \frac{k_G^2}{x_G}$$

Thus x_p is always greater than x_G but approaches x_G (i) if the distance x_G increases due to increased depth of immersion and (ii) if the inclination θ decreases.

For surfaces which can be divided into triangles, use can be made of the equimomental system (see Appendix A). For the vertical triangular surface shown in Figure 33.4, the area A is replaced by areas $A/3$ imagined concentrated at the mid-points of the sides.

If the depths of these areas below the surface are α, β and γ, the depth of the centre of pressure h_p is given by

$$h_p = \frac{\dfrac{A}{3}\alpha^2 + \dfrac{A}{3}\beta^2 + \dfrac{A}{3}\gamma^2}{\dfrac{A}{3}\alpha + \dfrac{A}{3}\beta + \dfrac{A}{3}\gamma}$$

$$= \frac{\alpha^2 + \beta^2 + \gamma^2}{\alpha + \beta + \gamma}$$

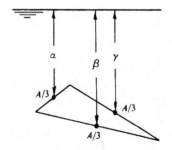

Figure 33.4

33.4 Forces on curved surfaces

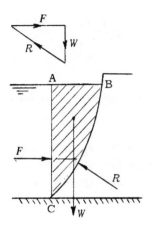

Figure 33.5

The force on a curved surface is obtained by considering the equilibrium of a wedge of fluid bounded by the curved surface and other suitably chosen plane surfaces.

For the curved surface of a dam, Figure 33.5, the water contained within the wedge ABC is in equilibrium under its weight W, the force F exerted on the face AC by the adjacent water, and the reaction of the curved surface BC. The magnitude and direction of the reaction R is obtained from the triangle of forces and its position is determined by the concurrency of R with the forces F and W. The force exerted *on* the dam is then equal and opposite to that exerted *by* the dam on the wedge ABC.

For a cylindrical or spherical surface, it is useful to note that, since the reaction R is normal to the surface, it must pass through the centre of curvature. Thus, for the sector gate shown in Figure 33.6, the wedge ABC is in equilibrium under the horizontal force F exerted on the face AB by the adjacent water, the upward force $V - W$ where V is the vertical reaction of the base BC and W is the weight of water enclosed in the wedge and the reaction R. The magnitude and direction of R is determined by the triangle of forces as before but it is unnecessary to determine the position of the vertical force $V - W$ since R must pass through the point O.

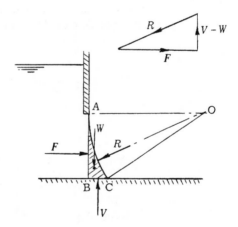

Figure 33.6

If such a gate is hinged at O, no torque will be required to overcome the fluid force when the gate is opened or closed.

When a surface is curved in one plane only, as in the foregoing examples, it is usual to consider the forces acting on a unit length of surface.

For the hemispherical body shown in Figure 33.7, the forces acting on it are the same as those acting on the liquid displaced, these being the horizontal force F acting on the face AB, the weight W of the liquid displaced and the reaction R on the curved surface, which passes through the centre O. It is again unnecessary to determine the position of the line of action of W.

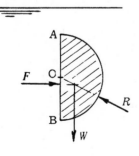

Figure 33.7

33.5 Surfaces subjected to gas pressure

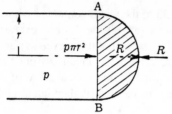

Figure 33.8

The density of a gas is small and so the pressure exerted on a plane or curved surface by a gas is assumed to be uniform. Thus for a plane surface

$$\text{force} = \text{pressure} \times \text{area}$$

and this acts at the centroid of the area.

For a curved surface such as the hemispherical end of the gas cylinder shown in Figure 33.8, the gas enclosed within the end is in equilibrium under the force on AB and the reaction of the curved surface,

i.e. $$R = p\pi r^2$$

A force equal and opposite to R, shown dotted, represents the force exerted *by* the gas *on* the surface.

33.6 The atmosphere

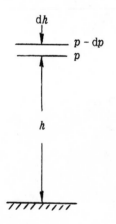

Figure 33.9

Air may be regarded as a perfect gas and when calculating the effect of altitude on pressure, it is assumed that the temperature falls at the rate of $0.0065°C/m$ (called the *lapse rate*). The relation between pressure and density is given by

$$p = c\rho^n$$

where c is a constant and n has the value 1.235.

Let the pressure at a height h above sea-level be p and let sea-level pressure, density and temperature be p_0, ρ_0 and T_0 respectively. If the pressure falls by dp as the height increases by dh, Figure 33.9,

$$dp = -\rho g\, dh$$

or

$$dh = -\frac{dp}{\rho g}$$

$$\frac{p}{\rho^n} = c \quad \text{so that} \quad \frac{1}{\rho} = \frac{c^{1/n}}{p^{1/n}}$$

Hence

$$dh = -\frac{c^{1/n}}{g}\left(\frac{dp}{p^{1/n}}\right)$$

Integrating

$$h = -\frac{c^{1/n}}{g}\left[\frac{p^{\frac{n-1}{n}}}{\frac{n-1}{n}}\right]_{p_0}^{p}$$

$$= \frac{n}{n-1}\frac{c^{1/n}}{g}\left[p_0^{\frac{(n-1)}{n}} - p^{\frac{(n-1)}{n}}\right]$$

But

$$c^{1/n} = \frac{p^{1/n}}{\rho} = \frac{p_0^{1/n}}{\rho_0}$$

$$\therefore h = \frac{n}{n-1}\left[\frac{p_0}{\rho_0 g} - \frac{p}{\rho g}\right]$$

$$= \frac{n}{n-1}\frac{p_0}{\rho_0 g}\left[1 - \frac{p}{p_0}\cdot\frac{\rho_0}{\rho}\right]$$

But

$$\frac{\rho_0}{\rho} = \left(\frac{p_0}{p}\right)^{1/n}$$

$$\therefore h = \frac{n}{n-1} \frac{p_0}{\rho_0 g} \left[1 - \left(\frac{p}{p_0}\right)^{\frac{(n-1)}{n}}\right] \qquad (33.4)$$

For air, $p_0 = 101.3$ kN/m² and $\rho_0 = 1.225$ kg/m³.

Hence, from equation (33.4),

$$h = \frac{1.235}{1.235 - 1} \times \frac{101.3}{1.225 \times 9.81} \left[1 - \left(\frac{p}{101.3}\right)^{\frac{(1.125-1)}{1.235}}\right]$$

from which $\quad p = 101.3(1 - 0.000\,022\,6h)^{5.255}$ kN/m² $\qquad (33.5)$

Worked examples

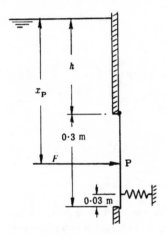

Figure 33.10

1. *A tank containing fuel of density 850 kg/m³ has an automatic device to prevent overflow; this consists of a rectangular door hinged at the upper edge and held closed by a spring, as shown in Figure 33.10. The door is 0.3 m deep and 0.2 m wide and the spring exerts a force of 1 kN.*

Calculate the level of the fuel above the hinge when the door is about to open.

Let the surface of the fuel be a height h above the hinge.

From equation (33.2),

force on door due to oil pressure = area × pressure at centroid

i.e. $\qquad F = (0.2 \times 0.3) \times (850 \times 9.81[h + 0.15])$

$$= 500(h + 0.15)\text{N}$$

From equation (33.3), depth of centre of pressure,

$$x_p = \frac{\text{second moment of area about surface}}{\text{first moment of area about surface}}$$

$$= \frac{\dfrac{bd^3}{12} + bd(h + 0.15)^2}{bd(h + 0.15)}$$

$$= \frac{0.3^2}{12(h + 0.15)} + (h + 0.15)$$

$$= \frac{0.007\,5}{h + 0.15} + (h + 0.15)$$

The door will commence to open when the moment of the spring force about the hinge is equal to that of the water force about the hinge,

i.e. $\qquad 1 \times 10^3(0.3 - 0.03) = 500(h + 0.15)\left\{\dfrac{0.007\,5}{h + 0.15} + (h + 0.15) - h\right\}$

from which $\qquad\qquad h = \underline{3.40 \text{ m}}$

2. *Each gate of a lock is 6 m high and 2 m wide, and is supported on hinges, situated 0.6 m from the top and bottom. The angle between the gates when they are closed is 140°. If the depths of water on the two sides are 5.4 m and 1.5 m, find the* *magnitude and position of the resultant force on each gate, the reaction between the gates and the magnitude and direction of the reactions at the hinges.*

The forces acting on one gate are as shown in Figure 33.11.

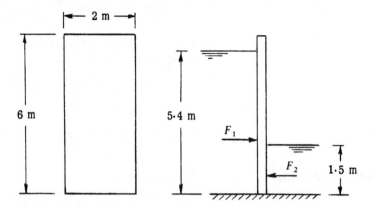

Figure 33.11

$$F_1 = \tfrac{1}{2}\rho g\, bh^2 \quad \text{from equation (33.1)}$$

$$= \tfrac{1}{2} \times 10^3 \times 9.81 \times 2 \times 5.4^2$$

$$= 286 \times 10^3 \text{ N}$$

Similarly, $$F_2 = \tfrac{1}{2} \times 10^3 \times 9.81 \times 2 \times 1.5^2$$

$$= 22 \times 10^3 \text{ N}$$

Thus the resultant force $F = 286 - 22 = 264$ kN

Each of these forces acts through the centre of pressure of their respective sides. If the resultant forces act at a height H above the base, then, taking moments about the base,

$$264H = 286 \times \frac{5.4}{3} - 22 \times \frac{1.5}{3}$$

$$\therefore H = \underline{1.91 \text{ m}}$$

The arrangement of the gates is shown in Figure 33.12. The force F is normal to the gate and the reaction R between the gates is normal to the line of contact. Hence, from symmetry, the reaction at the hinges is also R, acting through the point of intersection of the other two forces and inclined at 20° to the gate.

Equating forces normal to the gate,

$$F = 2R \sin 20°$$

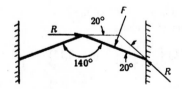

Figure 33.12

from which $R = \underline{386 \text{ kN}}$, acting at the same height above the base as F, i.e. 1.91 m.

If the reactions at the top and bottom hinges are R_1 and R_2 respectively,

then $R_1 + R_2 = R = 386$ kN   (1)

Taking moments about the base,

$$R_1 \times 5.4 + R_2 \times 0.6 = R \times 1.91 \qquad (2)$$

From equations (1) and (2), $R_1 = \underline{105 \text{ kN}}$ and $R_2 = \underline{281 \text{ kN}}$

3. *The water face of a dam is vertical for 6 m below the water surface and below this level, slopes at 30° to the vertical. The base is 20 m below the surface. Find the magnitude, direction and position of the resultant water thrust per metre width on the face of the dam.*

The arrangement is shown in Figure 33.13.

$$F_1 = \tfrac{1}{2}\rho g\, bh^2 \quad \text{from equation (33.1)}$$

$$= \tfrac{1}{2} \times 10^3 \times 9.81 \times 1 \times 6^2 \text{ N}$$

$$= 177 \text{ kN, acting at a depth of 4 m below the surface}$$

Depth of $G_2 = 6 + 7 = 13$ m

Length $BC = 14 \sec 30° = 16.17$ m

$$x_{G_2} = 13 \sec 30° = 15.01 \text{ m}$$

$$F_2 = \text{area} \times \text{pressure at } G_2$$

$$= 1 \times 16.17 \times 10^3 \times 9.81 \times 13 \text{ N}$$

$$= 2\,062 \text{ kN}$$

$$x_{P_2} = \frac{I_0}{Ax_{G_2}}$$

$$= \frac{\dfrac{1 \times 16.17^3}{12} + 1 \times 16.17 \times 15.01^2}{1 \times 16.17 \times 15.01} = 16.46 \text{ m}$$

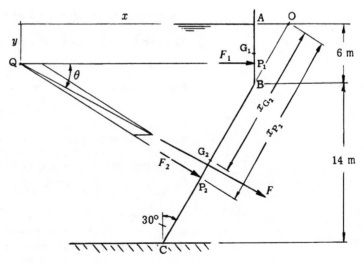

Figure 33.13

From the vector diagram at Q,

$$F = \underline{2\,217\text{ kN}}$$

and
$$\theta = \underline{27.72°}$$

By calculation or scale drawing, the coordinates of the point Q are

$$x = \underline{27.53\text{ m}}$$

and
$$y = \underline{4\text{ m}}$$

4. *The sector gate shown in Figure 33.14(a) is used to control the depth of water in a canal. The gate profile forms part of a circular arc of radius 2 m and 15 m wide. Determine the thrust on the gate.*

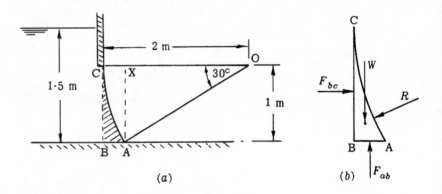

(a)   (b)

Figure 33.14

Consider the equilibrium of the wedge of water ABC shown in Figure 33.14(b).

Force on BC, F_{bc} = area × pressure at centroid

$$= (1 \times 15) \times (1 \times 10^3 \times 9.81) = 147\,000\text{ N}$$

Force on AB, F_{ab} = area × pressure at centroid

$$= (2 - 2\cos 30°) \times 15 \times (1.5 \times 10^3 \times 9.81) = 59\,300\text{ N}$$

Weight of liquid wedge ABC,

$$W = \text{weight of OXA} + \text{weight of XCBA} - \text{weight of OCA}$$

$$= 15 \times 10^3 \times 9.81 \left\{ \tfrac{1}{2} \times 2\cos 30° \times 1 + (2 - 2\cos 30°) \times 1 - \frac{\pi}{12} \times 2^2 \right\}$$

$$= 12\,800\text{ N}$$

Therefore resultant vertical force,

$$F_{ab} - W = 59\,300 - 12\,800 = 46\,500\text{ N}$$

The resultant force on the gate is equal and opposite to the reaction of the gate, R, given by

$$R = \sqrt{(F_{bc}^2 + (F_{ab} - W)^2)}$$

$$= \sqrt{(147\,000^2 + 46\,500^2)} = \underline{154\,200 \text{ N}}$$

R acts at an angle $\tan^{-1} \dfrac{46\,500}{147\,000} = 17.6°$ to the horizontal. The line of action of R must pass through the hinge O since it is normal to the circular surface. It therefore exerts no torque about O and the gate opening is unaffected by the depth of water.

5. *A stream is spanned by a bridge which is a single mass-concrete arch in the form of a circular arc of 5 m radius, the crown being 2.5 m above the springings. Measured in the direction of the stream, the overall width of the bridge is 6 m.*

During a flood, the water level rises to the level of the crown. Assuming that the arch remains watertight, calculate the force tending to lift the bridge from its foundations.

The upthrust on the bridge, F, is the same as the force which would be required to support a mass of water occupying the area shown shaded in Figure 33.15 if the bridge were removed.

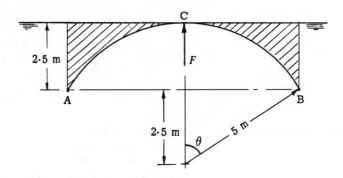

Figure 33.15

$$\text{Width of bridge } = 2\sqrt{(5^2 - 2.5^2)} = 8.66 \text{ m}$$

$$\cos\theta = \frac{2.5}{5} \qquad \therefore \theta = 60°$$

$$\therefore \text{ area of segment ABC} = \frac{120}{360} \times \pi \times 5^2 - \tfrac{1}{2} \times 8.66 \times 2.5$$

$$= 15.355 \text{ m}^2$$

$$\therefore \text{ shaded area} = 8.66 \times 2.5 - 15.355$$

$$= 6.295 \text{ m}^2$$

$\therefore$ weight of water displaced by arch

$$= 6.295 \times 6 \times 10^3 \times 9.81$$

i.e. $$F = 370\,500 \text{ N or } \underline{370.5 \text{ kN}}$$

6. *A diesel engine in a lorry has a power output of 200 kW at sea level, where the air density is 1.225 0 kg/m³, the pressure is 101.3 kN/m² and the temperature is 15°C. What will be the power output of the engine when crossing a mountain pass 5 000 m above sea level, assuming that the power output depends on the mass of air induced.*

From equation (33.5), $p = 101.3(1 - 0.000\,022\,6h)^{5.255}$

Hence, when h = 5 000 m, $p = 53.95$ kN/m²

At 5 000 m with a lapse rate of 0.006 5° C/m,

$$T = (273 + 15) - 5\,000 \times 0.006.5 = 255.5 \text{ K}$$

From the ideal gas equation (see Section 21.4)

$$pV = mRT \text{ where } R \text{ is the specific gas constant}$$

Hence $$m = \frac{pV}{RT}$$

At sea level, $$m = \frac{101.3V}{288R} = 0.352\frac{V}{R}$$

At 5 000 m, $$m = \frac{53.95V}{255.5R} = 0.211\frac{V}{R}$$

Therefore power at 5 000 m $$= \frac{0.211}{0.352} \times 200 = \underline{120 \text{ kW}}$$

Further problems

7. A closed cylindrical tank 0.6 m diameter and 1.8 m deep, with vertical axis, contains water to a depth of 1.2 m. Air to a pressure of 40 kN/m² above atmospheric is pumped into the cylinder. Determine the total normal force on the vertical wall of the tank and the distance of the centre of pressure from the base.

[149 kN; 0.855 m]

8. A tank contains a layer of oil 0.9 m deep resting on a layer of water 0.6 m deep. Calculate the total thrust on one side of the tank, which is 1.8 m wide, and find the depth at which it acts. The relative density of the oil is 0.9.

[18.23 kN; 0.494 m from base]

9. A sluice gate, 1 m wide and 2 m deep, has a mass of 500 kg and works in vertical guides. If the surface of the water on one side of the gate is 1 m above the upper edge of the gate, calculate the initial force required to lift the gate. The coefficient of friction between the gate and the guides is 0.4.   [20.6 kN]

10. A triangle ABC is immersed vertically in water with AB in the surface. Prove that the depth of the centre of pressure of the triangle is one-half the depth of C.

A square of side a is immersed vertically in a liquid of density ρ, with one vertex in the surface and a diagonal horizontal. Prove that the thrust on the square is $\rho g a^3/\sqrt{2}$ and find the depth of the centre of pressure of the square.   [0.825 a]

11. A circular opening, 0.6 m diameter, in the vertical side of a tank is closed by a disc which just fits the opening and can rotate about an axis lying along a horizontal diameter.

Calculate: (*a*) the force on the disc, and (*b*) the torque required to maintain the disc in equilibrium in the vertical position when the head of water above the axis of rotation is 0.8 m.

Show that this torque is independent of the head, provided that the disc is always submerged. [2.22 kN; 62.4 Nm]

12. The outlet of a horizontal rectangular channel, 1.2 m wide by 0.9 m deep is closed by a plane flap inclined at 40° to the vertical and hinged at the horizontal upper edge of the channel. The flap weighs 2.5 kN, the line of action of the weight being 0.35 m horizontally from the hinge.

To what height must the water level in the channel rise so as just to cause the flap to open? [0.33 m]

13. A channel of trapezoidal section is 2 m wide at the bottom. It contains water to a depth of 2.5 m and at the surface, the width is 4 m. The channel is closed by a dam, the water face of which is inclined at 30° to the vertical. Calculate the water force on the face of the dam and find the position of the centre of pressure.

[13.63 kN; 1.804 m]

14. A rectangular opening in the sloping side of a fresh water reservoir is 3 m deep by 2 m wide, the 2 m side being horizontal. The opening is filled by a door of uniform thickness, hinged at the top and the door is kept closed by its own weight and by a load *L* on the lever arm shown in Figure 33.16. The weight of the door is 45 kN. Calculate the load *L* so that the gate commences to open when the water level is 0.8 m above the top of the gate. [49.43 kN]

15. The vertical cross-section of a water tank in a barge is shown in Figure 33.17, the section being symmetrical about the vertical axis. Calculate the magnitude and position of the force on a vertical surface dividing two tanks in the barge when one tank is full and the other is filled to one quarter of its depth.

[3.09 MN acting at 4.75 m from the top]

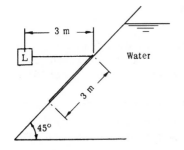

Figure 33.16

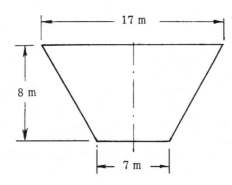

Figure 33.17

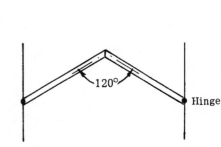

Figure 33.18

16. Figure 33.18 shows the gates of a lock which is 8 m wide. The gates make an angle of 120° with each other and each gate is hinged to the side of the lock. Determine the force on each hinge and the thrust between the gates when the depth of water above the gates is 12 m and that below the gates is 4 m.

[2.9 MN; 2.9 MN]

17. The sector gate shown in Figure 33.19 is used to control a water level. The circular arc has a radius of 4 m and the gate is 8 m wide. The gate is pivoted at the centre of the arc, the pivot being 2 m above the bottom of the gate. Determine the thrust on the pivot when the water depth is 2 m.

[167 kN at 20.35° to horizontal]

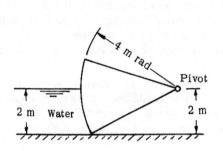

Figure 33.19   **Figure 33.20**

18. Figure 33.20 shows the vertical cross-section of a dam. The water face is part of a circular arc of radius 120 m, the centre of curvature being 30 m above the base of the dam. Calculate the magnitude and direction of the force on the 100 m wide face when the depth of water is 30 m.

[449 MN at 10.1° to the horizontal, through the centre of curvature]

19. Figure 33.21 shows the vertical cross-section of the side of a ship. Determine the force on the section AB which is 12 m long and has a circular profile of radius 10 m. The density of sea water is 1 026 kg/m^3.

[14.6 MN at 54.6° to horizontal, through centre of curvature]

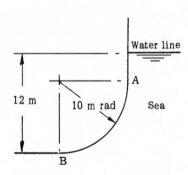

Figure 33.21

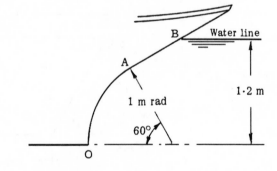

Figure 33.22

20. The stern of a yacht is shown in Figure 33.22, OA is a circular arc, AB is a straight line and the stern is 1.3 m wide. Determine the force on the stern for the water level shown, taking the density of sea water as 1 026 kg/m^3. (A partly graphical solution is recommended.)

[10.6 kN at 28° to the horizontal, intersecting OAB at 0.675 m below the water line]

21. A hinged hatch in a submarine is to be opened by compressed air. The hatch is circular with a diameter of 0.8 m and is inclined at 45° to the vertical. The centre of the hatch is 7 m below the sea surface and the hinge is at the uppermost point. The mass of the hatch is 95 kg and the barometer reads 101 kN/m². Calculate the air pressure required to open the hatch, taking the density of sea water as 1 026 kg/m³.

[175 kN/m²]

22. The pressure and temperature of air at sea level are 101 kN/m² and 288 K respectively. Determine the pressure, density and temperature at an altitude of 9 000 m, assuming a polytropic atmosphere for which $p/\rho^{1.235} = $ constant.

What lapse rate does this represent?

[30.7 kN/m²; 0.466 kg/m³; 229.5 K; 0.006 5 K/m]

23. A motorist decides to convert an aneroid barometer to an altimeter to use in his car on a mountain holiday. The barometer scale is marked at 900, 950 and 1 000 mb. What heights will correspond to the 900 and 950 mb readings assuming that at sea level, where the air density is 1.225 kg/m³, the average barometer reading is 1 000 mb?

[898 m; 432 m]

24. An airfield is situated on a plateau 3 000 m above sea level. For landings at this airfield, the altimeter has to be offset to allow for this elevation, the altimeter setting being determined by the pressure at local ground level. Assuming sea level conditions to be 101.32 kN/m², 15.2°C, 1.225 kN/m³, determine the pressure at the airfield.

[70.1 kN/m²]

34 Flow and flow measurement

34.1 Continuity equation

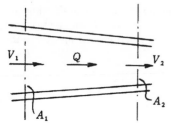

Figure 34.1

When a liquid is flowing along a pipe, the motion is normally turbulent and individual particles do not follow a regular path. The axial velocity is approximately uniform across the section except very close to the pipe wall and it is usual to assume a uniform velocity across the whole section.

Since liquids are assumed to be incompressible, the volume flow rate Q, in the pipe must be constant. Thus, in the tapered pipe shown in Figure 34.1,

$$Q = A_1V_1 = A_2V_2 \tag{34.1}$$

Equation (34.1) is known as the *continuity equation*.

34.2 Energy of a moving liquid

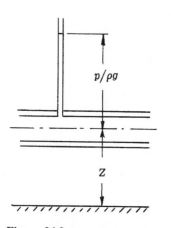

Figure 34.2

Consider a liquid at a pressure p, moving with a velocity V at a height Z above datum level, Figure 34.2.

If a tube were inserted in the top of the pipe, the liquid would rise up the tube a distance of $p/\rho g$ and this is equivalent to an additional height of liquid relative to datum level.

Thus potential energy of liquid per unit mass $= g\left(Z + \dfrac{p}{\rho g}\right)$

and kinetic energy of liquid per unit mass $= \dfrac{V^2}{2}$

Therefore total energy of liquid per unit mass $= g\left(Z + \dfrac{p}{\rho g} + \dfrac{V^2}{2g}\right)$

$$\tag{34.2}$$

Equation (34.2) represents the specific energy of the liquid. Each of the quantities in the brackets have the units of length and are termed *heads*. Thus Z is referred to as the *potential head* of the liquid, $p/\rho g$ as the *pressure head* and $V^2/2g$ as the *velocity head*. The expression

$Z + p/\rho g + V^2/2g$ is therefore the total head of the liquid and represents the energy per unit weight.

If there are no losses of energy between two sections of a pipe, Figure 34.3, and no energy changes due to work or heat transfer, the total energy remains constant,

i.e.
$$Z_1 + \frac{p_1}{\rho g} + \frac{V_1^2}{2g} = Z_2 + \frac{p_2}{\rho g} + \frac{V_2^2}{2g} \qquad (34.3)$$

This is known as *Bernoulli's equation.*

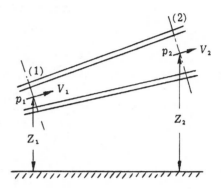

Figure 34.3

If a loss of head h occurs between sections (1) and (2), equation (34.3) becomes

$$Z_1 + \frac{p_1}{\rho g} + \frac{V_1^2}{2g} = Z_2 + \frac{p_2}{\rho g} + \frac{V_2^2}{2g} + h \qquad (34.4)$$

If, in addition, energy is added by a pump or deducted by a turbine, equation (34.3) becomes further modified to

$$Z_1 + \frac{p_1}{\rho g} + \frac{V_1^2}{2g} + \frac{w_{\text{in}}}{g} = Z_2 + \frac{p_2}{\rho g} + \frac{V_2^2}{2g} + \frac{w_{\text{out}}}{g} + h \qquad (34.5)$$

where w is the specific work in J/kg.

34.3 Venturi meter

Figure 34.4 shows a venturi meter, which consists of a constriction in a pipe with a manometer arranged to indicate the difference in pressure between the main and the throat.

Applying Bernoulli's equation to sections (1) and (2) and assuming no losses,

$$Z_1 + \frac{p_1}{\rho g} + \frac{V_1^2}{2g} = Z_2 + \frac{p_2}{\rho g} + \frac{V_2^2}{2g}$$

i.e.
$$\frac{p_1}{\rho g} - \frac{p_2}{\rho g} = \frac{V_2^2}{2g} - \frac{V_1^2}{2g} \qquad \text{if the meter is horizontal.}$$

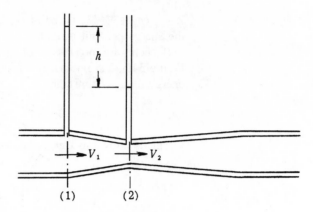

Figure 34.4

But
$$A_1 V_1 = A_2 V_2$$

$$\therefore V_2 = \frac{A_1}{A_2} V_1$$

$$\therefore \frac{p_1}{\rho g} - \frac{p_2}{\rho g} = h = \frac{V_1^2}{2g}\left(\frac{A_1}{A_2}\right)^2 - \frac{V_1^2}{2g} = \frac{V_1^2}{2g}\left(\frac{A_1^2 - A_2^2}{A_2^2}\right)$$

$$\therefore V_1 = \frac{A_2}{\sqrt{(A_1^2 - A_2^2)}}\sqrt{(2gh)}$$

$$\therefore Q = A_1 V_1 = \frac{A_1 A_2}{\sqrt{(A_1^2 - A_2^2)}}\sqrt{(2gh)}$$

$$= k\sqrt{h} \text{ since } A_1 \text{ and } A_2 \text{ are constants} \quad (34.6)$$

Due to friction in the tapered pipe, the pressure at the throat is slightly lower than the theoretical value, so that h becomes slightly larger, giving a value for Q which is too high. To correct this, the theoretical discharge is multiplied by a coefficient of discharge, c_d, which is usually about 0.97 or 0.98.

Thus
$$Q = c_d k\sqrt{h} \quad (34.7)$$

The difference in pressure between the main and the throat is usually measured with a mercury-and-water U-tube, Section 32.3(c), so that

$$h = x(S - 1)$$

If the venturi meter is inclined to the horizontal such that the sections (1) and (2) are at heights Z_1 and Z_2 above datum level, equation (34.6) becomes

$$Q = k\sqrt{(h - (Z_2 - Z_1))}$$

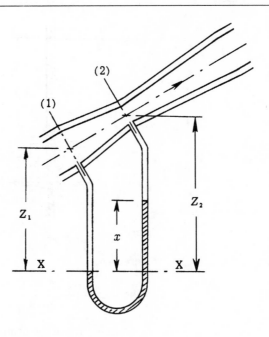

Figure 34.5

If the pressure difference is measured with a mercury-and-water U-tube, Figure 34.5, then, equating pressures at level XX,

$$\frac{p_1}{\rho g} + Z_1 = \frac{p_2}{\rho g} + (Z_2 - x) + Sx$$

i.e.
$$h - (Z_2 - Z_1) = x(S - 1)$$

Thus, for a given flow rate, the manometer reading is the same for an inclined meter as for a horizontal meter.

34.4 Small orifices

An orifice is a precisely made hole through which a liquid may flow. A *small* orifice is one in which the variation of head across the hole is small enough to be neglected.

Figure 34.6(*a*) shows a small sharp-edged orifice through which a liquid issues under a head h. The theoretical velocity of the jet is obtained by applying Bernoulli's equation to points (1) and (2),

i.e.
$$h + \frac{p_1}{\rho g} + 0 = 0 + \frac{p_2}{\rho g} + \frac{V^2}{2g}$$

But
$$p_1 = p_2 = \text{atmospheric pressure}$$

Hence
$$V = \sqrt{(2gh)}$$

Due to the component of the velocity perpendicular to the axis, the jet will contract after leaving the orifice and the section at which the jet becomes parallel is termed the *vena contracta*.

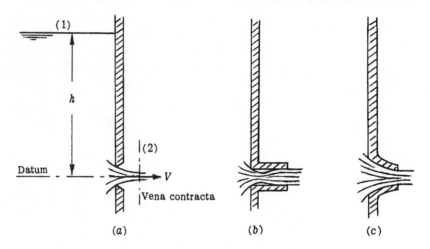

Figure 34.6

The ratio $\dfrac{\text{area of vena contracta}}{\text{area of orifice}}$ is called the *coefficient of contraction* (c_c).

Due to friction at the orifice, the velocity at the vena contracta is slightly less than the theoretical velocity and the ratio $\dfrac{\text{actual velocity of jet}}{\text{theoretical velocity of jet}}$ is called the *coefficient of velocity* (c_v).

If A is the orifice area, the discharge, or flow rate,

$$Q = c_c A \times c_v \sqrt{(2gh)}$$

$$= c_d A \sqrt{(2gh)} \text{ where } c_d = c_c c_v \qquad (34.8)$$

c_d is the *coefficient of discharge* which is the ratio $\dfrac{\text{actual discharge}}{\text{theoretical discharge}}$.

For a sharp-edged orifice, approximate values for c_c, c_v and c_d are 0.64, 0.97 and 0.62.

The value for c_d can be improved by fitting a pipe or mouthpiece to the orifice, as shown in Figures 34.6(*b*) and (*c*).

34.5 Large rectangular orifice

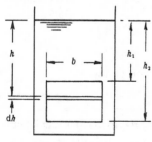

Figure 34.7

For the orifice shown in Figure 34.7, there will be an appreciable difference in pressure between the top and bottom edges. If the breadth of the orifice is b, the theoretical discharge through an elementary strip of thickness dh at a depth h below the surface is given by

$$dQ = b \, dh \times \sqrt{(2gh)}$$

$$\therefore Q = b\sqrt{(2g)} \int_{h_1}^{h_2} h^{1/2} dh$$

$$= \tfrac{2}{3} b \sqrt{(2g)}(h_2^{3/2} - h_1^{3/2})$$

Therefore actual discharge $= c_d \times \tfrac{2}{3} b \sqrt{(2g)}(h_2^{3/2} - h_1^{3/2})$ $\qquad (34.9)$

34.6 Orifice plate

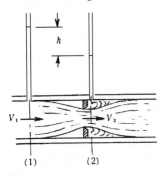

Figure 34.8

Figure 34.8 shows a sharp-edged orifice plate fitted into a pipe conveying a liquid. The flow will follow the pattern shown, forming a vena contracta just beyond the orifice before expanding to the pipe diameter again.

Applying Bernoulli's equation to sections (1) and (2),

$$\frac{p_1}{\rho g} + \frac{V_1^2}{2g} = \frac{p_2}{\rho g} + \frac{V_2^2}{2g}$$

Also $\qquad A_1 V_1 = A_2 V_2 \quad$ so that $\quad V_2 = \frac{A_1}{A_2} V_1$

Thus $\qquad \dfrac{p_1}{\rho g} - \dfrac{p_2}{\rho g} = h = \dfrac{V_1^2}{2g}\left(\dfrac{A_1^2 - A_2^2}{A_2^2}\right)$

Therefore $\quad Q = A_1 V_1 = \dfrac{A_1 A_2}{\sqrt{(A_1^2 - A_2^2)}}\sqrt{(2gh)} = k\sqrt{(h)}$ (34.10)

It is difficult to establish the exact position of the vena contracta and to determine its area, A_2, in relation to the known orifice area. The constant k is therefore determined experimentally, when it will incorporate the coefficient of discharge.

If h is small so that ρ is approximately constant, equation (34.10) may also be applied to compressible fluid flow. This type of orifice is used to meter flow in carburettors, fuel systems, control systems and engine test installations.

34.7 Pitot tube

A simple pitot tube consists of a tube, bent at right angles and tapered at the lower end, as shown in Figure 34.9. If such a tube is placed in an open channel, the liquid at the mouth of the tube is brought to rest (this point being termed a *stagnation point*) and liquid is forced up the tube above the level of the free surface.

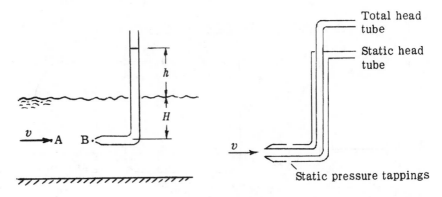

Figure 34.9　　　　　　　　　　**Figure 34.10**

Applying Bernoulli's equation to a point A in the free stream and a point B at the mouth of the pitot tube,

$$H + \frac{v^2}{2g} = H + h$$

or
$$v = \sqrt{(2gh)} \qquad (34.11)$$

If the pitot tube is used in a pipe under pressure, the reading h will represent both the velocity head and pressure head at the point under consideration and it is necessary to deduct the pressure head in order to obtain the velocity head only.

This is conveniently arranged by using a *pitot-static tube,* Figure 34.10. This consists of two concentric tubes, the inner one only being open at the lower end. The outer tube has radial holes drilled in it to measure the static pressure head, while the inner tube measures the total head. The two outlets are connected to a U-tube manometer of the type shown in Figure 32.7 and the difference in levels represents the velocity head only, from which the velocity can be deduced.

If the pitot tube is large enough to distort the flow, a calibration constant C may be required, so that

$$v = C\sqrt{(2gh)} \qquad (34.12)$$

Pitot tubes may also be used to measure gas velocity if the velocity is sufficiently low so that the density may be considered constant. They can also be used to determine the velocities of aircraft and ships.

34.8 Velocity distribution and flow rate

A pitot tube measures the velocity at a particular point in the section of a channel or pipe and since the velocity varies considerably over the cross-section, a series of pitot readings is necessary to determine the flow rate.

Figure 34.11 shows the type of variation which may be obtained by traversing the pitot tube along the line AA. The cross-section is then divided into convenient areas a_1, a_2, a_3, etc., Figure 34.12, and the velocity at the centre of each area is determined.

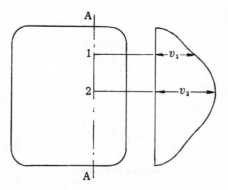

Figure 34.11

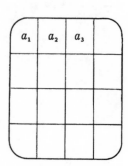

Figure 34.12

Then the flow rate

$$Q = a_1 v_1 + a_2 v_2 + a_3 v_3 + \ldots = \Sigma a v \qquad (34.13)$$

If the total cross-sectional area of the duct is A,

$$\text{the mean velocity } V = \frac{Q}{A} = \frac{\Sigma a v}{A} \qquad (34.14)$$

In the case of a circular pipe, the velocity profile is the same across any diameter and hence, for an annular ring of mean radius r and area Δa, Figure 34.13,

$$\Delta Q = v_r \times \Delta a,$$

where v_r is the velocity at radius r

Thus

$$Q = \Sigma v_r \Delta a \qquad (34.15)$$

and

$$V = \frac{\Sigma v_r \Delta a}{\pi r^2} \qquad (34.16)$$

The velocity at the boundary of any duct or pipe is zero.

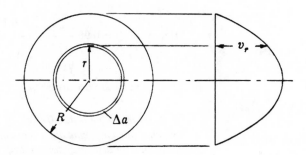

Figure 34.13

Worked examples

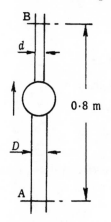

Figure 34.14

1. *Figure 34.14 shows part of the lubrication system of an engine. At A the pressure is atmospheric and at B the gauge pressure is 80 kN/m². The diameter of the suction pipe is twice that of the delivery pipe. The total head lost in the system is 440 mm of oil and point B is 0.8 m above A. The pump supplies a head of 11.3 m of oil and the oil has a density of 850 kg/m³.*

Determine the velocity of flow in the suction pipe.

From equation (34.5),

$$Z_a + \frac{p_a}{\rho g} + \frac{V_a^2}{2g} + \frac{w_{\text{in}}}{g} = Z_b + \frac{p_b}{\rho g} + \frac{w_{\text{out}}}{g} + h$$

Also

$$Q = A_a V_a = A_b V_b$$

$$\therefore V_b = V_a \left(\frac{D^2}{d^2} \right) = 4 V_a$$

Hence

$$0 + 0 + \frac{V_a^2}{2g} + 11.3 = 0.8 + \frac{80 \times 10^3}{850g} + \frac{(4V_a)^2}{2g} + 0 + \frac{440}{10^3}$$

$$\therefore \quad \frac{V_a^2}{2 \times 9.81}(16 - 1) = 0.46$$

from which $V_a = \underline{0.775 \text{ m/s}}$

2. *A horizontal venturi meter with a throat diameter of 50 mm is placed in a pipe of 120 mm diameter. The pipe conveys oil of density 800 kg/m³ and the difference in levels of mercury in a U-tube manometer is 220 mm. The coefficient of discharge of the meter is 0.98. Calculate the flow rate of oil.*

From Section 34.3, $Q = c_d \dfrac{A_1 A_2}{\sqrt{(A_1^2 - A_2^2)}} \sqrt{(2gh)}$

where h is the difference in head of liquid between the main and the throat. From equation (34.7),

$$h = \frac{p_1 - p_2}{\rho g} = \left(\frac{\rho_{\text{Hg}}}{\rho_{\text{oil}}} - 1 \right) x$$

$$= \left(\frac{13.6}{0.8} - 1 \right) \times 0.22 = 3.52 \text{ m}$$

Hence $Q = 0.98 \dfrac{\frac{\pi}{4} \times 0.12^2 \times \frac{\pi}{4} \times 0.05^2}{\sqrt{\left(\left(\frac{\pi}{4} \times 0.12^2 \right)^2 - \left(\frac{\pi}{4} \times 0.05^2 \right)^2 \right)}} \sqrt{(2 \times 9.81 \times 3.52)}$

$$= \underline{16.2 \times 10^{-3} \text{ m}^3/\text{s}}$$

3. *In an engine test, 0.04 kg/s of air flows through a 50 mm diameter pipe, into which is fitted a 40 mm diameter orifice plate. The density of air is 1.2 kg/m³ and the coefficient of discharge for the orifice is 0.63. The pressure drop across the orifice is measured by a U-tube manometer fitted with water.*
Calculate the manometer reading.

The arrangement is shown in Figure 34.15.

Mass flow rate, $\dot{m} = \rho Q$

i.e. $0.4 = 1.2Q$

$$\therefore \quad Q = 0.033\,3 \text{ m}^3/\text{s}$$

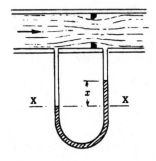

Figure 34.15

From equation (34.10), the theoretical volume flow rate is given by

$$Q = \frac{A_1 A_2}{\sqrt{(A_1^2 - A_2^2)}} \sqrt{(2gh)}$$

$$= \frac{\frac{\pi}{4} \times 0.05^2 \times \frac{\pi}{4} \times 0.04^2}{\sqrt{\left(\left(\frac{\pi}{4} \times 0.05^2 \right)^2 - \left(\frac{\pi}{4} \times 0.04^2 \right)^2 \right)}} \sqrt{(2 \times 9.81 \times h)}$$

$$= 0.007\,244 \sqrt{h}$$

Hence, actual discharge $= 0.63 \times 0.007\,244\sqrt{h} = 0.033\,3$

from which $\qquad\qquad h = 53.34$ m of air

Equating pressures at level XX in the U-tube,

$$1.2g \times 53.34 = 10^3 g \times x$$

$$\therefore x = 64 \times 10^{-3} \text{ m} = \underline{64 \text{ mm}}$$

4. *A tank is divided into two compartments by a vertical wall in which there is a rectangular orifice 0.8 m wide by 0.5 m deep to enable the oil in the tank to flow from one side to the other. Determine the discharge through the orifice when the level on one side is 3 m above the top of the orifice and that on the other side is 0.2 m below the top of the orifice. The coefficient of discharge is 0.65.*

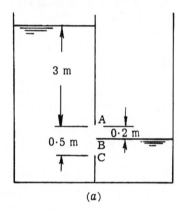 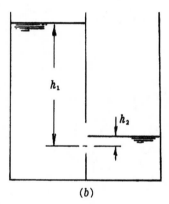

(a)   (b)

Figure 34.16

The upper part of the orifice AB, Figure 34.16 (*a*) permits free flow and the discharge is given by equation (34.9).

Hence $\qquad\qquad Q_{ab} = c_d \times \frac{2}{3}b\sqrt{(2g)}(h_2^{3/2} - h_1^{3/2})$

$$= 0.65 \times \tfrac{2}{3} \times 0.8(3.2^{3/2} - 3.0^{3/2})\sqrt{(2 \times 9.81)}$$

$$= 0.81 \text{ m}^3/\text{s}$$

The lower part, BC, is a 'drowned' orifice and the head causing flow at any point within BC is the difference in level, $h_1 - h_2$, Figure 34.16 (*b*), which is a constant, equal to 3.2 m.

Hence $\qquad\qquad Q_{bc} = c_d A\sqrt{(2gH)}$

$$= 0.65 \times 0.8 \times 0.3\sqrt{(2 \times 9.81 \times 3.2)}$$

$$= 1.25 \text{ m}^3/\text{s}$$

Therefore total discharge $= 0.81 + 1.25 = \underline{2.06 \text{ m}^3/\text{s}}$

5. (a) A pitot-static tube on the centre line of a 0.36 m diameter circular water pipe records a reading of 0.058 4 m of water. If the meter calibration coefficient is 1, what is the centre line velocity?

(b) The variation of velocity in the pipe is determined by a pitot-static traverse and the results are shown in the following table. Determine the quantity flowing and the mean velocity in the pipe.

Radius (m)	0	0.06	0.12	0.15	0.17	0.18
Velocity (m/s)	1.07	0.96	0.85	0.73	0.55	0

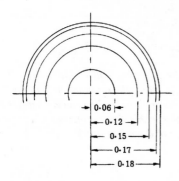

Figure 34.17

(a) From equation (34.12),

$$v = C\sqrt{(2gh)} = 1\sqrt{(2 \times 9.81 \times 0.058\,4)} = \underline{1.07 \text{ m/s}}$$

(b) The cross-section of the pipe is divided into five concentric rings of radii corresponding with the given information, Figure 34.17. The mean velocity for any ring is taken as the average of those at the boundaries of the ring. The following table is then compiled:

Mean velocity for ring, v_r	$\dfrac{1.07 + 0.96}{2}$	$\dfrac{0.96 + 0.85}{2}$	$\dfrac{0.85 + 0.73}{2}$	$\dfrac{0.73 + 0.55}{2}$	$\dfrac{0.55 + 0}{2}$
	= 1.015	= 0.905	= 0.790	= 0.640	= 0.275
Area of ring, Δa	0.011 3	0.033 9	0.025 4	0.020 2	0.011 0
$\Delta Q = v_r \Delta a$	0.011 5	0.030 7	0.020 1	0.012 9	0.003 0

$$Q = \Sigma \Delta Q = 0.078\,2 \text{ m}^3/\text{s}$$

$$V = \frac{Q}{A} = \frac{0.078\,2}{\pi \times 0.18^2}$$

$$= \underline{0.768 \text{ m/s}}$$

Alternative method

The given information is plotted and the cross-section is then divided into smaller rings of uniform width. The mean velocity for each ring is then obtained from the graph, Figure 34.18 on the next page.

Since the width of each ring is small, the area of a ring may be taken as $2\pi r \Delta r$, where r is the mean radius and Δr is the width.

Then $$\Delta Q = v_r \times 2\pi r \Delta r$$

and $$Q = 2\pi \Delta r \Sigma v_r r$$

The following table is then compiled:

r	0.01	0.03	0.05	0.07	0.09	0.11	0.13	0.15	0.17
v_r	1.05	1.00	0.97	0.94	0.91	0.87	0.81	0.73	0.55
$v_r r$	0.010 5	0.030 0	0.048 5	0.065 8	0.081 9	0.095 7	0.105 3	0.109 5	0.093 5

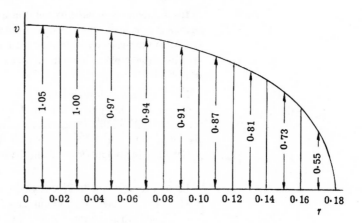

Figure 34.18

$$Q = 2\pi \Delta r \Sigma v_r r$$

$$= 2\pi \times 0.02 \times 0.641 = \underline{0.080\,6 \text{ m}^3/\text{s}}$$

The difference in the answers obtained is due to the approximations inherent in each method. Care is needed to be certain that the approximations made do not give inaccurate answers.

Further problems

6. Water flows downwards at the rate of 0.9 m³/min through a pipe which tapers from 100 mm diameter at A to 50 mm diameter at B, A being 1.5 m above B. The pressure in the pipe at A is 70 kN/m² above atmospheric. Assuming no loss of energy between A and B, find the pressure in the pipe at B.   [57.4 kN/m²]

7. A pipe carrying water tapers from 160 mm diameter at A to 80 mm diameter at B. A is 3 m above B. The pressure in the pipe is 100 kN/m² at A and 20 kN/m² at B, both measured above atmosphere. The flow is 4 m³/min and is in the direction A to B. Find the loss of energy, expressed as a head of water, between points A and B.

[2.745 m]

8. In a vertical pipe conveying water, pressure gauges are inserted at A and B where the diameters are 150 mm and 75 mm respectively. The point B is 1.8 m below A and the pressure at B is 15 kN/m² greater than at A. Determine the rate of flow down the pipe.

If the gauges at A and B are replaced by tubes filled with water and connected to a U-tube containing mercury of relative density 13.6, give a sketch showing how the levels in the two limbs of the U-tube differ and calculate the value of this difference.

[0.010 5 m³/s; 21.4 mm]

9. Oil of relative density 0.95 flows along a pipe from a point A to a point B. At A the pipe diameter is 100 mm and the pressure is 140 kN/m² absolute and at B, which is 2 m below A, the diameter is 150 mm and the pressure is 180 kN/m² absolute. Determine (a) the velocity of the oil at A, (b) the quantity of oil flowing along the pipe, (c) the total energy of the oil at A per unit mass, assuming that the horizontal through B is the datum level.   [7.49 m/s; 0.058 8 m³/s; 195 J]

10. A circular pipe has a fall of 4 m and narrows so that, at exit, its area is one quarter that at inlet. The pressure at inlet is 20 kN/m² above atmospheric and at exit, where the velocity is 15 m/s, the pressure is atmospheric. Calculate the inlet velocity and the mass flow rate of water through the pipe per unit cross-sectional area at inlet.   [2.81 m/s; 2 810 kg/s m²]

11. A turbine is installed in a vertical pipe line, 150 mm diameter. A gauge is connected to the pipe at a point A above the turbine and another at a point B below the turbine, the vertical distance between A and B being 1.2 m. When the discharge is 0.06 m³/s, the gauges read 400 kN/m² and 35 kN/m² respectively.

Calculate the output power from the turbine, assuming it to be 85% efficient.

[19.22 kW]

12. In a vertical pipe line conveying water, pressure gauges are inserted at A and B where the diameters are 150 mm and 75 mm respectively. The point B is 2.4 m below A and when the rate of flow down the pipe is 0.02 m³/s, the pressure at B is 12 kN/m² greater than at A. Assuming the losses in the pipe between A and B can be expressed as $k \cdot \dfrac{v^2}{2_g}$, where v is the velocity at A, find the value of k.

[3.02]

13. A horizontal pipe line 125 mm diameter, through which water is flowing, is fitted with a venturi meter having a throat diameter of 50 mm. Manometer tubes are led from the entrance and throat of the meter to a U-tube containing mercury which records a difference in the mercury levels of 250 mm. Working from first principles, and assuming a value of 0.98 for the coefficient of the meter, determine the flow of water in the pipe.   [0.015 3 m³/s]

14. A venturi meter is to be used to measure the flow of oil in a vertical pipe 150 mm in diameter. The flow is upward through the meter, which has a throat diameter of 50 mm. A mercury manometer connected between the throat and a point in the pipe below the throat reads 160 mm. The coefficient of discharge of the meter is 0.98 and the densities of oil and mercury are 850 kg/m³ and 13 600 kg/m³ respectively. Calculate the mass flow rate of oil.   [11.2 kg/s]

15. A venturi meter is tested with its axis horizontal and the flow is measured in a collecting tank. The pipe diameter is 75 mm, the throat diameter is 38 mm and the pressure difference is measured with a mercury-and-water differential manometer. When the manometer reads 270 mm, 1 100 kg of water were collected in 120 s. Determine the coefficient of discharge of the meter.   [0.96]

16. A cylindrical tank 2 m in diameter has a 50 mm diameter orifice in the base, for which the coefficient of discharge is 0.62.

(*a*) Water enters the tank at 0.009 m³/s. What is the depth in the tank when the level becomes steady?

(*b*) If the input flow rate is increased to 0.02 m³/s, at what rate does the water level rise when the depth is 4 m?   [2.8 m; 3 mm/s]

17. Two vertical sided locks are connected by a door 1 m square. The level in one lock is 10 m above the centre of the door and in the other 4 m above the centre. The coefficient of discharge of the door is 0.6. Calculate the flow rate when the door is opened wide.   [6.53 m³/s]

18. The air supply to a petrol engine passes through the carburettor choke tube 40 mm diameter. When the engine is running, the pressure at the throat of the choke tube is 7 N/m^2 below atmospheric. By treating air as an incompressible fluid (the pressure changes are small) and ignoring other losses, calculate the air/fuel ratio by mass used in the engine when the fuel flow rate is 1.25 kg/h. The density of air is 1.2 kg/m^3 and the coefficient of discharge for the choke is 0.95. The velocity of the air away from the choke tube may be assumed negligible.   [14:1]

19. The air supply to an engine on a test bed passes down a 180 mm diameter pipe fitted with an orifice plate 90 mm diameter. The pressure drop across the orifice is 80 mm of paraffin. The coefficient of discharge of the orifice is 0.62 and the densities of air and paraffin are 1.2 kg/m^3 and 830 kg/m^3 respectively. Calculate the mass flow rate of air to the engine.   [0.162 5 kg/s]

20. A pitot static tube is placed on the centre line of a 0.25 m diameter circular pipe carrying air of density 1.1 kg/m^3. The reading of the instrument, which has a calibration coefficient of 0.98, is 40 mm water. If the mean velocity in the pipe is 0.83 of the maximum velocity, determine the volume flow rate.   [1.066 m^3/s]

21. In a pitot static traverse of a circular duct of 0.5 m diameter the following results were obtained. Determine the volume flow rate of air and the mean velocity.

Radius (m)	0	0.05	0.10	0.15	0.20	0.25
Velocity (m/s)	29	28	26	24	14	0

[3.39 m^3/s, 17.3 m/s]

22. A pitot static tube is placed at a certain radius in a pipe to measure the velocity of flow. The pipe is 240 mm diameter and conveys 0.2 m^3/s of water. The tube is connected to a differential manometer which reads a pressure difference of 85 mm of mercury. Calculate the calibration coefficient for the pitot tube if it is to be used at this radius to indicate mean velocity.   [0.964 4]

35 Momentum of fluids

35.1 The momentum equation

The momentum equation is obtained from Newton's second law. When applied to fluid flow, it states that the sum of the external forces acting on a fluid in a given direction is equal to the rate of change of momentum of the fluid in that direction.

Referring to Figure 35.1, the force in the x direction exerted *by* the surface *on* the fluid is given by

$$F_x = \dot{m}(V_2 \cos \theta_2 - V_1 \cos \theta_1)$$

The force exerted *by* the jet *on* the surface is equal and opposite to the above force.

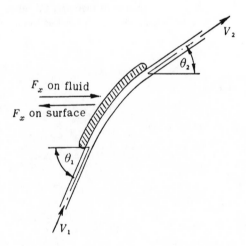

Figure 35.1

35.2 Impact of jets on stationary surfaces

When a jet of liquid impinges on a plate or vane, the force acting on the fluid to change its momentum is the reaction of the vane and the force exerted *by* the jet *on* the vane is equal and opposite to this reaction. In the cases considered below, it will be assumed that there is no friction between the liquid and the vane.

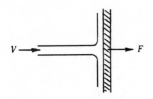

Figure 35.2

(a) Flat plate normal to jet

Figure 35.2 shows a jet of cross-sectional area A moving with a velocity V. After impact, the velocity normal to the plate is zero so that force on plate normal to jet

$$= \dot{m}(V - 0)$$

But $$\dot{m} = \rho A V$$

so that $$F = \rho A V^2 \tag{35.1}$$

(b) Flat plate inclined to jet

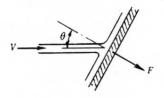

Figure 35.3

If the normal to the plate is inclined at an angle θ to the axis of the jet, Figure 35.3, the initial velocity normal to the plate is $V \cos \theta$. The final velocity normal to the plate is again zero, so that

$$F = \dot{m}(V \cos \theta - 0)$$

$$= \rho A V^2 \cos \theta \tag{35.2}$$

(c) Symmetrical curved vane

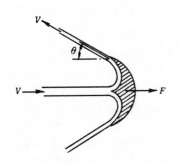

Figure 35.4

Figure 35.4 shows a curved vane of the type used in a Pelton wheel. If there is no friction as the liquid passes over the vane, the magnitude of the velocity remains unchanged. The initial velocity normal to the vane is V and the component of its final velocity normal to the vane is $-V \cos \theta$.

Hence $$F = \dot{m}(V - [-V \cos \theta])$$

$$= \rho A V^2 (1 + \cos \theta) \tag{35.3}$$

If the jet is completely reversed in direction, i.e. $\theta = 0$, then

$$F = 2\rho A V^2$$

(d) General case of curved vane

If the inlet and outlet velocities are V_1 and V_2 respectively, Figure 35.5, then in the horizontal direction,

$$F_x = \dot{m}(V_1 \cos \theta_1 + V_2 \cos \theta_2) \tag{35.4}$$

and in the vertical direction,

$$F_y = \dot{m}(V_1 \sin \theta_1 - V_2 \sin \theta_2) \tag{35.5}$$

The resultant force on the vane is given by

$$F = \sqrt{(F_x^2 + F_y^2)}$$

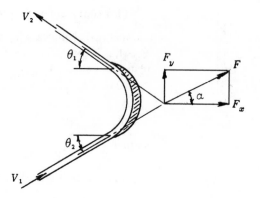

Figure 35.5

and the inclination to the horizontal is given by

$$\alpha = \tan^{-1} \frac{F_y}{F_x}$$

35.3 Impact of jets on moving plates

Figure 35.6

If the plate shown in Figure 35.6 is moving with velocity v in the direction of the jet, the mass flow rate of liquid impinging on the plate is given by

$$\dot{m} = \rho A(V - v)$$

The velocity of the jet is reduced from V to v after striking the plate, so that the change of momentum per second is $\dot{m}(V - v)$.

Thus $\qquad\qquad F = \dot{m}(V - v) = \rho A(V - v)^2 \qquad\qquad$ (35.6)

This is an impracticable case since the jet would need to be continually lengthened.

A practical example of impact on moving plates is the water wheel, Figure 35.7, where a series of flat plates is mounted radially on the rim of the wheel.

The whole of the water leaving the nozzle strikes the plates, although the change in velocity is still $(V - v)$

Hence $\qquad\qquad F = \dot{m}(V - v) \quad$ where $\quad \dot{m} = \rho A v$

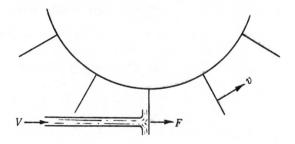

Figure 35.7

Work done per second, or power output $= \dot{m}(V - v)v$ (35.7)

Energy supplied per second, or power input $= \frac{1}{2}\dot{m}V^2$

Therefore efficiency, $\eta = \dfrac{\dot{m}(V - v)v}{\frac{1}{2}\dot{m}V^2}$

$$= \dfrac{2(V - v)v}{V^2}$$ (35.8)

For maximum efficiency, $\dfrac{d\eta}{dv} = 0$ from which $v = \dfrac{V}{2}$

Therefore maximum efficiency $= \dfrac{2\left(V - \dfrac{V}{2}\right)\dfrac{V}{2}}{V^2}$

$$= 0.5 \quad \text{or} \quad 50\%$$

35.4 Curved vanes and the Pelton wheel

If the flat plates of a water wheel are replaced by curved vanes, shown in plan view in Figure 35.8(a), the relative velocity of the jet to the vane at inlet is $(V - v)$ and, in the absence of friction, this will also be the velocity of the liquid relative to the vane at outlet.

Hence the absolute velocity of the liquid in the direction of motion of the vane is $v - (V - v)\cos\theta$, Figure 35.8(b).

Thus force on vane, $F = \dot{m}\{V - [v - (V - v)\cos\theta]\}$

$$= \dot{m}(V - v)(1 + \cos\theta)$$

Power output $= \dot{m}(V - v)(1 + \cos\theta)v$ (35.9)

Power input $= \frac{1}{2}\dot{m}V^2$

Hence $\eta = \dfrac{2(V - v)(1 + \cos\theta)v}{V^2}$ (35.10)

For maximum efficiency, $v = \dfrac{V}{2}$ as before and the maximum efficiency becomes

$$\eta_{\text{max}} = \tfrac{1}{2}(1 + \cos\theta)$$

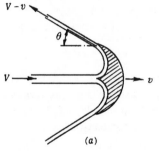

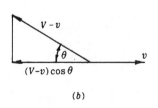

Figure 35.8

If $\theta = 0$, i.e. the jet is reversed in direction, the maximum efficiency is 100%, neglecting friction.

The above case is the basis of the Pelton wheel, Figure 35.9. The vertical wheel, W, carries a series of curved vanes, or buckets, B, attached radially to its circumference. The water enters through the nozzle, N, controlled by a needle-valve, and strikes the vanes in a direction tangential to the wheel, one half being deflected to either side. The lower edge of the vane is cut away so as not to impede the jet before it has come into the vertical position, Figure 35.10.

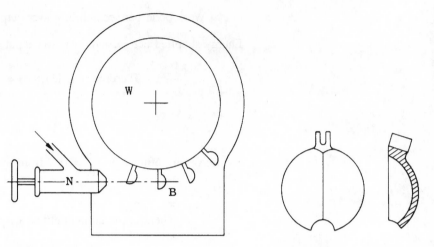

Figure 35.9   **Figure 35.10**

Due to friction, the relative velocity at outlet is slightly less than the relative velocity at inlet and the greatest angle through which the jet can be deflected is about 160°, otherwise the deflected jet will strike the back of the oncoming vane.

It is found that the maximum efficiency occurs when the vane speed is about 0.46 of the jet speed and the greatest efficiency available is about 85%.

35.5 Flow through turbines, fans, compressors and pumps

The Pelton wheel is a simple form of impulse turbine. There are also impulse steam and gas turbines, reaction water, steam and gas turbines, fans, compressors and pumps, all of which use the principle that the rate of change of momentum of the working fluid involves forces which give power

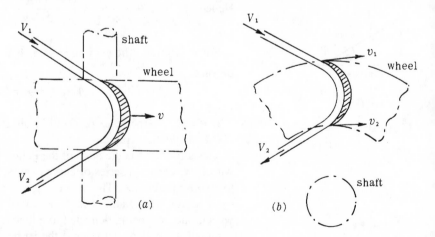

Figure 35.11

in the case of turbines and absorb power in the case of fans, compressors and pumps. In all these designs, the fluid is enclosed by a casing and the change of velocity as the fluid flows over the curved vanes is determined from relative velocity diagrams (see Chapter 41).

The machines are usually of the axial flow or radial flow type, as shown in Figures 35.11(a) and (b) respectively.

For each type, the relative velocity diagrams are constructed at inlet and outlet to the vanes. Referring to Figure 35.12, V_1 and V_2 are the absolute velocities of the fluid at inlet and outlet and v_1 and v_2 are the velocities of the vane tips. The relative velocities u_1 and u_2 are tangential to the vane tips so that the fluid enters and leaves the vane without shock.

The force on the vane,

$$F_x = \dot{m}(V_1 \cos\theta_1 + V_2 \cos\theta_2)$$

Torque on shaft,

$$T = \dot{m}(V_1 \cos\theta_1 r_1 + V_2 \cos\theta_2 r_2) \tag{35.11}$$

where r_1 and r_2 are the radii of the vane tips at the inlet and outlet respectively.

If ω is the angular velocity, then power output

$$= \dot{m}\omega(V_1 \cos\theta_1 r_1 + V_2 \cos\theta_2 r_2)$$

$$= \dot{m}(V_1 \cos\theta_1 v_1 + V_2 \cos\theta_2 v_2) \tag{35.12}$$

In the case of an axial flow machine, $v_1 = v_2 = v$, so that

$$\text{power output} = \dot{m}v(V_1 \cos\theta_1 + V_2 \cos\theta_2) \tag{35.13}$$

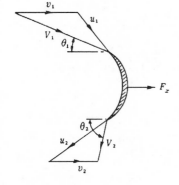

Figure 35.12

35.6 Forces on tapered pipes and bends

Figure 35.13 shows a tapered pipe in which the gauge pressure changes from p_1 to p_2 while the velocity changes from V_1 to V_2. The force exerted by the liquid on the pipe wall due to the change of momentum is $\dot{m}(V_1 - V_2)$, as before but there is an additional axial force $(p_1 A_1 - p_2 A_2)$ due to the pressure of the liquid.

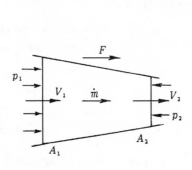

Figure 35.13

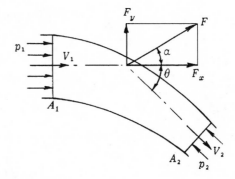

Figure 35.14

Thus total force on pipe,

$$F = \dot{m}(V_1 - V_2) + (p_1A_1 - p_2A_2) \qquad (35.14)$$

If the pipe turns through an angle θ between inlet and outlet, as shown in Figure 35.14, then horizontal force on bend,

$$F_x = \dot{m}(V_1 - V_2 \cos\theta) + (p_1A_1 - p_2A_2 \cos\theta)$$

and vertical force on bend,

$$F_y = \dot{m}(0 + V_2 \sin\theta) + (0 + p_2A_2 \sin\theta)$$

The resultant force, $F = \sqrt{(F_x^2 + F_y^2)}$

and $\qquad\qquad \alpha = \tan^{-1}\dfrac{F_y}{F_x}$

35.7 Jet and propeller propulsion

The aircraft and boat shown in Figure 35.15 move with velocity v while the fluid leaves with velocity V relative to the craft.

The change of momentum of the fluid is $\dot{m}(V - v)$ and hence the propelling force,

$$F = \dot{m}(V - v)$$

where $\dot{m}$ is the mass of fluid ejected per second.

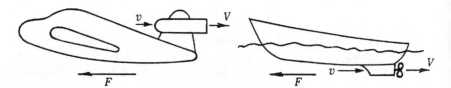

Figure 35.15

The propulsive efficiency is defined by

$$\eta = \frac{\text{work done on craft}}{\text{K.E. supplied to propellant}}$$

$$= \frac{Fv}{\tfrac{1}{2}\dot{m}(V^2 - v^2)}$$

$$= \frac{\dot{m}(V - v)v}{\tfrac{1}{2}\dot{m}(V^2 - v^2)} = \frac{2v}{V + v}$$

The efficiency rises as v rises but the thrust decreases and hence a compromise is needed.

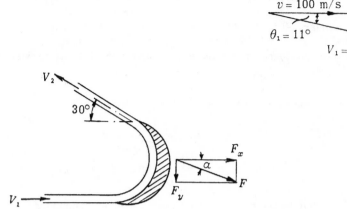

Figure 35.16   **Figure 35.17**

Worked examples

1. *A jet of water flows smoothly on to a stationary curved vane and turns through 150°C. The jet is 25 mm in diameter and is at atmospheric pressure. The jet leaves the vane with 80% of its entry velocity. If the flow rate is 0.01 m^3/s, determine the force on the vane.*

The vane is shown in Figure 35.16.

$$\text{Mass flow rate} = 0.01 \times 10^3 = 10 \text{ kg/s}$$

$$\text{Velocity of water entering vane, } V_1 = \frac{0.01}{\frac{\pi}{4} \times 0.025^2} = 20.4 \text{ m/s}$$

$$\text{Velocity of water leaving vane, } V_2 = 0.8 \times 20.4 = 16.3 \text{ m/s}$$

In the horizontal direction, change of momentum per second

$$= \dot{m}(V_1 - [-V_2 \cos 30°]) = \dot{m}(V_1 + V_2 \cos 30°)$$

i.e. $F_x = 10(20.4 + 16.3 \times 0.866) = 345 \text{ N}$

In the vertical direction, change of momentum per second

$$= \dot{m}(0 + V_2 \sin 30°)$$

i.e. $F_y = 10 \times 16.3 \times 0.5 = 81.5 \text{ N}$

The resultant force, $F = \sqrt{(345^2 + 81.5^2)} = \underline{354 \text{ N}}$

and the angle of the resultant to the horizontal is given by

$$\alpha = \tan^{-1} \frac{81.5}{345} = \underline{13.3°}$$

2. *Water is supplied to a Pelton wheel through a nozzle 25 mm diameter with velocity 60 m/s. The buckets move in a path of mean radius 0.3 m and deflect the water through 160°C. Calculate the speed of the wheel for maximum power and the value of this power (a) if the motion of the water over the buckets is assumed to be frictionless and (b) if friction reduces the velocity of the water as it passes over the vane by 10%.*

(*a*) For maximum efficiency, $v = \frac{1}{2}V = 30$ m/s   (from Section 35.4)

$$\therefore \omega = \frac{v}{R} = \frac{30}{0.3} = 100 \text{ rad/s}$$

$$\therefore N = \frac{60}{2\pi} \times 100 = \underline{955 \text{ rev/min}}$$

Maximum power $= \dot{m}(V - v)(1 + \cos\theta)v$ from equation (35.9)

$$= 10^3 \times \frac{\pi}{4} \times 0.025^2 \times 60(60 - 30)(1 + \cos 20°) \times 30$$

$$= 51\,400 \text{ W}   \text{ or }   \underline{51.4 \text{ kW}}$$

(*b*) If the exit velocity is reduced by 10%,

final momentum in original direction of jet $= -\rho AV \times 0.9\ V \cos\theta$

$$\therefore F = \rho AV^2(1 + 0.9\cos\theta)$$

The presence of the 0.9 in the formula for efficiency will not affect the condition for maximum efficiency, so that the speed for maximum efficiency is 955 rev/min, as before.

The power will be reduced in the ratio $\dfrac{1 + 0.9\cos\theta}{1 + \cos\theta}$

$$\text{so that maximum power} = \frac{1 + 0.9\cos 20°}{1 + \cos 20°} \times 51.4$$

$$= \underline{48.8 \text{ kW}}$$

3. *An impulse steam turbine has blades with equal inlet and outlet angles and the pressure across the blade is constant. Steam enters with a velocity of 270 m/s, inclined at 11° to the direction of motion of the blades, and leaves with 90% of the relative velocity at which it entered. Calculate the mass flow rate of steam if the turbine is to develop 300 kW at a blade speed of 100 m/s.*

Figure 35.17 shows the blade with the velocity triangles at inlet and outlet.
From the geometry of the inlet triangle, the velocity of the steam relative to the blade,

$$u_1 = 173 \text{ m/s}$$

and the angle it makes with the direction of motion is 17.3°.

The relative velocity of the steam at outlet,

$$u_2 = 0.9 \times 173 = 155.7 \text{ m/s}$$

Due to the symmetry of the blade, this also makes an angle of 17.3° with the direction of motion.

From the outlet triangle, $V_2 = 68$ m/s

and $\theta_2 = 43°$

Force on blade in direction of motion,

$$F = \dot{m}(V_1 \cos\theta_1 + V_2 \cos\theta_2)$$

$$= \dot{m}(270 \cos 11° + 68 \cos 43°) = 314\,\dot{m}$$

Power = force × velocity

i.e. $300 \times 10^3 = 314\,\dot{m} \times 100$

$$\therefore \dot{m} = \underline{9.56 \text{ kg/s}}$$

4. *Water at a constant gauge pressure of 70 kN/m² flows through a pipe at 3 m/s. The pipe, which is 200 mm diameter, makes a 30° bend. Determine the force on the bend.*

Mass flow rate, $\dot{m} = \rho A V$

$$= 10^3 \times \frac{\pi}{4} \times 0.2^2 \times 3 = 94.2 \text{ kg/s}$$

Let F_x and F_y be the horizontal and vertical components of the force on the bend, Figure 35.18(a).

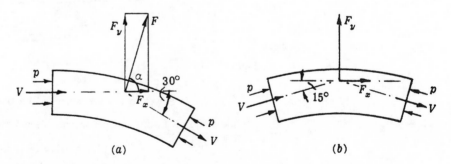

(a)   (b)

Figure 35.18

Then, as in Section 35.6,

$$F_x = \dot{m}(V - V\cos 30°) + (pA - pA\cos 30°)$$

$$= 94.2 \times 3(1 - 0.866) + 70 \times 10^3 \times \frac{\pi}{4} \times 0.2^2(1 - 0.866) = 323 \text{ N}$$

$$F_y = \dot{m}(0 + V\sin 30°) + (0 + pA\sin 30°)$$

$$= 94.2 \times 3 \times 0.5 + 70 \times 10^3 \times \frac{\pi}{4} \times 0.2^2 \times 0.5 = 1\,240 \text{ N}$$

$$\therefore \quad F = \sqrt{(323^2 + 1\,240^2)} = \underline{1\,280 \text{ N}}$$

and $\quad \alpha = \tan^{-1} \dfrac{1\,240}{323} = \underline{75°}$

Alternatively, due to the symmetry of the bend, Figure 35.18(*b*),

$$F_x = \dot{m}(V \cos 15° - V \cos 15°) + (pA \cos 15° - pA \cos 15°)$$

$$= 0$$

and $F_y = \dot{m}(V \sin 15° + V \sin 15°) + (pA \sin 15° + pA \sin 15°)$

$$= 2 \times 94.2 \times 3 \times 0.259 + 2 \times 70 \times 10^3 \times \frac{\pi}{4} \times 0.2^2 \times 0.259$$

$$= \underline{1\,280 \text{ N}}$$

This is directed along the axis of symmetry.

5. *A speedboat is propelled by a circular jet of fresh water. The water enters the system through a forward-facing intake and is pumped horizontally through the stern. At 8 m/s the resistance of the boat is 950 N and the power supply to the pump is 14 kW. The pump efficiency is 90%.*

Ignoring the difference in elevation between the intake and the jet, determine the mass flow rate of water, the jet diameter and the efficiency of the propulsion system.

Let $\dot{m}$ be the mass flow rate, v be the velocity of the boat and V be the velocity of the jet relative to the boat, Figure 35.19.

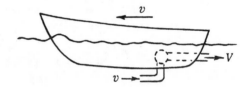

Figure 35.19

Power supplied to water $= 14 \times 0.9 = 12.6$ kW

Increase in K.E. of water per second $= \frac{1}{2}\dot{m}(V^2 - v^2)$

$$\therefore \quad 12.6 \times 10^3 = \frac{1}{2}\dot{m}(V^2 - 8^2)$$

or $\qquad\qquad\qquad\qquad 25\,200 = \dot{m}(V^2 - 64) \qquad\qquad\qquad (1)$

Propelling force on boat $= \dot{m}(V - v)$

i.e. $\qquad\qquad\qquad\qquad\qquad 950 = \dot{m}(V - 8) \qquad\qquad\qquad\qquad (2)$

From equations (1) and (2), $\quad \dfrac{25\,200}{950} = \dfrac{V^2 - 64}{V - 8} = V + 8$

$$\therefore \quad V = 18.5 \text{ m/s}$$

and $\qquad\qquad\qquad\qquad\qquad \dot{m} = \dfrac{950}{18.5 - 8} = \underline{90.5 \text{ kg/s}}$

If the jet diameter is d, then $\qquad \dot{m} = \rho A V = \rho \frac{\pi}{4} d^2 V$

i.e. $\qquad\qquad\qquad\qquad 90.5 = 10^3 \times \frac{\pi}{4} d^2 \times 18.5$

from which $\qquad\qquad\qquad\qquad d = \underline{0.079 \text{ m}}$

$$\text{Propulsion efficiency} = \frac{\text{output power}}{\text{input power}} = \frac{\text{thrust} \times \text{velocity}}{\text{energy supply to pump}}$$

$$= \frac{950 \times 8}{14 \times 10^3} = 0.543 \quad \text{or} \quad \underline{54.3\%}$$

Further problems

6. A jet of water 25 mm in diameter and having a velocity of 8 m/s strikes a flat plate. Calculate the force on the plate (*a*) if it is stationary, (*b*) if it moves in the same direction as the jet at 3 m/s.   [31.42 N; 12.28 N]

7. A 20 m/s wind blows against the side of a tanker in ballast. The tanker is 250 m long and has an average freeboard of 12 m. The density of air is 1.2 kg/m³. Calculate the normal force on the ship when the wind is abeam (i.e. at 90° to the ship's axis).   [1.44 MN]

8. A jet of water 50 mm diameter issues under a head of 25 m and strikes a fixed flat plate. Find the force exerted on the plate (*a*) when the plate is perpendicular to the axis of the jet, and (*b*) when the plate is inclined at 30° to the axis of the jet.   [963 N; 481.5 N]

9. A jet of water issues from a sharp-edged orifice, 80 mm diameter, in the vertical side of a tank, under a head of 3.5 m. The coefficients to contraction and velocity are 0.64 and 0.97 respectively. If a vertical plate is fixed in the path of the jet, find the force exerted on the plate.   [208 N]

10. A jet of water issuing from a nozzle under a pressure head of 40 m strikes normally against a fixed flat vane and exerts on it a force of 800 N. What must be the diameter of the nozzle?

If, instead of a single fixed vane, there were a series of flat vanes moving in the direction of the jet at a speed of 12 m/s, each normal to the jet, what force would then be exerted on the vanes and at what rate would work be done on the vanes by the jet? Comparing the work done with the energy of the jet, what efficiency does this represent?   [36 mm; 457.5 N; 5.49 kW; 49%]

11. A jet of water, 50 mm diameter, issues from a nozzle under a head of 60 m and strikes a series of moving vanes, each of which deflects the jet through an angle of 150°. If the velocity of the vanes in the direction of the jet is 10 m/s, calculate (*a*) the force acting on each vane, (*b*) the work done per second, (*c*) the efficiency.   [3.055 kN; 30.55 kW; 77%]

12. A jet of water 80 mm diameter, moving with a velocity of 25 m/s, flows horizontally on to a stationary curved vane which deflects it through 120°. Calculate the magnitude and direction of the resultant force on the vane.

If the jet impinges on a series of such vanes moving at 12 m/s in the direction of the jet, determine the force on the vanes in the direction of the jet. At what speed must the vanes move for maximum work transfer rate?   [5.45 kN at 30° to jet; 2.45 kN; 12.5 m/s]

13. A Pelton wheel works under an effective head of 700 m. The jet is 150 mm diameter, and the efficiency is 89%. The jet is deflected through an angle of 155°. Calculate the quantity of water discharged, the power and the speed in rev/min if the mean diameter of the wheel is 1 m.

[2.07 m^3/s; 12.66 MW; 834 or 1 410 rev/min]

14. The buckets of a Pelton wheel deflect a jet, having a velocity of 60 m/s, through an angle of 160°. Assuming that the velocity of the jet relative to the buckets is reduced by 15% as the jet water moves over them, find the efficiency of the wheel if the speed of the buckets is 27 m/s and, using this efficiency, calculate the diameter of the jet so that the wheel may develop 200 kW.   [89%; 51.5 mm]

15. A Pelton wheel water turbine has blades shaped as those in Figure 35.8(*a*). The blade speed is 14 m/s and the turbine receives 0.7 m^3/s of water from a reservoir with a water level 30 m above the jet. The blades deflect the jet through 160°. Neglecting friction in the pipe, determine the power output and the efficiency of the turbine.   [193.5 kW; 94%]

16. Water enters the blades of an impulse turbine with a velocity of 70 m/s in a direction inclined at 20° to the direction of motion of the blades. The velocity of the blades is 27 m/s and they are symmetrical in section, i.e., the inlet and outlet angles are equal. When the mass flow rate of water is 150 kg/s, determine the force on the blades and the power output.   [11.6 kN; 313 kW]

17. A 90° horizontal bend reduces a pipe from 200 mm to 100 mm in diameter. The gauge pressure at entry to the 200 mm end is 300 kN/m^2. If the flow rate through the pipe is 0.1 m^3/s, determine the magnitude of the force on the bend.

[10.2 kN]

18. A horizontal bend reduces a pipe from 150 mm diameter to 75 mm diameter and deflects the axis through 60°. The gauge pressure in the 150 mm diameter pipe is 350 kN/m^2 and the flow is 0.085 m^3/s. Find the forces on the bend parallel and normal to the 150 mm diameter pipe.   [5.386 kN; 2.092 kN]

19. The resistance to motion of a vessel in fresh water is 22 kN at 5 m/s. The vessel is propelled by water jet; the water enters through a forward-facing intake at the bow and is discharged at the stern. The propulsive efficiency is 0.8 and the mechanical efficiency of the jet pumps is 0.75. Determine the jet velocity, the jet exit area and the power required to drive the pumps.

[7.95 m/s; 0.938 m^2; 183.5 kW]

36 Viscosity

36.1 Introduction

The resistance to flow of a fluid is due to molecular cohesion, which results in a shearing action as layers of the fluid slide relative to each other. This resistance to shear stress is a measure of the *viscosity* of the fluid; simple experiments will show that viscosity is different for different fluids and that it varies with temperature. Thus oil is more viscous than water and its viscosity falls as the temperature rises.

36.2. Dynamic and kinematic viscosity

Most fluids have a viscosity which is independent of the rate of shearing and are known as *Newtonian fluids*. However there are fluids which are *non-Newtonian* and their special properties are shown in Figure 36.1(a).

(a) A Bingham plastic fluid can be stressed until it yields after which it behaves as a Newtonian fluid with constant viscosity.
(b) A pseudo-plastic fluid has a viscosity which decreases as the rate of shear increases. Clay, milk and blood behave in this way.
(c) A dilatent fluid has a viscosity which increases as the rate of shear increases. Concentrated solutions of sugar in water behave in this way.

Other fluids showing non-Newtonian behaviour include thixotropic fluids whose viscosity depends on the duration of the stress, and viscoelastic materials which exhibit both viscous and elastic properties depending on the rate of application of stress. Bitumen behaves in this way.

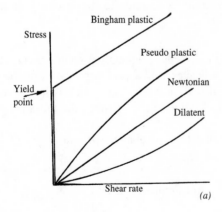
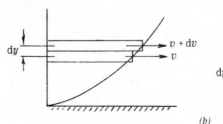
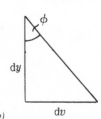

(a)   *(b)*

Figure 36.1

The study of non-Newtonian fluids is known as *rheology* but only Newtonian fluids are considered further in this book.

When a fluid flows smoothly over a stationary boundary, Figure 36.1(*b*), the layer in contact with the boundary is at rest and subsequent layers move with increasing velocities as the distance from the boundary increases. Thus there is a velocity gradient across the section of flow.

There is a shearing action between adjacent layers resulting in a loss of energy as the resisting force is overcome. If the velocity changes by dv in a distance dy perpendicular to the direction of flow, Figure 36.1(*b*), the viscous strain rate, $\phi = dv/dy$.

The viscous stress τ is the viscous resistance per unit area and the coefficient of viscosity μ is defined as the ratio viscous stress/viscous strain rate,

i.e.
$$\mu = \frac{\tau}{\phi} = \frac{\tau}{dv/dy}$$

or
$$\tau = \mu \frac{dv}{dy} \qquad (36.1)$$

The coefficient of viscosity is also called the *dynamic viscosity* to distinguish it from the *kinematic viscosity* v, defined by

$$v = \frac{\mu}{\rho} \qquad (36.2)$$

Kinematic viscosity is not a fundamental property of a fluid but the ratio μ/ρ occurs frequently in problems of fluid motion and it is found convenient to give it a separate name.

The units of μ are kg/m s   (1 poise $= 0.1$ kg/m s)

and the units of v are m^2/s   (1 stoke $= 10^{-4}$ m^2/s).

36.3 Measurement of viscosity

Viscosity is usually determined by a *viscometer*, which measures the time taken for a fixed volume of liquid to flow through a capillary tube. A calibration table then relates the time in seconds to the viscosity of the liquid.

36.4 Redwood viscometer

Figure 36.2 shows the Redwood viscometer, frequently used in Britain. The liquid under test is contained in a beaker with a capillary tube in the base. The liquid temperature is controlled by a water bath which is heated and stirred. The time taken for 50 ml of liquid to pass through the tubes is then observed.

For viscous flow through a round tube of diameter d and length l, the velocity of flow is given by

$$h_f = \frac{32\mu l V}{\rho g d^2} \qquad \text{from equation (37.5)}$$

where h_f is the head loss due to viscous friction.

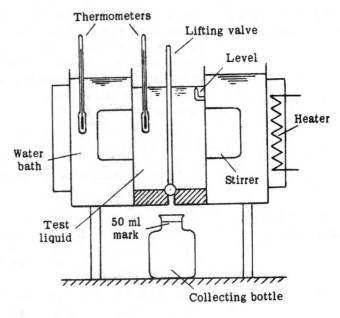

Figure 36.2

Thus quantity flowing,

$$Q = \frac{\pi}{4}d^2 \times V = \frac{\pi}{4}d^2 \times \frac{h_f \rho g d^2}{3\mu l}$$

or

$$Q \propto \frac{\rho}{\mu} \propto \frac{1}{\nu}$$

If t is the time for a fixed volume of liquid to pass, then $t \propto \dfrac{1}{Q}$, so that

$$t \propto \nu \qquad (36.3)$$

This equation is not quite correct since part of the available head is used to produce velocity rather than overcome friction losses in the tube, so that a more correct equation is

$$\nu = At - \frac{B}{t}, \quad \text{where } A \text{ and } B \text{ are constants.}$$

This equation is embodied in the conversion table provided with the equipment, relating time (Redwood seconds) with kinematic viscosity.

The Redwood No. 1 viscometer is suitable for viscosities less than $50 \times 10^{-4} \text{ m}^2/\text{s}$ and the Redwood No. 2 viscometer, having a larger diameter exit tube, is used for more viscous liquids. The time from the No. 2 viscometer is multiplied by 10 to convert to Redwood (No. 1) seconds.

36.5 British Standard U-tube viscometer

In the U-tube viscometer, Figure 36.3, the liquid under test flows from a reservoir through a capillary tube. The instrument is initially filled to the filling mark and placed vertically in a constant temperature water bath. The liquid is then forced by pressure to the top of the other limb and allowed to flow back through the capillary tube, being timed between the two marks shown. The kinematic viscosity is again proportional to the time taken to flow through the capillary tube,

$$v = Ct \tag{36.4}$$

The value of the constant C is supplied with the instrument.

36.6 Stokes's law and terminal velocity

When a small sphere of diameter d falls with velocity v through a liquid of viscosity μ, the force R opposing the motion is given by

$$R = 3\pi\mu vd \tag{36.5}$$

provided that $Re < 0.1$ (see Section 37.4).

This is known as Stokes's law.

If the sphere is released from rest, the velocity theoretically never reaches a limiting value but it effectively reaches a terminal velocity V in an extremely

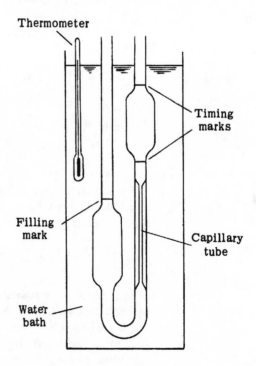

Figure 36.3

short time. The motion is then governed by the equilibrium equation,

downward force (weight) = upward force (buoyancy) + resistance

i.e.
$$\frac{\pi}{6}d^3\rho_s g = \frac{\pi}{6}d^3\rho g + 3\pi\mu V d$$

where ρ_s and ρ are the densities of the sphere and liquid respectively.

Hence
$$\mu = \frac{(\rho_s - \rho)gd^2}{18V} \tag{36.6}$$

The initial time taken to reach equilibrium conditions may be eliminated by measuring V from a point below the liquid surface.

36.7 Viscosity determination by falling ball

In the falling ball viscometer, Figure 36.4, the instrument is held vertically in a water bath and the velocity of the ball is determined by timing its motion between two marks. Equation (36.6) could then be used to determine the viscosity of the liquid but this assumes that the liquid surrounding the ball is of infinite extent.

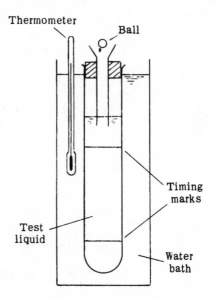

Figure 36.4

Due to the disturbance caused by the viscometer tube, equation (36.6) is multiplied by a correction factor F which varies with the ratio d/D, where D is the tube diameter and d the ball diameter.

Thus
$$\mu = \frac{(\rho_s - \rho)gd^2}{18V} \times F$$

The ball size is chosen to ensure that $V < 1$ cm/s and that $d/D < 0.1$. When $d/D = 0.1$, F is approximately 0.8.

Worked examples

1. *A piston, 50 mm diameter and 75 mm long, moves vertically in an open-ended lubricated cylinder with a radial clearance of 0.1 mm. When falling due to its own weight, the piston moves through 30 mm in 4.2 s at a uniform velocity. When a mass of 0.05 kg is added to the piston, it moves with uniform velocity through the same distance in 2.4 s. Calculate the viscosity of the oil and the mass of the piston*

$$\tau = \mu \frac{\mathrm{d}v}{\mathrm{d}y} \quad \text{from equation (36.1)}$$

If the mass of the piston is m, then

$$\tau = \frac{\text{axial force}}{\text{curved surface area of piston}}$$

$$= \frac{mg}{\pi \times 0.05 \times 0.075} = 84.88mg \text{ N/m}^2$$

$$\mathrm{d}v = \frac{0.03}{4.2} = 0.007\,143 \text{ m/s}$$

$$\mathrm{d}y = 0.1 \times 10^{-3} \text{ m}$$

$$\therefore 84.88 \ mg = \mu \times \frac{0.007\,143}{0.1 \times 10^{-3}}$$

$$\therefore mg = 0.841\,5\mu \tag{1}$$

When the additional 0.05 kg is added to the piston,

$$\tau = 84.88(m + 0.05)g$$

and $$\mathrm{d}v = \frac{0.03}{2.4} = 0.012\,5 \text{ m/s}$$

$$\therefore 84.88 \ (m + 0.05)g = \mu \times \frac{0.012\,5}{0.1 \times 10^{-3}}$$

$$\therefore (m + 0.05)g = 1.473\mu \tag{2}$$

Therefore, from equations (1) and (2),

$$\mu = \underline{0.777 \text{ kg/m s}}$$

and $$m = \underline{0.066\,6 \text{ kg}}$$

2. *A steel sphere, 1.6 mm diameter, has a mass of 16.6 mg. The ball falls at its terminal velocity in oil of density 950 kg/m^3, covering 300 mm in 34 s. Neglecting the wall effects of the enclosing vessel, determine the viscosity of the oil.*

$$V = \frac{0.3}{34} = 8.82 \times 10^{-3} \text{ m/s}$$

Assuming that Stokes's law is applicable,

$$\mu = \frac{(\rho_s - \rho)gd^2}{18V} \quad \text{from equation (36.6)}$$

$$\rho_s = \frac{\text{mass of sphere}}{\text{volume of sphere}}$$

$$= \frac{16.6 \times 10^{-6}}{\frac{\pi}{6} \times (1.6 \times 10^{-3})^3} = 7\,740 \text{ kg/m}^3$$

$$\therefore \mu = \frac{(7\,740 - 950) \times 9.81 \times (1.6 \times 10^{-3})^2}{18 \times 8.82 \times 10^{-3}}$$

$$= 1.074 \text{ kg/m s}$$

The validity of Stokes's law may now be verified, i.e. that $Re < 0.1$.

$$Re = \frac{\rho V d}{\mu} = \frac{950 \times 8.82 \times 10^{-3} \times 1.6 \times 10^{-3}}{1.074}$$

$$= \underline{0.012\,5}$$

Further problems

3. A piston 100 mm diameter and 200 mm long is enclosed in an open-ended lubricated cylinder containing oil of viscosity 0.8 kg/m s. The radial clearance between the piston and cylinder is 0.2 mm.

Calculate (*a*) the axial force required to move the piston at a uniform speed of 0.5 m/s;

(*b*) the torque required to rotate the piston at a uniform speed of 2 rad/s.
[126 N; 1.26 N m]

4. Assuming that Reynolds' number is 0.1, determine the settling velocity of the largest spherical particles of density $2\,000$ kg/m^3 which will fall from air of density 1.23 kg/m^3 and kinematic viscosity 15×10^{-6} m^2/s in accordance with Stokes's law. What is the mass of the particles?

[0.059 m/s; 1.7×10^{-8} g]

5. Assuming Stokes's law applies, find the time taken for a uniform suspension of spherical particles of density $3\,000$ kg/m^3 and diameter 0.02 mm to clear from water of depth 3 m. The viscosity of the water is 0.013 poise.

Check the validity of the assumption.
[2.485 h]

37 Real fluid flow

37.1 Introduction

In Chapter 34, the flow rates predicted by simple theory were not in agreement with measured values and the discrepancies were corrected by experimentally determined coefficients of discharge. These differences are due to the effects of viscosity, pipe walls, fittings, etc., and it is now necessary to determine the losses due to these causes.

37.2 Steady, unsteady and quasi-steady flow

If the velocity of flow at a point in a fluid stream does not vary with time, the flow is said to be *steady*, and if it does vary with time, the flow is *unsteady*. In some cases, such as the flow from a reservoir, the level falls very slowly as water is drawn off and the velocity in the discharge pipe may be treated as steady, although varying slightly with time. Such flow is termed *quasi-steady*.

37.3 Uniform and non-uniform flow

If the velocity of flow is constant at various points in a region, the flow is *uniform* but if it varies from point to point, it is *non-uniform*. Non-uniformity may occur in the direction of flow, such as in a tapering pipe, or normal to the direction of flow, such as in the cross-section of a pipe or duct.

37.4 Reynolds's experiment

If a duplicate of Reynolds's original experiment. Figure 37.1(*a*), is performed, in which a coloured dye is fed into a moving stream of water, it is found that at low speeds, the dye stream remains in a straight line, Figure 37.1(*b*), and the flow is described as *laminar* or *viscous*. However, at high speeds, the dye becomes mixed throughout the water. Figure 37.1(*c*), and the flow is described as *turbulent*. The speed at which the flow changes from laminar to turbulent is not clearly defined and there is a transition zone between the two patterns.

It is found that the change of pattern depends upon the velocity of flow V, the tube diameter d and the kinematic viscosity v, and the ratio $\dfrac{Vd}{v}$ $\left(\text{or } \dfrac{\rho V d}{\mu}\right)$ is known as the *Reynolds number (Re)*.

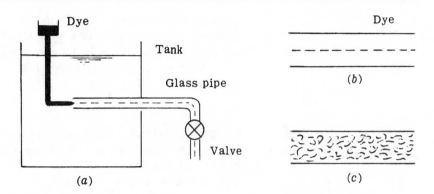

Figure 37.1

It is also found that the transition occurs at slightly higher values of *Re* when the velocity of flow is increased than when it is decreased. These values are termed the *upper critical number* and *lower critical number* respectively; the lower value is considered the more important and is that defined as the *critical Reynolds number*. In round pipes, flow is laminar for *Re* less than 2 000 and turbulent for *Re* greater than 4 000. The transition zone lying between these two values has flow which is unstable and indeterminate.

For liquids of low viscosity such as water, the velocity of flow at which the transition occurs is quite low and in most hydraulics problems, the flow will be turbulent.

37.5 Velocity distribution

If a pitot tube is traversed across the diameter of a pipe conveying a moving liquid, it is found that the flow is non-uniform. Figure 37.2 shows the velocity profiles for laminar and turbulent flow; for laminar flow, the profile is parabolic, as shown in Section 37.8 but for turbulent flow, the velocity is almost uniform, except in the vicinity of the walls. In both cases, the velocity at the pipe wall is zero, and it is therefore impossible to have uniform flow adjacent to a boundary.

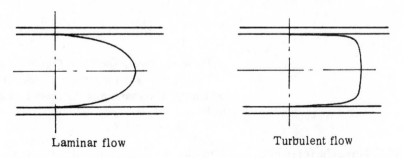

Laminar flow   Turbulent flow

Figure 37.2

To determine the flow rate for non-uniform velocity distribution, it is necessary to integrate the flow over small elements of the pipe section. For turbulent flow, the method was shown in Section 34.8 but for laminar flow, the flow rate can be determined by mathematical integration, as in Section 37.8.

37.6 Loss of head in pipes due to friction

Reynolds performed further experiments to determine the loss of head in a pipe for both laminar and turbulent flow. The method is shown in Figure 37.3 and it was found that the loss of head, h_f, over a length l increased as the velocity increased.

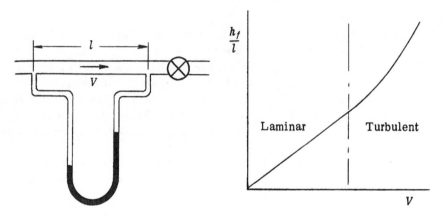

Figure 37.3   Figure 37.4

For laminar flow, the loss of head per unit length.

$$\frac{h_f}{l} \propto V$$

but for turbulent flow, the loss of head increased at a greater rate, given by

$$\frac{h_f}{l} \propto V^n$$

These results are shown in Figure 37.4.

The friction loss for turbulent water flow had already been investigated by Darcy, who found that the loss of head was approximately proportional to V^2.

37.7 Darcy's formula

Consider the flow of liquid through a pipe of diameter d and length l, Figure 37.5, and let the viscous stress at the wall be τ.

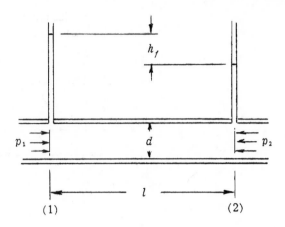

Figure 37.5

Equating forces on the liquid between sections (1) and (2),

$$p_1 \times \frac{\pi}{4}d^2 - p_2 \times \frac{\pi}{4}d^2 = \tau \times \pi dl$$

Thus the loss of head due to friction,

$$h_f = \frac{p_1 - p_2}{\rho g} = \frac{4\tau l}{\rho g d}$$

A dimensionless friction coefficient f is defined by the relation $f = \tau/\frac{1}{2}\rho V^2$ where V is the average velocity in the pipe,

so that
$$h_f = \frac{4\left(\frac{1}{2}\rho V^2 f\right)l}{\rho g d} = \frac{4flV^2}{2gd} \qquad (37.1)$$

This is known as *Darcy's formula*.

The loss of head per unit length, $\dfrac{4f}{d} \cdot \dfrac{V^2}{2g}$, is termed the *hydraulic gradient* since it represents the gradient on which the pipe must be laid to maintain a uniform pressure. The hydraulic gradient is denoted by i, so that

$$h_f = il \qquad (37.2)$$

37.8 Laminar flow through round pipes

Figure 37.6 shows the velocity distribution in a pipe of radius r and length l.

When the radius increases from y to $y + dy$, the velocity *decreases* by dv, so that viscous strain rate $= -dv/dy$.

Viscous resistance of a cylinder of radius y and length l

$$= 2\pi yl \times \tau = 2\pi yl \times -\mu\frac{dv}{dy} \quad \text{from Section 36.2.}$$

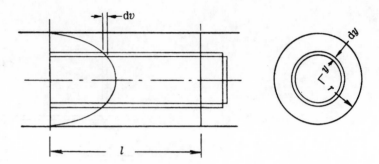

Figure 37.6

This resistance is equal to the difference in the forces on the ends of the cylinder,

i.e.
$$-2\pi y l \mu \frac{dv}{dy} = \rho g h_f \times \pi y^2 = \rho g i l \pi y^2$$

from which
$$\frac{dv}{dy} = -\frac{\rho g i}{2\mu} \cdot y$$

$$\therefore v = -\frac{\rho g i}{4\mu} y^2 + c$$

But $v = 0$ at $y = r$, so that
$$0 = -\frac{\rho g i}{4\mu} r^2 + c$$

from which
$$v = \frac{\rho g i}{4\mu}(r^2 - y^2) \qquad (37.3)$$

Equation (37.3) shows that the velocity distribution in laminar flow is parabolic. If V is the mean velocity, the flow rate is given by

$$Q = V \times \pi r^2 = \int_0^r \frac{\rho g i}{4\mu}(r^2 - y^2) \cdot 2\pi y \, dy$$

$$= \frac{\pi \rho g i}{2\mu} \int_0^r (r^2 y - y^3) \, dy$$

$$= \frac{\pi \rho g i r^4}{8\mu} \qquad (37.4)$$

Equation (37.4) is known as Poiseuille's equation

Thus
$$V = \frac{\rho g i r^2}{8\mu} = \frac{\rho g i d^2}{32\mu}$$

$$\therefore h_f = il = \frac{32\mu l V}{\rho g d^2} \qquad (37.5)$$

$$= \frac{32 l V^2}{g d \left(\dfrac{\rho V d}{\mu}\right)}$$

The term $\dfrac{\rho V d}{\mu}$ $\left(\text{or } \dfrac{V d}{v}\right)$ is called the *Reynolds number* and is denoted by Re.

Thus
$$h_f = \frac{32 l V^2}{g d (Re)}$$

Comparing this with the Darcy formula, equation (37.1),

$$h_f = \frac{4 f l V^2}{2 g d}$$

it will be seen that $\qquad f = \dfrac{16}{Re} \quad$ for laminar flow $\qquad$ (37.6)

37.9 Turbulent flow through round pipes

It is difficult to obtain a Reynolds number for water low enough to obtain laminar flow; thus the flow is generally turbulent, with a velocity profile as shown in Figure 37.2. It is therefore impracticable to analyse the motion analytically, as for laminar flow, and the friction coefficient f in Darcy's formula is obtained experimentally.

It is found that for turbulent flow in *smooth* pipes,

$$f = \frac{0.079}{(Re)^{1/4}} \qquad (37.7)$$

and substituting this in the Darcy formula gives $h_f \propto V^{1.75}$. However, it is usually sufficient to assume that $h_f \propto V^2$, i.e. that f is a constant.

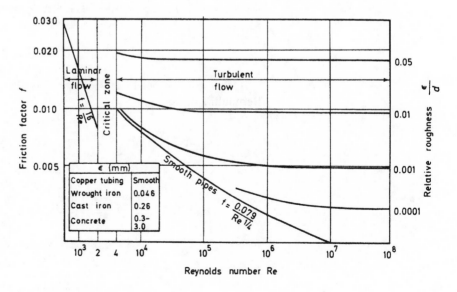

Figure 37.7

For *rough* pipes, the surface texture affects the value of f and reference may be made to the Moody diagram, Figure 37.7, which relates the values of f and Re for different degrees of roughness.

37.10 Losses due to sudden enlargement and contraction in turbulent flow

(a) Sudden enlargement

When the section of a pipe is suddenly enlarged, the pattern of flow is as indicated in Figure 37.8. As the velocity head is reduced, the pressure at section (2) is higher than at section (1) but experiments show that the pressure immediately beyond the enlargement is the same as that at section (1).

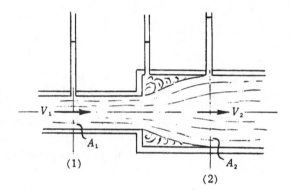

Figure 37.8

Net force acting from left to right = rate of change of momentum between sections (1) and (2)

i.e.
$$p_1 A_1 + p_1(A_2 - A_1) - p_2 A_2 = \dot{m}(V_2 - V_1)$$

i.e.
$$(p_1 - p_2)A_2 = \rho A_2 V_2(V_2 - V_1)$$

or
$$\frac{p_1}{\rho g} - \frac{p_2}{\rho g} = \frac{V_2}{g}(V_2 - V_1) \qquad \text{(a)}$$

If the loss of head between sections (1) and (2) is h_l, then from Bernoulli's equation,

$$\frac{p_1}{\rho g} + \frac{V_1^2}{2g} = \frac{p_2}{\rho g} + \frac{V_2^2}{2g} + h_l$$

Thus
$$\frac{p_1}{\rho g} - \frac{p_2}{\rho g} = \frac{V_2^2}{2g} - \frac{V_1^2}{2g} + h_l \qquad \text{(b)}$$

From (a) and (b)
$$\frac{V_2^2}{2g} - \frac{V_1^2}{2g} + h_l = \frac{V_2^2}{g} - \frac{V_1 V_2}{g}$$

$$\therefore h_l = \frac{V_1^2}{2g} - \frac{V_1 V_2}{g} + \frac{V_2^2}{2g}$$

$$= \frac{(V_1 - V_2)^2}{2g} \qquad (37.8)$$

Since $A_1V_1 = A_2V_2$, equation (37.8) may be expressed in the form

$$h_l = \left(\frac{A_2}{A_1} - 1\right)^2 \frac{V_2^2}{2g} \qquad (37.9)$$

(b) Sudden contraction

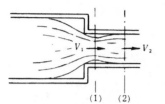

Figure 37.9

The flow pattern in a sudden contraction is shown in Figure 37.9, a vena contracta forming at section (1). The flow then expands to fill the pipe at section (2) and the only loss occurs in this expansion process. Thus, applying equation (37.9),

$$h_l = \frac{V_2^2}{2g}\left(\frac{A_2}{A_1} - 1\right)^2$$

$$= \frac{V_2^2}{2g}\left(\frac{1}{C_c} - 1\right)^2$$

where C_c is the coefficient of contraction.

By experiment, it is found that $\left(\dfrac{1}{C_c} - 1\right)^2$ is about 0.5, so that

$$h_l = 0.5\frac{V_2^2}{2g} \qquad (37.10)$$

An entry to a tank or exit from a pipe may be treated as an enlargement where A_2 is infinite, so that the loss is $V^2/2g$, i.e. all the kinetic energy at entry or exit is wasted.

The exit from a tank or entry to a pipe may be treated as a contraction, so that the loss is $0.5\dfrac{V^2}{2g}$ where V is the exit or entry velocity.

37.11 Other losses in turbulent pipe flow

In general, losses of head due to bends, valves, etc., are expressed as a fraction of the velocity head,

i.e. $\qquad h_l = k\dfrac{V^2}{2g}$ where k is a constant

Typical values of k are as follows:

Smooth bend	0.30	Angle valve (open)	5.0
Mitre bend	1.1	Gate valve (open)	0.19
90° elbow	0.9	Gate valve ($\frac{3}{4}$ open)	1.15
45° elbow	0.42	Gate valve ($\frac{1}{2}$ open)	5.6
Standard T	1.8	Conical enlargement (10°)	0.16
Globe valve (open)	10.0	Conical enlargement (25°)	0.55

In general terms, losses may be minimized by making changes in section or direction as gradual as possible. For example, a well-designed venturi meter involves practically no loss of head.

37.12 Total head and hydraulic gradient

Figure 37.10 shows a pipe line connecting two tanks or reservoirs. The total head at any section is represented by $Z + \dfrac{p}{\rho g} + \dfrac{V^2}{2g}$ and if this quantity is plotted along the pipe, it illustrates the distribution of the losses.

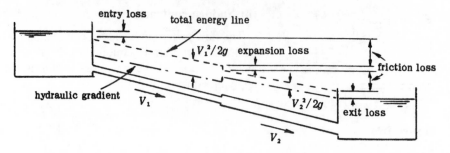

Figure 37.10

If the velocity head is subtracted from the total energy line, the remaining quantity $Z + \dfrac{p}{\rho g}$ represents the hydraulic gradient or pressure line. This shows the head available for distribution between pressure and elevation above datum and may be used to determine the pressure at points in a pipe line which varies in altitude.

Figure 37.11 shows a pipe line operating as a siphon over a hill. At A and B, the pressure is atmospheric and between these points, it is below atmospheric. The pressure should never be allowed to fall below about 20 kN/m^2 or 2 m, otherwise dissolved gases may evolve and flow will cease due to an air-lock.

In most pipe line problems, losses other than pipe friction are negligible.

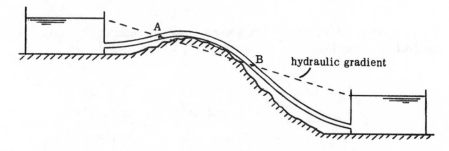

Figure 37.11

37.13 Combination of pipes

These are treated in the same way as combinations of electrical resistances.

(a) Pipes in series

For two pipes in series, Figure 37.12(a), the flow in each pipe is the same, i.e.
$$Q = A_1 V_1 = A_2 V_2$$
and the total loss of head is the sum of the losses in each pipe.

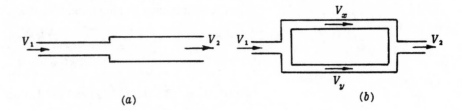

(a)   (b)

Figure 37.12

(b) Pipes in parallel

For two pipes in parallel, Figure 37.12(b), the flow in the main pipes is the sum of the flows in the two branches,

i.e. $$Q = A_1 V_1 = A_2 V_2 = A_x V_x + A_y V_y$$

and the loss of head in each branch must be the same.

(c) General case of branched pipes

Figure 37.13 shows a pipe from a reservoir A which divides into two branches to reservoirs B and C. It is required to find the rate of flow into each of the reservoirs B and C.

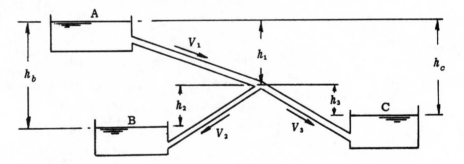

Figure 37.13

Neglecting all losses other than friction,

$$h_b = h_1 + h_2 = \frac{4f_1 l_1 V_1^2}{2gd_1} + \frac{4f_2 l_2 V_2^2}{2gd_2} \qquad (37.11)$$

$$h_c = h_1 + h_3 = \frac{4f_1 l_1 V_1^2}{2gd_1} + \frac{4f_3 l_3 V_3^2}{2gd_3} \qquad (37.12)$$

Also $$A_1 V_1 = A_2 V_2 + A_3 V_3 \qquad (37.13)$$

The unknown velocities V_1, V_2 and V_3 may be obtained from these equations and hence the discharges calculated.

Such problems may be simplified by writing

$$h_f = \frac{4flV^2}{2gd} = \frac{4flQ^2}{2 \times 9.81 \times d \times \left(\frac{\pi}{4}d^2\right)^2} \approx \frac{flQ^2}{3d^5}$$

when equations (37.11), (37.12) and (37.13) become

$$h_b = \frac{f_1 l_1 Q_1^2}{3d_1^5} + \frac{f_2 l_2 Q_2^2}{3d_2^5} \qquad (37.14)$$

$$h_c = \frac{f_1 l_1 Q_1^2}{3d_1^5} + \frac{f_3 l_3 Q_3^2}{3d_3^5} \qquad (37.15)$$

and
$$Q_1 = Q_2 + Q_3 \qquad (37.16)$$

The method of solving these equations is shown in Example 6.

37.14 Part-full pipes and open channel flow

From the Darcy formula, the hydraulic gradient for a pipe in which the only loss is due to friction is given by

$$i = \frac{h_f}{l} = \frac{4f}{d} \cdot \frac{V^2}{2g} = \frac{f}{\left(\dfrac{\frac{\pi}{4}d^2}{\pi d}\right)} \cdot \frac{V^2}{2g}$$

The term $\dfrac{\frac{\pi}{4}d^2}{\pi d}$ represents the ratio $\dfrac{\text{wetted cross-sectional area}}{\text{wetted perimeter}}$ and is termed the *hydraulic mean depth*. This is denoted by m, so that

$$i = \frac{f}{m} \cdot \frac{V^2}{2g}$$

Hence
$$V = \sqrt{\left(\frac{2g}{f} \cdot mi\right)} = c\sqrt{(mi)} \qquad (37.17)$$

This is known as the *Chezy formula* and may be used for pipes which are running part-full and for open channels provided that the appropriate value for m is used.

The value of c may be calculated from $\sqrt{\left(\dfrac{2g}{f}\right)}$ but for open channels it is more usually obtained from the Manning formula

$$c = \frac{m^{1/6}}{n}$$

where n, the roughness coefficient, varies from 0.01 for smooth surfaces to 0.03 for rivers.

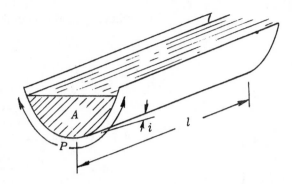

Figure 37.14

The parameter c has the units $m^{1/2}/s$ so that values for n and c depend upon the system of units used. The values of n quoted above are correct for S.I. units only.

The Chezy formula may be obtained from first principles by considering steady flow in a pipe or channel laid on a gradient i, Figure 37.14.

If the length of the channel is l, the wetted cross-section is A and the wetted perimeter is P, the component of the weight of liquid down the plane is equal to the frictional resisting force. For turbulent flow, the frictional resistance may be taken as proportional to the square of the velocity, so that the resistance per unit area, $R = kV^2$

Hence $$\rho g A l \times i = R \times P l = kV^2 \times P l$$

from which $$V = \sqrt{\left(\frac{\rho g}{k} \cdot \frac{A}{P} \cdot i \right)}$$

$$= c\sqrt{(mi)}$$

37.15 Power transmission in pipes with turbulent flow

Head represents energy per unit weight, as shown in Section 34.2. Thus the power P *available* for a flow rate Q under a head h is given by

$$P = \rho g Q h = \rho g A V h$$

For power transmitted a considerable distance by pipeline, the only significant loss of head is that due to friction. Thus, for a friction loss h_f, the *transmitted* power

$$P = \rho g A V (h - h_f)$$

and the transmission efficiency is given by

$$\eta = \frac{\text{power transmitted}}{\text{power available}}$$

$$= \frac{\rho g A V (h - h_f)}{\rho g A V h} = 1 - \frac{h_f}{h}$$

Assuming the friction coefficient f to be constant,

$$h_f = cV^2 \quad \text{where} \quad c = \frac{4fl}{2gd}$$

Thus

$$P = \rho gAV(h - cV^2)$$

$$= \rho gA(hV - cV^3)$$

For the power to be a maximum,

$$\frac{\mathrm{d}}{\mathrm{d}V}(hV - cV^3) = 0$$

i.e.

$$h - 3cV^2 = 0$$

or

$$h_f = \frac{h}{3}$$

For this condition, the transmission efficiency is $66\frac{2}{3}\%$.

Figure 37.15 shows the variation of P in terms of V. The value of V for maximum power is that which leads to $h_f = h/3$ and hence, for a given pipe length and roughness, the optimum pipe diameter is that which fulfils this relation.

Figure 37.15

37.16 Quasi-steady flow

As stated in Section 37.2, quasi-steady flow is unsteady but varying so slowly that it can be treated as steady.

Consider the case of a tank of uniform cross-sectional area A, Figure 37.16, with an orifice of area a in the base.

If the tank is filled to a depth H_1 and then allowed to discharge freely through the orifice, the outflow at any instant

$$Q = C_d a\sqrt{(2gh)}$$

where h is the instantaneous head over the orifice.

If the head falls a distance $\mathrm{d}h$ in time $\mathrm{d}t$, the reduction in volume of liquid in the tank is equal to the discharge through the orifice.

i.e.

$$A\,\mathrm{d}h = -C_d a\sqrt{(2gh)}\mathrm{d}t$$

The negative sign arises because h *decreases* as t *increases*.

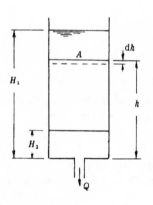

Figure 37.16

Thus

$$\mathrm{d}t = \frac{-A}{C_d a\sqrt{(2g)}} \cdot \frac{\mathrm{d}h}{h^{1/2}}$$

If T is the time taken for the head to fall from H_1 to H_2, then

$$\int_0^T \mathrm{d}t = \frac{-A}{C_d a\sqrt{(2g)}} \int_{H_1}^{H_2} \frac{\mathrm{d}h}{h^{1/2}}$$

from which

$$T = \frac{2A}{C_d a\sqrt{(2g)}}(\sqrt{H_1} - \sqrt{H_2}) \qquad (37.18)$$

If an experiment if performed in which the time T is noted for various values of H_2 as the tank discharges, a graph of T against $\sqrt{H_2}$ should be a straight line of slope $\dfrac{-2A}{C_d a \sqrt{(2g)}}$, from which C_d may be obtained.

As a further example of quasi-steady flow, consider the flow from reservoir 1 to reservoir 2, Figure 37.17.

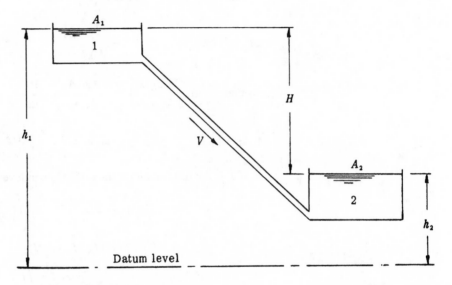

Figure 37.17

Let the surface areas of the reservoirs be A_1 and A_2 respectively, the cross-sectional area of the pipe be a and the velocity of flow be V. If the level in reservoir 1 falls by $\mathrm{d}h_1$ and that in reservoir 2 rises by $\mathrm{d}h_2$ in time $\mathrm{d}t$, then

$$Q = -A_1 \frac{\mathrm{d}h_1}{\mathrm{d}t} = A_2 \frac{\mathrm{d}h_2}{\mathrm{d}t} = aV$$

The difference in level, $\qquad H = h_1 - h_2$

Therefore, rate of change of level difference,

$$\frac{\mathrm{d}H}{\mathrm{d}t} = \frac{\mathrm{d}h_1}{\mathrm{d}t} - \frac{\mathrm{d}h_2}{\mathrm{d}t}$$

$$= \frac{\mathrm{d}h_1}{\mathrm{d}t}\left(1 + \frac{A_1}{A_2}\right) \quad \text{since} \quad \frac{\mathrm{d}h_2}{\mathrm{d}t} = -\frac{\mathrm{d}h_1}{\mathrm{d}t} \cdot \frac{A_1}{A_2}$$

$$\therefore \ Q = aV = \frac{-A_1}{1 + \dfrac{A_1}{A_2}} \cdot \frac{\mathrm{d}H}{\mathrm{d}t}$$

$$= -\frac{A_1 A_2}{A_1 + A_2} \cdot \frac{\mathrm{d}H}{\mathrm{d}t}$$

The losses of head between the two reservoirs due to friction and other causes are all multiples of $\dfrac{V^2}{2g}$ and so may be represented collectively by $c\dfrac{V^2}{2g}$.

Hence
$$H = c\frac{V^2}{2g}$$

or
$$V = \sqrt{\left(\frac{2gH}{c}\right)}$$

so that
$$a\sqrt{\left(\frac{2gH}{c}\right)} = -\frac{A_1 A_2}{A_1 + A_2} \cdot \frac{\mathrm{d}H}{\mathrm{d}t}$$

from which
$$\mathrm{d}t = -\frac{A_1 A_2}{a(A_1 + A_2)} \sqrt{\left(\frac{c}{2g}\right)} \frac{\mathrm{d}H}{\sqrt{H}}$$

If T is the time taken for the difference in level to change from H_1 to H_2,

$$\int_0^T \mathrm{d}t = -\frac{A_1 A_2}{a(A_1 + A_2)} \sqrt{\left(\frac{c}{2g}\right)} \int_{H_1}^{H_2} \frac{\mathrm{d}H}{\sqrt{H}}$$

i.e.
$$T = k(\sqrt{H_1} - \sqrt{H_2}) \qquad (37.19)$$

where
$$k = \frac{2A_1 A_2}{a(A_1 + A_2)} \sqrt{\left(\frac{c}{2g}\right)}$$

Worked examples

1. *Oil is pumped through a circular pipe 150 mm diameter and 500 m long and discharges at a level of 20 m above the pump. The oil has a density of 850 kg/m³ and a viscosity of 0.12 N s/m². Determine the power required to pump (a) 25 kg/s of oil, (b) 100 kg/s of oil. Neglect all losses other than pipe friction.*

(a)
$$V = \frac{\dot{m}}{\rho A} = \frac{25}{850 \times \frac{\pi}{4} \times 0.15^2} = 1.662 \text{ m/s}$$

$$\therefore Re = \frac{\rho V d}{\mu} = \frac{850 \times 1.662 \times 0.15}{0.12} = 1\,768$$

Hence the flow is laminar since $Re < 2\,000$.

Thus
$$f = \frac{16}{Re} \quad \text{from equation (37.6)}$$

$$= \frac{16}{1\,768} = 0.009\,05$$

$$\therefore h_f = \frac{4flV^2}{2gd} \quad \text{from equation (37.1)}$$

$$= \frac{4 \times 0.009\,05 \times 500 \times 1.662^2}{2 \times 9.81 \times 0.15} = 16.95 \text{ m}$$

Therefore head to be overcome $= 16.95 + 20 = 36.95$ m

Therefore power required $= \dot{m}gh$

$$= 25 \times 9.81 \times 36.95$$

$$= 9\,070 \text{ W} \quad \text{or} \quad \underline{9.07 \text{ kW}}$$

(b) $\qquad\qquad V = 4 \times 1.662 = 6.65$ m/s

and $\qquad\qquad Re = 4 \times 1\,768 = 7\,072$

Hence the flow is turbulent since $Re > 4\,000$.

Thus $\qquad\qquad f = \dfrac{0.079}{Re^{1/4}}$ from equation (37.7)

$$= \dfrac{0.079}{7\,072^{1/4}} = 0.008\,61$$

Thus $\qquad\qquad h_f = \dfrac{4 \times 0.008\,61 \times 500 \times 6.65^2}{2 \times 9.81 \times 0.15} = 259 \text{ m}$

$$\therefore h = 259 + 20 = 279 \text{ m}$$

Therefore power required $= 100 \times 9.81 \times 279$

$$= 273\,500 \text{ W or } \underline{273.5 \text{ kW}}$$

2. *At a sudden enlargement in a pipe carrying water, the diameter increases from 320 mm to 640 mm and at the enlargement, the hydraulic gradient rises by 0.12 m. Calculate the mass flow rate of water through the pipe.*

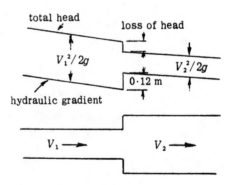

Figure 37.18

Figure 37.18 shows the total head and hydraulic gradient just before and after the enlargement. From the figure it will be seen that

$$\frac{V_1^2}{2g} = \frac{V_2^2}{2g} + 0.12 + \text{loss of head}$$

From equation (37.8), loss of head in the enlargement $= \dfrac{(V_1 - V_2)^2}{2g}$

Thus
$$\frac{V_1^2}{2g} = \frac{V_2^2}{2g} + 0.12 + \frac{(V_1 - V_2)^2}{2g}$$

But
$$A_1 V_1 = A_2 V_2$$

so that
$$V_2 = V_1 \left(\frac{d_1^2}{d_2^2}\right) = \frac{V_1}{4}$$

Thus
$$V_1^2 = \frac{V_1^2}{16} + 0.12 \times 2 \times 9.81 + \frac{9}{16} V_1^2$$

from which
$$V_1 = 2.51 \text{ m/s}$$

Hence
$$\dot{m} = \rho A_1 V_1$$
$$= 10^3 \times \frac{\pi}{4} \times 0.32^2 \times 2.51 = \underline{202 \text{ kg/s}}$$

3. *Two reservoirs, Figure 37.19, whose difference in level is 15 m are connected by a pipe PQR, 1 500 m in length, which has its highest point Q 1.7 m below the level in the upper reservoir. The pipe is 200 mm in diameter and* $f = 0.005$. *Find the length PQ so that the pressure at Q is 3 m below atmospheric.*

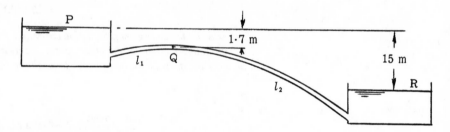

Figure 37.19

The losses of head at entry and exit are $0.5\dfrac{V^2}{2g}$ and $\dfrac{V^2}{2g}$ respectively.

Applying Bernoulli's equation to points P and Q, taking the level at R as datum and the pressure of the atmosphere as p,

$$\frac{p}{\rho g} + 15 = \left(\frac{p}{\rho g} - 3\right) + \frac{V^2}{2g} + 13.3 + \frac{4fl_1 V^2}{2gd} + 0.5\frac{V^2}{2g}$$

from which
$$4.7 = \frac{V^2}{2g}\left(1.5 + \frac{4 \times 0.005 l_1}{0.2}\right)$$

$$= \frac{V^2}{2g}\left(1.5 + \frac{l_1}{10}\right) \qquad (1)$$

Applying Bernoulli's equation to points P and R,

$$\frac{p}{\rho g} + 15 = \frac{p}{\rho g} + \frac{4f(l_1 + l_2)V^2}{2gd} + 0.5\frac{V^2}{2g} + \frac{V^2}{2g}$$

from which

$$15 = \frac{V^2}{2g} \left(1.5 + \frac{4 \times 0.005 \times 1\,500}{0.2} \right)$$

$$= 151.5 \frac{V^2}{2g} \qquad (2)$$

Thus, from equations (1) and (2),

$$\frac{15}{4.7} = \frac{151.5}{1.5 + \frac{l_1}{10}}$$

from which $\qquad l_1 = \underline{460 \text{ m}}$

4. *A pipe line comprising two pipes of the same length and diameter delivers water from a reservoir to a turbine. Each pipe is 4000 m in length and f = 0.006. Calculate the diameter of the pipes if the turbine is to develop 4 MW with a gross head of 170 m. All losses except pipe friction may be ignored.*

Each pipe delivers half the water to produce 2 MW and from Section 37.15, the maximum power will be achieved when the head lost is one-third of the available head.

i.e. $\qquad h_f = \frac{170}{3} = \frac{4flV^2}{2gd} = \frac{4 \times 0.006 \times 4\,000\ V^2}{2gd}$

from which $\qquad \dfrac{V^2}{d} = 11.6 \qquad (1)$

Power available $= \rho g A V (h - h_f)$

i.e. $\qquad 2 \times 10^6 = 10^3 \times 9.81 \times \frac{\pi}{4} d^2 V \left(170 - \frac{170}{3} \right)$

from which $\qquad V d^2 = 2.29 \qquad (2)$

Thus, from equations (1) and (2),

$$d^5 = \frac{2.29^2}{11.6} = 0.452$$

$$\therefore d = \underline{0.854 \text{ m}}$$

5. *A pipe line 2000 m in length has two pipes in parallel. One pipe is 0.4 m diameter and the other is 0.5 m diameter. The total flow rate is 0.8 m³/s and for both pipes, f = 0.006. Determine the loss of head in the system and the flow rate in each pipe.*

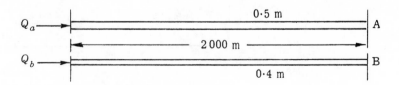

Figure 37.20

For pipes in parallel, the loss of head in each must be the same. Thus, referring to Figure 37.20,

$$h_f = \frac{4flV_a^2}{2gd_a} = \frac{4flV_b^2}{2gd_b}$$

$$\therefore V_a = V_b \sqrt{\left(\frac{d_a}{d_b}\right)} = V_b \sqrt{\left(\frac{0.5}{0.4}\right)} = 1.12\, V_b \qquad (1)$$

$$Q_a + Q_b = Q$$

i.e.
$$\frac{\pi}{4} \times 0.5^2 V_a + \frac{\pi}{4} \times 0.4^2 V_b = 0.8$$

i.e.
$$1.562\,5 V_a + V_b = 6.36 \qquad (2)$$

From equations (1) and (2),

$$V_a = 2.59 \text{ m/s} \quad \text{and} \quad V_b = 2.32 \text{ m/s}$$

$$\therefore h_f = \frac{4flV_a^2}{2gd_a}$$

$$= \frac{4 \times 0.006 \times 2\,000 \times 2.59^2}{2 \times 9.81 \times 0.5} = \underline{32.82 \text{ m}}$$

$$Q_a = \frac{\pi}{4} \times 0.5^2 \times 2.59 = \underline{0.51 \text{ m}^3/\text{s}}$$

and
$$Q_b = \frac{\pi}{4} \times 0.4^2 \times 2.32 = \underline{0.29 \text{ m}^3/\text{s}}$$

6. *Water flows from a reservoir through a pipe 150 mm diameter and 180 m long to a point below the surface of the reservoir where it branches into two pipes, each 100 mm diameter. One of the pipes is 48 m long discharging to atmosphere at a point 18 m below reservoir level and the other 60 m long discharging to atmosphere 24 m below reservoir level.*

Assuming that f = 0.008, calculate the discharge from each pipe, neglecting all losses other than friction.

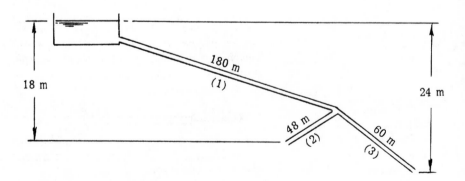

Figure 37.21

The system is shown in Figure 37.21.
From equations (37.14), (37.15) and (37.16),

$$18 = \frac{0.008 \times 180 \, Q_1^2}{3 \times 0.15^5} + \frac{0.008 \times 48 Q_2^2}{3 \times 0.10^5}$$

$$= 6320 \, Q_1^2 + 12\,800 \, Q_2^2 \tag{1}$$

$$24 = \frac{0.008 \times 180 Q_1^2}{3 \times 0.15^5} + \frac{0.008 \times 60 Q_3^2}{3 \times 0.10^5}$$

$$= 6320 \, Q_1^2 + 16\,000 \, Q_3^2 \tag{2}$$

and
$$Q_1 = Q_2 + Q_3 \tag{3}$$

From equation (1), $0.002\,85 = Q_1^2 + 2.025 Q_2^2$

or
$$\frac{0.002\,85}{Q_1^2} = 1 + 2.025 \, n^2 \quad \text{where } n = \frac{Q_2}{Q_1}$$

From equation (2), $0.003\,8 = Q_1^2 + 2.53 \, Q_3^2$

or
$$\frac{0.003\,8}{Q_1^2} = 1 + 2.53 \left(\frac{Q_3}{Q_1}\right)^2$$

$$= 1 + 2.53 \left(1 - \frac{Q_2}{Q_1}\right)^2 \quad \text{from equation (3)}$$

$$= 1 + 2.53(1 - n)^2$$

Thus
$$\frac{0.002\,85}{0.003\,8} = \frac{1 + 2.025 \, n^2}{1 + 2.53(1 - n)^2}$$

from which $n^2 + 30.7n - 13.35 = 0$

Thus
$$n = 0.425$$

$$\therefore \frac{0.002\,85}{Q_1^2} = 1 + 2.025 \times 0.425^2 = 1.365$$

$$\therefore Q_1 = \underline{0.045\,7 \text{ m}^3/\text{s}}$$

$$Q_2 = 0.425 \times 0.045\,7$$

$$= \underline{0.019\,43 \text{ m}^3/\text{s}}$$

7. *Figure 37.22 shows the cross-section of an open channel, the lower part being a semicircle. The channel is laid on a slope of 1 in 2500 and the Chezy constant c = 52. Calculate the volume flow rate when the maximum depth is 0.6 m and also the depth for this rate to be doubled.*

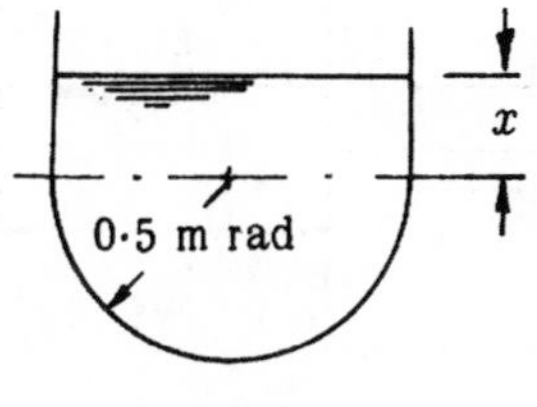

When the maximum depth is 0.6 m,

$$x = 0.1 \text{ m}$$

Therefore wetted area,
$$A = \frac{1}{2} \times \frac{\pi}{4} \times 1^2 + 1 \times 0.1$$

$$= 0.493 \text{ m}^2$$

0·5 m rad

Figure 37.22

Wetted perimeter, $\qquad P = \dfrac{\pi}{2} \times 1 + 2 \times 0.1 = 1.77 \text{ m}$

Hydraulic mean depth, $\qquad m = \dfrac{A}{P} = \dfrac{0.493}{1.77} = 0.279$

$$V = c\sqrt{(mi)} \quad \text{from equation (37.17)}$$

$$= 52\sqrt{\left(0.279 \times \dfrac{1}{2\,500}\right)}$$

$$= 0.548 \text{ m/s}$$

$$\therefore Q = AV = 0.493 \times 0.548 = \underline{0.271 \text{ m}^3/\text{s}}$$

In terms of x, $\qquad A = \dfrac{1}{2} \times \dfrac{\pi}{4} \times 1^2 + 1 \times x = 0.393 + x$

$$P = \dfrac{\pi}{2} \times 1 + 2 \times x = 1.57 + 2x$$

$$\therefore m = \dfrac{0.393 + x}{1.57 + 2x}$$

When the flow rate is doubled,

$$V = \dfrac{Q}{A} = \dfrac{2 \times 0.271}{0.393 + x}$$

$$\therefore \dfrac{0.542}{0.393 + x} = 52\sqrt{\left(\dfrac{0.393 + x}{1.57 + 2x} \times \dfrac{1}{2\,500}\right)}$$

from which $\qquad x = 0.494 \text{ m}$

Thus new maximum depth $= 0.5 + 0.494 = \underline{0.994 \text{ m}}$

8. *A rectangular tank 6 m by 6 m contains water to a depth of 5 m. The tank is connected by a horizontal pipe 300 m long and 75 mm diameter to a second rectangular tank 5 m by 5 m, containing water to a depth of 2 m. The friction coefficient for the pipe is 0.01. How long does it take to lower the depth in the first tank to 4 m?*

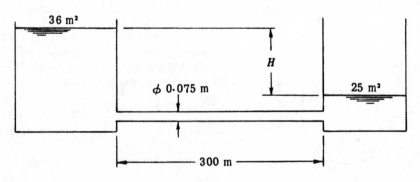

Figure 37.23

The arrangement of the tanks is shown in Figure 37.23.
From equation (37.19), $T = k(\sqrt{H_1} - \sqrt{H_2})$

where $$k = \frac{2A_1A_2}{a(A_1+A_2)}\sqrt{\left(\frac{c}{2g}\right)}$$

In this example, the losses are due to pipe friction and losses at entry and exit, so that

$$H = \left(\frac{4fl}{d} + 0.5 + 1\right)\frac{V^2}{2g}$$

$$= \left(\frac{4 \times 0.01 \times 300}{0.075} + 1.5\right)\frac{V^2}{2g}$$

$$= 161.5\frac{V^2}{2g}$$

i.e. $c = 161.5$

Initial head difference, $H_1 = 5 - 2 = 3$ m

When level in tank 1 falls by 1 m, level in tank 2 rises by $1 \times \dfrac{36}{25} = 1.44$ m.

∴ final head difference, $H_2 = 4 - 3.44 = 0.56$ m

Hence $$T = \frac{2 \times 36 \times 25}{\frac{\pi}{4} \times 0.075^2(36+25)}\left(\sqrt{\frac{161.5}{2 \times 9.81}}\right)(\sqrt{3} - \sqrt{0.56})$$

$$= 18\,850 \text{ s} \quad \text{or} \quad \underline{5.236 \text{ h}}$$

If losses other than friction are ignored, $c = 160$ and $T = 18\,760$, a difference of less than 0.5 %. In most cases where long pipes are involved, only friction losses are relevant.

Further problems

9. Oil of density 850 kg/m^3 and viscosity 0.13 kg/ms flows in a pipe of diameter 25 mm at a velocity of 10 m/s. Determine whether flow is laminar or turbulent.

If the critical Reynolds number is taken to be 2000, determine the mass flow rate through the pipe when flow ceases to be laminar.

[Laminar [$Re = 1\,634$]; 5.1 kg/s]

10. Water flows in a 6 mm diameter pipe at a mean velocity of 0.2 m/s; the viscosity of the water is 0.0013 N s/m^2. Determine the loss of head in a 50 m length of pipe and the power required to overcome the loss. [1.18 m; 0.0655 W]

11. A laminar flow of liquid, of viscosity μ, passes through a circular pipe of diameter d with velocity V. Show that the loss of pressure per unit length of pipe is 32 $V\mu/d^2$. It may be assumed that in laminar flow, $f = 16/Re$ where f is a friction constant defined by f = viscous shear stress/$\frac{1}{2}\rho V^2$.

Calculate the power required to force oil of viscosity 0.25 Ns/m^2 through a 100 mm diameter pipe of length 100 m at a velocity of 0.6 m/s. [226 W]

12. A fluid of density 800 kg/m^3 and viscosity 0.008 N s/m^2 flows along a smooth pipe, 8 mm diameter. The pipe is 4 m long and the mean velocity of the fluid is 3.5 m/s. Determine the friction factor and the loss of head in the pipe.

If the same flow rate of fluid is to pass along a smooth pipe of diameter (a) 4 mm, (b) 16 mm, determine the new loss of head in each case.

[0.010 9; 22.9 m; 365 m; 0.444 m]

13. Water at 20°C flows in a smooth pipe 8 m long and 50 mm diameter. The flow rate is 0.02 m^3/s. Determine the pressure change in the pipe when the pipe is inclined at 30° to the horizontal and the flow is (a) upwards, (b) downwards.

[138.2 kN/m^2; 59.8 kN/m^2]

14. A fluid is discharged from a tank through a pipe 700 m long. The first 200 m of the pipe is 80 mm in diameter and the remainder is 160 mm in diameter. The pipe outlet is 40 m below the inlet. The friction coefficient f for both portions of the pipe is 0.003 8, the loss of head at entry to the pipe is 0.5 $V^2/2g$ and the loss of head at exit is $V^2/2g$ where V is the relevant pipe velocity. Calculate the flow rate through the pipe, ignoring the loss at enlargement.   [0.021 7 m^3/s]

15. A vertical pipe AB conveys water. The diameter at A is 120 mm and at B, 3 m below A, it is 60 mm. The loss of head in a valve between A and B is of the form $kV^2/2g$, where V is the velocity at A. When the flow rate is 0.03 m^3/s upwards, the pressure at B is 12 kN/m^2 greater than that at A. Find the value of k.

[10.1]

16. A sudden enlargement occurs in a circular horizontal pipe which conveys 0.11 m^3/s of oil of density 850 kg/m^3. The enlargement is from 150 mm diameter to 210 mm and the pressure just before the enlargement is 0.1 MN/m^2. Calculate the power required to overcome the loss in the enlargement and the pressure after the enlargement.   [0.434 kW; 96.05 kN/m^2]

17. Water is siphoned from a tank by a bent pipe, 30 m long and 30 mm diameter. The highest point in the pipe is 2 m below the water level in the tank and the pipe discharges to atmosphere 15 m below the water level. If the highest point in the pipe is one-third of the way along it, determine the pressure at this point.

$f = 0.004\,8$ and entry and exit losses may be ignored.

[34 kN/m^2 below atmospheric]

18. A reservoir discharges into a second reservoir through a pipe 0.4 m diameter and 2 100 m long. The reservoirs are of equal depth and the difference in surface levels is 9 m. If the level of the pipe midway between the reservoirs is 17 m below the water level in the highest reservoir, calculate the gauge pressure at this point and the flow rate through the pipe. $f = 0.008$ [122.2 kN/m^2; 0.128 m^3/s]

19. A horizontal pipe 30 m long conveys 1.2 m^3/s of water for power. At entry to the pipe, a pump gives a head of 45 m. Determine the maximum power available at exit and the diameter of pipe required, ignoring all losses except friction. $f = 0.005$.

[353 kW; 0.343 m]

20. A 1 m diameter pipe 4 000 m long connects two reservoirs and its slope is 1 in 1 500. When the level of the reservoirs is low, the pipe runs part full and with a depth of 0.5 m, the flow rate is 0.3 m^3/s. Determine the value of f for the pipe and the flow rate when the pipe is full.

Consider also what happens when the pipe runs (a) one-quarter full, and (b) three-quarters full.   [0.005 6; 0.6 m^3/s]

21. Two pipes, one 50 mm diameter and one 100 mm diameter, each 30 m long, are connected in parallel between two reservoirs whose difference in level is 10 m. Find the flow rate through each pipe if $f = 0.008$ for both and entry and exit losses are to be included.   [0.006 m^3/s; 0.033 m^3/s]

22. Two reservoirs, having a constant difference in water level of 65 m, are connected by a straight pipe 0.2 m diameter and 4 km long. The pipe is tapped at a point distant 1.5 km from the upper reservoir and water is drawn off at the rate of 0.04 m^3/s.

If the friction coefficient f is 0.009, determine the rate at which water enters the lower reservoir, neglecting all losses except pipe friction.

Sketch the hydraulic gradient for the pipe.   [0.021 8 m^3/s]

23. A pipe line, 0.8 m diameter and 4 km long, delivers 0.8 m^3/s of water at 20°C. If the last 2 km of the line are replaced by two pipes in parallel, each 0.5 m diameter, what will be the percentage change in loss of head? For all pipes, f may be assumed to be twice the value for a smooth pipe. Losses other than that due to friction may be ignored.   [92% increase]

24. A reservoir A supplies two reservoirs B and C, the water surfaces of B and C being 20 m and 30 m respectively below that of A. A pipe 150 mm diameter is used for the first 270 m from A and this then branches into two pipes, each 100 mm diameter; the pipe to reservoir B is 65 m long while that to reservoir C is 100 m long.

Calculate the discharge to reservoirs B and C, taking $f = 0.007\,5$ for all pipes. Neglect all losses except those due to friction.   [0.017 m^3/s; 0.024 m^3/s]

25. A cylindrical water tank 0.8 m in diameter discharges through a sharp-edged circular orifice, 15 mm diameter, in the base of the tank. If the initial water level is 0.75 m above the base and the coefficient of discharge of the orifice is 0.62, calculate

(a) the time taken to empty the tank,

(b) the water level after 500 s.   [1 792 s; 0.39 m]

26. A vertical cylindrical tank, 0.3 m diameter, has a 19 mm diameter orifice in the base. The tank is filled with water to a depth of 1 m and allowed to discharge through the orifice. The table shows readings of time t at head h above the orifice.

h(mm)	1 000	920	835	750	670	580	500	420	335
t(s)	0	7.6	15.5	23.8	32.8	42.1	52.2	63.3	75.3

Determine the coefficient of discharge of the orifice.   [0.632]

27. The petrol tank drain plug falls out of the tank in a car, leaving a circular orifice 25 mm diameter in the tank base. The tank is rectangular in cross-section, 1 m × 0.2 m.

The motorist had just filled his tank with 66 litres of petrol. Ignoring the small amount of fuel consumed in motoring, determine how far the car will travel at 45 km/h before it runs out of petrol. The coefficient of discharge of the orifice is 0.6.   [2.2 km]

28. A cylindrical tank, 3 m diameter, is emptied through an open ended pipe fitted in the bottom of the tank. The pipe is 50 mm diameter and 30 m long, with its open end 10 m below the bottom of the tank. Taking the friction coefficient in the pipe as 0.007 and allowing for losses at entry and exit, find the time taken to lower the level in the tank from 4 m to 1 m.   [49.26 min]

29. A tank of constant cross-sectional area 8 m² is connected to another tank of constant cross-sectional area 4 m² by a pipe 50 mm diameter and 130 m long. The friction coefficient for the pipe is 0.01. Neglecting losses other than that due to friction, find the time taken for 2.25 m³ of water to flow from the large tank to the small tank if the initial level in the large tank is 2 m above that in the small tank.

[35.39 min]

38 Dimensional analysis

38.1 Introduction

The fundamental dimensions of engineering quantities are mass (M), length (L), time (T) and temperature (θ) and all equations must have the same dimensions on either side. This provides a method of determining the nature of a relationship which may be too complex for simple analysis. This *dimensional analysis* method will give only part of the solution and experimental work is necessary to complete it.

Although dimensional analysis may be used in the solution of problems in many fields, it is restricted here to fluid mechanics, in which temperature is not involved and the fundamental dimensions are then reduced to three: M, L and T. Table 38.1 shows the dimensions of quantities which commonly arise in fluid mechanics.

Table 38.1. Dimensions of quantities

Quantity	Dimensions	Quantity	Dimensions
Area	L^2	Pressure, stress	M/LT^2
Volume	L^3	Volume flow rate	L^3/T
Velocity	L/T	Work, energy	ML^2/T^2
Acceleration	L/T^2	Power	ML^2/T^3
Momentum	ML/T	Dynamic viscosity	M/LT
Force	ML/T^2	Kinematic viscosity	L^2/T
Angular velocity	$1/T$	Surface tension	M/T^2
Mass flow rate	M/T	Density	M/L^3

To illustrate the principle, consider the dimensions of the continuity equation

$$\dot{m} = \rho A V$$

Using the data in the table, the dimensions on the two sides are

$$\left[\frac{M}{T}\right] = \left[\frac{M}{L^3}\right] \left[L^2\right] \left[\frac{L}{T}\right] = \left[\frac{M}{T}\right]$$

Thus it can be seen that the dimensions on either side are the same.

38.2 Rayleigh's method of dimensional analysis

The variables considered to be involved in a problem are chosen and the equation is expressed in the form

$$X = \phi(X_1, X_2, X_3, \ldots)$$

$$= k(X_1^a, X_2^b, X_3^c, \ldots)$$

where X_1, X_2, X_3, etc., are the variables, ϕ represents 'some function of', k is a constant and a, b, c, d, etc., are unknown indices.

The dimensions of each of the variables are put into the equation and the corresponding dimensions in the two sides are equated, i.e.,

dimension of $X = $ [dimension of X_1]^a[dimension of X_2]^b

[dimension of X_3]$^c \ldots$

If all four fundamental dimensions are involved, four indices may be eliminated. Where there are not more than four variables, the complete form of the equation may be determined, except for numerical constants but when there are more than four variables, the resulting equation can only be expressed in terms of dimensionless groups and different, but equally valid, forms may be obtained by eliminating different indices.

Certain dimensionless groups occur frequently in hydraulics problems and are given special names such as Reynolds number and Froude number.

This method is illustrated in Examples 1 and 2 and in Sections 39.4 and 39.5. The method is also used in heat transfer studies, Chapter 31.

38.3 Buckingham's Π-theorem

The principle of dimensional analysis is also stated in Buckingham's Π-theorem as follows:

If a physical problem involves n variables, $X_1, X_2, \ldots, X_n$, then

$$\phi(X_1, X_2, \ldots, X_n) = 0 \qquad (38.1)$$

(for proof, see '*E. Buckingham*–Model Experiments and the Form of Empirical Equations', *Trans. ASME*, Vol. 37, 1915).

If p is the number of fundamental dimensions, such as M, L and T, in the problem, the n variables may also be arranged in $(n - p)$ independent dimensionless groups, each denoted by Π.

Then
$$\phi(\Pi_1, \Pi_2, \ldots, \Pi_{n-p}) = 0 \qquad (38.2)$$

Each Π term involves some of the X terms and ϕ represents 'some function of'.

The procedure is as follows:

(a) Select any p variables from the n available variables, which *must* include between them all the p fundamental dimensions.
(b) Select one other variable and form a dimensionless group from these $p + 1$ variables.
(c) Repeat (b) for each other variable in turn.

Thus, for example, select X_1, X_2 and X_3, containing between them M, L and T and combine with X_4 to form Π_1. Then combine X_1, X_2 and X_3 with X_5 to form Π_2 and so on up to X_n to form Π_{n-p}.

$$\Pi_1 = \phi_1(X_1, X_2, X_3, X_4)$$

$$\Pi_2 = \phi_2(X_1, X_2, X_3, X_5)$$

$$\vdots \qquad \vdots \qquad \vdots \qquad\qquad (38.3)$$

$$\Pi_{n-p} = \phi_{n-p}(X_1, X_2, X_3, X_n)$$

Equations (38.3) are solved individually as in Section 38.2, with the index for the variables $X_4, X_5, \ldots, X_n$ taken as unity in each case.

This method is illustrated in Examples 1 and 2. It can be seen that, although longer than Rayleigh's method, it is simpler for problems involving many variables since each group is produced individually.

38.4 Arrangement of dimensionless groups

By Rayleigh's method, it is possible to produce different dimensionless groups by varying the indices chosen to be eliminated. Similarly, by Buckingham's method, different dimensionless groups may be produced by varying the three chosen variables from the n available; this is of no significance since, because the groups are dimensionless, they may be multiplied or divided amongst themselves until the required arrangement is obtained. Any new groups formed in this way will still be dimensionless (see Examples 1 and 2).

In the case of Rayleigh's method, in which a given relationship such as

$$R = \rho V^2 D^2 \phi(Re, Fr) \quad \text{(see Section 39.5)}$$

is to be derived from the variables D, V, ρ, μ and g, the result may be obtained directly by eliminating the indices of the terms which appear outside the brackets, i.e. the indices of ρ, V and D. It may not always be possible to achieve this aim in multi-variable problems, so that some re-arrangement will still be needed but this choice minimizes the work.

A problem sometimes occurs with the Buckingham method when selecting the three initial variables. If it is possible to form a dimensionless group from the three chosen variables, the fourth variable which is added is not involved. This has an index of 1, which will automatically produce meaningless equations and hence it is necessary to change one of the three chosen variables to obtain a solution (see Example 1).

Worked examples

1. *A circular shaft of diameter D rotates at a speed N and is supported by a journal bearing with diametral clearance C. The viscosity of the lubricant is μ and the bearing pressure is p. Obtain an expression for the friction coefficient f, defined by the ratio friction force/bearing load, in terms of μ, N, p, D and C.*

(a) Rayleigh's method

$$f = \phi(\mu, N, p, D, C) = k(\mu^a N^b p^c D^d C^e)$$

Substituting dimensions,

$$0 = \left[\frac{M}{LT}\right]^a \left[\frac{1}{T}\right]^b \left[\frac{M}{LT^2}\right]^c [L]^d [L]^e$$

Equating powers of M: $0 = a + c$ (1)

Equating powers of L: $0 = -a - c + d + e$ (2)

Equating powers of T: $0 = -a - b - 2c$ (3)

Three indices may be eliminated from equations (1), (2) and (3). One solution is $a = b$, $c = -b$ and $e = -d$, so that

$$f = k(\mu^b N^b p^{-b} D^d C^{-d})$$

$$= k \left[\frac{D}{C}\right]^d \left[\frac{\mu N}{p}\right]^b$$

or $$f = \phi \left[\frac{D}{C}; \frac{\mu N}{p}\right]$$

(b) Buckingham's method

$$\phi(f, \mu, N, p, D, C) = 0$$

Three dimensionless groups will be formed from the six variables but one of these is the friction coefficient, f, leaving two further groups Π_1 and Π_2.

Thus $$\phi(\Pi_1, \Pi_2, f) = 0$$

or $$f = \phi(\Pi_1, \Pi_2)$$

Π_1 and Π_2 will be obtained from $\phi(\mu, N, p, D, C) = 0$.
Choose μ, N and D to involve the dimensions M, L and T.

Then $$\Pi_1 = \phi_1(\mu, N, D, p) = \mu^{a_1} N^{b_1} D^{c_1} p$$

Since each Π group is dimensionless, then, substituting dimensions:

$$0 = \left[\frac{M}{LT}\right]^{a_1} \left[\frac{1}{T}\right]^{b_1} [L]^{c_1} \left[\frac{M}{LT^2}\right]$$

Equating powers of M: $0 = a_1 + 1$

Equating powers of L: $0 = -a_1 + c_1 - 1$

Equating powers of T: $0 = -a_1 - b_1 - 2$

Hence $a_1 = -1$, $b_1 = -1$ and $c_1 = 0$

so that $$\Pi_1 = \mu^{-1} N^{-1} p = \frac{p}{\mu N}$$

Similarly $$\Pi_2 = \phi_2(\mu, N, D, C) = \mu^{a_2} N^{b_2} D^{c_2} C$$

$$\therefore 0 = \left[\frac{M}{LT}\right]^{a_2} \left[\frac{1}{T}\right]^{b_2} [L]^{c_2} [L]$$

Equating powers of M: $0 = a_2$

Equating powers of L: $0 = -a_2 + c_2 + 1$

Equating powers of T: $0 = -a_2 - b_2$

Hence $a_2 = 0,   b_2 = 0$ and $c_2 = -1$

so that $$\Pi_2 = D^{-1}C = \frac{C}{D}$$

Thus $$f = \phi(\Pi_1, \Pi_2) = \phi\left[\frac{p}{\mu N}; \frac{C}{D}\right]$$

It will be seen that, in this case, the groups are inverted compared with those obtained by Rayleigh's method. However, this is of no significance since, because they are dimensionless, they may be inverted without loss of validity.

This example may be used to demonstrate the problem that arises if the three chosen variables in the Buckingham method form a dimensionless group of their own. It can be seen from the results that μ, N and p could have been selected since they involve M, L and T but they form their own group.

Combining these variables with D, then

$$\Pi_1 = \phi_1(\mu, N, p, D) = \mu^{a_1} N^{b_1} p^{c_1} D$$

Substituting dimensions,

$$0 = \left[\frac{M}{LT}\right]^{a_1} \left[\frac{1}{T}\right]^{b_1} \left[\frac{M}{LT^2}\right]^{c_1} [L]$$

Equating powers of M, L and T respectively gives

$$0 = a_1 + c_1$$

$$0 = -a_1 - c_1 + 1$$

and $$0 = -a_1 - b_1 - 2c_1$$

The first two equations lead to the statement that $1 = 0$.

The results from this combination are therefore meaningless and it is necessary to change one of the three chosen variables.

2. *An orifice of diameter d is used to measure the rate of flow Q of a fluid of viscosity μ and density ρ along a pipe of diameter D. The pressure drop across the orifice is p. Show that*

$$Q = d^2 \left(\sqrt{\frac{p}{\rho}}\right) \phi\left[\frac{\mu}{d\sqrt{(\rho p)}}, \frac{D}{d}\right]$$

(a) Rayleigh's method

$$Q = \phi(d, \rho, \mu, D, p) = k(d^a \rho^b \mu^c D^d p^e)$$

Substituting dimensions,

$$\frac{L^3}{T} = [L]^a \left[\frac{M}{L^3}\right]^b \left[\frac{M}{LT}\right]^c [L]^d \left[\frac{M}{LT^2}\right]^e$$

Equating powers of M: $0 = b + c + e$

Equating powers of L: $3 = a - 3b - c + d - e$

Equating powers of T: $-1 = -c - 2e$

Since the required expression has the variables d, p and ρ as multipliers, it is necessary to eliminate the indices of these terms, i.e. a, e and b.

Thus $a = 2 - c - d$

$$e = \frac{1}{2} - \frac{c}{2}$$

and $b = -\frac{1}{2} - \frac{c}{2}$

Hence $Q = k\{d^{2-c-d}\rho^{-1/2-c/2}\mu^c D^d p^{1/2-c/2}\}$

$$= kd^2\left(\sqrt{\frac{p}{\rho}}\right)\left[\frac{\mu}{d\sqrt{(\rho p)}}\right]^c\left[\frac{D}{d}\right]^d$$

$$= d^2\left(\sqrt{\frac{p}{\rho}}\right)\phi\left[\frac{\mu}{d\sqrt{(\rho p)}};\frac{D}{d}\right]$$

(b) Buckingham's method

$$\phi(Q, d, \rho, \mu, D, p) = 0$$

There are six variables and three dimensions, which leads to three Π groups. Choose Q, d and ρ to involve M, L and T and combine μ, D and p in turn.

Then $\Pi_1 = \phi_1(Q, d, \rho, \mu) = Q^{a_1} d^{b_1} \rho^{c_1} \mu$

Equating dimensions,

$$0 = \left[\frac{L^3}{T}\right]^{a_1} [L]^{b_1} \left[\frac{M}{L^3}\right]^{c_1} \left[\frac{M}{LT}\right]$$

Thus, for M: $0 = c_1 + 1$

for L: $0 = 3a_1 + b_1 - 3c_1 - 1$

and for T: $0 = -a_1 - 1$

Hence $a_1 = -1,\quad b_1 = 1\quad$ and $\quad c_1 = -1$

so that $\Pi_1 = \frac{\mu d}{Q\rho}$

Similarly $\Pi_2 = \phi_2(Q, d, \rho, D) = Q^{a_2} d^{b_2} \rho^{c_2} D$

$$\therefore\ 0 = \left[\frac{L^3}{T}\right]^{a_2} [L]^{b_2} \left[\frac{M}{L^3}\right]^{c_2} [L]$$

Thus, for M: $0 = c_2$

for L: $0 = 3a_2 + b_2 - 3c_2 + 1$

and for T: $0 = -a_2$

Hence $\qquad\qquad a_2 = 0, \quad b_2 = -1 \quad \text{and} \quad c_2 = 0$

so that $\qquad\qquad \Pi_2 = \dfrac{D}{d}$

$$\Pi_3 = \phi_3(Q, d, \rho, p) = Q^{a_3} d^{b_3} \rho^{c_3} p$$

$$\therefore 0 = \left[\frac{L^3}{T}\right]^{a_3} [L]^{b_3} \left[\frac{M}{L^3}\right]^{c_3} \left[\frac{M}{LT^2}\right]$$

Thus, for M: $\qquad\qquad 0 = c_3 + 1$

for L: $\qquad\qquad 0 = 3a_3 + b_3 - 3c_3 - 1$

and for T: $\qquad\qquad 0 = -a_3 - 2$

Hence $\qquad\qquad a_3 = -2, \quad b_3 = 4 \quad \text{and} \quad c_3 = -1$

so that $\qquad\qquad \Pi_3 = \dfrac{d^4 p}{Q^2 \rho}$

Thus $\qquad\qquad \phi\left[\dfrac{\mu d}{Q\rho}; \dfrac{D}{d}; \dfrac{d^4 p}{Q^2 \rho}\right] = 0$

To give the required form, this needs rearrangement (see Section 38.4). The square root of Π_3 is $\dfrac{d^2}{Q}\left(\sqrt{\dfrac{p}{\rho}}\right)$ and dividing this into Π_1 gives $\dfrac{\mu}{d\sqrt{(\rho p)}}$ so that

$$\phi\left[\frac{\mu}{d\sqrt{(\rho p)}}; \frac{D}{d}; \frac{d^2}{Q}\left(\sqrt{\frac{p}{\rho}}\right)\right] = 0$$

or $\qquad\qquad \dfrac{Q}{d^2}\left(\sqrt{\dfrac{\rho}{p}}\right) = \phi\left[\dfrac{\mu}{d\sqrt{(\rho p)}}; \dfrac{D}{d}\right]$

from which $\qquad\qquad \underline{Q = d^2\left(\sqrt{\dfrac{p}{\rho}}\right)\phi\left[\dfrac{\mu}{d\sqrt{(\rho p)}}; \dfrac{D}{d}\right]}$

Further problems

3. Fluid of density ρ and viscosity μ flows through a circular pipe of diameter d at velocity V. Show that the resistance per unit length R is given by

$$R = \rho V^2 \phi\left[\frac{\rho V d}{\mu}\right]$$

4. Fluid of density ρ and viscosity μ flows at velocity V along a circular pipe of diameter d and length L. Show that the pressure loss due to friction p is given by

$$p = \rho V^2 \phi\left[\frac{\rho V d}{\mu}, \frac{L}{d}\right]$$

5. An orifice of diameter d discharges fluid under a head h in a gravitational field g. The fluid has density ρ and viscosity μ. Show that the quantity flowing Q may be expressed by

$$Q = d^2 \sqrt{(gh)}\phi\left[\frac{\mu}{\rho g d^{3/2}}, \frac{h}{d}\right]$$

6. A rectangular weir, breadth B, discharges fluid under a head h in a gravitational field g. The fluid has density ρ, viscosity μ and surface tension τ. Show that the quantity flowing per unit width is given by

$$\frac{Q}{B} = kh^{3/2}g^{1/2}\phi\left[\frac{\mu}{h^{3/2}g^{1/2}\rho}, \frac{\tau}{h^2g\rho}\right]$$

where k is a constant.

7. Show by dimensional analysis that the flow rate Q over a V-notch may be expressed by

$$Q = g^{1/2}h^{5/2}\phi\left[\frac{g^{1/2}h^{3/2}}{\nu}, \frac{gh^2\rho}{\tau}, \theta\right]$$

where h is the head over the notch vertex, ρ is the fluid density, ν is the fluid kinematic viscosity, τ is the surface tension of the fluid, θ is the angle of the notch and g is the gravitational acceleration.

8. The thrust of a propeller T depends on the propeller diameter D, the speed of rotation N, the speed of advance V, the fluid density ρ and the fluid viscosity μ. Show that the thrust may be expressed by

$$T = \rho D^2 V^2 \phi\left[\frac{\mu}{\rho DV}, \frac{ND}{V}\right]$$

9. Assuming that the wavemaking resistance R of a hydrofoil depends on the density of the fluid ρ, the size of the hydrofoil D, the velocity of the hydrofoil V and the gravitational acceleration g, show that

$$R = \rho V^2 D^2 \phi\left[\frac{V}{\sqrt{(gD)}}\right]$$

10. The friction power P dissipated in a water dynamometer is dependent on the brake wheel size D, the brake speed N, the fluid density ρ and the fluid viscosity μ. Show that the power may be expressed by

$$P = D^5 N^3 \rho\phi\left[\frac{\mu}{D^2N\rho}\right]$$

11. Obtain an expression for the resistance R to the motion of a ship, assuming that the important parameters are the size of the ship D, the speed V, the density of the sea ρ, the viscosity of the sea μ and the gravitational acceleration g.

$$\left[R = \rho L^2 V^2 \phi\left[\frac{\rho VD}{\mu}, \frac{V^2}{Dg}\right]\right]$$

39 Dynamic similarity

39.1 Similarity

In order to be able to predict the performance of a system from a larger or smaller model, it is necessary to consider the different types of similarity.

(i) *Geometric similarity* requires that the ratio of a length in the system to the corresponding length in the model is the same everywhere.

(ii) *Kinematic similarity* requires that the ratio of velocity and acceleration at any point in the system to the corresponding velocity and acceleration in the model is the same everywhere.

(iii) *Dynamic similarity* requires that the ratio of the force at any point in the system to the corresponding force in the model is the same everywhere.

Geometric similarity is simple to achieve but it may be shown that in order to obtain kinematic similarity, the conditions of dynamic similarity must also apply. Thus if dynamic similarity is achieved in geometrically similar situations, full similarity results.

39.2 Dynamic similarity

If surface tension and other small effects are neglected, the motion of a fluid is governed by a combination of viscous, gravitational and pressure forces. From Newton's second law, the vector sum, F, of the external forces acting on a body of mass m causes an acceleration a in the direction of the force, given by

$$F = ma$$

The quantity ma is called the inertia force. Thus for two fluid systems to be dynamically similar, the ratios $\dfrac{\text{inertia force}}{\text{viscous force}}$, $\dfrac{\text{inertia force}}{\text{gravitational force}}$ and $\dfrac{\text{inertia force}}{\text{pressure force}}$ must be equal. Since the inertia force is the sum of the viscous, gravitational and pressure forces, it is only necessary for two of these ratios to be equal to give dynamic similarity.

In fluid mechanics, two problems concerned are submerged bodies (submarines, pipes, pumps, aeroplanes, cars, etc.) where the whole of the surfaces under consideration are in contact with the fluid, and floating bodies, such as ships, which are only partly submerged.

For submerged bodies, gravitational forces are irrelevant and dynamic similarity is achieved if the ratio $\dfrac{\text{inertia force}}{\text{viscous force}}$ is equal for each system considered. For floating bodies, however, the ratio of both $\dfrac{\text{inertia force}}{\text{viscous force}}$ and $\dfrac{\text{inertia force}}{\text{gravitational force}}$ must be equal for each system.

If V represents velocity, D is a dimension specifying size and t is time,

then
$$\text{inertia force} = ma = \rho D^3 \times \frac{V}{t}$$

$$\text{viscous force} = \text{viscous stress} \times \text{ area}$$

$$= \tau D^2 = \mu \frac{V}{D} \cdot D^2 = \mu V D$$

and
$$\text{gravitational force} = mg = \rho D^3 g$$

The ratio
$$\frac{\text{inertia force}}{\text{viscous force}} = \frac{\rho D^3 \dfrac{V}{t}}{\mu V D} = \frac{\rho D}{\mu} \cdot \frac{D}{t} = \frac{\rho V D}{\mu}$$

This is the Reynolds number, Re.

The ratio
$$\frac{\text{inertia force}}{\text{gravitational force}} = \frac{\rho D^3 \dfrac{V}{t}}{\rho D^3 g} = \frac{V}{tg} = \frac{V}{\dfrac{D}{V} \cdot g} = \frac{V^2}{Dg}$$

$\dfrac{V}{\sqrt{(Dg)}}$ is called the Froude number (Fr), so that $\dfrac{V^2}{Dg}$ represents $(Fr)^2$. Thus, for geometrically similar submerged bodies, dynamic similarity requires the Reynolds number to be the same for each and for geometrically similar floating bodies, dynamic similarity requires both the Reynolds and Froude numbers to be the same for each.

39.3 Model testing

Dimensional analysis provides a method of predicting the performance of systems which are geometrically similar to systems for which the performance is known. This particular application, based on dynamic similarity, has wide uses but is here restricted to problems of resistance to motion of submerged and floating bodies. For this purpose, scale models are built so that any convenient length dimension may be used to specify size.

39.4 Submerged bodies

The resistance to motion, R, of a body completely submerged in an incompressible fluid will depend on the size of the body D, its velocity V and the density ρ and viscosity μ of the liquid. Using the method of dimensional analysis,

$$R = \phi(D, V, \rho, \mu)$$

$$= k(D^a V^b \rho^c \mu^d)$$

Substituting dimensions,

$$\left[\frac{ML}{T^2}\right] = [L]^a \left[\frac{L}{T}\right]^b \left[\frac{M}{L^3}\right]^c \left[\frac{M}{LT}\right]^d$$

Equating powers of M: $1 = c + d$

Equating powers of L: $1 = a + b - 3c - d$

Equating powers of T: $-2 = -b - d$

One solution is $a = 2 - d$, $b = 2 - d$ and $c = 1 - d$

so that

$$R = k(D^{2-d} V^{2-d} \rho^{1-d} \mu^d)$$

$$= k\rho V^2 D^2 \left[\frac{\rho V D}{\mu}\right]^{-d}$$

$\dfrac{\rho V D}{\mu}$ is the Reynolds number of the flow past the body, Re,

so that

$$R = \rho V^2 D^2 \phi(Re) \tag{39.1}$$

or

$$\frac{R}{\rho V^2 D^2} = \phi(Re) \tag{39.2}$$

For a series of geometrically similar bodies, dynamic similarity will be achieved if Re is the same for each, when the unknown function ϕ becomes irrelevant. Thus, when Re is the same for each, $R/\rho V^2 D^2$ will be the same for each, from equation (39.2).

If suffix m refers to the model and suffix p refers to the prototype, then for dynamic similarity,

$$(Re)_p = (Re)_m$$

i.e.

$$\left(\frac{\rho V D}{\mu}\right)_p = \left(\frac{\rho V D}{\mu}\right)_m \tag{39.3}$$

It is usual to make model tests in wind tunnels. The tunnel velocity, or *corresponding speed*, is given by

$$V_m = V_p \left(\frac{D_p}{D_m}\right) \left(\frac{\rho_p}{\rho_m}\right) \left(\frac{\mu_m}{\mu_p}\right) \tag{39.4}$$

For these conditions, the resistance of the prototype is given by

$$\left(\frac{R}{\rho V^2 D^2}\right)_p = \left(\frac{R}{\rho V^2 D^2}\right)_m$$

from which

$$R_p = R_m \left(\frac{\rho_p V_p^2 D_p^2}{\rho_m V_m^2 D_m^2}\right) \qquad (39.5)$$

Submarines, cars and aeroplanes at low speeds (where compressibility effects may be neglected) are examples of submerged bodies which may be treated by this method.

Pipes may also be considered as submerged bodies. If loss of head is to be predicted,

resistance $R = \rho g h A$ where A is the pipe cross-sectional area

so that the equation $\dfrac{R_p}{R_m} = \dfrac{\rho_p V_p^2 D_p^2}{\rho_m V_m^2 D_m^2}$ becomes

$$\frac{h_p}{h_m} = \frac{V_p^2}{V_m^2} \qquad (39.6)$$

This relation is only valid for geometrically similar pipes and if the lengths of pipes considered are not in the correct ratio, it must be modified according to the ratio of the actual pipe length to the geometrically similar length,

i.e.

$$\frac{h_p}{h_m} = \frac{V_p^2}{V_m^2} \times \frac{\text{actual length of pipe}}{\text{geometrically similar length of pipe}}$$

$$= \frac{V_p^2}{V_m^2} \times \frac{L_p}{L_m(D_p/D_m)}$$

$$= \frac{V_p^2 D_m L_p}{V_m^2 D_p L_m} \qquad (39.7)$$

If the pipes are of equal length, $L_p = L_m$, so that

$$\frac{h_p}{h_m} = \frac{V_p^2 D_m}{V_m^2 D_p}$$

These results may also be obtained from the Darcy formula,

i.e.

$$h_p = \frac{4 f_p L_p V_p^2}{2 g D_p} \quad \text{and} \quad h_m = \frac{4 f_m L_m V_m^2}{2 g D_m}$$

At the same Reynolds number with both pipes of the same material,

$$f_p = f_m$$

so that

$$\frac{h_p}{h_m} = \frac{V_p^2 D_m L_p}{V_m^2 D_p L_m}, \quad \text{as before.}$$

39.5 Floating bodies

The resistance to motion of the hull of a ship consists of two parts, (a) that due to the viscous effects of the water, and (b) the wavemaking effects as water is lifted by the hull against gravity. If surface tension effects are ignored, the dimensional analysis of Section 39.4 may be repeated with the gravitational acceleration g as an additional factor.

$$R = \phi(D, V, \rho, \mu, g) = k(D^a V^b \rho^c \mu^d g^e)$$

Substituting dimensions,

$$\frac{ML}{T^2} = [L]^a \left[\frac{L}{T}\right]^b \left[\frac{M}{L^3}\right]^c \left[\frac{M}{LT}\right]^d \left[\frac{L}{T^2}\right]^e$$

Equating powers of M: $1 = c + d$

Equating powers of L: $1 = a + b - 3c - d + e$

Equating powers of T: $-2 = -b - d - 2e$

One solution is $a = 2 - d + e$, $b = 2 - d - 2e$ and $c = 1 - d$.

so that

$$R = k(D^{2-d+e} V^{2-d-2e} \rho^{1-d} \mu^d g^e)$$

$$= k\rho V^2 D^2 \left[\frac{\rho V D}{\mu}\right]^{-d} \left[\frac{V^2}{Dg}\right]^{-e}$$

$$= k\rho V^2 D^2 \left[\frac{\rho V D}{\mu}\right]^{-d} \left[\frac{V}{\sqrt{(Dg)}}\right]^{-2e}$$

$\dfrac{V}{\sqrt{(Dg)}}$ is the Froude number of the flow, Fr, so that

$$R = \rho V^2 D^2 \phi(Re, Fr) \tag{39.8}$$

or

$$\frac{R}{\rho V^2 D^2} = \phi(Re, Fr)$$

For dynamic similarity, Re must be the same for the model and prototype and Fr must be the same for the model and prototype. Model tests are conducted in water tanks so that the fluid properties are substantially the same for the model and prototype.

Then, for equality of Reynolds numbers,

$$V_m = V_p \left(\frac{D_p}{D_m}\right)$$

and for equality of Froude numbers,

$$V_m = V_p \sqrt{\left(\frac{D_m}{D_p}\right)}$$

These two conditions are incompatible and so dynamic similarity cannot be achieved.

If R_v is the viscous resistance and R_w is the wavemaking resistance,

$$R = R_v + R_w$$

The viscous resistance is approximately proportional to the wetted area and the square of the velocity, so that

$$R_v = \tfrac{1}{2}\rho C_D A V^2 \tag{39.9}$$

where C_D is a dimensionless drag coefficient.

The value of C_D is determined by experiment on a flat plate of the same material and surface finish, so that for the model,

$$R_{v_m} = \tfrac{1}{2}\rho C_D A_m V_m^2$$

and for the prototype,

$$R_{v_p} = \tfrac{1}{2}\rho C_D A_p V_p^2$$

The wavemaking resistance is given by equation (39.8) when μ is omitted from the analysis,

i.e.

$$R_w = \rho V^2 D^2 \phi(Fr)$$

Tests are then conducted at the corresponding speed given by the Froude equality,

i.e.

$$V_m = V_p \sqrt{\left(\frac{D_m}{D_p}\right)} \tag{39.10}$$

so that

$$\left(\frac{R_w}{\rho V^2 D^2}\right)_m = \left(\frac{R_w}{\rho V^2 D^2}\right)_p$$

or

$$\frac{R_{w_p}}{R_{w_m}} = \frac{\rho_p V_p^2 D_p^2}{\rho_m V_m^2 D_m^2} = \frac{\rho_p D_p^3}{\rho_m D_m^3} \tag{39.11}$$

The total resistance, R_p, is then obtained from

$$\frac{R_p - R_{v_p}}{R_m - R_{v_m}} = \frac{\rho_p D_p^3}{\rho_m D_m^3} \tag{39.12}$$

where R_m is the total model resistance measured in the test at the speed given by equation (39.10).

In the case of a hydrofoil, which is considered to have no viscous resistance, the resistance is given directly by equation (39.11).

39.6 Drag and lift coefficients

Submerged bodies are so common that it is convenient to express the resistance parallel to the fluid flow as a *drag force* and to use a drag coefficient C_D, as in Section 39.5, to determine this force, given by

$$R_D = \tfrac{1}{2}\rho C_D A V^2 \tag{39.13}$$

But from equation (39.1),

$$R = \rho V^2 D^2 \phi(Re)$$

so that C_D is a function of Re. Thus, if C_D is found for a model, it will apply also to a geometrically similar system under dynamically similar conditions.

In a similar way, the lift force, R_L, normal to the direction of motion may be expressed in terms of a lift coefficient C_L, given by

$$R_L = \tfrac{1}{2}\rho C_L A V^2 \qquad (39.14)$$

C_L is also a function of Re and so has the same value for geometrically similar systems under dynamically similar conditions.

The value of A in equations (39.13) and (39.14), equivalent to the D^2 term in equation (39.1), is chosen by convention for common applications, i.e. wing plan area for aircraft and frontal area for cars.

Equations (39.13) and (39.14) may be rearranged to give

$$C_D = \frac{R_D/A}{\tfrac{1}{2}\rho V^2} \quad \text{and} \quad C_L = \frac{R_L/A}{\tfrac{1}{2}\rho V^2}$$

The term $\tfrac{1}{2}\rho V^2$ is called the dynamic pressure of the fluid at velocity V and represents, from Section 34.7, the increase in pressure when the fluid is brought to rest. Thus, at the *stagnation point* S on the submerged body shown in Figure 39.1, the free stream pressure is increased by $\tfrac{1}{2}\rho V^2$. Elsewhere on the surface, the free stream pressure will be reduced or increased, depending on whether the local velocity is greater than, or less than, the free stream velocity.

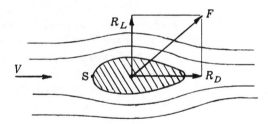

Figure 39.1

The resultant force, F, on the body, which might be a turbine blade, aerofoil or propeller, may be resolved into the drag force R_D and lift force R_L, respectively parallel and normal to the direction of flow, to enable the performance of the body to be conveniently analysed.

39.7 Model testing problems

It is sometimes found that the achievement of dynamic similarity is only possible if the fluid or its density is changed; this is to avoid excessive corresponding speed requirements when testing small models.

If the performance of an aircraft, to fly at 250 m/s, is examined by means of a $\tfrac{1}{5}$ scale model, tested in air at the same pressure and temperature as the real aircraft operating conditions, then ρ and μ are the same for the model and prototype.

For dynamic similarity,

$$\left(\frac{\rho V D}{\mu}\right)_m = \left(\frac{\rho V D}{\mu}\right)_p$$

But $\rho_m = \rho_p$ and $\mu_m = \mu_p$, so that $V_m D_m = V_p D_p$. Since $D_p = 5 D_m$, then $V_m = 5 V_p$ and so the model test needs to be conducted at 1 250 m/s. At such high speeds, other effects occur which would render the experimental results useless and so, to avoid this problem, the ambient pressure in model tests may be increased to increase ρ_m and give a smaller value for V_m. The tests may also be conducted in an entirely different fluid of greater density.

39.8 Experimental results

Dimensional analysis will not produce a complete answer to a problem and experiments are necessary to supply further data. In order to use such experimental results for all geometrically similar systems, they must be presented graphically in non-dimensional form.

Suppose that for a particular problem, dimensional analysis shows that

$$\Pi_1 = \phi(\Pi_2, \Pi_3, \Pi_4)$$

where Π_1, Π_2, Π_3, and Π_4 are non-dimensional groups.

Experiments are performed so that the values of these groups can be determined. The results are then displayed as graphs of Π_1 against Π_2 with Π_4 constant and Π_3 as contour, as shown in Figure 39.2. To produce such a set of results would need a considerable number of tests but the graphs are then available to predict the performance of any geometrically similar system. Such graphs would be used for a range of pumps or propellers.

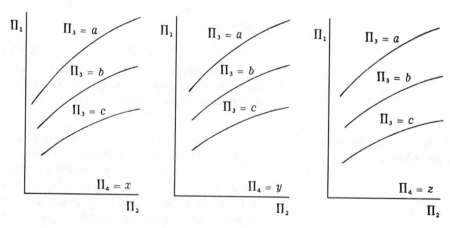

Figure 39.2

Worked examples

1. *The power requirements of a small submarine cruising submerged in sea water at a speed of 15 knots are to be determined using a $\frac{1}{10}$ th scale model in a high density wind tunnel. The model drag is measured as 300 N when the tunnel pressure is 6 bar and the tunnel temperature is 27°C.*

Determine the tunnel velocity and the power required for the submarine.

For sea water, $\rho = 1\,030\,kg/m^3$ and $\mu = 1.12 \times 10^{-3}\,Ns/m^2$.

For air at 27°C, 6 bar, $\rho = 7.06\,kg/m^3$ and $\mu = 18.46 \times 10^{-6}\,Ns/m^2$.

1 knot = 6 080 ft/h = 0.514 m/s.

From equation (39.4), the corresponding speed of the model,

$$V_m = V_p \left(\frac{D_p \rho_p \mu_m}{D_m \rho_m \mu_p} \right)$$

$$= 15 \times 0.514 \left(\frac{10 \times 1\,030 \times 18.46 \times 10^{-6}}{1 \times 7.06 \times 11.12 \times 10^{-3}} \right)$$

$$= \underline{186 \text{ m/s}}$$

At the corresponding speed, $R_p = R_m \left(\frac{\rho_p V_p^2 D_p^2}{\rho_m V_m^2 D_m^2} \right)$

$$= 300 \left(\frac{1\,030 \times [15 \times 0.514]^2 \times 10^2}{7.06 \times 186^2 \times 1^2} \right)$$

$$= 7\,600 \text{ N}$$

Power required to overcome resistance

$$= 7\,600 \times (15 \times 0.514)$$

$$= 58\,700 \text{ W or } \underline{58.7 \text{ kW}}$$

2. *A scale model of a ship is 4 m long and is towed through a water tank at 2.3 m/s. The wetted area of the ship is 500 m² and its length is 50 m. What is the power requirement of the ship, if the model resistance at the corresponding speed is 56 N?*

For both model and ship, the drag friction coefficient C_D is 0.0021. It may be assumed that the water in the tank has the same properties as the water in which the ship operates, which has a density of 1 026 kg/m³.

For a floating body, corresponding speeds are obtained from equality of Froude numbers,

$$V_p = V_m \sqrt{\left(\frac{D_p}{D_m} \right)} \qquad \text{from equation (39.10)}$$

$$= 2.3 \sqrt{\left(\frac{50}{4} \right)} = 8.14 \text{ m/s}$$

Surface area of model $= 500 \times \left(\frac{4}{50} \right)^2 = 3.2 \text{ m}^2$

Viscous drag, $\qquad R_v = \frac{1}{2}\rho C_D A V^2$ from equation (39.9)

At 8.14 m/s, $\qquad R_{v_p} = \frac{1}{2} \times 1\,026 \times 0.002\,1 \times 500 \times 8.14^2 = 35\,600$ N

At 2.5 m/s, $\qquad R_{v_m} = \frac{1}{2} \times 1\,026 \times 0.002\,1 \times 3.2 \times 2.3^2 = 18.2$ N

Thus, from equation (39.12)

$$\frac{R_p - R_{v_p}}{R_m - R_{v_m}} = \frac{\rho_p D_p^3}{\rho_m D_m^3}$$

i.e. $$\frac{R_p - 35\,600}{56 - 18.2} = \frac{1\,026 \times 50^3}{1\,026 \times 4^3}$$

from which $\qquad\qquad R_p = 109\,500$ N

Therefore power required $= 109\,500 \times 8.14$

$$= 893\,000 \text{ W} \quad \text{or} \quad \underline{893 \text{ kW}}$$

3. *A pipe 50 m long and 20 mm diameter carries water at 10°C. The water flows with a velocity of 5 m/s and the loss of head in the pipe is 60 m. A pipe of similar material is to carry oil at the corresponding speed, the oil having a density of 850 kg/m³ and a viscosity of 0.032 kg/m s. Calculate the pressure difference between the ends of the oil pipe, which is 75 m long and 1.5 m diameter.*
For water at 10°C, $\rho = 10^3 kg/m^3$ and $\mu = 1.3 \times 10^{-3}$ kg/m s.

At the corresponding speeds, the Reynolds numbers must be equal,

i.e. $$\left(\frac{\rho V D}{\mu}\right)_{\text{water}} = \left(\frac{\rho V D}{\mu}\right)_{\text{oil}}$$

i.e. $$\frac{1\,000 \times 5 \times 0.02}{1.3 \times 10^{-3}} = \frac{850 \times V_{\text{oil}} \times 1.5}{0.032}$$

$$\therefore V_{\text{oil}} = 1.93 \text{ m/s}$$

From equation (39.7), $$\frac{h_{\text{oil}}}{h_{\text{water}}} = \frac{V_{\text{oil}}^2 D_{\text{water}} L_{\text{oil}}}{V_{\text{water}}^2 D_{\text{oil}} L_{\text{water}}}$$

$$\therefore h_{\text{oil}} = 60 \times \frac{1.93^2 \times 0.02 \times 75}{5^2 \times 1.5 \times 50}$$

$$= 0.178 \text{ m}$$

$$\therefore p = \rho g h$$

$$= 850 \times 9.81 \times 0.178 = \underline{148 \text{ N/m}^2}$$

Further problems

4. In order to determine the drag of an aircraft designed to fly at 150 m/s, a $\frac{1}{25}$ scale model is tested in a wind tunnel at 29 atmospheres. The drag of the model at the corresponding speed is 180 N. Assuming that the viscosity of air is independent of pressure and that the tunnel temperature is equal to the air temperature at which the aircraft will operate, determine the power required by the aircraft.   [783 kW]

5. A $\frac{1}{5}$ scale model of an automobile component with frontal area 100 cm^2 is tested in water to determine its drag. The automobile moves at 18 m/s and at the corresponding speed in the water test, the model drag was 7 N. Calculate the power required to overcome the component drag and determine the component drag coefficient.

For air, $\rho = 1.412$ kg/m^3 and $\mu = 1.599 \times 10^{-5}$ Ns/m^2.

For water, $\rho = 1\,000$ kg/m^3 and $\mu = 1\,002 \times 10^{-6}$ Ns/m^2.   [22.7 W, 0.55]

6. A hydrofoil craft is to travel at 30 m/s in sea water. The foils are 10 m in length and in tests conducted at the corresponding speed in a fresh water tank, the scale model 0.5 m long had a resistance of 5 N. Determine the power required for the hydrofoil craft.

For sea water, $\rho = 1\,030$ kg/m^3 and for fresh water, $\rho = 1\,000$ kg/m^3,

[1 236 kW]

7. A ship is to be built 150 m in length with a wetted surface area of 3 000 m^2. The ship is to operate at 20 knots. A $\frac{1}{40}$ scale model is towed at the corresponding speed in a tank of fresh water and the model resistance is 19 N. For both model and ship, the flat plate drag coefficient may be assumed to be 0.002 5 (based on S.I. units). Determine the power requirement for the ship.

For sea water, $\rho = 1\,026$ kg/m^3 and $\mu = 11.2 \times 10^{-4}$ Ns/m^2.

For fresh water, $\rho = 1\,000$ kg/m^3 and $\mu = 11.5 \times 10^{-4}$ Ns/m^2.

1 knot = 0.514 m/s.   [12 750 kW]

8. Determine the power required to drive a ship 100 m in length with a wetted area of 1 300 m^2 at 11 m/s in fresh water. The resistance in fresh water at the same temperature of a $\frac{1}{10}$ scale model run at corresponding speed is 500 N. The drag coefficient for flat plates may be obtained from the relation $C_D = 0.074/Re^{0.2}$.

For fresh water, $\rho = 1\,000$ kg/m^3 and $\mu = 1\,136 \times 10^{-6}$ Ns/m^2.   [4 560 kW]

9. The velocity of flow of water through an 80 mm diameter pipe is 5 m/s and the loss of head at this speed is 48 m of water per 100 m of pipe. A pipe of similar material, 400 mm diameter, carries air at 300 K at the corresponding speed. What will be the loss of head, in metres of water, per 100 m of pipe?

For air at 300 K, $\mu = 18.46 \times 10^{-6}$ Ns/m^2, $\rho = 1.177$ kg/m^3.

For water, $\mu = 1\,136 \times 10^{-6}$ Ns/m^2, $\rho = 1\,000$ kg/m^3.   [0.088 m]

40 Turbomachinery

40.1 Introduction

In a hydraulic turbomachine, liquid flows over curved blades, or through passages between curved blades, on a rotor. The force exerted on the blades derives from the rate of change of momentum of the liquid and this in turn produces a torque on the rotor shaft. The power developed is then the product of torque and angular speed.

When power is produced by a turbomachine, it is called a turbine and when power is absorbed to raise pressure, it is called a pump. Similar machines exist for compressible fluids but these are beyond the scope of this book.

40.2 Dimensional analysis applied to turbomachinery

In turbomachines, the flow is extremely turbulent and viscous effects are small. Thus viscosity and Reynolds number are absent in dimensional analysis applied to turbo-machines and the relevant variables are pressure (p), shaft power (P), efficiency (η), volume flow rate (Q), fluid density (ρ) and runner diameter (D). Pressure may be expressed in terms of the head ($p = \rho g H$) and gH is preferred as a variable.

These variables are not all independent and it is usually considered that gH, P and η are dependent on the remainder.

$$gH = f_1(Q, N, \rho, D) \tag{40.1}$$

$$P = f_2(Q, N, \rho, D) \tag{40.2}$$

and
$$\eta = f_3(Q, N, \rho, D) \tag{40.3}$$

From equation (40.1)

$$gH = k(Q^a N^b \rho^c D^d)$$

Substituting dimensions,

$$\left[\frac{L^2}{T^2}\right] = \left[\frac{L^3}{T}\right]^a \left[\frac{1}{T}\right]^b \left[\frac{M}{L^3}\right]^c [L]^d$$

Equating powers of M: $0 = c$

Equating powers of L: $2 = 3a - 3c + d$

Equating powers of T: $-2 = -a - b$

Three indices may be eliminated from these equations and one solution

is
$$b = 2 - a, \quad c = 0 \quad \text{and} \quad d = 2 - 3a,$$

so that
$$gH = k(Q^a N^{2-a} D^{2-3a})$$

$$= kN^2 D^2 \left[\frac{Q}{ND^3}\right]^a$$

or
$$\frac{gH}{N^2 D^2} = \phi\left[\frac{Q}{ND^3}\right] \tag{40.4}$$

Similarly, from equations (40.2) and (40.3), it is found that

$$\frac{P}{\rho N^3 D^5} = \phi\left[\frac{Q}{ND^3}\right] \tag{40.5}$$

and
$$\eta = \phi\left[\frac{Q}{ND^3}\right] \tag{40.6}$$

For a range of geometrically similar turbines or pumps, dynamic similarity is achieved if $\dfrac{Q}{ND^3}$ is the same for each. At this condition, the values of $\dfrac{gH}{N^2 D^2}$, $\dfrac{P}{\rho N^3 D^5}$ and η will also be the same for each machine in the range. These parameters are found by experiment for a particular design of machine *at the point of maximum efficiency*; these will be the

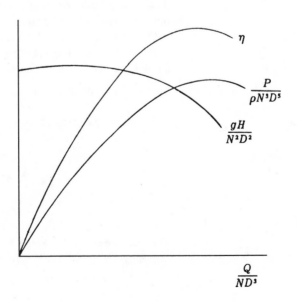

Figure 40.1

same for all similar machines and are given the following names:

Flow coefficient, or discharge number, $C_Q = \dfrac{Q}{ND^3}$ (40.7)

Head coefficient, $C_H = \dfrac{gH}{N^2D^2}$ (40.8)

Power coefficient, $C_P = \dfrac{P}{\rho N^3 D^5}$ (40.9)

These non-dimensional parameters may be used to plot the performance characteristics of a turbomachine, which will apply to all geometrically similar machines. The relation between these parameters is shown in Figure 40.1.

Equations (40.4), (40.5) and (40.6) show that in any range of machines operating at corresponding points, doubling the speed will double the discharge, quadruple the head and multiply the power eight-fold.

40.3 Specific speed

The specific speed of a turbomachine is the speed at which a geometrically similar machine would need to run to produce unit output from unit input at maximum efficiency. This speed has a typical value for each different design of machine and is a useful parameter for selecting the optimum type for given performance requirements.

For dynamic similarity, the dimensionless groups $\dfrac{gH}{N^2D^2}$, $\dfrac{Q}{ND^3}$ and $\dfrac{P}{\rho N^3 D^5}$ must be the same for each machine and, as stated in Section 38.4, these groups may be multiplied or divided amongst themselves to produce other dimensionless groups, as required. These new groups must also be the same for dynamically similar machines.

For a *turbine*, the specific speed is defined as the speed at which a similar turbine would generate an output of 1 kW under a head of 1 m. This is determined by eliminating the variable D from the groups $\dfrac{gH}{N^2D^2}$ and $\dfrac{P}{\rho N^3 D^5}$, i.e. between C_H and C_P.

Thus
$$\frac{\left(\dfrac{P}{\rho N^3 D^5}\right)^{1/2}}{\left(\dfrac{gH}{N^2D^2}\right)^{5/4}} = \frac{NP^{1/2}}{\rho^{1/2} g^{5/4} H^{5/4}}$$

This ratio must be the same for the actual turbine and the specific turbine. Denoting the variables for the specific turbine by suffix s,

$$\frac{N_s P_s^{1/2}}{\rho^{1/2} g^{5/4} H_s^{5/4}} = \frac{NP^{1/2}}{\rho^{1/2} g^{5/4} H^{5/4}}$$

But $P_s = 1$ kW and $H_s = 1$ m and so, assuming ρ to be the same for each,

$$N_s = \frac{NP^{1/2}}{H^{5/4}} \tag{40.10}$$

For a *pump*, the specific speed is defined as the speed at which a similar pump would deliver an output of 1 m³/s against a head of 1 m. This is determined by eliminating the variable D from the groups $\dfrac{gH}{N^2D^2}$ and $\dfrac{Q}{ND^3}$, i.e. between C_H and C_Q.

Thus

$$\frac{\left(\dfrac{Q}{ND^3}\right)^{1/2}}{\left(\dfrac{gH}{N^2D^2}\right)^{3/4}} = \frac{NQ^{1/2}}{g^{3/4}H^{3/4}}$$

This ratio must be the same for the actual pump and the specific pump. Denoting the variables for the specific pump by suffix s,

$$\frac{N_sQ_s^{1/2}}{g^{3/4}H_s^{3/4}} = \frac{NQ^{1/2}}{g^{3/4}H^{3/4}}$$

But $Q_s = 1$ m³/s and $H_s = 1$ m, so that

$$N_s = \frac{NQ^{1/2}}{H^{3/4}} \tag{40.11}$$

The specific speeds for turbines and pumps depend on the systems of units employed and would have different values if power, discharge and head were expressed in horsepower and feet units rather than S.I. Units. (Although specific speeds are normally quoted in rev/min, the units appropriate to S.I. definitions are actually rev min⁻¹kW^{1/2}m⁻⁵/⁴ for turbines and rev min⁻¹s⁻¹/²m³/⁴ for pumps.)

40.4 Choice of machine

The specific speed of a turbomachine depends only on the shape and might be termed a 'shape factor'. The value is determined from coefficients defined at the point of maximum efficiency and thus the *optimum* choice of a machine for particular performance requirements is restricted. The specific speed for these requirements is obtained from equations (40.10) or (40.11) and this then determines the most suitable type of machine, the only other variable being the size necessary to provide the required output.

The 'shape' of a turbine or pump is determined by two factors: the degree of reaction and the flow direction.

40.5 Degree of reaction

When a liquid passes through a machine, both the kinetic and pressure energy of the liquid may change; the distribution of the change between kinetic and pressure energy defines the *degree of reaction*. When there is no change in static pressure across the runner, the degree of reaction is zero and the machine is an impulse design. There are no impulse pumps but impulse turbines, known as Pelton wheels, are used (see Section 35.4).

In an impulse turbine, the available head is converted to kinetic energy through fixed nozzles and the high velocity stream of water passes over blades, or buckets, where the change of momentum produces the power. The runner is not full of water; when the water falls from a bucket, it remains empty until it meets the next jet and the pressure is atmospheric throughout. The specific speed of a Pelton wheel is low ($N_s < 40$) and this design is used when a large head of water is available.

Turbines and pumps in which there is a change of kinetic and pressure energy in the runner have a degree of reaction and are called reaction machines. The degree of reaction is determined by the shape of the runner and the blade angles. All such machines *must run full of water* in order to achieve the static pressure change.

Reaction turbines include Francis and Kaplan (or propeller) types which offer a range of specific speeds. Pumps include centrifugal, axial and mixed flow designs which again offer a range of specific speeds. These various types are described briefly in the following sections, while centrifugal pumps are considered in more detail in Chapter 41.

40.6 Reaction turbines

The arrangement of a turbine plant is shown in Figure 40.2. Water is supplied to the turbine from a reservoir and thence, via a draft tube, to a tail race. The object of the draft tube is to keep the turbine full of water and it is

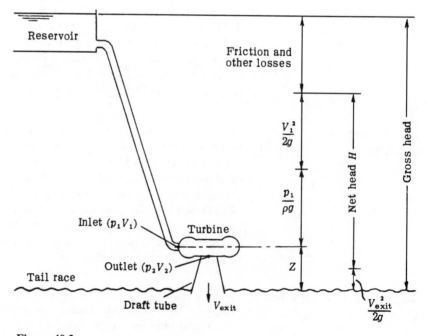

Figure 40.2

divergent to reduce the final velocity of the water, thus keeping the loss of kinetic energy at exit, $\dfrac{V^2_{exit}}{2g}$ to a minimum. The height of the draft tube, Z, is limited by the need to keep the outlet pressure, p_2, at the runner above about 2.5 m of water absolute, otherwise bubbles of vapour will be released which will damage the turbine. This phenomenon is known as cavitation.

Some of the gross head is lost due to pipe friction, entry loss, bends, etc., and some is lost in kinetic energy at exit; the remaining head is available to produce power.

Thus net head, H = gross head − losses − exit K.E.

40.7 Types of reaction turbine

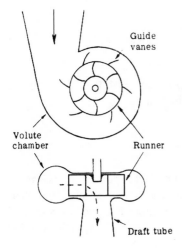

Figure 40.3

(a) Radial flow turbine

The principle of the Francis radial flow turbine is shown in Figure 40.3. Water enters a spiral volute chamber and then flows radially through stationary pivoted guide vanes; these direct the water, ideally without shock, on to the moving vanes attached to the runner. The water flows radially inwards through the vanes but leaves the runner axially to enter the draft tube. Francis turbines have medium specific speeds ($N_s = 40 - 350$) and are suited to heads between 20 and 200 m.

(b) Mixed flow turbines

The runner is generally similar to that of the radial flow turbine but is designed to give partly radial and partly axial flow. Figure 40.4 shows variations in the flow direction for different runners, together with typical specific speeds.

It will be seen that the specific speed increases as the flow becomes more axial and less radial.

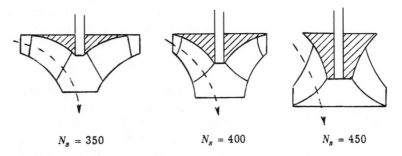

Figure 40.4

(c) Axial flow turbine

In a Kaplan axial flow or propeller turbine, Figure 40.5, the water enters a spiral volute chamber and then flows radially through stationary pivoted guide vanes. It is then turned into the axial direction *before* passing through the runner, which is similar to a propeller.

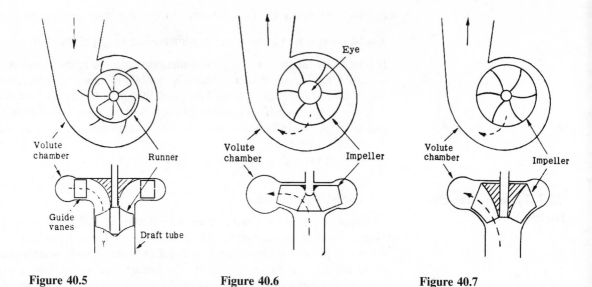

Figure 40.5   Figure 40.6   Figure 40.7

Kaplan turbines have a high specific speed ($N_s = 430 - 750$) and are suited to low net heads, i.e. between 3 and 25 m.

40.8 Types of pump

(a) Radial flow pump

The principle of a radial flow or centrifugal pump is shown in Figure 40.6. Water enters the impeller (or rotor) eye axially and then flows radially outwards through the blade passages. The specific speed of a centrifugal pump may be up to 100 but when $N_s < 30$, a set of diffuser blades is fixed round the rotor to improve efficiency.

Centrifugal pumps give relatively high heads with a low flow rate.

(b) Mixed flow pumps

In a mixed flow pump, Figure 40.7, the water enters axially and then passes through an impeller which gives it a partly radial flow at exit. As with turbines, the transition from axial to radial flow is gradual and the specific speed and general performance lie between those for centrifugal and axial flow pumps. Specific speeds range from 100 to 200.

(c) Axial flow, or propeller, pump

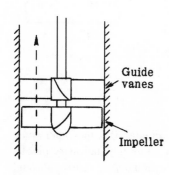

Figure 40.8

The principle of action is shown in Figure 40.8. Water flows axially through a propeller-type runner and then through stationary guide vanes.

The specific speed of a propeller pump is high ($N_s > 200$) and this design is suitable for cases where a high flow rate with a low head is required.

40.9 Machine choice based on specific speed

It has been seen that the optimum choice of machine type for a given performance is dictated by its specific speed and this information is summarized in Figures 40.9 and 40.10, which show the range of specific speeds for different types of turbine and pump respectively.

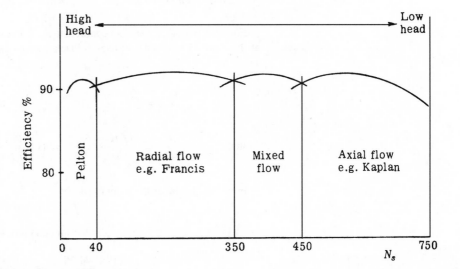

Figure 40.9

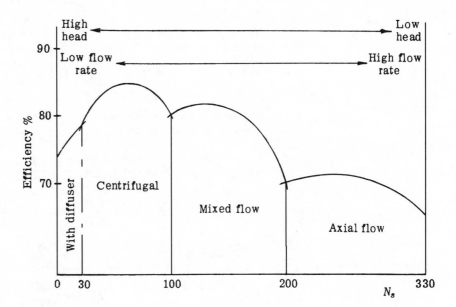

Figure 40.10

Worked examples

1. *A Francis turbine is required to develop about 4500 kW at maximum efficiency when running at 110 rev/min with a head of 14 m. A scale model is tested with a head of 3.7 m and at maximum efficiency, the speed was 210 rev/min, the flow rate was 1.45 m³/s and the power output was 44.25 kW.*

Determine the ratio of sizes of turbine and model, the specific speed, the flow rate required through the turbine and its power output and efficiency.

From Section 40.3, the head coefficient C_H is the same for the turbine and model at the point of maximum efficiency. Denoting model by suffix m and turbine by suffix t,

$$C_H = \left(\frac{gH}{N^2D^2}\right)_m = \left(\frac{gH}{N^2D^2}\right)_t \quad \text{from equation (40.8)}$$

so that

$$\frac{D_t}{D_m} = \sqrt{\left(\frac{H_t \times N_m^2}{H_m \times N_t^2}\right)} = \sqrt{\left(\frac{14 \times 210^2}{3.7 \times 110^2}\right)} = \underline{3.714}$$

From the results of the model test,

$$N_s = \frac{NP^{1/2}}{H^{5/4}} \quad \text{from equation (40.10)}$$

$$= \frac{210 \times 44.25^{1/2}}{3.7^{5/4}} = \underline{272.2 \text{ rev/min}}$$

The flow coefficient C_Q is also the same for model and turbine at the point of maximum efficiency,

i.e.

$$C_Q = \left(\frac{Q}{ND^3}\right)_m = \left(\frac{Q}{ND^3}\right)_t \quad \text{from equation (40.7)}$$

so that

$$Q_t = Q_m \times \frac{N_tD_t^3}{M_mD_m^3} = 1.45 \times \frac{110}{210} \times 3.714^3 = \underline{38.94 \text{ m}^3/\text{s}}$$

Similarly, the power coefficient C_P is the same for model and turbine at the point of maximum efficiency,

i.e.

$$C_P = \left(\frac{P}{\rho N^3D^5}\right)_m = \left(\frac{P}{\rho N^3D^5}\right)_t \quad \text{from equation (40.9)}$$

so that

$$P_t = P_m \times \frac{N_t^3D_t^5}{N_m^3D_m^5} = 44.25 \times \left(\frac{110}{210}\right)^3 \times 3.714^5 = \underline{4490 \text{ kW}}$$

Alternatively,

$$N_s = \frac{N_tP_t^{1/2}}{H_t^{5/4}}$$

i.e.

$$272.2 = \frac{110P^{1/2}}{14^{5/4}}$$

from which

$$P = \underline{4490 \text{ kW}}$$

The efficiency is given by

$$\eta = \frac{\text{power output}}{\text{power input}} = \frac{P}{\rho g Q H}$$

$$= \frac{4\,490 \times 10^3}{10^3 \times 9.81 \times 38.94 \times 14} = 0.839\,6 \quad \text{or} \quad \underline{83.96\%}$$

2. *A small rotary pump with an impeller 0.2 m diameter runs at maximum efficiency and discharges 0.065 m³/s of water against a head of 4.27 m when running at 1 800 rev/min. What is the specific speed?*

A similar pump is to deliver 0.92 m³/s of water to a condenser. If the impeller diameter is 0.4 m, at what speed should this pump run and what will be the head produced?

From equation (40.11),

$$N_s = \frac{N Q^{1/2}}{H^{3/4}}$$

$$= \frac{1\,800 \times 0.065^{1/2}}{4.27^{3/4}} = \underline{154.5 \text{ rev/min}}$$

If both pumps are to operate at maximum efficiency, the flow coefficient C_Q will be the same for each.

Denoting the small pump by suffix 1 and the large pump by suffix 2,

$$\frac{Q_1}{N_1 D_1^3} = \frac{Q_2}{N_2 D_2^3} \quad \text{from equation (40.7)}$$

$$\therefore N_2 = N_1 \frac{Q_2}{Q_1} \left(\frac{D_1}{D_2}\right)^3$$

$$= 1\,800 \times \frac{0.92}{0.065} \times \left(\frac{0.2}{0.4}\right)^3 = \underline{3\,185 \text{ rev/min}}$$

Similarly, the head coefficient C_H will be the same for each,

i.e.

$$\frac{g H_1}{N_1^2 D_1^2} = \frac{g H_2}{N_2^2 D_2^2} \quad \text{from equation (40.8)}$$

$$\therefore H_2 = H_1 \left(\frac{N_2 D_2}{N_1 D_1}\right)^2$$

$$= 4.27 \times \left(\frac{3\,185 \times 0.4}{1\,800 \times 0.2}\right)^2 = \underline{53.48 \text{ m}}$$

Note: The specific speed for the large pump is given by

$$N_s = \frac{3\,185 \times 0.92^{1/2}}{53.48^{3/4}} = \underline{154.5 \text{ rev/min}}$$

This agrees with that for the small pump.

From Figure 40.10, a suitable type of pump would be a mixed flow design.

Further problems

3. A quarter-scale turbine model is tested under a head of 11 m. The full-scale turbine is to use a head of 30 m and to run at 430 rev/min. At what speed should the model test be made?

If, during the test, the model produces 100 kW at the point of maximum efficiency, using a flow of 1.08 m³/s of water. What will be the power output of the full-scale turbine and the flow rate required at the point of maximum efficiency?

[1 042 rev/min; 7.196 kW; 28.52 m³/s]

4. A water turbine is to develop 5 MW at 120 rev/min using a head of 32 m. A scale model test conducted at 210 rev/min with a head of 3.6 m uses a flow rate of 0.22 m³/s at the point of maximum efficiency.

Determine the specific speed and efficiency of the machines, the flow rate for the turbine and the power output of the model.

What type of turbine is appropriate to this performance?

[111.5 rev/min; 89.2%; 17.85 m³/s; 6.93 kW; Francis]

5. A hydro-electric power station is to use turbines which have a specific speed of 188 rev/min at 82% efficiency. The head and quantity of water available are 60 m and 33 m³/s respectively. The generator design requires the turbines to run at 480 rev/min. Determine the number of turbines to be installed and the power output of the station. What type of turbine would be required?

[4; 15.93 MW; Francis]

6. A pump running at 1 400 rev/min delivers 0.03 m³/s at the point of maximum efficiency against a head of 33.5 m. A geometrically similar pump has linear dimensions 50% greater and operates at 1 200 rev/min. Determine the head produced by the larger pump at the point of maximum efficiency and the discharge at this point.

What is the specific speed of this range of pumps and the relative power consumption of the two pumps?   [55.37 m; 0.087 m³/s; 17.41 rev/min; 4.78]

7. The efficiency, η, of a rotodynamic pump depends on the speed, N, the impeller diameter, D, the fluid density, ρ, the flow rate, Q, the head, expressed as gH and the power, P. Show that

$$\eta = \phi \left[\frac{P}{\rho N^3 D^5}; \frac{Q}{ND^3}; \frac{gH}{N^2 D^2} \right]$$

A pump supplying 0.3 m³/s of water develops a head of 20 m at 2 000 rev/min and requires a power supply of 100 kW. For dynamically similar conditions at the point of maximum efficiency, what are the rotational speed, the head and flow rate when the power supply is reduced by 30%?   [1 776 rev/min; 15.77 m; 0.27 m³/s]

8. The power required to drive a pump at the design point is 6 kW at 1 465 rev/min. The discharge is then 25 kg/s against a head of 15.25 m.

Determine the size ratio of another similar pump which would operate at the point of maximum efficiency from a motor which produces 4.7 kW at 2 500 rev/min. What would be the head and discharge from this pump under these conditions?

[0.691; 21.2 m; 14.1 kg/s]

9. A pump for use in an irrigation scheme is to raise 4 500 m³/h of water against a head of 2.5 m with the pump working at maximum efficiency.

Tests on a geometrically similar pump of smaller size gave the following results at a constant speed of 1 000 rev/min.

Quantity (m³/s)	0.045	0.057	0.062	0.068	0.076	0.079	0.083	0.091
Head (m)	14.0	13.7	13.4	12.95	12.2	11.73	11.0	8.84
Power (kW)	10.07	11.41	11.86	12.31	12.76	12.98	13.20	13.50

Determine the specific speed of the series and the speed at which the irrigation pump should run. What type of pump should be used?

[42.23 rev/min; 75.1 rev/min; centrifugal]

41 Centrifugal pumps

41.1 Introduction

The centrifugal pump is one of the most common types of turbomachine and is now analysed in more detail. A similar analysis could, however, be applied to axial flow pumps, turbines, fans, etc., with appropriate essential differences.

In Chapter 35, the momentum equation is used to determine the force on a curved blade and this forms the basis of the action of a turbomachine. It is common to resolve the absolute velocity of the water into components in the radial, tangential and axial direction; the force in any direction then derives from the change in momentum in that direction.

The velocity component tangential to the rotor circumference is called the *velocity of whirl*; the force in the whirl direction is that which gives rise to the torque and power and is therefore the most important force considered in turbomachine analysis.

41.2 The Euler equation

In this analysis, the following notation is used:

Absolute velocity at inlet	V_1
Absolute velocity at outlet	V_2
Relative velocity at inlet	u_1
Relative velocity at outlet	u_2
Tangential velocity of blade tip at inlet	$v_1 (= \omega r_1)$
Tangential velocity of blade tip at outlet	$v_2 (= \omega r_2)$
Velocity of whirl at inlet	v_{w_1}
Velocity of whirl at outlet	v_{w_2}
Radial velocity at inlet	v_{r_1}
Radial velocity at outlet	v_{r_2}

Figure 41.1 shows the relative velocity triangles at inlet and outlet of a typical blade of a radial flow (outward) pump impeller. The directions of u_1 and u_2 are tangential to the blade tips for smooth flow; the directions of v_1 and v_2 are tangential to the impeller circumference at the blade tips.

The torque about the shaft axis, exerted on the water, is given by the rate of change of the angular momentum of the water about that axis.

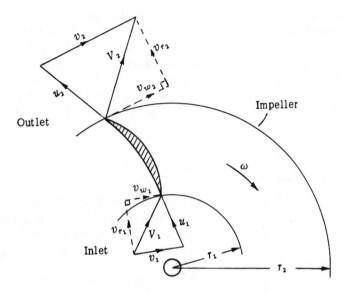

Figure 41.1

Momentum of water at inlet/sec $=$ mass of water/sec $\times$ tangential velocity

$$= \rho Q \times v_{w_1}$$

$\therefore$ moment of momentum (or angular momentum) of water/sec

$$= \rho Q v_{w_1} r_1$$

Similarly angular momentum of water/sec at outlet

$$= \rho Q v_{w_2} r_2$$

$\therefore$ torque $=$ change of angular momentum/sec

i.e. $$T = \rho Q(v_{w_2} r_2 - v_{w_1} r_1) \qquad (41.1)$$

This is known as the *Euler equation*.

The power input, $P = T\omega = \rho Q(v_{w_2} r_2 - v_{w_1} r_1)\omega$

But $\omega r_1 = v_1$ and $\omega r_2 = v_2$, so that

$$P = \rho Q(v_{w_2} v_2 - v_{w_1} v_1) \qquad (41.2)$$

Equation (41.2) represents the energy input per second but in pump analysis, it is often more convenient to work in terms of the head produced, rather than the power. The head H represents the energy per unit weight (see Section 34.2) and so

$$P = \rho g Q H$$

Thus $$H = \frac{\rho Q(v_{w_2} v_2 - v_{w_1} v_1)}{\rho Q g} = \frac{v_{w_2} v_2 - v_{w_1} v_1}{g} \qquad (41.3)$$

Equation (41.3) represents the *Euler head*.

41.3 Application to centrifugal pumps

The general arrangement of a centrifugal pump is shown in Figure 40.6. Water enters the casing in an axial direction but is turned through 90° before entering the impeller. It then passes radially through the blades to enter the volute chamber.

At inlet to the impeller, the water velocity has no component in the tangential direction, i.e. the velocity of whirl v_{w_1} is zero and the velocity triangle is right-angled. Hence equations (41.2) and (41.3) reduce to

$$P = \rho Q v_{w_2} v_2 \tag{41.4}$$

and

$$H = \frac{v_{w_2} v_2}{g} \tag{41.5}$$

The relative velocity triangles for this case are shown in Figure 41.2.

The Euler head is greater than the actual head produced, due to losses. The actual head produced comprises the suction head H_s, the discharge head H_d, the friction losses in the suction and delivery pipes, H_{f_s} and H_{f_d} and the discharge velocity head $\frac{v_d^2}{2g}$. This head is known as the *manometric head*, H_m, since it is the head which would be registered by a manometer connected across the pump at inlet and outlet if the inlet and outlet pipes were the same diameter.

Thus

$$H_m = H_s + H_d + H_{f_s} + \frac{v_d^2}{2g}$$

The distribution of these heads is shown in Figure 41.3.

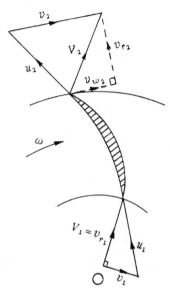

Figure 41.2

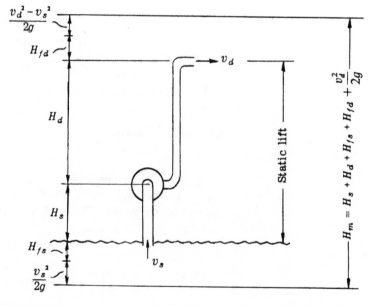

Figure 41.3

The ratio of the manometric head to the Euler head is called the manometric efficiency,

i.e.
$$\eta_m = \frac{H_m}{\dfrac{v_{w_2} v_2}{g}} = \frac{g H_m}{v_{w_2} v_2} \tag{41.6}$$

The difference between $\dfrac{v_{w_2} v_2}{g}$ and H_m is made up of hydraulic losses such as friction, bends, eddies, etc., and the effectiveness of the volute casing or diffuser in converting the impeller exit velocity to pressure energy. In addition to these hydraulic losses, there are also mechanical losses such as bearing friction and the overall efficiency of the pump will be less than the manometric efficiency.

41.4 Forward facing, radial and backward facing impeller blades

Figure 41.2 shows a pump impeller with backward facing blades but there are also designs with radial and forward facing blade directions at exit. Figure 41.4 shows these arrangements, from which it will be seen that, for the same values of u_2 and v_2, v_{w_2} (and hence the Euler head) is greatest with forward facing blades and least with backward facing blades.

In Figure 41.4, θ represents the outlet angle of the water relative to the whirl direction. Then, in all cases,

$$v_{w_2} = v_2 - v_{r_2} \cot\theta,$$

$\cot\theta$ being negative when θ is greater than $90°$.

Thus the Euler head,

$$\frac{v_{w_2} v_2}{g} = \frac{v_2}{g}(v_2 - v_{r_2}\cot\theta)$$

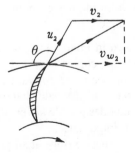 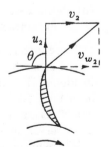 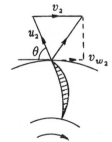

Forward facing blades   Radial blades   Backward facing blades

Figure 41.4

If the outlet flow area is A_2, then $v_{r_2} = \dfrac{Q}{A_2}$, which is a constant. Also, $v_2 = \omega r_2$, so that

$$\text{Euler head} = A\omega^2 - B\omega Q \cot\theta \qquad (41.7)$$

where A and B are constants.

This equation is represented graphically in Figure 41.5 for forward facing, radial and backward facing blades at a constant speed ω, with allowance for the sign of $\cot\theta$.

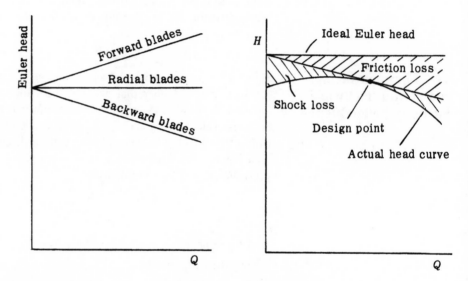

Figure 41.5   Figure 41.6

The actual head–flow rate curves will be modified by

(a) friction effects, which increase as Q increases, and
(b) shock losses away from the design point, when the liquid flow is not at the correct inlet angle.

These effects are shown in Figure 41.6, for the case of radial blades.

The actual head–flow rate curves for the three types of blade are shown in Figure 41.7. It will be seen that forward-facing blades give a greater head for a given flow rate and it would therefore be expected that pumps would generally be of this design. However, this is not so – the *slopes* of these curves are of great importance.

Where the curve has a negative slope at all points, such as with backward facing blades, there is only one flow rate for any value of H and the flow is stable but for a curve having positive and negative slopes, there are two values of Q for some values of H. This can lead to unstable operation, especially when pumps are used in parallel and so pumps with backward facing blades are preferred.

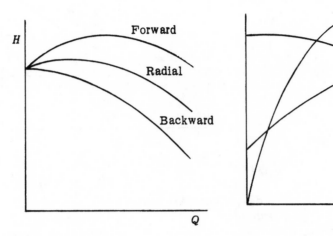

Figure 41.7   Figure 41.8

41.5 Performance characteristics

The parameters of turbomachine performance, developed in Section 40.2, are head, power and efficiency, each being a function of the flow rate, Q. Figure 41.8 shows the relations of these quantities for a centrifugal pump with backward facing blades, running at constant speed. The design point is that corresponding with maximum efficiency; it is at this point that the specific speed is evaluated, in order to assist in the selection of a pump for a particular application.

41.6 System characteristics and matching

A pump is installed in a system to give a static lift, h, between suction and delivery points but, in addition, it will also have to overcome losses due to friction, bends, valves, etc. These losses are all proportional to V^2 and hence proportional to Q^2 for a given cross-sectional area of flow. Thus a *system characteristic* at constant speed may be expressed in the form

$$H = h + kQ^2 \qquad (41.8)$$

where H is the required output head and k is a constant.

This relation is shown in Figure 41.9.

Having selected a pump from specific speed considerations, the actual running point is determined from the combination of the pump and system characteristics; this is known as *matching* and is illustrated in Figure 41.10. The matching point is at the intersection of the two characteristic curves which should ideally correspond with the maximum efficiency point for the pump.

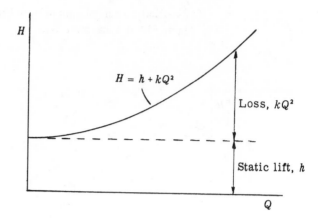

Figure 41.9

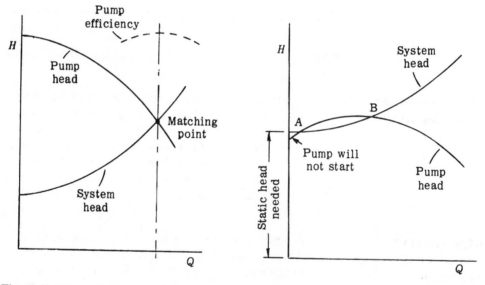

Figure 41.10

Figure 41.11

In the case of forward facing blades, the pump characteristic initially rises with increase of flow and it is possible that the system characteristic may intersect it at two points, Figure 41.11. Thus there will be two operating points, A and B, and variations in the speed of the driving motor or in the delivery head may cause the discharge to oscillate between these positions, leading to unstable running.

This possibility is greatly increased when two unstable pumps are operating in parallel, discharging into a common delivery pipe; a change in the operating point in one pump may cause the second pump to start oscillating, which in turn will affect the running of the first pump. It is therefore necessary that pumps operating in parallel should have stable characteristics.

A further problem can arise if the system head is greater than the pump head at zero discharge. The pump will then be unable to start unless the discharge pipe is partly emptied to reduce the opposing head.

Worked examples

1. *A centrifugal pump is required to deliver 0.75 m³/s of water against a manometric head of 16 m. At outlet, the relative velocity of the water is inclined at 120° to the tangential velocity of the rotor (i.e. backward facing blades) and the radial velocity is 0.27 times the rotor tangential velocity. The manometric efficiency is 85% and the width of the impeller at exit is 0.1 times the outside diameter.*

Determine the outside diameter of the impeller and the speed of rotation.

From equation (41.6),
$$\eta_m = \frac{gH_m}{v_{w_2}v_2}$$

$$\therefore v_{w_2} = \frac{9.81 \times 16}{0.85v_2} = \frac{184.66}{v_2} \qquad (1)$$

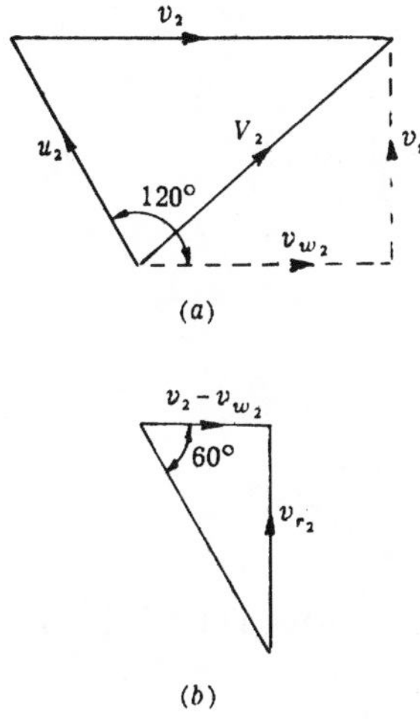

Figure 41.12(a) shows the relative velocity diagram at exit.
From Figure 41.12(b),

$$\tan 60° = \frac{v_{r_2}}{v_2 - v_{w_2}}$$

But $v_{r_2} = 0.27v_2$,

$$\therefore 1.732 = \frac{0.27v_2}{v_2 - v_{w_2}}$$

from which
$$v_{w_2} = 0.844v_2 \qquad (2)$$

Hence, from equations (1) and (2),

$$v_2 = 14.79 \text{ m/s}$$

But
$$v_2 = \frac{\pi DN}{60}$$

from which
$$DN = \frac{60 \times 14.79}{\pi} = 282.47 \qquad (3)$$

Quantity flowing,
$$Q = \text{exit area} \times v_{r_2}$$

i.e.
$$0.75 = (\pi D \times 0.1D) \times (0.27 \times 14.79)$$

from which
$$D = \underline{0.773 \text{ m}}$$

From equation (3),
$$N = \frac{282.47}{0.773}$$

$$= \underline{365.4 \text{rev/min}}$$

(a)

(b)

Figure 41.12

2. *A centrifugal pump running at 1000 rev/min gave the following results in a test:*

Discharge (m³/s)	0	0.075	0.151	0.227	0.302	0.359
Head (m)	23.0	22.5	21.8	19.8	14.3	0

(a) The pump is to be connected to 300 mm diameter suction and delivery pipes with the discharge point 13 m above sump level. The total length of the suction and delivery pipes is 75 m and the entry loss is equivalent to a further 7 m of pipe. If the friction coefficient for the pipes is 0.006, calculate the discharge from the pump.

(b) If the speed is reduced to 800 rev/min, determine the new flow rate, assuming the efficiency to be unchanged.

(a) Pipe velocity, $\quad V = \dfrac{Q}{\dfrac{\pi}{4} \times 0.3^2} = 14.147Q$

Effective length of pipe $= 75 + 7 = 82$ m

$\therefore$ friction loss in pipe $= \dfrac{4flV^2}{2gd}$

$$= \frac{4 \times 0.006 \times 82(14.147Q)^2}{2 \times 9.81 \times 0.3}$$

$$= 66.92Q^2$$

Exit loss $= \dfrac{V^2}{2g} = \dfrac{(14.147Q)^2}{2 \times 9.81} = 10.2Q^2$

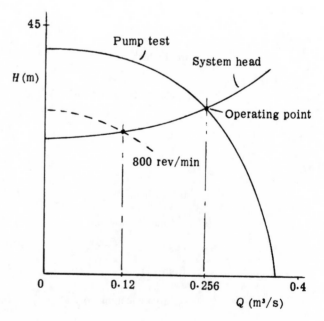

Figure 41.13

$$\text{System characteristic,} \quad H = \text{static lift} + \text{losses}$$

$$= 13 + (66.92 + 10.2)Q^2$$

$$= 13 + 77.12Q^2$$

For the values of Q used in the results, corresponding values of H are as follows:

H(m):   13   13.44   14.76   16.97   20.03   22.94

The graphs of pump head and system head against discharge are shown in Figure 41.13, from which the intersection gives a discharge of 0.256 m³/s.

(b) At a given efficiency, $\dfrac{Q}{ND^3}$ is a constant

$$\therefore Q_2 = Q_1 \times \frac{N_2}{N_1} \text{ since } D \text{ is unchanged}$$

Also $\dfrac{gH}{N^2D^2}$ is a constant,

$$\therefore H_2 = H_1 \times \frac{N_2^2}{N_1^2}$$

The pump characteristic at 800 rev/min can now be obtained by multiplying the original discharge values by 0.8 and the corresponding head values by 0.8^2.
The results are as follows:

Discharge (m³/s)	0	0.06	0.121	0.182	0.242	0.287
Head (m)	14.7	14.4	14.0	12.7	9.2	0

This curve is plotted on the original diagram and intersects the system head curve, which remains unchanged, at a discharge of 0.12 m³/s

Further problems

3. A centrifugal pump delivers 0.012 m³/s of water when the pressure difference across the pump is 0.33 MN/m² and the speed is 2 100 rev/min. The manometric efficiency is 0.83. If the impeller is 0.2 m in diameter and the exit width is 0.012 m, determine the blade exit angle.   [21.75°]

4. A centrifugal pump is to be used in a system having a head characteristic given by $H = (12 + 7\,410Q^2)$ m where Q is the discharge in m³/s. In a trial on the pump, the following results were obtained:

Speed (rev/min)	1 215	1 212	1 210	1 202	1 200	1 195	1 190
Head (m)	25.6	27.0	26.8	26	24.3	20	13.1
Discharge (m³/s)	0	0.011	0.018	0.025	0.032	0.041	0.049

Correct the table to a speed of 1 200 rev/min and determine the pump discharge at 1 200 rev/min with the system.   [0.036 5 m³/s]

5. In a test on a centrifugal pump the following results were obtained:

Speed (rev/min)	2 160	2 140	2 130	2 120	2 105	2 120
Head (m)	34.44	37.19	34.14	29.41	21.34	14.33
Discharge (m³/s)	0	0.007 1	0.011 8	0.016 7	0.023 2	0.028 6
Power (kW)	–	4.77	5.80	6.56	7.16	7.16

Plot curves of head and efficiency against quantity flowing for a speed of 2 100 rev/min and hence determine the power requirement at 2 100 rev/min when used in a system with a head demand, $H = (13.4 + 59\,300Q^2)$ m where Q is the quantity flowing in m³/s.   [6.35 kW]

6. A centrifugal pump is required to discharge 0.57 m³/s of water with a head of 13 m when rotating at 750 rev/min. The manometric efficiency is 0.8 and the radial velocity is 3 m/s. The loss of head in the pump is $0.026\,2V^2$ where V is the impeller absolute exit velocity in m/s.
Determine the impeller diameter, exit width and blade exit angle.
[0.375 m; 0.16 m; 35.7°]

7. A centrifugal pump delivers 0.22 m³/s with a head of 25 m at 1 500 rev/min. The manometric efficiency is 75% and the loss of head in the system is $0.033V^2$ where V is the absolute velocity at impeller exit in m/s. The exit area of the pump is $1.2D^2$ where D is the impeller diameter. Determine the blade exit angle and the impeller diameter.   [26.3°; 0.266 m]

8. A centrifugal pump impeller has an external diameter of 0.3 m and an outlet area of 0.11 m². The backward facing blades make an angle of 145° to the whirl direction. Pressure gauges at the same level above the sump in the suction and delivery lines close to the pump read heads of water of 3.7 m below atmospheric pressure and 21 m above atmospheric pressure respectively when the pump delivers 0.204 m³/s of water at 1 200 rev/min. The power input is 71.6 kW and pipe friction losses are 3 m of water.
Determine the manometric efficiency and the overall efficiency of the pump.
[79.4%; 61%]

9. A centrifugal pump delivers 0.94 m³/s of water against a head of 25 m of water. The outer diameter of the impeller is 1 m and the inner diameter is 0.5 m. The pump speed is 600 rev/min. The radial flow velocity is constant at 6 m/s and the pump efficiency is 0.75. Find the inlet and outlet impeller blade angles relative to the whirl direction, the inlet and outlet blade widths and the power input to the pump if the mechanical efficiency is 0.9.   [21°; 15.9°; 0.1 m; 0.05 m; 342 kW]

10. The results of a trial on a centrifugal pump are as follows:

Discharge (m³/s)	0	0.255	0.481	0.750	0.990
Head (m)	132	121	106	78	31

Two such pumps deliver water into a pipeline 0.6 m diameter and 1 000 m long against a static head of 50 m. The friction coefficient in the pipe is 0.005. Neglecting velocity head, determine the discharge if the pumps are coupled (a) in parallel, and (b) in series.   [1.35 m³/s; 0.98 m³/s]

42 Hydraulic machines

42.1 Introduction

This chapter deals with equipment which uses liquid pressure to perform various tasks. It includes positive displacement pump designs but excludes centrifugal pumps which are considered separately in Chapter 41.

42.2 Reciprocating pumps

A reciprocating pump for incompressible fluids uses a piston and cylinder mechanism. Liquid is drawn into the pump cylinder through an inlet valve as the piston moves back, then, as the piston moves forward the inlet valve closes and the liquid is raised in pressure until the delivery valve opens when the required pressure is reached, Figure 42.1(*a*). Valve operation is usually automatic, for example, the delivery valve opens when the pump pressure exceeds that in the delivery pipe and the liquid is pushed out by the piston.

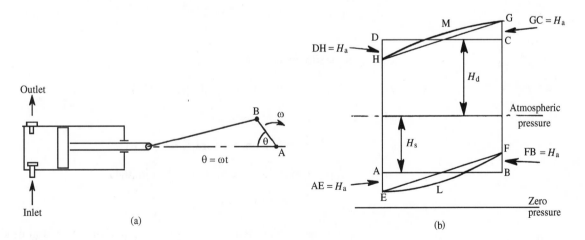

Figure 42.1

The ideal working cycle is represented by the rectangle ABCD in Figure 42.1(*b*). The lines EF and GH show the pressure changes due to acceleration and deceleration of the fluid and there are also friction losses in the suction pipe shown on line ELF and in the delivery pipe GMH. The pump speed must not be such that the pressure at E allows dissolved gases or vapour to form bubbles which will cause damage when the bubbles collapse as pressure rises. This is called *cavitation*.

Pressure variation due to acceleration and deceleration in the suction pipe may be analysed. In order to do so the pump piston acceleration is required. This is discussed in Section 12.5, equation (12.4) but it is usual to simplify the piston motion by assuming simple harmonic motion (Section 18.2). Using the latter method with an angular velocity ω of the crank AB, Figure 42.1(a), it is shown that

piston velocity $= \omega r \sin \omega t = \omega r \sin \theta$

and

piston acceleration $= \omega^2 r \cos \omega t = \omega^2 r \cos \theta$

The velocity v and acceleration f of the liquid in the pipe is therefore

$$v = (A/a)\omega r \sin\ \theta \tag{42.1}$$

$$f = (A/a)\omega^2 r \cos\ \theta \tag{42.2}$$

where A is the piston area and a is the pipe area.

If the length of the suction pipe is l, the mass of water in the pipe is $\rho a l$ then let p_a be the acceleration pressure and since force = mass × acceleration

$$p_a \cdot a = (\rho a l) \times ((A/a)\omega^2 r \cos\theta)$$

$$\therefore\ p_a = \rho l (A/a)\omega^2 r.\cos\theta$$

Now p_a is maximum when $\theta = 0°$,

i.e.   maximum pressure $p_a = \rho l (A/a)\omega^2 r$

and p_a is a minimum when $\theta = 180°$,

i.e.   minimum pressure $p_a = -(\rho l (A/a)\omega^2 r$

But $p_a = \rho g H_a$, hence the maximum acceleration head

$$H_a = (l/g)(A/a)\omega^2 r \tag{42.3}$$

The minimum head in the suction pipe will be the sum of the suction head H_s and the acceleration head H_a and this must be less than that causing bubbles to form in the suction pipe.

In general in a suction pipe,

$$H_s + H_a < 8 \text{ m of water} \tag{42.4}$$

A similar analysis for the delivery pipe yields

$$H_a < 8 \text{ m water } + H_d \qquad (42.5)$$

which is less likely to cause cavitation problems.

It is also possible to analyse the effects of friction losses in the pipes from a knowledge of the liquid velocity in the pipe (equation (42.1):

$$v = (A/a)\omega r \sin\theta$$

and the use of Sections 37.7, 37.8 and 37.9. This involves quite lengthy calculations and is not demonstrated here. Effectively the parabolic areas EFLE and HGMH in Figure 42.1(b) represent the extra work required to overcome pipe friction losses. Since the velocity of the water is zero at entry to and exit from the pipe ($\theta = 0°$ and $180°$) there is no effect on the minimum heads.

At high speeds, the acceleration will be greater and an air cushion chamber is fitted on the suction side to smooth fluctuations and reduce the risk of cavitation. Similarly, on the delivery side, an air cushion chamber will smooth pulsating flow. Figure 42.2 shows the arrangement of air cushion chambers. The effect of the cushion chamber is to reduce the amount of liquid being accelerated to that between each chamber and the pump.

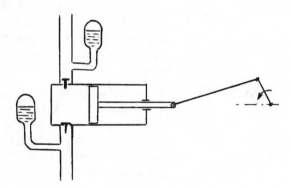

Figure 42.2

42.3 Reciprocating pump performance

Pump delivery increases with speed as would be expected but is reduced by *slip* which is the difference between the swept volume and delivered volume. A volumetric efficiency is defined by

$$\eta_{\text{VOL}} = \frac{\text{delivered volume}}{\text{swept volume}} = \frac{Q}{Q_0} = \frac{Q_0 - Q_L}{Q_0} = 1 - \frac{Q_L}{Q_0} \qquad (42.6)$$

where Q is the delivered volume, Q_L is the slip and Q_0 is (swept volume × speed).

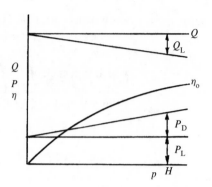

Figure 42.3

The overall efficiency of the pump is given by

$$\eta_0 = \frac{\rho Q g H}{P} \tag{42.7}$$

where P is the power input to the pump, H is the head and ρ is the fluid density.

Overall efficiency rises as head increases but at large heads the volume flow rate falls as slip increases. In general both η_{VOL} and η_0 are in the range 80–90% (Figure 42.3).

Double acting pumps are arranged so that as the piston moves forward raising the pressure and expelling liquid the space behind the piston is filled with fresh liquid. If the piston rod area is neglected, delivery is doubled. Two sets of valves will be required.

42.4 Multi-cylinder pumps

One disadvantage of the reciprocating pump which is not apparent in the performance characteristics is that the delivery is not steady. A single cylinder pump will deliver the pulsating flow shown in Figure 42.4(*a*). This problem may be overcome by using a multi-cylinder pump, and Figure 42.4(*b*) shows the effect of a three-cylinder pump with cranks 120 degrees apart.

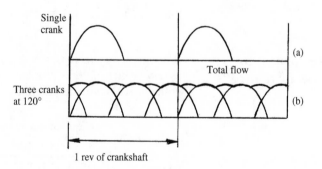

Figure 42.4

42.5 Rotary positive displacement pumps

These comprise a group of pumps which use close meshing or sliding of rotating parts to move liquid from the low pressure supply side to the high pressure delivery side. Some common designs include gear pumps, lobe pumps and sliding vane pumps (Figure 42.5). The performance characteristics of such pumps are similar to those shown in Figure 42.3. The values of η_0 and η_{VOL} for rotary positive displacement pumps are similar to those quoted above for reciprocating pumps and choice will be decided on the suitability of the pump to the particular application.

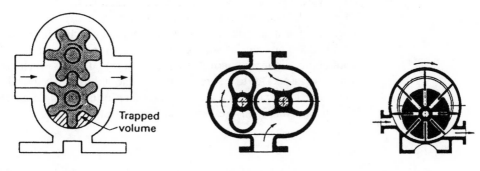

Figure 42.5

42.6 Hydraulic accumulators

The hydraulic accumulator is used for temporary storage of high pressure water (Figure 42.6). It consists of a vertical cylinder with a sliding ram around which are attached circumferential containers filled with heavy material. Water is pumped into the cylinder and lifts the ram and heavy

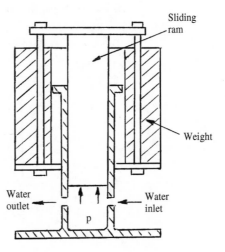

Figure 42.6

material until the cylinder is full. When the machine served by the accumulator requires the high pressure water it is passed to the working cylinder of the machine. An accumulator is also useful for overcoming the fluctuating nature of reciprocating pump flow to allow a steady power supply at constant pressure.

The maximum energy or capacity stored by an accumulator is the product of the accumulator pressure p and the accumulator volume V.

42.7 Hydraulic intensifiers

The hydraulic intensifier is a device for increasing the pressure of a quantity of water. In order to achieve this, a larger amount of low pressure water is required. The intensifier has a fixed ram through which high pressure water is supplied to a machine. A sliding ram is fitted over the fixed ram and the outer wall of this sliding ram moves inside a fixed cylinder into which low pressure water is passed. Due to the area difference, the sliding ram is forced down over the lower fixed ram, increasing the pressure in the sliding ram. The sliding ram passes this high pressure water to the machine in use (Figure 42.7).

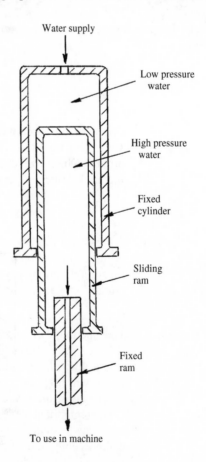

Figure 42.7

When the sliding ram reaches the bottom of the stroke, valves are operated to allow the sliding ram to move up to the beginning of the stroke and repeat the process; however some *double acting* designs are made.

42.8 Hydraulic presses and jacks

The principle of the hydraulic press or jack is simple; a small force f applied to a small diameter piston of area a in a cylinder will generate a pressure f/a. If this system is connected to a large diameter cylinder containing a piston of area A, the pressure will exert a large force $F = Af/a$ in a hydraulic press or lift large weights with a hydraulic jack (Figure 42.8). Whether in press or jack form, there must be a means of repeating the action. In the case of a press this may be achieved by the operation of return rams or the press may be a *double acting* design. The hydraulic jack must also have a means of lowering the load, often by means of a controlled *leak* of the fluid to a sump so that the lifted load does not fall too quickly. Presses are sometimes coupled to an intensifier to achieve greater load capability.

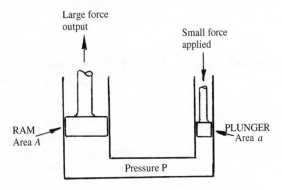

Figure 42.8

Presses of this type were originally designed by Joseph Bramah at the end of the eighteenth century and enabled a great step forward in engineering capability, particularly in lifting large civil engineering structures used in bridges. Other uses include lock gates, cranes, lifts, etc.

42.9 Hydraulic brakes

The operation of hydraulic vehicle brakes is similar to that of the press (Figure 42.9). A large movement of the vehicle brake pedal raises the pressure in the 'master cylinder'. This pressure is transmitted to the larger area of the wheel cylinders which operate the brake shoes or disc pads forcing them against the brake drums or brake discs, respectively.

One important difference in this application is to be certain that there is always enough fluid in the system to avoid inward air leaks. Air is a compressible fluid which would lead to inefficient braking, perhaps causing an accident. For this reason the master cylinder is connected to a brake

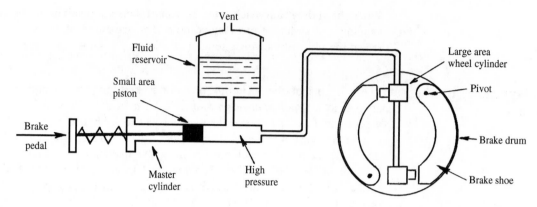

Figure 42.9

fluid reservoir and some designs involve retaining a small pressure in the hydraulic system so that fluid leaks are outward which is safer than air leaking inwards.

42.10 Conclusion

It should be clear from these examples that the use of hydraulic machines to generate large forces has many engineering uses.

Worked examples

1. *A reciprocating pump has a stroke of 300 mm and a bore of 0.44 m. Water is to be lifted through a total height of 12 m. The pump is driven by an electric motor at 70 rev/min and delivers 0.052 m³/s of water. Determine the percentage slip and the power required to drive the pump if the overall efficiency of the system is 95%.*

Swept volume of piston $= \dfrac{\pi}{4} \times 0.44^2 \times 0.3 = 0.045\,6 \text{ m}^3$

At 70 rev/min, ideal volume flow rate $= 0.045\,6 \times \dfrac{70}{60} = 0.053\,2 \text{ m}^3/\text{s}$

Actual delivered volume $= 0.052 \text{ m}^3/\text{s}$

Hence slip $= (0.053\,2 - 0.052) = 0.001\,2 \text{ m}^3/\text{s}$

Percentage slip $= \dfrac{0.001\,2}{0.053\,2} \times 100 = 2.256\%$

Ideal power is the work rate of lifting 0.052 m³/s of water through 12 m.

Ideal power $= 0.052 \text{ m}^3/\text{s} \times 1\,000 \text{ kg/m}^3 \times 12 \text{ m} \times 9.81 \text{ m/s}^2 = 6\,121.4 \text{ W}$

Actual power $= 6\,121.4 \times \dfrac{1}{0.95} = 6\,443.5 \text{ W}$

2. *A single-acting reciprocating pump runs at 28 rev/min. The pump has a piston of diameter 125 mm and a stroke of 300 mm. The suction pipe is 10 m in length and the diameter is 75 mm. Calculate the acceleration head at the beginning of the suction stroke. The pump centre line is 3 m above the water level in the river used*

as a sump. Assuming simple harmonic motion (SHM), determine the minimum head of water in the pump and compare the result with equation (42.4).

In Figure 42.1(*b*) the minimum head is at point E when the crank angle $\theta = 0°$. Hence from equation (42.3), the maximum acceleration head is given by

$$H_a = (l/g)(A/a)\omega^2 r = \frac{10}{9.81}\left[\frac{125}{75}\right]^2 \left[\frac{2\pi 28}{60}\right]^2 0.15$$

$$= 3.65 \text{ m water.}$$

Atmospheric pressure $= 101325 \text{ N/m}^2$

$$= 101\,325/(1\,000 \times 9.81) = 10.329 \text{ m water}$$

Minimum head in pump $= H_{\text{atmos}} - H_s - H_a = 10.329 - 3 - 3.65$

$$= \underline{3.679 \text{ m water.}}$$

Equation (42.4) requires $H_s + H_a < 8$ m water; i.e. $(3 + 3.65) < 8$ which is satisfied so that no cavitation should occur on the suction side of the pump.

3. *A single-acting reciprocating pump has a vertical delivery pipe 30 m in length and 75 mm in diameter. The pump piston diameter is 140 mm and the stroke is 0.6 m. Calculate the maximum pump speed so that there is no cavitation in the delivery stroke. Assume the pump diameter is negligible compared with the delivery pipe length.*

Assume that the piston motion is SHM.

The maximum acceleration head is at the point H in Figure 42.1(*b*) where crank angle θ is $0°$.

From equation (42.3) applied to the delivery pipe

$$H_a = (l/g)(A/a)\omega^2 r = \frac{30}{9.81}\left[\frac{140}{75}\right]^2 \omega^2 \left[\frac{0.6}{2}\right]$$

$$H = 3.197\omega^2$$

But from equation (42.5) the limiting condition is $H_a < (8 + H_d)$, i.e.

$$3.197\omega^2 < (8 + 30)$$

$$\therefore \omega_{\text{max}}^2 = (38/3.197) = 11.886$$

i.e. $$\omega_{\text{max}} = 3.448$$

If the maximum speed is N rev/min,

$$\frac{2\pi N}{60} = 3.448.$$

Hence the maximum speed is $\underline{32.92 \text{ rev/min.}}$

Further problems

4. (a) A fluid of specific gravity 0.8 is raised through a total height of 20 m by a single acting reciprocating pump. The bore of the pump is 0.15 m and the stroke is 0.35 m. The pump is driven at 36 rev/min by an electric motor. When tested the pump delivered 13 m³/h. Calculate the percentage slip and the power required from the electric motor if the mechanical efficiency of the system is 93%.

[2.67%; 609.5 W]

(b) If the pump in part (a) is replaced by a double-acting pump running at 40 rev/min, what would be the fluid flow rate if the bore and stroke are the same as the single acting pump?

What assumptions are required to be able to solve this problem.   [28.9 m³/h]

NOTE. Pipe friction losses are not to be considered in problems 5, 6, 7 and 8. All answers assume SHM.

5. (a) A single-acting reciprocating pump running at 30 rev/min. has a piston diameter of 110 mm and a stroke of 250 mm. The vertical suction pipe is 52 mm in diameter and 8 m in length. The pump centre line is 3.5 m above the level from which water is drawn. Calculate the minimum head of water in the pump and comment on the result by considering equation (42.4).

(b) What actions could be taken to solve the problem?

[(a) 2.328 m water; change H_s, l_s or A/a]

6. A single-acting reciprocating pump has a suction pipe 5 m long and 0.15 m in diameter. The water level in the sump is 1.5 m above the suction pipe entry. The pump has a piston diameter of 0.25 m and the stroke is 0.75 m.

(a) What is the flow rate of water through the pump when running at 25 rev/min with 2% slip?
(b) Check the cavitation criteria for safe running.
(c) How much power is required if the total lift is 5.5 m and the mechanical efficiency of the system is 92%?
(d) What is the maximum speed at which the pump can be run before cavitation may occur?   [0.015 m³/s; 7.14 < 8; 880 W; 27.8 rev/min]

7. A single-acting reciprocating water pump has a vertical delivery pipe 55 mm in diameter and 35 m in length. (The lift of the water above the pump centre line may be assumed to be 35 m.) The pump has a plunger diameter of 120 mm and a stroke of 0.475 m.

(a) Calculate the maximum speed of the pump so that there is no cavitation in the delivery pipe.
(b) What is the maximum flow rate of water from the delivery pipe if the slip is 2%?
(c) If the suction head is 7 m and the mechanical efficiency is 96%, calculate the power required for this delivery.   [31.18 rev/min; 0.002 74 m³/s; 1.175 kW]

8. A single-acting reciprocating pump supplies water to a factory through a vertical suction pipe 8 m in length and 0.18 m in diameter. The vertical delivery pipe is 10 m in length and 0.15 m in diameter. The sump level is 5 m below the pump centre line. The pump piston diameter is 0.3 m and the stroke is 0.6 m. At what speed may the pump run to avoid cavitation in both suction and delivery strokes and what will be the flow rate of water delivered with 2% slip?

Calculate the power required to drive the pump if the overall efficiency of the system is 94%.   [20 rev/min; 0.013 9 m³/s; 2 038.7 W]

Appendix A
Moments of area and inertia

First moment of area

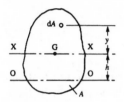

Figure A.1

Figure A.1 shows a plane surface of area A and XX is an axis passing through the centroid G.

The first moment of area of an element dA about XX is defined as $dA \times y$ and the total first moment for the whole area is given by $\int dA \times y$.

If distances above XX are regarded as positive, those below must be regarded as negative and *the total first moment about any axis passing through the centroid is zero.*

Second moment of area

The second moment of area of element dA about axis XX is defined as $dA \times y^2$ and the total second moment for the whole area is then given by $\int dA \times y^2$; this is denoted by I. Second moments of areas below XX do not cancel those above XX since squares of negative distances are positive.

The units of I are m^4, cm^4, mm^4, etc and it should be noted that $1\ m^4 = 10^8\ cm^4 = 10^{12}\ mm^4$.

Theorem of parallel axes

Referring to Figure A.1, let OO be an axis parallel to XX and distance h from it. Then second moment of area of element dA about OO

$$= dA(y+h)^2$$

$$= dA(y^2 + 2yh + h^2)$$

Therefore total second moment of area about OO,

$$I_{OO} = \int dA \times y^2 + 2h \int dA \times y + h^2 \int dA$$

$\int dA \times y^2 = I_{XX}$, $\int dA \times y$ is the first moment of area about XX, which is zero since this axis passes through the centroid G, and $\int dA$ is the total area A.

Hence
$$I_{OO} = I_{XX} + Ah^2 \tag{A.1}$$

Care must be exercised in using this relationship; it is not possible to transfer I values from any axis to any other axis, only from an axis through the centroid to another axis or vice versa.

Theorem of perpendicular axes

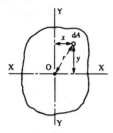

Figure A.2

Let the element dA be at distances x, y and r, respectively from axes YY, XX and O (perpendicular to the plane), Figure A.2,

$$I_{XX} = \int dA \times y^2$$

$$I_{YY} = \int dA \times x^2$$

$$\therefore I_{XX} + I_{YY} = \int dA \times y^2 + \int dA \times x^2$$

$$= \int dA(y^2 + x^2)$$

$$= \int dA \times r^2$$

$\int dA \times r^2$ represents the second moment of area about an axis perpendicular to the plane, which is termed the polar second moment, denoted by J.

Hence
$$I_{XX} + I_{YY} = J \tag{A.2}$$

Equimomental system

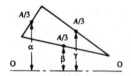

Figure A.3

For triangular figures, it can be shown that for the determination of first and second moments of area, the area A of the triangle is equivalent to three areas, each $A/3$, considered concentrated at the mid-points of the three sides. Since these areas are imagined concentrated at points, they have no second moments about their own axes.

Thus, for the triangle of area A, shown in Figure A.3, first moment of area about OO

$$= \frac{A}{3}\alpha + \frac{A}{3}\beta + \frac{A}{3}\gamma$$

and second moment of area about OO

$$= \frac{A}{3}\alpha^2 + \frac{A}{3}\beta^2 + \frac{A}{3}\gamma^2$$

This equivalent system, known as an *equimomental system*, may be extended to any area which can be divided into triangles (see Example 1).

Radius of gyration

The radius of gyration of a section is the distance from the axis at which the area may be imagined concentrated to give the same value of I. This is denoted by k, so that

$$I = Ak^2$$

or
$$k = \sqrt{\left(\frac{I}{A}\right)} \tag{A.3}$$

Worked examples

(a)

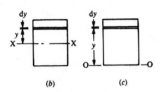

(b)   (c)

Figure A.4

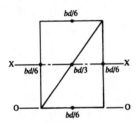

Figure A.5

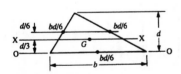

Figure A.6

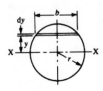

Figure A.7

1. *Determine I_{XX} and I_{OO} for the rectangle shown in* Figure A.4(a).

The area of an elementary strip of thickness dy, Figure A.4(b), is $b\,dy$.

$$\therefore I_{XX} = \int_{-d/2}^{d/2} b\,dy \times y^2 = b\left[\frac{y^3}{3}\right]_{-d/2}^{d/2} = \frac{bd^3}{12} \qquad (A.4)$$

In Figure A.4(c), $\quad I_{OO} = \int_0^d b\,dy \times y^2 = b\left[\frac{y^3}{3}\right]_0^d = \frac{bd^3}{3} \qquad (A.5)$

Alternatively, from the theorem of parallel axes,

$$I_{OO} = I_{XX} + Ah^2$$

$$= \frac{bd^3}{12} + bd \times \left(\frac{d}{2}\right)^2 = \frac{bd^3}{3}$$

These results could alternatively be obtained by the use of the equimomental system. Dividing the rectangle into two triangles, the area of each being $bd/2$, the equimomental system is as shown in Figure A.5. One third of the area of each triangle, i.e. $bd/6$, is placed at the centre of each side and so there will be an element $bd/3$ at the centre of the diagonal, $bd/6$ coming from each triangle.

Then $\qquad\qquad I_{XX} = 2 \times \frac{bd}{6} \times \left(\frac{d}{2}\right)^2 = \frac{bd^3}{12},$

the elements lying on XX being at zero distance from that axis.

Similarly $\quad I_{OO} = \frac{bd}{6} \times d^2 + \left(\frac{bd}{6} + \frac{bd}{3} + \frac{bd}{6}\right) \times \left(\frac{d}{2}\right)^2 = \frac{bd^3}{3}$

the element lying on OO having no second moment about that axis.

2. *Determine I_{XX} and I_{OO} for the triangle shown in* Figure A.6.

Using the equimomental system, as in Example 1,

$$I_{XX} = 2 \times \frac{bd}{6} \times \left(\frac{d}{6}\right)^2 + \frac{bd}{6} \times \left(\frac{d}{3}\right)^2 = \frac{bd^3}{36} \qquad (A.6)$$

$$I_{OO} = 2 \times \frac{bd}{6} \times \left(\frac{d}{2}\right)^2 = \frac{bd^3}{12} \qquad (A.7)$$

3. *Determine I_{XX} for the circle shown in* Figure A.7.

The area of an elementary strip parallel to XX is $b\,dy$ but b is a variable and must be expressed in terms of r and y.

By Pythagoras' theorem,

$$\left(\frac{b}{2}\right)^2 + y^2 = r^2$$

from which $\qquad\qquad b = 2\sqrt{(r^2 - y^2)}$

Hence
$$I_{XX} = \int_{-r}^{r} 2\sqrt{(r^2 - y^2)}\,dy \times y^2$$

$$= \frac{\pi r^4}{4}$$

(a trigonometrical substitution $y = r\sin\theta$ must be used to evaluate this integral).

If the diameter of the circle is d, then $r = d/2$.

Hence
$$I_{XX} = \frac{\pi}{4}\left(\frac{d}{2}\right)^4 = \frac{\pi}{64}d^4 \qquad (A.8)$$

4. *Determine J for the circle shown in* Figure A.8.

To obtain the second moment about an axis O perpendicular to the plane, consider the annular element, radius x and thickness dx.

Figure A.8

Area of element $= 2\pi x\,dx$

$$\therefore J = \int_0^r 2\pi x\,dx \times x^2 = 2\pi\left[\frac{x^4}{4}\right]_0^r = \frac{\pi r^4}{2} \quad \text{or} \quad \frac{\pi d^4}{32} \qquad (A.9)$$

Referring to Figure A.9.

$$I_{XX} + I_{YY} = J, \quad \text{from the theorem of perpendicular axes.}$$

But
$$I_{XX} = I_{YY}, \quad \text{by symmetry}$$

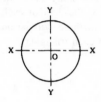

Figure A.9

so that
$$I_{XX} = \frac{J}{2} = \tfrac{1}{2} \times \frac{\pi d^4}{32} = \frac{\pi d^4}{64}$$

This is a much simpler method of obtaining I_{XX} than that of Example 3.

5. *Determine the radius of gyration about axis XX, passing through the centroid for: (a) a rectangle,* Figure A.4; *(b) a circle,* Figure A.7.

(a)
$$k = \sqrt{\left(\frac{I}{A}\right)} = \sqrt{\left(\frac{\dfrac{bd^3}{12}}{bd}\right)} = \frac{d}{2\sqrt{3}} \qquad (A.10)$$

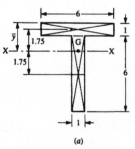

(a)

Dimensions in cm

(b)
$$k = \sqrt{\left(\frac{I}{A}\right)} = \sqrt{\left(\frac{\dfrac{\pi d^4}{64}}{\dfrac{\pi d^2}{4}}\right)} = \frac{d}{4} \qquad (A.11)$$

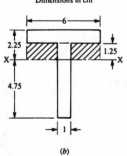

(b)

Figure A.10

6. *Determine I_{XX} for the T-section shown in* Figure A.10(a).

First find the position of the centroid G.
Taking moments about the upper edge,

$$6 \times 1 \times \tfrac{1}{2} + 6 \times 1 \times 4 = (6 \times 1 + 6 \times 1)\bar{y}$$

$$\therefore \bar{y} = 2.25 \text{ cm}$$

From the theorem of parallel axes,

$$I_{XX} = \left\{ \frac{6 \times 1^3}{12} + 6 \times 1 \times 1.75^2 \right\} + \left\{ \frac{1 \times 6^3}{12} + 1 \times 6 \times 1.75^2 \right\}$$

$$= \underline{55.25 \text{ cm}^4} \quad \text{or} \quad \underline{55.25 \times 10^{-8} \text{ m}^4}$$

Alternatively, the second moment of area of a rectangle about an edge is $bd^3/3$, from equation (A.5). Therefore, referring to Figure A.10(b),

$$I_{XX} = \frac{6 \times 2.25^3}{3} - \frac{5 \times 1.25^3}{3} + \frac{1 \times 4.75^3}{3}$$

$$= \underline{55.25 \text{ cm}^4}$$

Moment of inertia

Moment of inertia is analogous with second moment of area, being second moment of mass; it is also denoted by I but has units of kg m^2.

For the body shown in Figure A.11, the moment of inertia of a particle of mass dm about an axis through the centre of gravity G, perpendicular to the plane of the body is $dm \times r^2$. Hence for the whole body,

$$I_G = \int dm \times r^2$$

Similarly, the moment of inertia about a parallel axis through O,

$$I_O = \int dm \times l^2$$

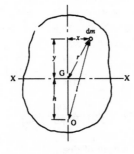

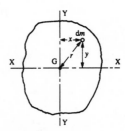

Figure A.11   Figure A.12

Theorem of parallel axes

$$I_O = \int dm \times l^2 = \int dm\{x^2 + (y + h)^2\}$$

$$= \int dm\{x^2 + y^2 + 2yh + h^2\}$$

$$= \int dm\{r^2 + 2yh + h^2\}$$

$$= \int dm \times r^2 + 2h \int dm \times y + h^2 \int dm$$

$\int dm \times r^2$ is I_G, $\int dm \times y$ is the first moment of the mass about axis XX, which is zero since XX passes through G, and $\int dm$ is the total mass, m.

Hence
$$I_O = I_G + mh^2 \qquad \text{(A.12)}$$

Theorem of perpendicular axes

Referring to Figure A.12,
$$I_{XX} = \int dm \times y^2$$
and
$$I_{YY} = \int dm \times x^2$$
$$\therefore I_{XX} + I_{YY} = \int dm(y^2 + x^2)$$
$$= \int dm \times r^2$$
$$= I_G \qquad \text{(A.13)}$$

Radius of gyration

The radius of gyration of a body about an axis is the distance from that axis at which the mass may be considered concentrated to give the same value of I. This is denoted by k, so that

$$I = mk^2$$

or
$$k = \sqrt{\left(\frac{I}{m}\right)} \qquad \text{(A.14)}$$

Worked examples

Figure A.13

7. *Find the moment of inertia and radius of gyration of a uniform disc of mass m and radius r about the axis through O, perpendicular to its plane,* Figure A.13.

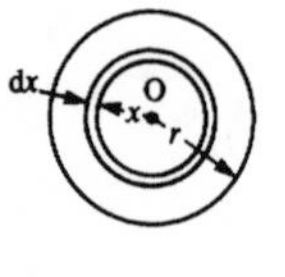

The mass represented by unit area of disc is $m/\pi r^2$. Therefore mass of elementary ring of radius x and thickness dx

$$= \frac{m}{\pi r^2} \times 2\pi x\, dx = \frac{2m}{r^2} x\, dx$$

$\therefore$ moment of inertia of ring $= \dfrac{2m}{r^2}x\, dx \times x^2 \ = \dfrac{2m}{r^2}x^3\, dx$

$\therefore$ total moment of inertia of disc $\qquad = \dfrac{2m}{r^2}\displaystyle\int_0^r x^3\, dx$

$$= \frac{2m}{r^2}\left[\frac{x^4}{4}\right]_0^r$$

i.e.
$$I = \frac{mr^2}{2} \qquad \text{(A.15)}$$

Also
$$I = mk^2$$

so that
$$k = \frac{r}{\sqrt{2}} \qquad \text{(A.16)}$$

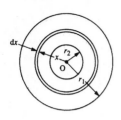

Figure A.14

8. *Find the moment of inertia and radius of gyration of an annular ring of mass m, outer radius r_1 and inner radius r_2 about the axis through O, perpendicular to its plane, Figure A.14.*

$$\text{Mass of elementary ring} = \frac{m}{\pi(r_1^2 - r_2^2)} \times 2\pi x \, dx$$

$$= \frac{2m}{r_1^2 - r_2^2} x \, dx$$

$$\therefore \text{ moment of inertia of ring} = \frac{2m}{r_1^2 - r_2^2} x^3 \, dx$$

$$\therefore \text{ total moment of inertia} = \frac{2m}{r_1^2 - r_2^2} \int_{r_2}^{r_1} x^3 \, dx$$

$$= \frac{m}{2(r_1^2 - r_2^2)} (r_1^4 - r_2^4)$$

i.e.

$$I = \frac{m(r_1^2 + r_2^2)}{2} \tag{A.17}$$

Also

$$I = mk^2$$

$$\therefore k = \sqrt{\left(\frac{r_1^2 + r_2^2}{2} \right)} \tag{A.18}$$

9. *Find the moment of inertia of a sphere, mass m and radius r, about an axis through its centre.*

Let the density of the material be ρ. Then mass of elementary disc of radius y, Figure A.15,

$$= \rho \times \pi y^2 \, dx$$

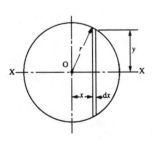

Figure A.15

$\therefore$ moment of inertia of disc about XX

$$= \rho \pi y^2 \, dx \times \frac{y^2}{2}$$

$$= \tfrac{1}{2} \rho \pi y^4 \, dx$$

But

$$y^2 = r^2 - x^2$$

$$\therefore \text{ moment of inertia of disc} = \tfrac{1}{2} \rho \pi (r^2 - x^2)^2 \, dx$$

$$\therefore \text{ total moment of inertia of sphere} = 2 \int_0^r \tfrac{1}{2} \rho \pi (r^4 - 2r^2 x^2 + x^4) \, dx$$

$$= \rho \pi \left[r^4 x - \frac{2r^2 x^3}{3} + \frac{x^5}{5} \right]_0^r$$

$$= \frac{8}{15} \rho \pi r^5$$

But
$$m = \rho \times \frac{4\pi r^3}{3}$$

$$\therefore I = \frac{8}{15}\rho\pi r^5 \times \frac{m}{\frac{4}{3}\rho\pi r^3} = \frac{2}{5}mr^2 \qquad (A.19)$$

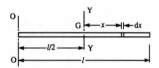

Figure A.16

10. *Determine I_{YY}, I_{OO}, k_{YY} and k_{OO} for a thin uniform rod of mass m and length l, Figure A.16.*

$$\text{Mass of element of length } dx = \frac{dx}{l} \times m$$

$$\therefore \text{ moment of inertia of element about YY} = \frac{dx}{l}m \times x^2$$

$$\therefore \text{ total moment of inertia about YY} = \frac{m}{l} \times 2 \int_0^{l/2} x^2 \, dx$$

$$= \frac{2m}{l}\left[\frac{x^3}{3}\right]_0^{l/2}$$

i.e.
$$I_{YY} = \frac{ml^2}{12} \qquad (A.20)$$

From the theorem of parallel axes,

$$I_{OO} = I_{YY} + m\left(\frac{l}{2}\right)^2$$

$$= \frac{ml^2}{12} + \frac{ml^2}{4} = \frac{ml^2}{3} \qquad (A.21)$$

$$k_{YY} = \sqrt{\left(\frac{I_{YY}}{m}\right)} = \frac{l}{2\sqrt{3}} \qquad (A.22)$$

$$k_{OO} = \sqrt{\left(\frac{I_{OO}}{m}\right)} = \frac{l}{\sqrt{3}} \qquad (A.23)$$

Details of the steel sections dimensions and properties in Table A.1 have been taken from the *Structural Steelwork Handbook* and are reproduced in this publication by permission of the British Constructional Steelwork Association Ltd. Copies of this complete publication, which contains the Safe Load Tables, can be obtained from the BCSA at 4 Whitehall Court, Westminster, London SW1A 2ES.

The copyright of these extracts belongs to the BCSA and they may not be re-copied in any form or stored in a retrieval system without BCSA's permission.

In Table A.1, 'moment of inertia' can be read as 'second moment of area, I' and 'elastic modulus' as 'modulus of section Z'.

Table A.1. Extract from universal beams: dimensions and properties

| Serial size (mm) | Mass per metre (kg) | Moment of inertia | | Radius of gyration | | Elastic modulus | | Area of section (cm²) |
		Axis x − x (cm⁴)	Axis y − y (cm⁴)	Axis x − x (cm)	Axis y − y (cm)	Axis x − x (cm³)	Axis y − y (cm³)	
457 × 152	82	36 160	1 093	18.6	3.24	1 555	142.5	104.4
	74	32 380	963	18.5	3.18	1 404	126.1	94.9
	67	28 522	829	18.3	3.12	1 248	109.1	85.3
	60	25 464	794	18.3	3.23	1 120	104.0	75.9
	52	21 345	645	17.9	3.11	949.0	84.61	66.5
406 × 178	74	27 279	1 448	17.0	3.91	1 322	161.2	94.9
	67	24 279	1 269	16.9	3.85	1 186	141.9	85.4
	60	21 520	1 108	16.8	3.82	1 059	124.7	76.1
	54	18 576	922	16.5	3.67	922.8	103.8	68.3
406 × 152	74	26 938	1 047	16.9	3.32	1 294	136.2	94.8
	67	23 798	908	16.7	3.26	1 155	118.8	85.3
	60	20 619	768	16.5	3.18	1 011	100.9	75.8
406 × 140	46	15 603	500	16.3	2.92	775.6	70.26	58.9
	39	12 408	373	15.9	2.75	624.7	52.61	49.3
381 × 152	67	21 276	947	15.8	3.33	1 095	122.7	85.4
	60	18 632	814	15.7	3.27	968.4	106.2	75.9
	52	16 046	685	15.5	3.21	842.3	89.96	66.4
356 × 171	67	19 483	1 278	15.1	3.87	1 071	147.6	85.3
	57	16 038	1 026	14.9	3.77	894.3	119.2	72.1
	51	14 118	885	14.8	3.71	794.0	103.3	64.5
	45	12 052	730	14.6	3.58	684.7	85.39	56.9
356 × 127	39	10 054	333	14.3	2.60	570.0	52.87	49.3
	33	8 167	257	14.0	2.48	468.7	40.99	41.7
305 × 165	54	11 686	988	13.1	3.80	751.8	118.5	68.3
	46	9 924	825	13.0	3.74	646.4	99.54	58.8
	40	8 500	691	12.9	3.67	659.6	83.71	51.4
305 × 127	48	9 485	438	12.5	2.68	611.1	69.94	60.8
	42	8 124	367	12.4	2.63	530.0	58.99	53.1
	37	7 143	316	12.3	2.58	470.3	51.11	47.4
305 × 102	33	6 482	189	12.5	2.13	414.6	37.00	41.8
	28	5 415	153	12.2	2.05	350.7	30.01	36.3
	25	4 381	116	11.8	1.92	287.5	22.85	31.4
254 × 146	43	6 546	633	10.9	3.39	504.3	85.97	55.0
	37	5 544	528	10.8	3.34	433.1	72.11	47.4
	31	4 427	406	10.5	3.19	352.1	55.53	39.9
254 × 102	28	4 004	174	10.5	2.19	307.6	34.13	36.2
	25	3 404	144	10.3	2.11	264.9	28.23	32.1
	22	2 863	116	10.0	2.02	225.4	22.84	28.4
203 × 133	30	2 880	354	8.71	3.05	278.5	52.85	38.0
	25	2 348	280	8.53	2.94	231.1	41.92	32.3

Appendix B
Differential equations

Differential equations of the type

$$\frac{d^2x}{dt^2} + 2\mu\frac{dx}{dt} + \omega^2 x = f(t)$$

may be written in the form

$$\frac{d^2x}{dt^2} + 2\mu\frac{dx}{dt} + \omega^2 x = 0 + f(t)$$

and the solution consists of two parts:

(a) the value of x which satisfies the equation

$$\frac{d^2x}{dt^2} + 2\mu\frac{dx}{dt} + \omega^2 x = 0,$$

this being called the *complementary function;*

(b) the value of x which satisfies the equation

$$\frac{d^2x}{dt^2} + 2\mu\frac{dx}{dt} + \omega^2 x = f(t),$$

this being called the particular integral.

(a) Complementary function

(i) $\dfrac{d^2x}{dt^2} + \omega^2 x = 0$

The solution is $x = A\cos\omega t + B\sin\omega t$

(ii) $\dfrac{d^2x}{dt^2} + 2\mu\dfrac{dx}{dt} + \omega^2 x = 0$

The solution is $x = e^{-\mu t}(A\cos\sqrt{(\omega^2 - \mu^2)}t + B\sin\sqrt{(\omega^2 - \mu^2)}t)$

(b) Particular integrals

(i) $$\frac{d^2x}{dt^2} + \omega^2 x = C$$

The solution is $\quad x = \dfrac{C}{\omega^2}$

(ii) $$\frac{d^2x}{dt^2} + 2\mu\frac{dx}{dt} + \omega^2 x = C$$

The solution is $$x = \frac{C}{\omega^2}$$

(iii) $$\frac{d^2x}{dt^2} + \omega^2 x = C\cos pt$$

The solution is $$x = \frac{C\cos pt}{\omega^2 - p^2}$$

(iv) $$\frac{d^2x}{dt^2} + 2\mu\frac{dx}{dt} + \omega^2 x = C\cos pt$$

The solution is $$x = \frac{C\cos(pt - \alpha)}{\sqrt{(4\mu^2 p^2 + (\omega^2 - p^2)^2)}}$$

where $$\alpha = \tan^{-1}\frac{2\mu p}{\omega^2 - p^2}$$

Index